Potential Pulses: Genetic and Genomic Resources

Potential Pulses: Genetic and Genomic Resources

Edited by

Rahul Chandora

ICAR-National Bureau of Plant Genetic Resources-Regional Station, Shimla-171004, Himachal Pradesh, India

T. Basavaraja

ICAR-Indian Institute of Oilseeds Research, Rajendranagar, Hyderabad – 500 030, India

and

Aditya Pratap

ICAR-Indian Institute of Pulses Research, Kanpur - 208 024, India

⟨⟨◉⟩⟩ CABI

CABI is a trading name of CAB International

CABI
Nosworthy Way
Wallingford
Oxfordshire OX10 8DE
UK

Tel: +44 (0)1491 832111
E-mail: info@cabi.org
Website: www.cabi.org

CABI
200 Portland Street
Boston
MA 02114
USA

Tel: +1 (617)682-9015
E-mail: cabi-nao@cabi.org

The views expressed in this publication are those of the author(s) and do not necessarily represent those of, and should not be attributed to, CAB International (CABI). Any images, figures and tables not otherwise attributed are the author(s)' own. References to internet websites (URLs) were accurate at the time of writing.

CAB International and, where different, the copyright owner shall not be liable for technical or other errors or omissions contained herein. The information is supplied without obligation and on the understanding that any person who acts upon it, or otherwise changes their position in reliance thereon, does so entirely at their own risk. Information supplied is neither intended nor implied to be a substitute for professional advice. The reader/user accepts all risks and responsibility for losses, damages, costs and other consequences resulting directly or indirectly from using this information.

CABI's Terms and Conditions, including its full disclaimer, may be found at https://www.cabidigitallibrary.org/terms-and-conditions.

A catalogue record for this book is available from the British Library, London, UK.

Library of Congress Cataloging-in-Publication Data

Names: Chandora, Rahul, editor. | Basavaraja, T, editor. | Pratap, Aditya, 1976- editor.
Title: Potential pulses : genetic and genomic resources / edited by Rahul Chandora, T Basavaraja, and Aditya Pratap
Description: Boston, MA : CAB International, 2024 | Includes bibliographical references and index | Summary: "Potential pulses include adzuki bean, rice bean, faba bean, cow pea, etc. and have potential to become climate-smart crops for a sustainable future. These underutilized crops offer numerous benefits, including mitigating the effects of climate change and improving nutritional security"-- Provided by publisher
Identifiers: LCCN 2024019888 (print) | LCCN 2024019889 (ebook) | ISBN 9781800624634 (hardback) | ISBN 9781800624641 (ebk) | ISBN 9781800624658 (epub)
Subjects: LCSH: Legumes.
Classification: LCC SB177.L45 P68 2024 (print) | LCC SB177.L45 (ebook) | DDC 633.3--dc23/eng/20240709
LC record available at https://lccn.loc.gov/2024019888
LC ebook record available at https://lccn.loc.gov/2024019889

ISBN-13: 9781800624634 (hardback)
         9781800624641 (ePDF)
         9781800624658 (ePub)

DOI: 10.1079/9781800624658.0000

Commissioning Editor: Rebecca Stubbs
Editorial Assistant: Emma McCann
Production Editor: Rosie Hayden

Typeset by Straive, Pondicherry, India

Contents

Foreword

The importance of grain legumes in global food and nutritional security, crop diversification, soil amelioration and environmental sustainability is well recognized. Their nutritional richness and adaptability to varied agro-ecosystems contribute significantly to the global agriculture, promoting especially the plant-based food systems. Owing to their properties, promoting widespread cultivation and consumption of food legumes can be one of the major drivers of the first three of the Sustainable Development Goals (SDGs), namely eliminating poverty, ending hunger and establishing good health and wellbeing. Among the grain legumes, the potential food legumes, also called underutilized or minor food legumes, hold a special significance as they are niche specific, can withstand harsher environments and provide an additional opportunity for the diversification of micro-climate cropping systems and diets.

A phenomenal economic growth has taken place in Asia and the Pacific in the past few decades, which has resulted in significant improvements in the socio-economic conditions in the region, positively impacting the lifestyle and living situations of the people. This has also resulted in expanding the food habits of the population and the evolution of a class of consumers preferring healthier and plant-based foods over animal products. In such a scenario, potential food legumes offer an obvious advantage due to their high and quality protein, low fat, low glycaemic index and plenty of antioxidants. Low input requirement during their cultivation, including an all-time scarce resource, water, also makes them a good choice for the smallholder and subsistence farmers. However, unfortunately, the focus has remained confined to a few major legumes until now channelling ample funding, research and policy support in the past three to four decades, marginalizing underutilized yet potential food legumes. Nevertheless, potential legumes present a substantial promise for improving food security, meeting dietary demands, and fostering agricultural growth. These offer a plethora of alternative protein crops, which are also resilient to changing climates and have an inherent fortified ability to withstand biotic and abiotic challenges. They also offer a viable avenue for crop diversification, contributing significantly to agricultural sustainability and a nation's overall economic prosperity. The nutritional analysis of minor pulses revealed their high nutrient density, sometimes even more than the major legumes for a few elements, encompassing crucial micronutrients, dietary fibre, resistant starch, proteins and bioactive compounds. It has been observed that the protein content of the minor pulses varies from 8.6% to 43%, surpassing even that of chickpea and soybean. While crops such as winged bean outdo all pulses with the highest protein content, its amino acid composition is comparable to soybean. Likewise, the African yam bean boasts of diverse proximate compositions, including carbohydrates, protein, ash, fat and fibre. Many of the minor pulses, such as winged bean, runner bean and cowpea, flaunt multiple edible parts within the same plant, making available a range of nutrient sources. The potential pulses are not only inexpensive, but can also be grown in some of the areas where no other food legumes or even cereals can be grown. Therefore, there is a need to advocate strongly for the cause of potential pulses and promote them for healthier diets and the sustainability of the environment.

This book explores the potential and scope of underutilized pulses to greater depths and is very timely and well thought of in promoting this goldmine for global food and nutrition security. This volume, edited by three promising pulses researchers, offers a comprehensive and extensive treatise on the biology, ecology and breeding of the potential legumes besides providing an insight into the innovative developments that have changed our understanding of these neglected crops. Chapters on different minor pulses have been contributed by some of the best workers of these crops and have globally acknowledged research contributions. I am sure that the book will be widely read and the information contained in each individual chapter will be tremendously helpful to the researchers, students, teachers and extension workers.

TR Sharma
DDG (Crop Sciences)
ICAR, New Delhi

Preface

_____

Welcome to *Potential Pulses: Genetic and Genomic Resources*, a comprehensive exploration of some of the world's most promising yet often overlooked crops. Pulses such as African yam bean, cowpea, adzuki bean, rice bean, runner bean, lima bean, faba bean, and grass pea have been dietary staples for centuries, valued for their nutritional richness and adaptability to various agroecological zones. However, despite their significant potential, these crops have not received the same level of research attention as other major pulses. Consequently, their cultivation and use have waned, leading to their classification as 'orphan', 'neglected', and 'underutilized' crops.

This book aims to change that narrative. By bringing together the knowledge and insights of leading experts from around the globe, *Potential Pulses: Genetic and Genomic Resources* seeks to unlock the potential of these vital crops. The 16 chapters within this volume delve into the genetic diversity, genomic tools, and biotechnological innovations essential for enhancing the productivity and resilience of these pulses. Each chapter provides a thorough examination of the nutritional values, usage patterns, origins, distribution, evolutionary relationships, roles in global food security, crop improvement strategies, cultivation practices, and future perspectives of these crops. Detailed tables and diagrams are included to facilitate a deeper understanding of the topics covered.

Our goal is to shine a spotlight on the untapped potential of these crops and inspire renewed interest and research within the scientific community and beyond. This book is designed to be an invaluable resource for researchers, agronomists, policymakers, and anyone dedicated to advancing sustainable agriculture and food security.

We extend our heartfelt gratitude to the contributors who have shared their expertise and passion, making this book a reality. Their collaborative efforts have resulted in a rich and diverse body of work that we believe will significantly advance our understanding of potential pulses and their role in sustainable agriculture and food security. We also express our sincere thanks to CAB International (UK) for their steadfast support and commitment to promoting scientific knowledge.

It is our hope that *Potential Pulses: Genetic and Genomic Resources* will serve as an essential reference and a source of inspiration for ongoing and future research in the realm of pulse crops.

Rahul Chandora, T. Basavaraja and Aditya Pratap
Editors,
Potential Pulses: Genetic and Genomic Resources

List of Contributors

Michael T. Abberton, Genetic Resources Centre, International Institute of Tropical Agriculture, Ibadan-200136, Nigeria

Lynn Abou-Khater, International Center for Agricultural Research in the Dry Areas (ICARDA), Beirut 1108-2010, Lebanon

Daniel. B. Adewale, Department of Crop Science and Horticulture, Federal University Oye-Ekiti, Ikole-Ekiti campus-370105, Nigeria; daniel.adewale@fuoye.edu.ng

Nadège Adoukè Agbodjato, Laboratoire de Biologie et de Typage Moléculaire en Microbiologie (LBTMM), Université d'Abomey-Calavi (UAC), 05 BP 1604 Cotonou, Bénin

Atabong Paul Agendia, Department of Plant Biology, Faculty of Science, University of Yaoundé I, Yaoundé, Cameroon

Lamine Baba-Moussa, Laboratoire de Biologie et de Typage Moléculaire en Microbiologie (LBTMM), Université d'Abomey-Calavi (UAC), 05 BP 1604 Cotonou, Bénin; laminesaid@yahoo.fr

Olubukola Oluranti Babalola, Food Security and Safety Focus Area, Faculty of Natural and Agricultural Sciences, North-West University, Private Mail Bag X2046, Mmabatho, South Africa; olubukola.babalola@nwu.ac.za

Rind Balech, International Center for Agricultural Research in the Dry Areas (ICARDA), Beirut 1108-2010, Lebanon

P.S. Basavaraj, ICAR-National Institute of Abiotic Stress Management, Baramati- 413115, Maharashtra, India; bassuptl@gmail.com

T. Basavaraja, ICAR-Indian Institute of Pulses Research, Kanpur, India; basu86.gpb@gmail.com

Ingudam Bhupenchandra, ICAR-KVK, Tamenglong, ICAR-RC-NEH Region, Manipur Centre-795147, India

K.M. Boraiah, ICAR-National Institute of Abiotic Stress Management Malegaon, Baramati, Maharashtra, India; bors_km@yahoo.co.in

Kanishka Chandora, ICAR-National Bureau of Plant Genetic Resources-Regional Station, Shimla-171004, Himachal Pradesh, India

Rahul Chandora, ICAR-National Bureau of Plant Genetic Resources-Regional Station, Shimla-171004, Himachal Pradesh, India

Debasis Chattopadhyay, DBT-National Institute of Plant Genome Research, New Delhi-110067, India

Khushwant B. Choudhary, Division of Plant Improvement and Pest Management, ICAR-Central Arid Zone Research Institute (CAZRI), Jodhpur, 342 003, India

Durand Dah-Nouvlessounon, Laboratoire de Biologie et de Typage Moléculaire en Microbiologie (LBTMM), Université d'Abomey-Calavi (UAC), 05 BP 1604 Cotonou, Bénin

Raviprakash Govindrao Dani, Gen Scan Inc. Global Consultancy, Houston, Texas, USA; Namangan Engineering and Technology Institute, Namangan, Uzbekistan

M. Devindrappa, ICAR-Central Institute for Cotton Research, Nagpur- 441108, Maharashtra, India

Isabel M.G. Figari, Technological Institute of Production, Lima, Peru

Lakshmi Gangavati, Department of Genetics and Plant Breeding, University of Agricultural Sciences, Dharwad-580005, Karnataka, India; lakshmirg8@gmail.com

Gayacharan, ICAR-National Bureau of Plant Genetic Resources, New Delhi-110012, India; gayacharan@icar.gov.in

Rameshwar Baliram Ghorade, Dr. Panjabrao Deshmukh Krishi Vidyapeeth, Akola Maharashtra, 444104, India

S. Gurumurthy, ICAR-National Institute of Abiotic Stress Management, Baramati-413115, Maharashtra, India

Hanamant M. Halli, ICAR-National Institute of Abiotic Stress Management Malegaon, Baramati, Maharashtra, India

C.B. Harisha, ICAR-National Institute of Abiotic Stress Management, Baramati-413115, Maharashtra, India

Pravin Vishwanathrao Jadhav, Dr. Panjabrao Deshmukh Krishi Vidyapeeth, Akola Maharashtra, 444104, India; jpraveen26@yahoo.co.in

Shailesh Kumar Jain, Rajasthan Agricultural Research Institute - Sri Karan Narendra Agriculture University, Durgapura, Jaipur-302 018, Rajasthan, India

Sandeep Jaiswal, ICAR-RC-NEH Region, Umiam, Meghalaya-793103, India

Krishna Kumar Jangid, ICAR-National Institute of Abiotic Stress Management, Baramati-413115, Maharashtra, India

D.C. Joshi, ICAR-Vivekananda Parvatiya Krishi Anusandhan Sansthan, Almora-263601, India

Gopal Katna, CSK, Himachal Pradesh Krishi Vishvavidyalaya, Palampur, 176062, Himachal Pradesh, India; gkatna@gmail.com

Vikas Khandelwal, ICAR-All India Coordinated Research Project on Pearl Millet, Mandor, Jodhpur-342304, India

Amit Kumar, ICAR-RC-NEH Region, Umiam, Meghalaya-793104, India

Mithlesh Kumar, Agricultural Research Station, Mandor, Agriculture University, Jodhpur-342304, Rajasthan, India; mithleshgenetix@gmail.com

Sushil Kumar, Department of Agricultural Biotechnology, Anand Agricultural University (AAU), Anand, 388110, India; sushil254386@gmail.com

Fouad Maalouf, International Center for Agricultural Research in the Dry Areas (ICARDA), Beirut 1108-2010, Lebanon; f.maalouf@cgiar.org

C. Mahadevaiah, ICAR-Indian Institute of Horticultural Research Hesaraghatta Lake Post, Bangalore-560089, India

Hans Raj Mahla, Division of Plant Improvement and Pest Management, ICAR-Central Arid Zone Research Institute (CAZRI), Jodhpur, 342003, India

Arpita Mahobia, Dr Panjabrao Deshmukh Krishi Vidyapeeth, Akola Maharashtra, 444104, India

Moti Lal Mehriya, Agricultural Research Station, Mandor, Agriculture University, Jodhpur-342 304, Rajasthan, India

Gopika Krishna Mote, Dr Panjabrao Deshmukh Krishi Vidyapeeth, Akola Maharashtra, 444104, India

Balwant Sayasrao Mundhe, Dr Panjabrao Deshmukh Krishi Vidyapeeth, Akola Maharashtra, 444104, India,

Umakanta Ngangkham, ICAR-RC-NEH Region, Manipur Centre-795004, India

Francis Ajebesone Ngome, Institute of Agricultural Research for Development, Yaoundé, Cameroon

Martial Nounagnon, Laboratoire de Biologie et de Typage Moléculaire en Microbiologie (LBTMM), Université d'Abomey-Calavi (UAC), 05 BP 1604 Cotonou, Bénin

K.K. Pal, ICAR-National Institute of Abiotic Stress Management Malegaon, Baramati, Maharashtra, India

Swarup K. Parida, DBT-National Institute of Plant Genome Research, New Delhi-110067, India

Nandan L. Patil, Department of Genetics and Plant Breeding, University of Agricultural Sciences, Dharwad-580005, Karnataka, India

W.S. Philanim, ICAR-RC-NEH Region, Umiam, Meghalaya-793103, India; philanim09@gmail.com

Aliza Pradhan, ICAR-National Institute of Abiotic Stress Management Malegaon, Baramati, Maharashtra, India

Aditya Pratap, ICAR-Indian Institute of Pulses Research, Kanpur-208024, India

Vinita Ramtekey, ICAR-Indian Institute of Seed Science, Mau-275103, Uttar Pradesh, India; and Plant Breeding Department, University of Bonn, Kirschallee 1, Bonn, Germany; s79vramt@uni-bonn.de

Kirti Rani, Indian Council of Agricultural Research – National Bureau of Plant Genetic Resources, Regional Station, Jodhpur-342 003, Rajasthan, India

Ramya Rathod, Professor Jayashankar Telangana State Agriculture University, Rajendranagar, Hyderabad-500030, Telangana, India

Aditya Vishnudas Rathod, Dr Panjabrao Deshmukh Krishi Vidyapeeth, Akola Maharashtra, 444104, India

A. Amarender Reddy, School of Crop Health Policy Support Research, ICAR-National Institute of Biotic Stress Management (ICAR-NIBSM), Raipur, India

K. Sammi Reddy, ICAR-National Institute of Abiotic Stress Management, Baramati-413115, Maharashtra, India

Gautier Roko, Laboratoire de Biologie et de Typage Moléculaire en Microbiologie (LBTMM), Université d'Abomey-Calavi (UAC), 05 BP 1604 Cotonou, Bénin

Sanjay Bapu Sakhare, Dr. Panjabrao Deshmukh Krishi Vidyapeeth, Akola Maharashtra, 444104, India

Rumki Ch. Sangma, ICAR-RC-NEH Region, Umiam, Meghalaya-793103, India

Shravani Sanyal, School of Crop Health Policy Support Research, ICAR-National Institute of Biotic Stress Management (ICAR-NIBSM), Raipur, India

Parul Sharma, Punjab Agricultural University, Ludhiana-141004, India

Ramavtar Sharma, Division of Plant Improvement and Pest Management, ICAR-Central Arid Zone Research Institute (CAZRI), Jodhpur, 342 003, India; ramavtar.cazri@gmail.com

Siddhanath Shendekar, Mahatma Phule Krishi Vidyapeeth, Rahuri, Ahmednagar, Maharashtra, India

Chetan Shinde, Mahatma Phule Krishi Vidyapeeth, Rahuri, Ahmednagar, Maharashtra, India

Umesh Dnyaneshwar Shinde, Dr. Panjabrao Deshmukh Krishi Vidyapeeth, Akola Maharashtra, 444104, India

Amit Kumar Singh, ICAR-National Bureau of Plant Genetic Resources, New Delhi-110012, India

Binay Kumar Singh, ICAR-RC-NEH Region, Umiam, Meghalaya-793103, India

N. Raju Singh, ICAR-RC-NEH Region, Umiam, Meghalaya-793103, India

Carine Nono Temegne, Department of Plant Biology, Faculty of Science, University of Yaoundé I, Yaoundé, Cameroon; carine.temegne@facsciences-uy1.cm

Elena Alexandrovna Torop, Voronezh State Agrarian University, Voronezh, St. Michurina, 1, Russia, 394087

Letngam Touthang, ICAR-RC-NEH Region, Umiam, Meghalaya-793103, India

Anupam Tripathi, Acharya Narendra Deva University of Agriculture and Technology, Kumarganj, Ayodhya- 224229, Uttar Pradesh, India

Kuldeep Tripathi, ICAR-National Bureau of Plant Genetic Resources, New Delhi-110012, India

Esaïe Tsoata, Department of Plant Biology, Faculty of Science, University of Yaoundé I, Yaoundé, Cameroon

Sagar Laxman Zanjal, Dr Panjabrao Deshmukh Krishi Vidyapeeth, Akola Maharashtra, 444104, India

1 Leveraging the Potential of Lesser-known Pulses for the Sustainability of Future Food Systems

T. Basavaraja[1]*, Aditya Pratap[1] and Rahul Chandora[2]

[1]*ICAR-Indian Institute of Pulses Research, Kanpur, India;* [2]*ICAR-National Bureau of Plant Genetic Resources, Regional Station, Shimla, India*

Abstract

The rapid growth of the global population presents a daunting challenge to agriculture and natural resources, especially in the face of rising food demand. Hunger remains a crucial issue in various regions, particularly in South Asia and sub-Saharan Africa, where high rates of hunger and malnutrition persist. Despite improvements in agricultural practices, increasing hunger in 18 countries since 2015 highlights the need for innovative solutions. This book explores the potential of underutilized legume crops, often neglected in agriculture, to enhance food and nutritional security. With the world heavily reliant on a few staple crops, diversifying the food supply is crucial. These underutilized pulses, including adzuki bean, Bambara groundnut, African yam bean and winged bean, among others, offer rich nutritional profiles, high protein content and resilience to changing climates. Through genetic and genomic resources, including advanced sequencing technologies and marker-assisted selection, we can expedite the development of climate-resilient and nutritionally rich crop varieties. Harnessing the genetic diversity of these underutilized pulses is crucial in addressing food insecurity, improving agricultural sustainability and promoting diversified, nutrient-rich diets, offering a promising pathway to ensure global food security amidst the challenges posed by a growing population and changing climates.

Keywords: Potential legumes, orphan crops, genetic resources, underutilized, pulse genomics, sustainability

1.1 Introduction

Addressing the impact of the escalating global population on agriculture and natural resources in the 21st century requires a nuanced understanding of the challenges at hand. The 2023 Global Hunger Index (GHI) score of 18.3, though classified as moderate, highlights the persistent nature of hunger, particularly in 34 countries, with South Asia and sub-Saharan Africa facing the highest burden (Kousar *et al.*, 2021; von Grebmer *et al.*, 2023). Despite incremental improvements, 18 countries have seen an increase in hunger since 2015, underscoring the urgency of addressing food and nutritional security, especially with the projected global population of 10 billion by 2050 (Singh *et al.*, 2022).

Compounding these challenges are limited agricultural land, which is further declining, changing climates, and the impact of abiotic and biotic factors on crop productivity (Mayes *et al.*, 2012; Devaux *et al.*, 2020). The scarcity of genetic

*Corresponding author: basu86.gpb@gmail.com

© CAB International 2024. *Potential Pulses: Genetic and Genomic Resources* (eds R. Chandora, T. Basavaraja and A. Pratap)
DOI: 10.1079/9781800624658.0001

diversity further complicates matters, hindering the development of enhanced varieties of orphan crops. Innovative solutions are imperative to tackle these multifaceted challenges.

1.1.1 Pulses for food and nutritional security

The threat posed by climate-change-induced erratic weather patterns to global food production necessitates a shift from the overreliance on a narrow range of crop genetic variability. With only 30 domesticated plant species dominating human diets, there is a pressing need to diversify beyond these key crops (Fghire *et al.*, 2022; Ali and Bhattacharjee, 2023). This diversification becomes critical to address food insecurity comprehensively.

The decline in crop diversity, unhealthy dietary habits and sedentary lifestyles have led to a dual nutritional crisis affecting around 2 billion individuals globally. This crisis causes malnutrition and obesity-related non-communicable diseases (Sreenivasulu and Fernie, 2022). The past decade has seen substantial research on alternative crops, particularly underutilized potential legumes. These crops, often overlooked in terms of research, investment and commercial output, hold significant promise in delivering valuable food, nutrition and economic benefits.

The importance of agrobiodiversity, which refers to the variety of crop species in a system, is gaining recognition for its role in sustainability, socio-economic resilience, and human health improvement (Dwivedi *et al.*, 2017; Tian *et al.*, 2021; Regmi *et al.*, 2023). Neglected and underutilized crop species are emerging as focal points owing to their potential to enhance nutritional value, promote agrobiodiversity, sustain hostile environs and withstand environmental stresses. Integrating these species into food production systems can enhance the resilience and diversity of food production with a reduced environmental footprint, and their integration into diets can mitigate non-communicable and diet-related diseases.

Despite the potential benefits, mainstreaming neglected crops faces obstacles such as insufficient research, low public awareness, inadequate government policies and limited public/private investments. Recent efforts have focused on genetic conservation and enhancement of these crops to unlock their full potential.

1.1.2 Smart food crops: a path to sustainable agriculture

The focus on underutilized crops, aptly named 'smart food crops', aims to enhance food quality, reduce reliance on a limited set of staple crops, preserve cultural diet diversity, and explore climate-resilient alternatives (Ali and Bhattacharjee, 2023). The potential legume crops possess the capacity to thrive in diverse environmental challenges and changing climates due to their exceptional nitrogen-fixing capability. Beyond environmental adaptability, these crops boast higher nutritional content, making them instrumental in addressing nutrition deficiencies and ensuring food security in underserved regions (Joshi *et al.*, 2020; Ayilara *et al.*, 2022). As we navigate the complexities of global food security, investing in the potential of underutilized crops emerges as a critical avenue for sustainable and resilient agricultural practices.

1.2 Potential Pulse Crops: Nutrient Status, Consumption Pattern and Usage

The current global landscape is characterized by a grave situation marked by widespread hunger and significant loss of lives owing to malnutrition. The escalating shift towards vegetarian protein diets on a global scale has intensified the demand for alternative protein sources. In response, legumes have emerged as a viable and economically beneficial option to meet protein requirements (FAO, 2021; Singh *et al.*, 2022). These crops are highly esteemed globally for providing nutritional protein sustainably and cost-effectively, positioning them as the second most important dietary source after cereal grains (Nayak *et al.*, 2022). The pressing needs in the current context revolve around sustainable methods to enhance crop resilience to changing climates, fortify their ability to withstand biotic and abiotic challenges, and concurrently promote large-scale production and nutritional diversity. Underutilized pulses present substantial potential for improving food security, meeting dietary demands and fostering agricultural growth. They offer a viable avenue for crop diversification, contributing significantly to agricultural sustainability and a nation's overall economic prosperity (Mayes *et al.*, 2012; Joshi *et al.*, 2020; Fghire *et al.*, 2022). A meticulous nutritional analysis of underutilized pulse crops unveils their high nutrient density, encompassing crucial micronutrients, dietary fibre, resistant starch, proteins and bioactive compounds (Table 1.1). The protein content of untapped pulses varies from 8.6% to 43%, surpassing that of soybeans (Ayilara *et al.*, 2022). Among these pulses, the winged bean seed stands out with the highest protein percentage (27–43%), and its amino acid composition is comparable to soybeans, with methionine and cysteine as the limiting amino acids. Horse gram seed boasts a protein level ranging from 18.5% to 22.5% (Savithramma and Shambulingappa, 1996). Crude protein in cowpea ranges from 25.38% to 27.56% (Gondwe *et al.*, 2019). Recent studies on African yam bean grains indicate diverse proximate compositions, including carbohydrates, protein, ash, fat and fibre (George *et al.*, 2020).

Underutilized pulses offer unique potential for addressing diverse nutritional needs, catering to various preferences. Many of these legumes feature multiple edible parts within the same plant, allowing for a range of nutrient sources. For example, winged bean provides edible components such as leaves, flowers, green pods, seeds and tuberous roots, all rich in protein and suitable for raw or cooked bean consumption (Ayilara *et al.*, 2022; Dwivedi *et al.*, 2023).

Cowpea, known for its versatility, includes young leaves, growing points, green pods and green seeds used as vegetables in South Africa. Its fresh leaves, ranked among the top four vegetables in various African and Asian countries, are well-suited for cultivation in high-rainfall agroecologies, aiding in reducing malnutrition (Mekonnen *et al.*, 2022). Hyacinth bean serves multiple purposes as a pulse, vegetable, livestock feed, green manure, and even a decorative or medicinal herb (Naeem *et al.*, 2020). Similarly, runner bean in South-east Europe is grown for dry seeds and consumed as immature green pods (Sinkovič *et al.*, 2019). Despite being underutilized, these legumes are recognized for producing delicious meals for domestic consumption, although consumption rates remain below their

nutritional potential. Common processing methods involve boiling after removing foreign ingredients and soaking in water, indicating opportunities for increased integration into everyday diets (Agyekum *et al.*, 2023). Potential pulses have high amounts of amino acids, minerals, proteins, vitamins, dietary fibres and some beneficial bioactive components, often on par with or even superior to those found in domesticated main pulses. These potential pulses are not only inexpensive, but they also provide an excellent source of protein, especially during times when the consumption of animal products has been minimized because of potential health risks. As a result, there has been a significant increase in the promotion of plant-based protein by researchers as an approach to fulfilling the population's nutritional needs.

1.3 Harnessing Genetic and Genomic Resources for Climate-resilient Agriculture

1.3.1 Genetic resources

Germplasm collections serve as repositories of genetic diversity, encompassing phenotypic variations in agronomic features. Diverse genetic resources in crops play a crucial role in crop development and breeding efforts to improve resilience, yield potential, nutritional content and adaptability to changing environmental conditions (Silva *et al.*, 2019; Ulian *et al.*, 2020; Sserumaga *et al.*, 2021; Kanishka *et al.*, 2024).

Understanding phenotypic variations, assessing population structure and diversity, and characterizing genotype × environment interactions (GEI) in germplasm collections aid in identifying genetically varying, stable and agriculturally desirable germplasm. The International Institute of Tropical Agriculture (IITA) gene bank, for instance, houses a collection of 15,003 cultivated cowpea accessions from 89 countries, with a core collection of 2062 accessions developed based on geographical, agronomical and botanical descriptions (Boukar *et al.*, 2019). Sources of new traits continue to be discovered in cowpea germplasm, and the traits are being characterized at high genetic precision with the new genotyping methods available for detecting marker-trait

Table 1.1. An overview of potential pulse crops' nutritional profiles (per 100 g) (Dwivedi *et al.*, 2023) CC BY 4.0 DEED.

Potential pulses	Carbohydrate (%)	Crude protein (%)	Fat (%)	Energy (Kcal/100 g)	Crude fibre (%)	Ca (mg)	P (mg)	Fe (mg)
Adzuki bean	28.5–60.7	16.3–29.2	0.3–1.3	329	12.7	66	381	5
Bambara groundnut	53–69	17–25	6.5–8.5	1609	5–12	30–128	81–563	2–9
Grass pea	48–52	18–34	0.7–2.8	362	3.9–6.0	220–370	350–640	6.9–8.7
Horse gram	50–60	18.5–31.2	0.6–2.6	321	4.3–25.0	244–312	311–443	5.9–7.4
Faba bean	57.8–81.0	12.9–22.9	1.0–5.3	348	2.0–10.9	103–183	345–392	15
Hyacinth bean	29.6–66.3	20.5–35.5	0.3–9.7	344–383	3.7–14	94–132	317–428	1.7–9.4
Lima bean	49.4–77.34	8.6–30.3	0.5–5.9	338	2–16	68.7–81.0	4.3–11	91.6–128
Moth bean	61.5–66.0	22.0–26.0	1.6	343	–	1144–150	231–489	10.8–15.1
Rice bean	50–70	14–26	0.5–2.3	347	3.6–5.6	111–598	124–568	3.7–9.2
Tepary bean	65–69	21–25	0.9–1.2	360	2.1–3.1	28	450	7.1–8.3
Runner bean	62–83	27–32	1.0–1.8	372–458	1.2–1.8	72–138	59–559	10.3–12.0
Winged bean	12.7–42.2	27–43	13.9–26.7	409	3.4–27.0	102–850	310–637	4.9–6.0
Cowpea	50–60	23–32	5.4–11.2	336	3.9–10.6	85–93	438–498	10–11

associations, described in Akohoue *et al.* (2020). Tepary bean is a prominent legume crop with unique genes that can be introduced into common bean or *Phaseolus* species. The high levels of diversity reported in wild *Phaseolus acutifolius* and *Phaseolus parvifolius* suggest they contain beneficial genes for domesticated tepary bean genetic improvement (Mhlaba *et al.*, 2018). Lima bean (*Phaseolus lunatus* L.), one of the five domesticated *Phaseolus* bean crops, has extensive ecological adaptations from Mexico to Argentina. Garcia *et al.* (2021) successfully assembled a chromosome-level genome for the lima bean by combining long-read and short-read sequencing methods and using a dense genetic map derived from a biparental population. The high genetic variability of runner bean (*Phaseolus coccineus*) is valuable for breeding objectives, especially as a source of disease resistance and cold tolerance. Because *P. coccineus* germplasm has not been characterized, however, it cannot be used as a donor species for inter-specific hybridization (Guerra-García *et al.*, 2022). About 4685 runner bean accessions are documented worldwide, with the Austrian Agency for Health and Food Safety in Linz, Austria, maintaining 2674 accessions through the European Cooperative Programme for Plant Genetic Resources (Schwember *et al.*, 2017). Likewise, major moth bean *ex-situ* collections are maintained in the Indian National Gene Bank, ICAR–National Bureau of Plant Genetic Resources (NBPGR), New Delhi, India, with a total of 3422 accessions, which have been utilized in extensive characterization and evaluation programmes to identify superior germplasm accessions such as IC36245, IC36555, IC36667, IC36577 and IC36604. The genetic stocks MH 65, MH 34/66, MH 45 and MH 66 exhibited the most promising traits, making them potentially beneficial candidates for integration into the moth bean varietal improvement programme (Chandora *et al.*, 2023).

Certain legumes, such as winged bean (*Psophocarpus tetragonolobus*), lima bean (*P. lunatus*), hyacinth bean (*Lablab purpureus*) and Bambara groundnut (*Vigna subterranea*), have received attention as potentially underutilized legumes within the diverse range of legume species cultivated in tropical countries. This recognition is due to their hardiness, immense nutritional qualities and notably high protein content, particularly in their seeds (Cheng *et al.*, 2019). Bambara

groundnut landraces exhibit highly significant variations in morpho-agronomic traits. Additionally, positive relationships are found between yield and traits associated with yield, indicating an opportunity for further enhancement of agronomic attributes in these landraces (Khan *et al.*, 2021; Uba *et al.*, 2021). Likewise, a total of 94 grass pea accessions were evaluated for three qualitative and 19 quantitative traits in lowland (Antalya, Turkey) and highland (Isparta, Turkey) environments. There were substantial variations between genotypes for all agronomic traits in lowland environments. The highest biological yields were found in GP104 and GP145, with 22.5 and 82.4 g values in the lowland and highland, respectively (Arslan *et al.*, 2022). The diversity found in the world's most extensive collection of grass pea and their relatives maintained at the ICARDA seed bank in Syria has been utilized by the ICARDA scientists. Extensive Lathyrus collections are conserved in France, and the NBPGR is in India, Bangladesh and Chile (Singh *et al.*, 2013).

1.3.2 Genomic resources

Molecular markers have become essential tools for evaluating germplasm, exploring genetic diversity, mapping genes and employing marker-assisted selection (MAS) to enhance crop quality. Although genomic research on underutilized pulses has lagged behind major pulse crops, recent breakthroughs in genome sequencing using next-generation sequencing (NGS) technology have accelerated progress (Varshney *et al.*, 2009).

The development of genomic resources for underutilized pulses has progressed more recently compared to major pulse crops such as chickpea, pigeon pea, lentil, mung bean and urd bean. Initial efforts focused on crop/germplasm domestication, collection and evaluation, utilizing various marker systems in genetic diversity investigations. This led to the rapid identification of molecular markers, including those linked to quantitative trait loci (QTL) associated with desirable characteristics in underutilized pulses (Table 1.2).

With the advancement of sequencing technology, single nucleotide polymorphisms (SNPs) have become the predominant molecular markers owing to their unbiased nature, especially when exploring closely related untapped pulse species (Feuillet *et al.*, 2011; Cheng *et al.*, 2019; Sahruzaini

Table 1.2. Details on germplasm, DNA markers, and primary output from molecular-based diversity studies of untapped pulses (modified from Dwivedi *et al.*, 2023) CC BY 4.0 DEED.

Germplasm no.	No. of markers	Output	Reference
African yam bean (*Sphenostylis stenocarpa*)			
169	1789 SNPs	Genetically differentiated sub-populations and a strong correlation between phenotype and genotype-based matrices are observed	Shitta *et al.* (2022)
93	3722 SNPs	Highly predictive genotypic–phenotypic diversity relationships	Aina *et al.* (2021)
77	AFLPs	Out of 227 AFLP bands, 59 were polymorphic, with genetic distances between 0.048 and 0.842. Four distinct clusters were identified, unrelated to geographical origin	Adewale *et al.* (2015)
Adzuki bean (*Vigna angularis*)			
261	110 SSR	North and South China accessions sufficiently differentiated	Chen *et al.* (2015)
96	26 SSR	The wild relatives of Chinese and Japanese origins possess enough similarity	Liu *et al.* (2014)
176	85 SSR	Higher allelic diversity in the wild than in cultivated germplasm	Wang *et al.* (2012)
Bambara groundnut (*Vigna subterranea*)			
100	5927 DArTseq SNPs	Two distinct clusters, with cluster I containing TVSu-1897, the remaining in cluster II based on DArTseq SNPs and stress tolerance index	Odesola *et al.* (2023)
100	5925 SNPs	Higher genetic diversity among Nigerian accessions	Osundare *et al.* (2023)
93	2286 SNPs + morpho-agronomic traits	Two heterotic groups and unique accessions with specific characteristics	Majola *et al.* (2022)
270	3343 DArT SNPs	Greater diversity among accessions from diverse regions	Uba *et al.* (2021)
78	19 SSR	Higher within landrace diversity than between landraces; significant gene flow among the landraces	Minnaar-Ontong *et al.* (2021)
96	32 ISSR	Moderate to high levels of genetic differentiation	Khan *et al.* (2021)
Grass pea (*Lathyrus sativus*)			
400	56 SSR	Highly polymorphic and diverse germplasm	Rahman *et al.* (2022)
22	31 SSR	A few genetically diverse germplasms with low β-ODAP identified	Arslan *et al.* (2020)
118	18 EST-SSR	Sufficiently differentiated high- and low-β-ODAP (neurotoxin) accessions	Gupta *et al.* (2018)
283	30 SSR	Wild relatives clustered separately from cultigens	Wang *et al.* (2015)
Horse gram (*Macrotyloma uniflorum*)			
58	150 SSR	Three to four distinct genetic stocks	Kumar *et al.* (2020a)
48	117 SSR	Within a population, genetic variance is greater than between populations	Kaldate *et al.* (2017)
360	33 SSR and 24 morphological descriptors	Two distinct gene pools with higher levels of accessions variability	Chahota *et al.* (2017)

Continued

Table 1.2. Continued.

Germplasm no.	No. of markers	Output	Reference
Cow pea (*Vigna ungiculata*)			
217	886 DArTseq	High level of differentiation among populations and morphotypes; eight distinct clusters	Kafoutchoni *et al.* (2021)
281	493 SNPs	Distinct subpopulation structures differentiated by seed coat colour and ecological regions	Akohoue *et al.* (2020)
Hyacinth bean (*Lablab purpureus*)			
166	2460 SNPs	Four distinct subpopulations differentiated by ecogeographic regions	Njaci *et al.* (2023)
142	1000 SNPs	Substantial among accessions diversity than within accession variance	Muktar *et al.* (2021)
65	9320 DArTseq-based SNPs and 15,719 SilicoDart markers	Lower discrimination; higher within-population variance than among the population	Sserumaga *et al.* (2021)
Lima bean (*Phaseolus lunatus*)			
183 landraces	12 SSR and 7 morphological descriptors	Sufficient discrimination and introgression between Andean and Mesoamerican gene pools	Silva *et al.* (2019)
46	73 ISSR	There is higher genetic diversity in Mayan lowlands than in Mayan highland landraces; Mayan culture shaped diversity	Camacho-Pérez *et al.* (2018)
Rice bean (*Vigna umbellata*)			
Sequence 440 landraces	1,400,862 SNPs	Identified loci harbouring orthologues of FUL (FRUITFULL), FT (FLOWERING LOCUS T), and PRR3 (PSEUDO-RESPONSE REGULATOR 3) contribute to the adaptation of rice bean from its low latitude centre of origin towards higher latitudes, and the landraces which pyramid early-flowering alleles for these loci display maximally short flowering times	Guan *et al.* (2022)
32 accessions	300 genic SSR markers	A total of 3011 genic SSRs were identified as potential molecular markers; of these loci, 23 primer pairs were polymorphic among 32 rice bean accessions	Chen *et al.* (2016)
353 accessions	2,145,937 SNPs	Constructed a rice bean pangenome size of 679.32 Mb by GWAS analysis	Francis *et al.* (2023)
Tepary bean (*Phaseolus acutifolius*)			
Wild *P. acutifolius* accession	2,247,877 SNPs	Reasonable heat stress tolerance and reduced disease resistance gene acquisition are signs of tepary bean adaptation to arid and hot environments	Moghaddam *et al.* (2021)
156 accessions	24 SNPs	High-throughput genotyping has facilitated large-scale SNP detection, resulting in the development of molecular markers with related sequence information	Gujaria-Verma *et al.* (2016)
87 genotypes	15,645 SNPS	A total of 90 flanking candidate genes were identified using 1-kb genomic windows centred in each associated SNP marker	López-Hernández *et al.* (2023)

Continued

Table 1.2. Continued.

Germplasm no.	No. of markers	Output	Reference
422 cultivated accessions	53,676 SNPs	Genome-wide association studies identified loci and candidate genes controlling biotic stress resistance, including quantitative trait loci for weevil resistance, common bacterial blight, Fusarium wilt, and bean common mosaic necrosis virus	Bornowski *et al.* (2023)
Runner bean (*Phaseolus coccineus*)			
242 accessions	42,548 SNPs	Identified 24 SNPs associated with domestication, 13 with cultivar diversification, and eight with natural selection	Guerra-García *et al.* (2017)
113 genotypes	1190 SNPs	The genome-wide association analysis resulted in 18 high-quality SNPs that were subsequently used for the calculation of an estimated heat tolerance	Bomers *et al.* (2022)
237 accessions	79,286 SNPs	Introgression from wild to domesticated populations was detected, which might contribute to the recovery of the genetic variation	Guerra-García *et al.* (2022)
Moth bean (*Vigna aconitifolia*)			
428 accessions	9078 SNPs	Analysed genetic relationships among the *Vigna* species using SNP markers	Yadav *et al.* (2023)
240 accessions	1287 SSRs and 5606 transcripts	Genes associated with moisture stress tolerance in moth bean have been discovered	Tiwari *et al.* (2018)
F2 population (188)	172 SSR markers	As a genetic linkage map was being developed, 50 QTLs and three genes linked to 20 traits related to domestication were identified	Yundaeng *et al.* (2019)
Winged bean (*Psophocarpus tetragonolobus*)			
Two accessions (PI 491423 & PI 639033)	12,956 SSRs and 5190 SNPs	Gene discovery and marker development	Vatanparast *et al.* (2016)
F2 population (86)	1384 SNPs	The development of the first genetic linkage map for winged bean was accomplished	Chankaew *et al.* (2022)
Six accessions	9682 genic SSR	Developed assembly and validated genic SSR markers	Wong *et al.* (2017)

AFLP, amplified fragment length polymorphism; EST, expressed sequence tag; DArT, diversity array technology; ISSR, inter-simple sequence repeat; SNP, single nucleotide polymorphism; SSR, simple sequence repeat.

et al., 2020). Notably, substantial levels of synteny (44 syntenic blocks) have been identified between *P. vulgaris* and *Glycine max*, providing valuable insights for comparative genomic and genome selection research in *Phaseolus* species. A comprehensive database, Phaseolus Genes (2016), has been established, encompassing legacy markers and genomics-based markers such as simple sequence repeats (SSRs), SNPs and indels, primarily focusing on the common bean. In the case of lima bean, the chloroplast genome was sequenced for the first-time using Illumina sequencing technology, revealing its structure and organization (Tian *et al.*, 2021). Genomic studies by Garcia *et al.* (2021) and Wisser *et al.* (2021) resulted in a 623 Mb annotated assembly for lima bean, showcasing the combined efforts of sequencing, linkage and comparative analysis. The recent development of a tepary reference genome (Moghaddam *et al.*, 2021) provides valuable insights into linked loci conferring tolerance to biotic stresses (Bornowski *et al.*, 2023), given its unique breeding traits such as drought, heat, and salt stress tolerance (Gujaria-Verma *et al.*, 2016; Mhlaba *et al.*, 2018; Moghaddam *et al.*, 2021).

The genus *Vigna*, comprising five subgenera and over 100 wild species, has played a crucial role in domesticating seven distinct crops. Whole-genome sequences for these crops, including mung bean, black gram, rice bean, moth bean, cowpea, Bambara groundnut and adzuki bean, offer valuable genomic resources for achieving genetic progress in other *Vigna* species (Kang *et al.*, 2014; Yadav *et al.*, 2023). Comparative genomic studies have revealed substantial genome synteny among moth bean, mung bean, adzuki bean, rice bean and cowpea (Yundaeng *et al.*, 2019). Comprehensive sets of high-quality SNPs have been identified in rice bean and moth bean accessions through genotyping by sequencing (GBS) approaches (Suranjika *et al.*, 2023; Yadav *et al.*, 2023). Genome assemblies for moth bean have been generated, providing insights into stress-inducible genes associated with moisture stress tolerance (Tiwari *et al.*, 2018). Winged bean, despite its vast potential, has not received extensive research attention for molecular tool development supporting breeding programmes. However, recent efforts by Vatanparast *et al.* (2016) and Chankaew *et al.* (2022) have significantly expanded genetic resources for winged bean, including the identification of SSRs and high-confidence SNPs. The development of the first genetic linkage map and the identification of QTLs further emphasizes the importance of genomic resources for elucidating complex trait architectures in underutilized pulse crops. The genomic resources established in major pulse crops offer an excellent platform for strengthening the genetic enhancement of potential pulse crops. Leveraging these resources enables researchers and breeders to expedite the enhancement of traits in underutilized pulse crops, thereby making substantial contributions to the overarching objectives of bolstering food security and promoting agricultural sustainability.

The integration of genetic and genomic resources has propelled our comprehension and refinement of underutilized pulse crops to unprecedented heights. These resources serve as the bedrock for expediting progress in breeding initiatives, thereby aligning with the enduring objectives of fortifying global food security and fostering sustainable agricultural practices. As we continue to delve into research and foster collaborative efforts, the potential of underutilized pulses becomes increasingly unlocked, solidifying their pivotal role in establishing resilient food systems amidst the challenges of a changing climate.

References

Adewale, B.D., Vroh-Bi, I., Dumet, D.J., Nnadi, S., Kehinde, O.B. *et al.* (2015) Genetic diversity in African yam bean accessions based on AFLP markers: Towards a platform for germplasm improvement and utilization. *Plant Genetic Resources* 13, 111–118. DOI: 10.1017/s1479262114000707

Agyekum, P.B., Dombrowski, J., Lutterodt, H.E. and Ofosu, I.W. (2023) Consumption patterns and usage of selected underutilized legumes in a Ghanaian community. *Legume Science* p.e202.

Aina, A., Garcia-Oliveira, A.L., Ilori, C., Chang, P.L., Yusuf, M. *et al.* (2021) Predictive genotype-phenotype relations using genetic diversity in African yam bean (*Sphenostylis stenocarpa* (Hochst. ex. A. Rich) Harms). *BMC Plant Biology* 21, 547. DOI: 10.1186/s12870-021-03302-0

Akohoue, F., Achigan-Dako, E.G., Sneller, C., Deynze, A.V. and Sibiya, J. (2020) Genetic diversity, SNP-trait associations and genomic selection accuracy in a West African collection of Kersting's groundnut [*Macrotyloma geocarpum* (Harms) Maréchal & Baudet]. *PLoS ONE* 15, e0234769. DOI: 10.1371/journal.pone.0234769

Ali, A. and Bhattacharjee, B. (2023) Nutrition security, constraints, and agro-diversification strategies of neglected and underutilised crops to fight global hidden hunger. *Frontiers in Nutrition* 10:1144439. DOI: 10.3389/fnut.2023.1144439

Arslan, M., Basak, M., Aksu, E., Uzun, B. and Yol, E. (2020) Genotyping of low β-ODAP grass pea (*Lathyrus sativus* L.) with EST-SSR markers Braz. *Brazilian Archives of Biology and Technology* 63, e20190150. DOI: 10.1590/1678-4324-2020190150

Arslan, M., Yol, E. and Türk, M. (2022) Disentangling the genetic diversity of grass pea germplasm grown under lowland and highland conditions. *Agronomy* 12(10), 2426.

Ayilara, M.S., Abberton, M., Oyatomi, O.A., Odeyemi, O. and Babalola, O.O. (2022) Potentials of underutilized legumes in food security. *Frontiers in Soil Science* 2. DOI: 10.3389/fsoil.2022.1020193

Bomers, S., Sehr, E.M., Adam, E., von Gehren, P., Hansel-Hohl, K., Prat, N. and Ribarits, A. (2022) Towards heat tolerant runner bean (*Phaseolus coccineus* L.) by utilizing plant genetic resources. *Agronomy* 12(3), p.612.

Bornowski, N., Hart, J.P., Palacios, A.V., Ogg, B., Brick, M.A. *et al.* (2023) Genetic variation in a tepary bean (*Phaseolus acutifolius* A. Gray) diversity panel reveals loci associated with biotic stress resistance. *The Plant Genome* p.e20363.

Boukar, O., Belko, N., Chamarthi, S., Togola, A., Batieno, J. *et al.* (2019) Cowpea (*Vigna unguiculata*): Genetics, genomics and breeding. *Plant Breeding* 138(4), 415–424.

Camacho-Pérez, L., Martínez-Castillo, J., Mijangos-Cortés, J.O., Ferrer-Ortega, M.M., Baudoin, J.P. and Andueza-Noh, R.H. (2018) Genetic structure of Lima bean (*Phaseolus lunatus* L.) landraces grown in the Mayan area. *Genetic Resources and Crop Evolution* 65, 229–241. DOI: 10.1007/s10722-017-0525-1

Chahota, R.K., Shikha, D., Rana, M., Sharma, V., Nag, A. *et al.* (2017) Development and characterisation of SSR markers to study genetic diversity and horse gram germplasm population structure (*Macrotyloma uniflorum*). *Plant Molecular Biology Reporter* 35, 550–561. DOI:10.1007/s11105-017-1045-z

Chandora, R., Basavaraja, T. and Rana, J.C. (2023) Moth bean (*Vigna aconitifolia*): a minor legume with major potential to address global agricultural challenges. *Frontiers in Plant Science* 14, p.1179547.

Chankaew, S., Sriwichai, S., Rakvong, T., Monkham, T., Sanitchon, J. *et al.* (2022) The first genetic linkage map of winged bean [*Psophocarpus tetragonolobus* (L.) DC.] and QTL mapping for flower-, pod-, and seed-related traits. *Plants (Basel)* 11(4), 500. DOI: 10.3390/plants11040500.

Chen, H., Liu, L., Wang, L., Wang, S., Wang, M.L. and Cheng, X. (2015) Development of SSR markers and assessment of genetic diversity of adzuki bean in the Chinese germplasm collection. *Molecular Breeding* 35, 191. DOI: 10.1007/s11032-015-0383-5

Chen, H., Chen, X., Tian, J., Yang, Y., Liu, Z. *et al.* (2016) Development of gene-based SSR markers in rice bean (*Vigna umbellata* L.) based on transcriptome data. *PloS One* 11(3), p.e0151040.

Cheng, A., Raai, M.N., Zain, N.A.M., Massawe, F., Singh, A. and Wan-Mohtar, W.A.A.Q.I. (2019) In search of alternative proteins: unlocking the potential of underutilized tropical legumes. *Food Security* 11, 1205–1215.

Devaux, A., Goffart, J.-P., Petsakos, A., Kromann, P., Gatto, M. *et al.* (2020) Global food security, contributions from sustainable potato agri-food systems. In: Campos, H. and Ortiz, O. (eds) *The Potato Crop.* Springer, Cham, Switzerland, pp. 3–35. DOI: 10.1007/978-3-030-28683-5_1

Dwivedi, S.L., Chapman, M.A., Abberton, M.T., Akpojotor, U.L. and Ortiz, R. (2023) Exploiting genetic and genomic resources to enhance productivity and abiotic stress adaptation of underutilized pulses. *Frontiers in Genetics* 14, 1193780.

FAO (2021) The state of food security and nutrition in 2021: the world is at a critical juncture. FAO, Rome.

Feuillet, C., Leach, J.E., Rogers, J., Schnable, P.S. and Eversole, K. (2011) Crop genome sequencing: Lessons and rationales. *Trends in Plant Science* 16(2), 77–88.

Fghire, R., Anaya, F., Lamnai, K. and Faghire, M. (2022) Alternative crops as a solution to food security under climate changes. In: Chatoui, H., Merzouki, M., Moummou, H., Tilaoui, M., Saadaoui, N. and Brhich, A. (eds) *Nutrition and Human Health.* Springer, Cham, Switzerland. DOI: 10.1007/978-3-030-93971-7_7

Francis, A., Singh, N.P., Singh, M., Sharma, P., Kumar, D. *et al.* (2023) The ricebean genome provides insight into Vigna genome evolution and facilitates genetic enhancement. *Plant Biotechnology Journal* 21(8), 1522.

Garcia, T., Duitama, J., Zullo, S.S., Gil, J., Ariani, A. *et al.* (2021) Comprehensive genomic resources related to domestication and crop improvement traits in Lima bean. *Nature Communications* 12, 702. https://doi.org/10.1038/s41467-021-20921-1

George, T.T., Obilana, A.O. and Oyeyinka, S.A. (2020) The prospects of African yam bean: past and future importance. *Heliyon* 6, e05458. DOI: 10.1016/j.heliyon.2020.e05458

Gondwe, T.M., Alamu, E.O., Mdziniso, P. and Maziya-Dixon, B. (2019) Cowpea (*Vigna unguiculata* (L.) Walp) for food security: an evaluation of end-user traits of improved varieties in Swaziland. *Science Reports* 9(1), 15991. DOI: 10.1038/s41598-019-52360-w.

Guan, J., Zhang, J., Cong, D., Zhang, Z., Yu, Y. *et al.* (2022) Genomic analyses of rice bean landraces reveal adaptation and yield-related loci to accelerate breeding. *Nature Communications* 13, 5707. DOI: 10.1038/s41467-022-33515-2

Guerra-García, A., Suárez-Atilano, M., Mastretta-Yanes, A., Delgado-Salinas, A. and Piñero, D. (2017) Domestication genomics of the open-pollinated scarlet runner bean (*Phaseolus coccineus* L.). *Frontiers in Plant Science* 8, 1891.

Guerra-García, A., Rojas-Barrera, I.C., Ross-Ibarra, J., Papa, R. and Piñero, D. (2022) The genomic signature of wild-to-crop introgression during the domestication of scarlet runner bean (*Phaseolus coccineus* L.). *Evolution Letters* 6(4), 295–307. DOI: 10.1002/evl3.285

Gujaria-Verma, N., Ramsay, L., Sharpe, A.G., Sanderson, L.-A., Debouck, D.G. *et al.* (2016) Gene-based SNP discovery in tepary bean (*Phaseolus acutifolius*) and common bean (*P. vulgaris*) for diversity analysis and comparative mapping. *BMC Genomics* 17, 239. DOI: 10.1186/s12864-016-2499-3

Gupta, P., Udupa, S.M., Gupta, D.S., Kumar, J. and Kumar, S. (2018) Population structure analysis and determination of neurotoxin content in a set of grass pea (*Lathyrus sativus* L.) accessions of Bangladesh origin. *Crop Journals* 6, 435–442. DOI: 10.1016/j.cj.2018.03.004

Joshi, B.K., Shrestha, R., Gauchan, D. and Shrestha, A. (2020) Neglected, underutilized, and future smart crop species in Nepal. *Journals of Crop Improvement* 34, 291–313.

Kafoutchoni, K.M., Agoyi, E.E., Agbahoungba, S., Assogbadjo, A.E. and Agbangla, C. (2021) Genetic diversity and population structure in a regional collection of Kersting's groundnut (*Macrotyloma geocarpum* (Harms) Maréchal & Baudet). *Genetic Resource Crop Evolution* 68, 3285–3300. DOI: 10.1007/s10722-021-01187-4

Kaldate, R., Rana, M., Sharma, V., Hirakawa, H., Kumar, R. *et al.* (2017) Development of genome-wide SSR markers in horse gram and their use for genetic diversity and cross transferability analysis. *Molecular Breeding* 37, 103. DOI: 10.1007/s11032-017-0701-1

Kang, Y.J., Kim, S.K., Kim, M.Y., Lestari, P., Kim, K.H. *et al.* (2014) Genome sequence of mungbean and insights into evolution within *Vigna* species Yang. *Nature Communications* 5, 5443. DOI: 10.1038/ncomms6443

Kanishka, R.C., Jamir, M., Theunuo, S., T, B., Verma, H., Chandora, R. (2024) On the road to a sustainable and climate-smart future: recent advancements in genetics and genomics of pulse crops in the hills. In: Gahlaut, V., Jaiswal, V. (eds) *Genetics and Genomics of High-Altitude Crops*. Springer, Singapore, pp.1–45. https://doi.org/10.1007/978-981-99-9175-4_1

Khan, M.M.H., Rafii, M.Y., Ramlee, S.I., Jusoh, M. and Mamun, M.A. (2021) Genetic analysis and selection of Bambara groundnut (*Vigna subterranea* [L.] Verdc.) landraces for high yield revealed by qualitative and quantitative traits. *Scientific Reports* 11, 7597. DOI: 10.1038/s41598-021-87039-8

Kousar, S., Ahmed, F., Pervaiz, A. and Bojnec, S. (2021) Food insecurity, population growth, urbanization and water availability: The role of government stability. *Sustainability* 13(22), 12336. DOI: 10.3390/su132212336

Kumar, R., Kaundal, S.P., Sharma, V., Sharma, A., Singh, G. *et al.* (2020) Development of transcriptome-wide SSR markers for genetic diversity and structure analysis in *Macrotyloma uniflorum* (Lam.) Verdc. *Physiology Molecular Biology Plants* 26, 2255–2266. DOI: 10.1007/s12298-020-00898-9

Liu, C.Y., Fan, B.J., Cao, Z.M., Su, Q.Z., Wang, Y. *et al.* (2014) Genetic diversity analysis of wild adzuki bean germplasm and its relatives by using SSR markers. *Acta Agronomica Sinica (China)* 40, 174. DOI: 10.3724/SP.J.1006.2014.00174

López-Hernández, F., Burbano-Erazo, E., León-Pacheco, R.I., Cordero-Cordero, C.C., Villanueva-Mejía, D.F., Tofiño-Rivera, A.P. and Cortés, A.J. (2023) Multi-environment genome-wide association studies of yield traits in common bean (*Phaseolus vulgaris* L.) × tepary bean (*P. acutifolius* A. Gray) interspecific advanced lines in humid and dry Colombian Caribbean Subregions. *Agronomy* 13(5), 1396. DOI: 10.3390/agronomy13051396

Majola, N.G., Gerrano, A.S., Amelework, A., Shimelis, H. and Swanevelder, D. (2022) Genetic diversity and population structure analyses of South African Bambara groundnut (*Vigna subterranea* [L.] Verdc) collections using SNP markers. *South African Journal of Botany* 150, 1061–1068. DOI: 10.1016/j.sajb.2022.09.008

Mayes, S., Massawe, F.J., Alderson, P.G., Roberts, J.A., Azam-Ali, S.N. and Hermann, M. (2012) The potential for underutilised crops to improve the security of food production. *Journal of Experimental Botany* 63 (3), 1075–1079. DOI: 10.1093/jxb/err396

Mekonnen, T.W., Gerrano, A.S., Mbuma, N.W. and Labuschagne, M.T. (2022) Breeding of vegetable cowpea for nutrition and climate resilience in Sub-Saharan Africa: progress, opportunities, and challenges. *Plants* 11(12), 1583.

Mhlaba, Z.B., Mashilo, J., Shimelis, H., Assefa, A.B. and Modi, A.T. (2018) Progress in genetic analysis and breeding of tepary bean (*Phaseolus acutifolius* A. Gray): A review. *Scientia Horticulturae* 237, 112–119.

Minnaar-Ontong, A., Gerrano, A.S. and Labuschagne, M.T. (2021) Assessment of genetic diversity and structure of Bambara groundnut [*Vigna subterranea* (L.) Verdc.] landraces in South Africa. *Scientific Reports* 11, 7408. DOI: 10.1038/s41598-021-86977-7

Moghaddam, S.M., Oladzad, A., Koh, C., Ramsay, L., Hart, J.P. *et al.* (2021) The tepary bean genome provides insight into evolution and domestication under heat stress. *Nature Communications* 12(1), 2638.

Muktar, M.S., Sartie, A., Teressa, A., Habte, E. and Jones, C.S. (2021) Genetic diversity among and within accessions of a lablab (*Lablab purpureus*) collection maintained in the ILRI forage genebank. *Paper presented at the Joint XXIV International Grassland and XI Rangeland 2021 Congress, Nairobi, Kenya, 25–29 October 2021*. ILRI, Nairobi.

Naeem, M., Shabbir, A., Ansari, A.A., Aftab, T., Khan, M.M.A. and Uddin, M. (2020) Hyacinth bean (*Lablab purpureus* L.)–An underutilised crop with future potential. *Scientia Horticulturae* 272, 109551.

Nayak, S.P., Lone, R.A., Fakhrah, S., Chauhan, A., Sarvendra, K. and Mohanty, C.S. (2022) Mainstreaming underutilized legumes for providing nutritional security. In: Bhat, R. (ed.) *Future Foods*. Academic Press, Elsevier, London, pp. 151–163.

Njaci, I., Waweru, B., Kamal, N., Muktar, M.S., Fisher, D. *et al.* (2023) Chromosome-level genome assembly and population genomic resource to accelerate orphan crop lablab breeding. *Nature Communications* 14, 1915. DOI: 10.1038/s41467-023-37489-7

Odesola, K.A., Olawuvi, O.J., Paliwal, R., Ovatomi, O.A. and Abberton, M.T. (2023) Genome-wide association analysis of phenotypic traits in Bambara groundnut under drought-stressed and non-stressed conditions based on DArTseq SNP. *Frontiers Plant Science* 14, 1104417. DOI: 10.3389/fpls.2023.1104417

Osundare, O.T., Akinyele, B.O., Odiyi, A.C., Paliwal, R., Oyatomi, O.A. and Abberton, M.T. (2023) Genetic diversity and population structure of some Nigerian accessions of Bambara groundnut (*Vigna subterranea* (L.) Verdc.,) using DArT SNP markers. *Genetic Resources and Crop Evolution* 70(3), 887–901.

Rahman, M.M., Quddus, M.R., Ali, M.O., Liu, R., Li, M. *et al.* (2022) Genetic diversity of Lathyrus sps collected from different geographical regions. *Molecular Biology Reports* 49, 519–529. DOI: 10.1007/s11033-021-06909-6

Regmi, B., Kunwar, S., Acharya, T.D. and Gyawali, P. (2023) Potential of underutilized grain crops in the Western Mountains of Nepal for food and nutrient security. *Agriculture* 13(7), 1360.

Sahruzaini, N.A., Rejab, N.A., Harikrishna, J.A., Khairul Ikram, N.K., Ismail, I., Kugan, H.M. and Cheng, A. (2020) Pulse crop genetics for a sustainable future: Where we are now and where we should be heading. *Frontiers in Plant Science* 11, 531.

Savithramma, D.L. and Shambulingappa, K.G. (1996) Genetic divergence studies in horse gram *Macrotyloma uniflorum* (L.) Verdec. *Mysore Journal of Agriculture Science* 30, 223–229.

Schwember, A.R., Carrasco, B. and Gepts, P. (2017) Unravelling agronomic and genetic aspects of runner bean (*Phaseolus coccineus* L.). *Field Crops Research* 206, 86–94.

Shitta, N.S., Unachukwu, N., Edemodu, A.C., Abebe, A.T., Oselebe, H.O. and Abtew, W.G. (2022) Genetic diversity and population structure of an African yam bean (*Sphenostylis stenocarpa*) collection from IITA GenBank. *Scientific Reports* 12, 4437. DOI: 10.1038/s41598-022-08271-4

Silva, R.N.O., Lopes, A.C.A., Gomes, R.L.F., Pádua, J.G. and Burle, M.L. (2019) There is a high diversity of cultivated lima beans (*Phaseolus lunatus*) in Brazil, consisting of one Andean and two Mesoamerican groups with strong introgression between the gene pools. *Genetics Molecular Research* 18, GMR18441. DOI:10.4238/gmr18441

Singh, M., Upadhyaya, H.D. and Bisht, I.S. (2013) Introduction. In: Singh, M., Upadhyaya, H.D. and Bisht, I.S. (eds) *Genetic and Genomic Resources of Grain Legume Improvement*. Elsevier, London, pp. 1–10. DOI: 10.1016/b978-0-12-397935-3.00001-3

Singh, R.K., Sreenivasulu, N. and Prasad, M. (2022) Potential of underutilized crops to introduce the nutritional diversity and achieve zero hunger. *Functional Integrated Genomics* 22, 1459–1465. https://doi.org/10.1007/s10142-022-00898-w

Sinković, L., Pipan, B., Vasić, M., Antić, M., Todorović, V. *et al.* (2019) Morpho-agronomic characterisation of runner bean (*Phaseolus coccineus* L.) from south-eastern Europe. *Sustainability* 11(21), 6165.

Sreenivasulu, N. and Fernie, A.R. (2022) Diversity: current and prospective secondary metabolites for nutrition and medicine. *Current Opinion Biotechnology* 74, 164–170. DOI: 10.1016/j.copbio.2021.11.010.

Sserumaga, J. P., Kavondo, S. I., Kigozi, A., Kiggundu, M., Namazzi, C. *et al.* (2021) Genome-wide diversity and structure variation among lablab [*Lablab purpureus* (L.) Sweet] accessions and their implication in a forage breeding program. *Genetic Resource Crop Evolution* 68, 2997–3010. DOI: 10.1007/s10722-021-01171-y

Suranjika, S., Pradhan, S., Kalia, R.K. and Dey, N. (2023) De novo assembly of the whole genome of Moth bean (*Vigna aconitifolia*), an underutilised Vigna species of India. *bioRxiv* 023.05.18.540937.

Tian, J., Isemura, T., Kaga, A., Vaughan, D.A. and Tomooka, N. (2013) Genetic diversity of the rice bean (*Vigna umbellata*) gene pool assessed by SSR markers. *Genome* 56(12), 717–727.

Tian, S., Lu, P., Zhang, Z., Wu, J.K., Zhang, H. and Shen, H. (2021) Chloroplast genome sequence of Chongming lima bean (*Phaseolus lunatus* L.) and comparative analyses with other legume chloroplast genomes. *BMC Genomics* 22, 194. DOI: 10.1186/s12864-021-07467-8

Tiwari, B., Kalim, S., Tyagi, N., Kumari, R., Bangar, P. *et al.* (2018) Identification of genes associated with stress tolerance in moth bean [*Vigna aconitifolia* (Jacq.) Marechal], a stress hardy crop. *Physiology and Molecular Biology of Plants* 24, 551–561.

Uba, C.U., Oselebe, H.O., Tesfaye, A.A. and Abtew, W.G. (2021) Genetic diversity and population structure analysis of Bambara groundnut (*Vigna subterrenea* L) landraces using DArT SNP markers. *PLoS One* 16, e0253600. DOI: 10.1371/journal.pone.0253600

Ulian, T., Diaz-Granados, M., Pironon, S., Padulosi, S., Liu, U. *et al.* (2020) Unlocking plant resources to support food security and promote sustainable agriculture. *Plants, People, Planet* 2(5), 421–445.

Varshney, R.K., Nayak, S.N., May, G.D. and Jackson, S.A. (2009) Next-generation sequencing technologies and their implications for crop genetics and breeding. *Trends in Biotechnology* 27(9), 522–530.

Vatanparast, M., Shetty, P., Chopra, R., Doyle, J.J., Sathyanarayana, N. and Egan, A.N. (2016) Transcriptome sequencing and marker development in winged bean (*Psophocarpus tetragonolobus*; Leguminosae). *Scientific Reports* 6, 29070. DOI: 10.1038/srep29070

von Grebmer, K., Bernstein, J., Geza, W., Ndlovu, M., Wiemers, M. *et al.* (2023) Global Hunger Index: The Power of Youth in Shaping Food Systems. Welthungerhilfe (WHH), Bonn; Concern Worldwide, Dublin.

Wang, F., Yang, T., Burlyaeva, M., Li, L., Jiang, J. *et al.* (2015) Genetic diversity of grasspea and its relative species revealed by SSR markers. *PLoS ONE* 10, e0118542. DOI: 10.1371/journal.pone.0118542

Wang, L., Cheng, X., Wang, S. and Tian, J. (2012) Analysis of an applied core collection of adzuki bean germplasm by using SSR markers. *Journal of Integrative Agriculture* 11, 1601–1609. DOI:10.1016/S2095-3119(12)60163-4

Wisser, R.J., Oppenheim, S.J., Ernest, E.G., Mhora, T.T., Dumas, M.D. *et al.* (2021) Genome assembly of a Mesoamerican derived variety of lima bean: a foundational cultivar in the Mid-Atlantic USA. *G3 Genes|Genomes|Genetics* 11 (11), jkab207. DOI: 10.1093/g3journal/jkab207

Wong, Q.N., Tanzi, A.S., Ho, W.K., Malla, S., Blythe, M. *et al.* (2017) Development of gene-based SSR markers in winged bean (*Psophocarpus tetragonolobus* (L.) DC.) for diversity assessment. *Genes* 8(3), 100.

Yadav, A.K., Singh, C.K., Kalia, R.K., Mittal, S., Wankhede, D.P. *et al.* (2023) Genetic diversity, population structure, and genome-wide association study for the flowering trait in a diverse panel of 428 moth bean (*Vigna aconitifolia*) accessions using genotyping by sequencing. *BMC Plant Biology* 23, 228. DOI: 10.1186/s12870-023-04215-w

Yundaeng, C., Somta. P., Amkul, K., Kongjaimun, A., Kaga, A. and Tomooka, N. (2019) Construction of genetic linkage map and genome dissection of domestication-related traits of moth bean (*Vigna aconitifolia*), a legume crop of arid areas. *Molecular Genetic Genomics* 294(3), 621–635. DOI: 10.1007/s00438-019-01536-0

2 African Yam Bean (*Sphenostylis stenocarpa*)

Daniel B. Adewale¹* and Michael T. Abberton²

¹*Department of Crop Science and Horticulture, Federal University Oye-Ekiti, Ikole-Ekiti campus, Nigeria;* ²*Genetic Resources Centre, International Institute of Tropical Agriculture, Ibadan, Nigeria*

Abstract

The increasing focus on the inclusion of orphan crops into dietary regimens is a significant trend observable in the area of nutrition. Little information is available on the present state of genetic resource repositories for this group of crops. Historically, most of these resources have evolved as indigenous landraces within diverse traditional and cultural communities. Therefore, it is very important to identify niches that host them for documentation, which could guide subsequent exploration. African yam bean (*Sphenostylis stenocarpa* Hochst Ex. A. Rich) Harms is a prominent and foremost tuberous legume of humid tropical Africa. It is an orphan crop with no breeding record. Some African yam bean (AYB) germplasm resources have been conserved in national and international institutes, and some still exist as unconserved landraces. Within national and international conservation domains, ≤600 AYB accessions are conserved. These are inadequate numbers for capturing the total intraspecific variability within the species. Through reports, locations where AYB landraces were obtained for research were documented. Furthermore, institutes that host some accessions were documented. The Genetic Resources Centre (GRC) at the International Institute for Tropical Agriculture (IITA), Ibadan, Nigeria, holds the largest (450 resources) conservation of the crop, the majority of which are from Nigeria. Areas where AYB landraces exist as indigenous crops in Nigeria cover 22 states, 445 local government areas, and diverse ethnicities, climate, soil, relief, vegetation, etc. This study further documented reported polymorphic traits in AYB and provided insights and strategies to overcome constraints in AYB. Some landraces have proven significant in past research but have yet to be conserved. The need to source them for conservation, utilization and genetic improvement is imperative.

Keywords: African yam bean, genetic resources, geographical distribution, germplasm, landraces, niches, Nigeria

2.1 Introduction

Accessions (already conserved genetic resources) and landraces (genetic resources in the hands of cultural/traditional custodians) represent one of the greatest treasures to humanity. The different genetic resources host utilizable diversity for nutrition, better taste, increased yield, disease/pest resistance, climate resilience, reduced agricultural inputs and improvement (Vodouhe *et al.*, 2007; McCouch *et al.*, 2020). Conservation of the genetic resources of most orphan crops is rare in international and national agricultural institutes because they are not 'mandate' crops. However,

---

*Corresponding author: daniel.adewale@fuoye.edu.ng

© CAB International 2024. *Potential Pulses: Genetic and Genomic Resources*
(eds R. Chandora, T. Basavaraja and A. Pratap)
DOI: 10.1079/9781800624658.0002

the less or non-preferred crop species may be the most relevant to solving future problems such as sustainable food, feed, fibre and fuel productivity (Stifel, 1990). The quantity of research on African yam bean (*Sphenostylis stenocarpa* (Hochst Ex. A. Rich) Harms) is increasing. The reported nutritional and health benefits of the crop compared well with, and are superior to, most common legumes (Okigbo, 1973; Rachie, 1973; NRC, 1979, 2006; Nwokolo, 1996; Adewale and Nnamani, 2022) for humans and livestock. The identified benefits in African yam bean (AYB) have significantly promoted the campaign for its cultivation and utilization.

Now that the global environmental boundaries for food production have been exceeded (Gulisano *et al.*, 2019), further reliance on a few crops to meet the global food supply will not be sustainable. All innovations that relegate the incorporation of other crops in the global food system will continually support the loss of biodiversity, degradation of agricultural land, lowering water quality, etc. Moreover, the continuous impactful trend of climate change will lead to further genetic erosion, agronomic, ecological, nutritional and economic risks, and so on. The food security of the future does not only depend on the conservation of plant genetic resources in international and national seed banks, but also the genetic resources in farmers' hands and the traditional culinary knowledge of the food products (Ramprasad and Clements, 2016). Therefore, securing food for the future may depend on the knowledge of the different landraces, places and sources of their indigeneity and availability for access.

Whilst the farmers hold the genetic resources for traditional crops, the potential of such genetic material remains unknown, the threat to its germplasm loss would be much stronger and genetic value addition would be impossible. Aside from knowledge of species taxonomy, identifying the ecological distribution of crops and their wild relatives is critical (Ng, 1990). Ng (1990) stated that a thorough review of the literature and past exploration records (individual or institutional) is needed as an essential priority to guide the subsequent collection of plant genetic resources, because continuous germplasm collection and conservation will remain a primary priority in plant breeding. Most diversity studies on AYB revealed that the present gene pool of

the crop is still inadequate; the present study proposed to unveil niches, especially within Nigeria, where more genetic resources of the species can be further sourced for conservation. Because data on orphan crops (including AYB) are undocumented in agricultural production statistics, Foyer *et al.* (2016) suggested the use of systematic inventories for the documentation of a legume germplasm collection. The present chapter achieves this for AYB in Nigeria, and partially for Ghana, to identify national segments hosting AYB genetic resources from the literature.

2.2 Origin, Distribution and Botanical Description

African yam bean is native to Africa, primarily indigenous and endemic to west and central Africa. Ethiopia hosts the highest genetic resource diversity of the crop in the wild (Kay, 1987; Nyanayo and Nyigifa, 2011). According to Adesoye and Nnadi (2011), the species was disseminated from Ethiopia to other parts of Africa. The humid tropics of Africa, which includes western and central regions, have sustainably provided the most enabling environment for the crop to thrive since its introduction. Potter (1992) identified Cameroon, Cote d'Ivoire, the Democratic Republic of Congo, Ghana, Nigeria and Togo as major countries where the crop is cultivated. The wild and cultivated relatives of the African yam bean exist in tropical Africa (Kay, 1987; Nyanayo and Osuji, 2007; Moteetee and Van Wyk, 2012). Potter (1992) noted that the cultivation and production of AYB in Nigeria were much higher than those of other countries, and research articles on the crop from Nigeria greatly exceeded those of other countries in the producing regions.

2.2.1 African yam bean – a legume for humid tropical regions

African yam bean is among the few legumes well adapted to Africa's hot and humid environment, along with lima bean (*Phaseolus lunatus* L.) (Snapp *et al.*, 2018). The high rainfall belt of Nigeria, which includes the south, south-east, south-west and middle belt (mainly the Southern

Guinea Savanna), is where AYB is endemic as an indigenous crop in Nigeria. An extensive survey study by Saka *et al.* (2004) and Nnamani *et al.* (2017) for south-west and south-east Nigeria, respectively, ascertained the ecological domain for AYB genetic resources to be in the rainforest belt of Nigeria. Amoatey *et al.* (2000) and Adjei *et al.* (2023) made different collections of AYB landraces in the rainforest regions of Ghana. These further reveal that the crop is indigenous to the humid tropical region of Africa and corroborate the earlier assertion of Potter and Doyle (1992) and Okpara and Omaliko (1995) that the crop is mainly grown in high rainfall regions.

According to Saka *et al.* (2004), most minor legumes such as lima bean, African yam bean, pigeon pea and Bambara groundnut are intercropped with major crops on an average land range of 0.2–0.4 ha. Their study also revealed that minor legumes occupy less than 10% of cultivated land area in south-west Nigeria. A survey of underutilized legume cultivation in Nigeria showed that their production is mostly among older (≥50 years) farmers, whose production amounts to 80% of the estimated total (Saka *et al.*, 2004; Nnamani *et al.*, 2017). Fig. 2.1 shows the zone where the genetic resources of African yam bean can be sourced in Nigeria, and the information in Klu *et al.* (2001), Saka *et al.* (2004) and Nnamani *et al.* (2017) reflects the risk for the gradual disappearance of plant genetic resources of African yam bean in Africa.

2.3 Crop Gene Pool, Evolutionary Relationships and Systematics

2.3.1 Assurances of extant plant genetic resources as landraces

Landraces account for most of the planting materials for crop production worldwide, especially in developing nations where peasant agriculture supplies more than 70% of the food. Landraces are the primary planting material for rural farming

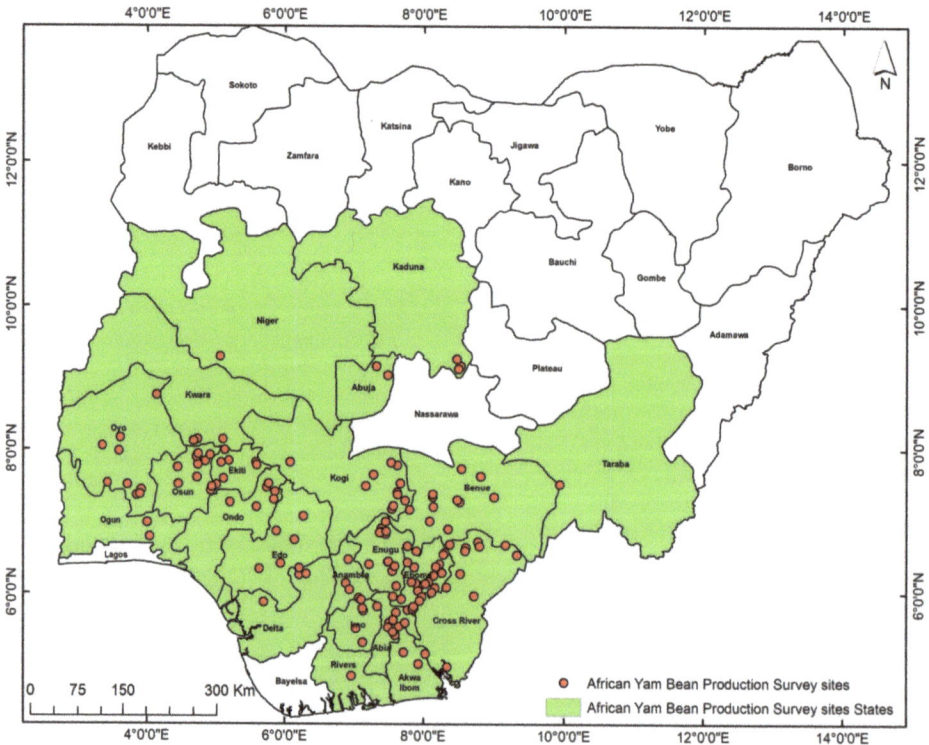

Fig. 2.1. Distribution pattern of African yam bean in Nigeria. (Author's own figure.)

families (Poinetti, 2005), ensuring that their genetic resources are in existence and available in the growing areas. Rural farmers in Nigeria still hold sizeable and promising quantities of genetic resources for African yam bean (Akande, 2008; Adewale *et al.*, 2012; Nnamani *et al.*, 2017). Informally, seeds of AYB landraces are saved and exchanged among farmers or sold in local/rural markets by vendors, primarily women, in various geo-cultural regions of Nigeria.

Within the rainforest/humid tropical regions of West Africa, AYB landraces still exist as:

- morphologically distinct species recognized by local farmers with a distinct name;
- possessing genetic integrity with distinct variation;
- a genetically dynamic crop; and
- morphotypes or ecotypes for specific culinary values, with a significant performance under different conditions, resistance to pests, quality, etc. (Kingsbury, 2009).

Landraces are dynamic cultivated crops with clear links to historical/traditional origin, having unique identities, being distinct in utility by farmers' preference, associated with some set of farmers' practices, hosting genetic diversity, adaptable to varied edapho-climatic environments, and so on (Villa *et al.*, 2005; Casañas *et al.*, 2017; Azeez *et al.*, 2018). Table 2.1 shows the natural phenotypic diversity in colour and colour pattern among seeds of AYB obtained from many sources, especially from markets and rural farmers in various locations in Nigeria. This provides support that African yam bean landraces are within Nigeria. Generally, as shown in Table 2.1, sources of AYB genetic resources include local markets, farmers, vendors, State Ministries of Agriculture, research institutes and experimental fields. Furthermore, the extent of diversity in the seed base, on testa colour description, was huge; landraces with mono-colorations varied from white to black, and different forms of marbling/mosaic coloration were also observed in the many reports (Table 2.1). Local markets and rural farmers are Nigeria's most reliable sources for African yam bean landraces. Every landrace gradually becomes a distinct genetic entity through interactions with many edapho-climatic factors, including farming systems and cultural practices. This particular way of development is strongly associated with a specific geographical location. Such variants do not easily lose their genetic identity owing to natural selection, and they exhibit good 'fitness' for sustainable productivity in their ecology, according to Kingsbury (2009). The current germplasm of AYB is still very low, coupled with a narrow genetic base (Akande *et al.*, 2009; Ojuederie *et al.*, 2014). For instance, the morpho-biochemical and genomic diversity studies (Popoola *et al.*, 2011, 2017; Adewale *et al.*, 2012, 2015; Ojuederie *et al.*, 2014, 2015; Abdulkareem *et al.*, 2015; Animasaun *et al.*, 2021; Shitta *et al.*, 2021, 2022) were not able to identify genetic resource(s) or candidate gene(s) for some of the major constraints limiting the commercial cultivation and wide utilization of the crop. This implies that the optimum genetic variability of the crop is yet to be captured.

2.4 Germplasm Collection, Characterization, Evaluation and Conservation

There are records of active explorations for AYB genetic resources and conservation in Nigeria by national and international programmes. Efforts for creating AYB landrace collections were equally recorded for Ghana, but their conservation status is unclear.

2.4.1 Historic exploration of African yam bean in Nigeria and conservation status

The genetic resources of AYB have been explored throughout parts of Nigeria by the Genetic Resources Centre (GRC), International Institute of Tropical Agriculture (IITA), Ibadan, Nigeria, and the National Centre for Genetic Resources and Biotechnology (NACGRAB), Ibadan, Nigeria. Moreover, underutilized legume breeders in the Institute of Agricultural Research and Training (IAR&T), Ibadan, Nigeria, explored AYB as well as other underutilized legumes in south-west Nigeria (Akande, 2008). Furthermore, and quite recently, collections of some AYB landraces were created in south-east Nigeria by scientists from Ebonyi State University, Abakaliki, Nigeria (Nnamani *et al.*, 2017). Abberton *et al.* (2022) and Nnamani *et al.* (2017) reported 450 and 34 accessions of African yam bean, respectively, to

Table 2.1. Sources with testa descriptions of some African yam bean landraces obtained in various locations in Nigeria. (Author's own table.)

Sr. No.	Places of collection	Seed sources	Testa descriptions	Documenting references
1	Umuahia, Abia state	Market	Brown	Ajayi (2011)
2	Within Lat.5° and 9°N and Long.3° and 7°E	Farmers	NS	Akande (2009)
3	Oba-Adesida, Akure, Iyere-Owo, Oja-Oba, Ibadan	Market	NS	Akinlabi et al. (2015)
4	South-east and south-west Nigeria (Owerri, Umuahia, Akungba-Akoko, Oye Igbo-Eze, Orba, Ogbete, Akure, Isiukwuato, Ubani and Enugu)	Market	Creamy white and brown	Akinyosoye et al. (2017)
5	Ika-Ika Qua	NS	NS	Alozie et al. (2009)
6	Obudu and Obanliku, Cross River state	Farmers	NS	Anya and Ozung (2019)
7	Obukpa, via Nsukka	Market	Brown	Anyika et al. (2009)
8	Ishan area, Edo state	Farmers	Black	Azeke et al. (2005)
9	Oye Igbo-Eze, Orba	Market	Black, marble and white	Egbujie and Okoye (2019)
10	Ogbete, Enugu	Market	NS	Enujiugha et al. (2012)
11	IAR&T, Ibadan	NS	NS	Esan and Fasasi (2013)
12	Owerri	Market	NS	Ibeawuchi et al. (2007)
13	Akpan Andem, Uyo	Market	NS	Igbabul et al. (2015)
14	Auchi, Ekpoma, Sabongida-Ora, Igueben, Igbanke, Benin City	Farmers and markets	Black, brown, light grey	Ikhajiagbe and Mensah (2012)
15	Umuahia	Vendor	NS	James et al. (2020)
16	Ubani, Umuahia, Enugu	NS	NS	Machuka and Okeola (2000)
17	Makurdi, Onitsha, Umudike, Ibadan, Iseyin	NS	NS	Moyib et al. (2008)
18	Fadan Ninzo, Gwantu, Randa, Kaduna state	Farmers	NS	Ndidi et al. (2014)
19	Akungba-Akoko	Market	Cream	Ngwu et al. (2014)
20	Within the coordinate covering Lat. 4° 30′ to 7° 00′ N and Long. 5° 30′ and 9° 30′ E	Farmers and vendors	White/grey, light to dark brown, purple-black, various mosaic features	Nnamani et al. (2018) and (2019)
21	Ogige market, Nsukka	Market	NS	Nwankwo et al. (2021)
22	Ankpa, Ekwashi, Giver, Tarka, Eha-Alumona, Isiukwato, Ikwuano, Umuahia, Ilaro	NS	White, brown, dark brown	Nwofia et al. (2014)
23	Owerri	Market	NS	Nwosu (2013)
24	Port Harcourt	Experimental field	NS	Nyananyo and Nyingifa (2011)
25	Makurdi, Oju, Adoka, Gboko and Ado, Benue state	NS	Black, brown, light grey, red and golden stripe	Ogbaji and Okeh (2021)
26	Akure	Market	White and brown	Ohaegbulam et al. (2018)
27	Owerri	Market	NS	Okereke et al. (2018)
28	Umuahia	Market	NS	Okoye et al. (2010)

Continued

Table 2.1. Continued.

Sr. No.	Places of collection	Seed sources	Testa descriptions	Documenting references
29	Ibadan	Experimental field	Brown	Olatunde and Adarabioyo (2010)
30	Ede, Osun state	Farmers and vendors	NS	Olayiwola et al. (2012)
31	Oke-Ozzi	Market	NS	Onoja and Obizoba (2009)
32	Eha-Amufu and Ugboloko	NS	NS	Onyeke et al. (2013)
33	New Market, Aba	Market	NS	Opiriari and Ogwo (2014)
34	Ayedun-Ekiti, Ekiti state	Farmers	White, light brown with black stripes, reddish-brown, reddish-brown with black stripes, light brown and black with light brown stripes	Oshodi et al. (1995)
35	Etitiama-Nkporo, Etitiulo-Bende, Abam, Amauwom, Alayi, Itu, Ngor-Okpuala, Akaeze, Ishiagu, Nsukka, Ankpa	NS	NS	Osuagwu and Nwofia (2014)
36	Bodija, Ibadan, Oyo state	Market	Brown	Raji et al. (2014)
37	ADP, Anyigba, Kogi state	Ministry of Agriculture	NS	Sale et al. (2016)
38	Isiukwuato, Ogbete, Enugu	Market	NS	Uchegbu and Amulu (2015)
39	Ogbete, Ubani, Umuahia	Market	NS	Ukom et al. (2019)
40	17 locations within Abia, Benue, Ebonyi, Enugu, Imo, Kogi, Taraba states	NS	NS	Usoroh et al. (2019)
41	Port Harcourt	Market	Light grey	Wokoma and Aziagba (2001)

ADP, Agricultural Development Programme; NS, not specified; IAR&T, Institute of Agricultural Research and Training.

be intact under conservation. The majority of the AYB collections in the GRC and IITA were from Nigeria. However, Ghana and the Democratic Republic of Congo have representative accessions (Abberton et al., 2022). Through an inventory, NACGRAB and IAR&T have AYB accessions intact in their repositories (Table 2.2). Only 564 AYB accessions are intact in the gene banks of national and international institutes, and only 258 are duplicated (Table 2.2). Talabi et al. (2022) described the conservation status of AYB: 10% are under long-term conservation, 49.15% are under medium-term conservation and 0.1% are in short-term conservation.

Moreover, 37.87% exist as seed collection, but AYB genetic resources do not exist under cryopreservation, field collections and DNA collections.

2.4.2 African yam bean germplasm in Nigeria

Only recent AYB collections by GRC and IITA have precise passport data. Most initial collections lack traceable information on the collection locations within Nigeria and elsewhere. Furthermore, in the literature, AYB collections by

Table 2.2. Institutions holding some genetic resources of African yam bean. (Author's own table.)

Sr. No.	Institutions	Quantity	References	Status and remark	Quantity duplicated in Svalbard Global Seed Vault, Norway
1	Genetic Resources Centre, International Institute of Tropical Agriculture, Ibadan, Nigeria	≥450	Abberton *et al.* (2022)	Intact	224 accessions
2	Institute of Agricultural Research and Training (IAR&T), Ibadan, Nigeria	≥40	Inventory	Lost	No
3	National Centre for Genetic Resources and Biotechnology (NACGRAB), Ibadan, Nigeria	≥80	Inventory	Intact	No
4	Germplasm Conservation Unit, Biotechnology Research and Development Centre, Ebonyi State University Abakaliki, Nigeria	34	Nnamani *et al.* (2017)	Intact and duplicated at Germplasm Screening Laboratory, Department of Crop Production and Protection, Obafemi Awolowo University, Ile-Ife, Nigeria.	No
5	Landraces in the hands of individuals in Nigerian communities	Unknown		Seed conservation depends on rare communal and individual modes of seed storage, available facilities, selection pressure, etc.	No

Machuka and Okeola (2000) and Akande *et al.* (2009) were skewed to south-east and south-west Nigeria, respectively. The lack of precise passport data makes it very difficult to trace locations where collections of AYB germplasm have been made earlier, the availability of which would have provided the necessary information for identifying further areas for unduplicated collections. Beyond the two major regions identified from Machuka and Okeola (2000) and Akande *et al.* (2009), the present review identified additional locations outside east and west Nigeria where scientists obtained landraces of AYB for various research purposes. Based on the available literature, leading to the generation of Fig. 2.1, a collection of AYB genetic resources has been made in 22 states of Nigeria. The 22 states comprised 445 local governments housing many ethno-cultural settings. The utilization of the crop will differ in the various communities. Therefore, subsequent exploration should incorporate the documentation of various ethnical uses such as food, feed, medicine, etc. The identified areas were from the south to the north, reaching the middle belt, traversing east to the west, and covering the forest region to the wetter savannas (Fig. 2.1). The identified covering areas, according to Eludoyin *et al.* (2014), were the tropical wet (coastal/swamp/mangrove), tropical wet and dry (rainforest) and tropical savanna (derived, Guinea, Sudan and Sahel savannas).

2.4.3 African yam bean genetic resources in Ghana

Notable regions of cultivation and production of African yam bean in Ghana include Ho-West, Nkwanta, where 22 AYB landraces were collected in villages around Abrewankor, Abodome, Bonakye, Bonto, Chali, Gbadzeme, Galiba, Kebukuu, Mehwe-me-Nyame, Nana Yaw, Nkwanta, and so on (Amoatey *et al.*, 2000). These areas cut across flood plains, lowlands, highlands, mountains, forests, savannas, and cool-to-hot environments. In a most recent work by Adjei *et al.* (2023), 15 landraces were collected from farmers in Ghana's Upper West and Volta regions. Ho-west, Nkwanta, Upper West and Volta are humid tropical regions in Ghana with high rainfall and

forest vegetation; by inference, Ghana also has some germplasm of AYB in the high rainfall regions.

2.5 Use of Germplasm in Crop Improvement by Conventional Approaches

The discovery of crops with remarkable market or end-user potential should drive focused genetics and breeding programmes to resolve its unappealing constraints (Mayes *et al.*, 2012), ultimately leading to cultivar development.

2.5.1 Importance of African yam bean and its pre-breeding status

The potential of AYB is invaluable for the two economic products (grain and tuber), which have food, feed, medicinal, nutritional and security benefits. African yam bean is an excellent food legume with low fat and high protein, fibre, minerals and vitamins. Its potential for enhancing soil fertility management through crop rotation has remained influential in Nigeria's traditional agriculture and farming systems (Saka *et al.*, 2004), although there is little research in this area. Bernado (2010) identified factors for breeding success to include a definitive understanding of the pollination mechanism and breeding system of the species, awareness of the forms of sterility status of the species (cytoplasmic or genic), availability of fertility restorer gene, polyploidy status, etc. There is no report of sterility and polyploidy in AYB. The crop possesses large and complete flowers exhibiting approximately 90% selfing and 10% outcrossing breeding mechanisms (Ojuederie and Balogun, 2017; Adewale and Adegbite, 2018; Adewale and Amazue, 2018). With access to genetic diversity and reproductive biology information, the generation of breeding population and cultivar development in AYB through conventional introgression is possible.

2.5.2 Unveiled intraspecific variabilities in African yam bean genetic resources

The identified variation in testa colour and patterns of the different landraces across the various ecologies in Table 2.3 is naturally apparent,

Table 2.3. List of qualitative traits and proportion of polymorphism reported on major agro-morphological diversity studies on African yam bean. (Author's own table.)

Sr. No.	Variables	Types	Adewale et al. (2012)	Ojuederie et al. (2015)	Abdulkareem et al. (2015)	Aina et al. (2020)	Shitta et al. (2021)
Binary variables (%)							
1	Pigmentation of plant parts	Non-pigmented	45.44	37.51	58.33	NA	NA
		Pigmented	54.55	62.49	41.67	NA	NA
2	Testa colour variegation	Non-variegated	22.07	45.01	NA	NA	NA
		Variegated	77.92	54.99	NA	NA	NA
3	Seed cavity ridges on pods	Present	52.27	95.05	NA	74.01	28.99
		Absent	47.53	4.99	NA	25.99	71.01
4	Compression on the seed	Compressed	NA	NA	33.33	NA	NA
		Not compressed	NA	NA	66.67	NA	NA
5	Pod dehiscence	Non-shattering	61.14	NA	NA	NA	77.51
		Shattering	38.85	NA	NA	NA	22.49
6	Splitting of testa	Absent	26.58	NA	NA	NA	NA
		Present	59.49	NA	NA	NA	NA
Ordinal variables (%)							
7	Eye colour pattern (ECP)	ECP0	26.58	NA	NA	NA	NA
		ECP1	3.78	42.49	NA	11.99	NA
		ECP 2	48.09	4.99	NA	12	NA
		ECP 3	7.63	NA	NA	38.00	NA
		ECP 4	10.13	42.5	NA	38.01	NA
		ECP 5	2.52	0	NA	0	NA
		ECP 6	1.25	2.49	NA	0	NA
		ECP 7	NA	4.99	NA	NA	NA
8	Testa basal color	Grey/white	11.4	2.49	8.33	NA	17.93
		Pale green	NA	2.5	NA	NA	NA
		Cream	2.52	NA	8.33	NA	NA
		Brown	58.27	55.01	50	NA	60
		Dark brown	1.27	NA	NA	NA	20
		Purple/black	1.27	2.49	8.33	NA	2.07
		Mosaic/variegated	25.31	35.01	25	NA	20.71

9	Seed shapes	Round	13.89	NA	25	NA	57.4
		Oval	21.54	85	41.67	NA	21.89
		Oblong	21.53	7.51	33.33	NA	NA
		Rhomboid	43.02	2.5	NA	NA	NA
		Globular	NA	4.99	NA	NA	NA
10	Basal colour of variegated seeds	Non-variegated	73.42	59.99	NA	NA	NA
		Cream	6.32	15	NA	NA	NA
		Brown	8.85	25.01	NA	NA	NA
		Black	11.41	NA	NA	NA	NA
11	Pattern of testa variegation (PTV)	PTV0	73.42	72.49	NA	NA	NA
		PTV1	3.81	NA	NA	NA	NA
		PTV2	13.93	27.51	NA	NA	NA
		PTV3	8.85	17.22	NA	NA	NA

and it is equally an inference for the presence of qualitative variations in African yam bean germplasm. The colour variation ranges from white through very light brown to completely black (Fig. 2.2). Genetic resources with variegated colour patterns equally exist with different background colours. Others, according to Nnamani *et al.* (2017), were milky seeds with brown and black hilum. Genetic resources collected from five communities in Benue state were characterized based on seed colour into black, brown, light grey, red and golden stripe (Ogbaji and Okeh, 2021). Many earlier workers (Oshodi *et al.*, 1995; Machuka and Okeola, 2000; Ikhajiagbe and Mensah, 2012) employed the colour and colour pattern of the testa for the morphological characterization of AYB.

Pigmentation is another evident morphological feature on the vines of African yam bean; it could be absent or present with proportional intensities (Table 2.3). Binary qualitative traits are also revealed for pod and seed. Pods could be ridge-like or uniformly smooth on the surface; moreover, pods of some accessions or landraces may or may not shatter when dried (Table 2.3). Testa could be monocoloured or mosaic with different patterns; some could be compressed while others are uniform. Furthermore, the testa of some accessions or landraces could split open while others could be intact (Table 2.3). Further qualitative variations were observed and classified by ordinal descriptive scales. Depending on the different number of genetic materials used by the five referenced workers in Table 2.3, significant (based on proportion) variations exist for eye colour pattern, testa basal colour, seed shape, basal colour of variegated seeds and pattern of testa variation in AYB (Table 2.3).

Fig. 2.2. Conspicuous variation in seeds of some African yam bean accessions. (Author's own figure.)

The identification of variations in this study offers the opportunity for selection criteria to improve pod, seed and vegetative characteristics.

Eight reported research works on the agromorphological diversity assessment of African yam bean with a total of 39 quantitative (discrete and continuous) variables were considered in Table 2.4. The number of genetic materials used ranged from 10 (Nwofia *et al.*, 2014) to 169 (Shitta *et al.*, 2021). The mean and coefficient of range were used to assess the level of variability of the 39 variables (Table 2.4). Means for each of the 39 traits in Table 2.4 differed depending on the study; differences in the sample size, type of genetic resources used and locations where the studies were conducted could be valid reasons for the non-similarity in the means. The coefficient of range values ≥50% depicts a huge variation for quantitative traits. The reports from the eight studies identified significant variation in leaf-related traits (Table 2.4). According to Potter and Doyle (1994), phenotypic expression of leaf/leaflet dimensions in African yam bean is under genetic rather than environmental influence. A determined look into this may lead to the initiation of breeding for physiological features, including vigour, which can positively enhance grain or tuber yield.

Furthermore, in Table 2.4, one or more of the authors (Popoola *et al.*, 2011; Adewale *et al.*, 2012; Nwofia *et al.*, 2014; Abdulkareem *et al.*, 2015; Ojuederie *et al.*, 2015; Aina *et al.*, 2020; Animasaun *et al.*, 2021; Shitta *et al.*, 2021) identified 15 pod and seed related traits with ≥50% coefficient of range. The traits and their mean ranges from Table 2.4 were: pod/plant (14.5–62.61), peduncle length (9.85–24.99 cm), pod length (18.05–23.67 cm), pod weight/plant (68.4–79.45 g), pods/peduncle (2.6–4.54), individual pod weight (5.88 g), pod beak length (1.05 cm), seed weight/plant (29.55–52.4 g), seed weight/pod (3.73–4.95 g), seed width (5.87–8.05 mm), seed thickness (5.46–9.2 mm), number of seeds/pod (10.95–18.43), locules/pod (12.5–17.68), shelling percentage (32.9–42.98 %) and seed yield/ha (221.5–415.5 kg). Significant variation exists in many of the quantitative traits of AYB. The identified variations in all the quantitative traits could be due to the influence of the inherent genes (G), the environment (E) or the interaction of both. Furthermore, multilocational (Adewale *et al.*, 2017, 2021) and multi-year (Akinyosoye *et al.*, 2017; Aremu *et al.*, 2020) grain yield evaluation has been attempted on African yam bean in Nigeria, leading to the identification of significant G × E interaction. The need for further analysis involving partitioning of the G × E interaction for stability performance of the tested genotypes for the different unpredictable environments has been implicated and recommendations of genotypes for specific environments have been made accordingly.

2.6 Integration of Genomic in African Yam Bean Genetic Resources for Crop Improvement

Understanding the cytology of a crop is a basic fundamental biology for crop genetic improvement because chromosomes are directly responsible for trait inheritance. Studies on the cytology of AYB have been scanty, but all related studies affirmed that the species is diploid. Notably, the reports of Adesoye and Nnadi (2011) and Popoola *et al.* (2011) identified variation in chromosome counts to include 18, 20, 22 and 24; however, both works agreed that 2n = 22 is the most predominant chromosome number. Biotechnological approaches have been employed in AYB research: plant tissue culture and diversity analysis using molecular markers. Calli have been induced from root, stem and leaf explants (Akande *et al.*, 2009), direct organogenesis was obtained from the embryo, leaf, cotyledonary node and shoot tips (Adesoye *et al.*, 2012), and *in vitro*, morphogenic responses were obtained in two AYB accessions (Ogunsola *et al.*, 2016). An earlier review by Oluwole *et al.* (2021) identified and listed the use of the following molecular marker-based diversity analysis: RAPD (Moyib *et al.*, 2008; Popoola *et al.*, 2017), AFLP (Ojuederie *et al.*, 2014; Adewale *et al.*, 2015), cowpea derived SSR (Shitta *et al.*, 2022), ISSR (Nnamani *et al.*, 2019; Animasaun *et al.*, 2021) and SNP (Oluwole *et al.*, 2020; Aina *et al.*, 2021; Shitta *et al.*, 2022). The species diversity analysis (morpho-agronomic and genomic) has been quite substantial; all identified significant diversity, with high genetic distances, high heterozygosity, and so on within the different AYB populations. Most of the genetic resources used were obtained from GRC and IITA; by inference, the

Table 2.4. List of quantitative traits, mean (M) and coefficient of ranges (CoR) depicting polymorphism as reported in major agro-morphological diversity studies on African yam bean. (Author's own table.)

Sr. No.	Morphological variables	Popoola et al. (2011); n = 25 M	CoR	Adewale et al. (2012); n = 80 M	CoR	Nwofia et al. (2014); n = 10 M	CoR	Abdulkareem et al. (2015); n = 15 M	CoR	Ojuederie et al. (2015); n = 40 M	CoR	Aina et al. (2020); n = 50 M	CoR	Animasaun et al. (2021); n = 21 M	CoR	Shitta et al. (2021); n = 169 M	CoR
	Vegetative data before the reproductive stage																
1	Days to seedling emergence	6.28	24.31														
2	Hypocotyl length at 2 weeks (cm)	14.47	20.55														
3	Days to primary leaflets emergence	11.83	13.92														
4	Terminal leaflet length (cm)	12.66	24.71					10.93	17.18			8.55	32.16				
5	Terminal leaflet width (cm)	5.37	19.93					4.32	19.37			3.3	36.36				
6	Leaf area (cm²)							35.85	32.01								
7	Internode length (cm)	13.32	19.95					11.58	37.41			12.1	57.85				
8	Petiole length (cm)	5.77	21.32					7.96	18.99			4.1	26.83				
9	Number of leaves at flowering inception							121.13	17.79								
10	Stem girth (cm) at flowering inception							0.96	42.20								
11	Number of branches at flowering inception							12.72	27.48								
	Flowers																
12	Days to peduncle initiation									75.19	10.31						
13	Days to flowering					95.33	6.81										
14	Days to 50% flowering	85.92	14.45	80	12.5					106.57	8.59	82		150.02	6.63		
15	Standard petal length (cm)	2.91	8.17														

No.	Trait	Values
16	Standard petal width (cm)	3.73, 15.25
17	Flowers/peduncle	11.04, 32.40
	Pods	
18	Days to pod formation	147.76, 9.57, 156.65, 7.27, 148.5, 11.11, 154.19, 7.08
19	Pod filling period	51.31, 10.71
20	Days to pod maturity	154.64, 2.42
21	Pods/plant	62.61, 67.46, 14.5, 86.21, 15.83, 80.61, 15.5, 93.54, 17.54, 51.94
22	Peduncle length (cm)	17.84, 34.13, 17.02, 18.46, 9.85, 91.87, 24.99, 14.64
23	Pod length (cm)	23.67, 41.93, 22..5, 34.22, 20.73, 37.17, 18.05, 71.19, 22.36, 21.67
24	Pod weight/ plant (g)	79.45, 63.73, 68.4, 98.09
25	Pods/ peduncle	4.54, 29.01, 2.6, 61.53
26	Individual pod weight (g)	5.88, 57.45
27	Filled pod/plant	27.12, 36.32
28	Pod beak length (cm)	1.05, 61.91
	Seeds and seed yield	
29	Seed weight/plant (g)	29.78, 6.15, 82.11, 29.55, 68.38, 52.4, 98.85, 18.3, 16.98
30	Seed weight/pod (g)	27.77, 36.4, 3.73, 55.67, 4.95, 73.74, 6.7, 21.81
31	100 seed weight (g)	25, 28.80, 20.33, 24, 33.33
32	Seed length (mm)	8.69, 9.38, 8.8, 18.18, 8.58, 13.79, 9.05, 40.33, 19.69, 30.74
33	Seed width (mm)	6.79, 5.87, 6.85, 10.94, 6.77, 13.11, 8.05, 54.04, 7.97, 7.16
34	Seed thickness (mm)	6.98, 5.46, 6.66, 11.94, 9.2, 56.52, 6.28, 9.09
35	Seeds/pod	15.82, 49.85, 10.95, 58.91, 13.78, 13.06, 11.55, 24.62, 12.5, 84.00, 5.6, 21.49, 5.92, 9.87
36	Locules/pod	17.68, 45.51, 12.65, 56.52, 13.89, 18.90, 12.5, 84.00, 18.43, 18.75
37	Seed set percentage (%)	89.25, 7.16, 83.6, 18.18, 82.71, 18.79, 80, 25.00, 14.25, 31.96
38	Shelling percentage (%)	32.9, 67.47, 42.98, 55.90
39	Seed yield/ha (kg)	291.72, 415.5, 49.57, 68.38, 221.5, 93.81

CoR, Coefficient of range. CoR was estimated as $(\text{Value}_{(max)} - \text{Value}_{(min)}) / (\text{Value}_{(max)} + \text{Value}_{(min)}) \times 100$.

centre holds quite a large diversity of species. However, beyond diversity studies, some nutritional (Oluwole *et al.*, 2020) and agro-morphological (Aina *et al.*, 2021; Shitta *et al.*, 2022) traits have been linked to the genome in genome-wide analysis. There is no report or documentation of the AFB genome size and sequence, new varieties of AFB, transgenic AFB, use of marker-assisted breeding methods for improvement, etc. The availability of a complete whole genome sequence would significantly improve marker discovery and precise detection of various quantitative trait loci (QTL) positions in the AYB genome (Paliwal *et al.*, 2020). Molecular markers associated with specific traits in AYB need to be developed, their growing cycle needs to be shortened, growth habit needs to be modified, pod-shattering needs to be addressed, etc., through genetic engineering.

2.7 Erosion of Genetic Resources from Traditional Areas and Limitations in Germplasm Use

The rapid urban migration and simultaneous increase in the world population, according to Baldermann *et al.* (2016), has become a serious challenge to food supply chains, causing losses in farming and arable lands and increasing the number of consumers and their demands. The complexity continually poses a reverse trend to plant genetic resource conservation, especially for underutilized crops.

2.7.1 Realities of losses of plant genetic resources

The disappearance of some species before being documented and studied is an ecological reality needing addressing. More species are becoming endangered and becoming extinct. Fowler (1994) reported a proportional loss of different crops such as maize (91%), pea (94%) and tomato (81%) within 100 years in the USA; Kothari (1997) documented the disappearance of thousands of rice, wheat, cotton, minor millets, different pulses, etc. High-yielding varieties have massively substituted many landraces (Pionetti, 2005), and about 95% of traditional rice varieties are no longer in cultivation in Sri Lanka since the early 1970s (Thanthriarachchi *et al.*, 2007). The consequences of the loss of these germplasms

include the irreversible loss of genetic resources and the increased risk of sustainable crop productivity. This directly undermines human welfare, food and nutritional security. Therefore, saving genetic resource diversity and stemming the trend of further losses to extinction should be a continuing project for the genetic resource rescue of crop species (Moe *et al.*, 2012; Padulosi *et al.*, 2013).

2.7.2 Resilience of African yam bean genetic resources despite neglect and underutilization

African yam bean tolerates wide edapho-climatic variations, which may have supported its existence despite neglect. Ecotype development through evolution is a continuous process; it forms variants and makes them fit or adaptable to different environments and agroecologies. Many other factors have been identified to favour the resilience of the crop through the ages, one of which is the high nutritional content of whole food. According to Nwokolo (1996), AYB was notably significant as food and nutritional relief for the Igbo refugees during the Nigerian Civil War. Its preference above other legumes, especially by the aged folks in the rural areas, must have enhanced its seed preservation and exchange for cultivation; its economic companionship with yam (*Dioscorea* spp.) in the field and in home gardens leads to continuous annual cropping and production. Furthermore, its survival amidst the neglect and pest pressure could be tightly linked to the presence of lectins in the grains and the ability of the seeds to maintain viability beyond one year under ambient storage conditions. Lectins were identified in the seeds of AYB by Omitogun *et al.* (1999) and Machuka and Okeola (2000), with the identified content determined to be potent enough as a lethal dose against most storage legume pests, including bruchids. The above factors are among those that may have prevented the outright extinction of the species in Nigeria and Africa.

2.8 Positive Utilization of the Discovered Genetic Resources in African Yam Bean

So far, available literature on AYB diversity is quite comprehensive, fairly optimal and capable of providing a working knowledge on the phylogeny and taxonomy of the species. The identification of sufficient genetic variation and their use is the most critical step to genetic advances in crop improvement (Baenziger *et al.*, 2006; Carena, 2009; McCouch *et al.*, 2020). The improvement programme of AYB through recombination techniques (hybridization among parents) and further variation inducement (through mutation, polyploidy, etc.) to generate elite germplasm is still quiescent. Promising genetic materials of AYB are available as landraces and accessions (Table 2.5). Most of the reported AYB genetic resources were from the Genetic Resources Centre, IITA, Ibadan, Nigeria. The present work identified some landraces and accessions with significant performances for various agronomic and nutritional traits (Table 2.5). Okoye *et al.* (2019) reported that cadmium contamination did not affect the stem diameter of TSs 93 (an AYB accession), normal days to emergence of TSs 96 (another AYB accession) were not delayed after planting and the grain yield reduction of TSs 92 was not significant under cadmium toxicity. This supports the comment by Bhattarai and Subudhi (2018) that genomic variation among varieties was substantial in determining environmental stress. The investigation by Okoye *et al.* (2019) provides a hint at the possibility of making varietal selections based on the varied resistance of different cultivars to adverse edaphic, climatic and other environmental conditions.

2.8.1 Assimilate sharing for grain and tuber yield in African yam bean

Although the physiology of tuberization in AYB is still unclear and the factors responsible for and the pattern of assimilate sharing between the two notable sinks is yet to be unravelled, Snapp *et al.* (2018) remarked that trade-off in harvest index is real in grain legumes. For African yam bean, with the exception of Ene-Obong and Okoye (1992), who hinted that the relationship between grain and tuber production in AYB is inverse, other workers (Adewale *et al.*, 2012; Ojuederie *et al.*, 2015) have reported seasonal or annual inconsistencies in tuber production of the crop. A good understanding of this is necessary for a holistic yield improvement programme for the crop.

2.8.2 Meal quality improvement and breeding for pest resistance complexes

Many anti-nutritional and nutritional components are present in AYB seeds and tubers. Both, according to Cullis and Kunert (2016), are closely associated with complex unique biochemistry to provide nutrients for humans and livestock and biochemical defence against pests and pathogens. Specific biochemistry has identified the development and evolution of variations in enzyme inhibitors, lectins, polyphenolics, phytates, oxalate, tannins, etc, in AYB. Although these are notably associated with reduced nutritional value, digestibility and bioavailability of some minerals, their presence could confer tolerance and resistance to the grain legume pests. Rapid introgression of important genes in grain legume breeding could be challenging (Foyer *et al.*, 2016). The same difficulty is anticipated in breeding for pest resistance and improved meal quality in AYB. However, the application of appropriate modern tools may address the conflict.

2.8.3 Lectin-based product development

Typically, the proportion of lectin in the total protein of most leguminous grains ranges from 0.1 to 5% (Machuka and Okeola, 2000). Where the proportion of lectin in the seed protein is higher, the presence of the biochemical compound would actively be involved in the defence mechanism of the seed at storage against pests. The proportion of lectin in some AYB landraces was 25–31% (Machuka and Okeola, 2000). Omitogun *et al.* (1999) and Machuka and Okeola (2000) remarked that the lectin in African yam bean is highly potent against bruchids and some field pests of *Vigna unguiculata*. Therefore, the isolation and purification of lectin in AYB can become an industrial engagement in product development to promote organic agriculture.

2.8.4 Modelling the growth habit of African yam bean

Staking is an obligate practice in AYB cultivation because it is a climber, and the leaves of the plant need good exposure to solar radiation for optimum photosynthesis and increased yield.

Table 2.5. Genetic resources of African yam bean with promising genetic potential documented in the literature. (Author's own table.)

Sr. No.	Landraces in some localities	Identified potential(s)	Reference(s)
1	UM951 (from Umuahia)	Lowest lectin	Machuka and Okeola (2000)
2	EN954 (from Enugu)	Highest lectin	Machuka and Okeola (2000)
3	AYB-11 (Ankpa, Kogi state), AYB-04 (Amauwom, Abia state), AYB-03 (Abam, Abia stste) and AYB-07 (Ngor-Okpuala, Imo state)	Tolerant to spent oil and engine oil	Osuagwu and Nwofia (2014)
4	A light brown landrace sample from Ayedun Ekiti, Ekiti state	High: crude protein content (> 25%) and carbohydrate (>63%)	Adeyeye (1997)
5	Landraces from Eha-Amufu, Enugu state and Ugbokolo, Benue state	Tolerance to *Meloidogyne incognita* infection	Onyeke and Akueshi (2012)
6	AYB-10 (from Umuahia, Abia State), AYB-2 (from Eha-Alumona, Enugu State), AYB-4 (from Giver, Benue State), AYB-3 (from Ekwashi, Benue State)	Respectively attained flowering earlier at 88 days with the longest (28.5cm) pod length, pod filling within 47 days, highest (33.5g) 100 seed weight, highest grain yield/ha (701.5kg)	Nwofia *et al.* (2014)
7	Landraces from the middle belt (Benue, Kogi and Taraba) and southeastern (Abia, Enugu and Imo) states of Nigeria	High seed protein content	Usoroh *et al.* (2019)
8	Landraces from Ishan, Edo state	The black and the marble landraces had a much higher quality for tempeh production than the white	Azeke *et al.* (2005)

Accessions in GRC, IITA

1	TSs 93 and TSs 90	Low susceptibility to cadmium toxicity	Okoye *et al.* (2019)
2	TSs 82, TSs 87 and TSs 44	Long seed length (10.45mm), seed width (7.59mm) and seed thickness (7.79mm) respectively	Adewale *et al.* (2010)
3	TSs 117 and TSs 154	Long pod length (≥30cm) and 100 seed weight (34.1g)	Adewale *et al.* (2012); Ojuederie *et al.* (2015)
4	TSs 86 and TSs 115	Thicker stem girth with a corresponding higher number of leaves	Animasaun *et al.* (2021)
5	TSs 1, TSs 2, TSs 91, TSs 111, TSs 6A and TSs 30B	Early flowering and pod maturity	Popoola *et al.* (2011); Aina *et al.* (2020); Animasaun *et al.* (2021)
6	TSs 63 and TSs 56	Tolerance to *Meloidogyne incognita* infection	Onyeke and Akueshi (2012)
7	TSs 33, TSs 24, TSs 82 and TSs 116	High and stable grain yield in many locations	Adewale *et al.* (2017)
8	TSs 118, TSs 12, TSs 148, TSs 61 and TSs 69	High and stable grain yield in one location over many years	Aremu *et al.* (2020)

Continued

Table 2.5. Continued.

Sr. No.	Landraces in some localities	Identified potential(s)	Reference(s)
9	TSs 107	The highest (15.9% and 68.7%) protein and carbohydrate in the tuber, respectively	Ojuederie and Balogun (2019)
10	TSs 107 and TSs 140	Tubers with low antinutritional factors	Ojuederie and Balogun (2019)
11	TSs 95	High nitrogen fixer, good for marginal soils	Ohanmu *et al.* (2018)
12	44C and TSs 49	The highest (66.5% and 14.5%) for carbohydrate and crude protein in the tuber	Konyeme *et al.* (2020)

A trial on the indispensability of staking for AYB led to a very high significant yield reduction (Okpara and Omaliko, 1997). A search for non-climbing variants is recommended. Mutation breeding would, however, be the next option if a non-climbing variant is not available in the germplasm. Gulisano *et al.* (2019) obtained completely determinate lines of *Lupinus mutabilis* upon induced mutation with ethyl methanesulfonate (EMS). Similar success could be attained in AYB.

2.8.5 Reduction of cooking time in African yam bean

Agunbiade and Longe (1999) stated that the structural arrangements of the amylose and amylopectin constituents affect the swelling properties of starch in legumes. This has also been identified as a significant factor in determining the cooking conditions for legumes. The starch granules of AYB were more strongly held together by crystalline intermolecular bonds, which poses resistance to swelling by hydration (Agunbiade and Longe, 1999). Their report further revealed that AYB equally had a very high water–oil absorption index. The characterization of a wide germplasm for starch structural arrangements and water–oil absorption index would be a good starting point to assess diversity to counter the hard-to-cook problems in AYB.

2.9 Cultural Practices

As an orphan crop, AYB awareness is poor, and the crop is mostly cultivated by farmers mostly aged above 50 years (Saka *et al.*, 2004). Although the crop produces two economic products – grain and tubers – the crop is propagated sexually through the seeds. The subterranean tuber of the AYB does not possess buds or eyes. Hence, it is not a vegetative propagule for propagation. Day-to-seedling emergence from the seed ranges between 5 to 7 days. AYB is a climber that produces vigorous vines and branches that twine clockwise on available erect plants. The growth habit made it a stake-dependent crop. In the field, it is usually in companionship with other crops, especially yam (*Dioscorea* spp) and cassava in West Africa (Okpara and Omaliko, 1997). Since it exhibits a twining growth habit and is an underutilized crop, its cultivation as a sole component and in commercial quantities is very rare.

2.10 Conclusion

Legumes in the humid tropics are particularly susceptible to biotic stress, especially from pest pressure and diseases (Snapp *et al.*, 2018). However, genetic resources for AYB with a candidacy to tackle the biotic stress and other constraints will be available because AYB is indigenous to the humid environment. Practically, genes for fitness must have evolved owing to eco-physiological interactions of the species with the environment over the years. This chapter identified two repositories of the status of AYB genetic resources: accessions and landraces. Accessions are safe, but the landraces need to be rescued. Some landraces of AYB were identified that have specific utilities in Nigeria, for example, those with low lectin content capable of producing highly nutritious meals for humans and livestock, and those with high lectin content, capable of becoming a great resource against legume storage pests (Machuka and Okeola, 2000). Furthermore, some yet-to-be-conserved AYB germplasm was identified for tolerance to spent oil and engine oil (Okoye *et al.*, 2019), high content of crude protein and carbohydrate (Nwokolo, 1996; Adeyeye *et al.*, 1999; Usoroh *et al.*, 2019), tolerance to *Meloidogyne incognita* (Onyeke and Akueshi, 2012), shorter days to flowering and pod maturation, high grain yield and yield determining traits (Nwofia *et al.*, 2014; Usoroh *et al.*, 2019), and proven and reliable suitability for tempeh (Azeke *et al.*, 2005). The array of identified important characteristics among these AYB landraces is enormous and capable of promoting cultivar development within the species. The genetic base of the available AYB as accessions in some institutions is still very narrow (Akande *et al.*, 2009). The record of the availability of AYB landraces in human custody in Nigeria was ascertained but is yet to be quantified. The usefulness of landraces remains unknown until the plant breeders can access them. Increased awareness and utilization of AYB hinges on developing improved varieties targeted to overcome the notable constraints; such an achievement needs to attract policy consideration, increased multidisciplinary research, wider cultivation, and further utilization.

Acknowledgements

We would like to acknowledge Mr Tunrayo Alabi, IITA, Ibadan, Nigeria, for generating the map from information in the literature.

References

Abberton, M., Paliwal, R., Faloye, B., Marimagne, T., Moriam, A. and Oyatomi, O. (2022) Indigenous African orphan legumes: potential for food and nutrition security in SSA. *Frontiers in Sustainable Food Systems* 6:708124. DOI: 10.3389/fsufs.2022.708124

Abdulkareem, K.A., Animasaun, D.A., Oyedeji, S. and Olabanji, O.M. (2015) Morphological characterisation and variability study of African yam beans (*Sphenostylis stenocarpa* (Hochst ex A. Rich)). *Global Journal of Pure and Applied Sciences* 21, 21–27.

Adesoye, A.I. and Nnadi, N.C. (2011) Mitotic chromosome studies of some accessions of African yam bean *Sphenostylis stenocarpa* (Hochst. Ex. A. Rich.) Harm. *African Journal of Plant Science* 5, 835–841.

Adesoye, A.I., Emese, A. and Olayode, O.M. (2012) In vitro regeneration of African yam bean (*Sphenostylis stenocarpa* (Hochst Ex. A. Rich.) Harms by direct organogenesis. *Kasetsart Journal - Natural Science* 46, 592–602.

Adewale, B.D. and Adegbite, E.A. (2018) Investigation of the breeding mechanism of African yam bean [Fabaceae] (*Sphenostylis stenocarpa* Hochst. Ex. A. Rich) Harms. *Notulae Scientia Biologicae* 10(2), 199–204.

Adewale, B.D. and Amazue, E.U. (2018) Floral maturation indices of African Yam Bean (*Sphenostylis stenocarpa* Hochst Ex. A. Rich) Harms (Fabaceae). *Notulae Scientia Biologicae* 10, 102–106.

Adewale, B.D. and Nnamani, C.V. (2022) Introduction to food, feed, and health wealth in African yam bean, a locked-in African indigenous tuberous legume. *Frontiers in Sustainable Food Systems* 6:726458. DOI: 10.3389/fsufs.2022.726458

Adewale, B.D., Kehinde, O.B., Aremu, C.O., Popoola, J.O. and Dumet, D.J. (2010) Seed metrics for genetic and shape determinations in African yam bean [Fabaceae] (*Sphenostylis stenocarpa* Hochst. Ex. A. Rich.) Harms. *African Journal of Plant Science* 4(4), 107–115.

Adewale, B.D., Dumet, D.J., Vroh-Bi, I., Kehinde, O.B., Ojo, D.K., Adegbite, A.E. and Franco, J. (2012) Morphological diversity analysis of African yam bean (*Sphenostylis stenocarpa* Hochst. Ex a. rich.) Harms and prospects for utilization in germplasm conservation and breeding. *Genetic Resources and Crop Evolution* 59(5), 927–936. DOI: 10.1007/s10722-011-9734-1

Adewale, B.D., Vroh-Bi, I., Dumet, D.J., Nnadi, S., Kehinde, O.B. *et al.* (2015) Genetic diversity in African yam bean accessions based on AFLP markers: towards a platform for germplasm improvement and utilization. *Plant Genetic Resources* 13(2), 111–118. DOI: 10.1017/S1479262114000707

Adewale, B.D., Ojo, D.K. and Abberton, M. (2017) GGE Biplot application for adaptability of African yam bean grain yield to four agro-ecologies in Nigeria. *African Crop Science Journal* 25, 333–347. DOI: 10.4314/acsj.v25i3.7

Adewale, B.D., Aremu, C.O., Alake, C.O. and Ige, S.A. (2021) Multi-year genetic estimates of quantitative traits and intra-specific diversity of some *Sphenostylis stenocarpa* accessions for reliable selection and characterization. *Agricultural Research* 10, 165–174. DOI: 10.1007/s40003-020-00496-x

Adeyeye, E., Oshodi, A.A. and Ipinmoroti, K.O. (1999) Fatty acid composition of six varieties of dehulled African yam bean (*Sphenostylis stenocarpa*) flour. *International Journal of Food Science and Nutrition* 50, 357–365. DOI: 10.1080/096374899101094

Adjei, R.R., Donkor, E.F., Santo, K.G., Adarkwah, C., Boateng, A.S., Afreh, D.N. and Sallah, E. (2023) A preliminary evaluation of variability, genetic estimates, and association among phenotypic traits of African yam bean landraces from Ghana. *Advances in Agriculture* 1996255. DOI: 10.1155/2023/1996255

Agunbiade, S.O. and Longe, O.G (1999) Essential amino acid composition and biological quality of yam bean *Sphenostylis stenocarpa* (Hoechst Ex. A. Rich.) Harms. *Nahrung* 43, 22–24.

Aina, A.I., Ilori, C.O., Ukoabasi, O.E., Olaniyi, O., Potter, D. and Abberton, M.T. (2020) Morphological characterisation and variability analysis of African yam bean (*Sphenostylis stenocarpa* Hochst. Ex. A. Rich) Harms. *International Journal of Plant Research* 10(3), 45–52.

Aina, A., Garcia-Oliveira, A.L., Ilori, C., Chang, P.L., Yusuf, M. *et al.* (2021) Predictive genotype-phenotype relations using genetic diversity in African yam bean (*Sphenostylis stenocarpa* (Hochst. Ex. A. Rich) Harms). *BMC Plant Biology* 21, 547. DOI: 10.1186/s12870-021-03302-0

Ajayi, A.O. (2011) Sustainable dietary supplements: an analytical study of African yam bean-*Sphenostylis sternocarpa* and corn-*Zea mays*. *European Journal of Experimental Biology* 1(4), 189–201.

Akande, S.R. (2008) Germplasm characterization of African yam bean (*Sphenostylis stenocarpa*) from southwest Nigeria. *Plant Genetic Resources Newsletter* 154, 25–29.

Akande, S.R., Balogun, M.O. and Ogunbodede, B.A. (2009) Effects of plant growth regulators and explant types on callus formation in African yam bean (*Sphenostylis stenocarpa* (Hochst. Ex A. Rich) Harms). *Kasetsart Journal - Natural Science* 43, 442–448.

Akinlabi, A.A., Kehinde, I. and Afolabi, C.G. (2015) Report of *Sclerotinia sclerotiorum* as the causal organism of the leaf spot and stem blight disease of African yam bean (*Sphenostylis stenocarpa*). *Journal of Applied Biosciences* 90, 8408–8412.

Akinyosoye, S.T., Adetumbi, J.A., Amusa, O.D., Agbeleye, A., Anjorin, F., Olowolafe, M.O. and Omodele, T. (2017) Bivariate analysis of the genetic variability among some accessions of African Yam Bean (*Sphenostylis stenocarpa* (Hochst ex A. Rich) Harms. *Acta Agriculturae Slovenica* 109(3), 493–507. DOI: 10.14720/aas.2017.109.3.02.

Alozie, Y.E., Udofia, U.S., Lawal, O. and Ani, I.F. (2009) Nutrient composition and sensory properties of cakes made from wheat and African yam bean flour blends. *Journal of Food Technology* 7(4), 115–118.

Amoatey, H.M., Klu, G.Y.P., Bansa, D., Kumaga, F.K., Aboagye, L.M. *et al.* (2000) African yam bean (*Sphenostylis stenocarpa*) a neglected crop in Ghana. *Western African Journal of Applied Ecology* 1, 53–60.

Animasaun, D., Adikwu, V., Alex, G., Akinsunlola, T., Adekola, O. and Krishnamurthy, R. (2021) Morpho-agronomic traits variability, allelic polymorphism and diversity analysis of African yam bean: towards improving utilization and germplasm conservation. *Plant Genetic Resources* 19(3), 216–228. DOI: 10.1017/S1479262121000253

Anya, M.I. and Ozung, P.O. (2019) Proximate, mineral and anti-nutritional compositions of raw and processed African yam bean (*Sphenostylis stenocarpa*) seeds in Cross River state, Nigeria. *Global Journal of Agricultural Sciences* 18, 19–29.

Anyika, J.U., Obizoba, I.C. and Ojimelukwe, P. (2009) Effect of food intake on weight, liver weight and composition in rat fed dehulled African yam bean and Bambara groundnut supplemented with sorghum or crayfish. *Pakistan Journal of Nutrition* 8(4), 500–504.

Aremu, C.O., Ige, S.A., Ibirinde, D., Raji, I., Abolusoro, S. *et al.* (2020) Assessing yield stability in African yam bean (*Sphenostylis stenocarpa*) performance using year effect. *Open Agriculture* 5, 202–212.

Azeez, M.A., Adubi, A.O. and Durodola, F.A. (2018) Landraces and Crop Genetic Improvement. In: Grillo, O. (ed.) *Rediscovery of Landraces as a Resource for the Future*. IntechOpen Ltd, London, pp. 1–21.

Azeke, M.A., Fretzdorff, B., Buening-Pfaue, H., Holzapfel, W. and Betsche, T. (2005) Nutritional value of African yam bean (*Sphenostylis stenocarpa*): improvement by lactic acid fermentation. *Journal of the Science of Food and Agriculture* 85, 963–970. DOI: 10.1002/jsfa.2052

Baenziger, P.S., Russell, W.K., Graef, G.L. and Campbell, B.T. (2006) Improving lives: 50 years of crop breeding, genetics, and cytology. *Crop Science* 46, 2230–2244.

Baldermann, S., Blagojevic, L., Frede, K., Klopsch, R., Neugart, S. *et al.* (2016) Are neglected plants the food for the future? *Critical Reviews in Plant Sciences* 35(2), 106–119. DOI: 10.1080/07352689.2016.1201399

Bernado, R. (2010) *Breeding for Quantitative Traits in Plants.* Stemma Press, Minnesota. pp. 6–10.

Bhattarai, U. and Subudhi, P.K. (2018) Identification of drought responsive QTLs during vegetative growth stage of rice using a saturated GBS-based SNP linkage map. *Euphytica* 214(2), 38.

Carena, M.J. (2009) *Handbook of Plant Breeding. Volume 3 – Cereals.* Springer–Verlag, New York. DOI: 10.1007/978-0-387-72297-9

Casañas, F., Simó, J., Casals, J. and Prohens, J. (2017) Toward an evolved concept of landrace. *Frontiers in Plant Science* 8, 145. DOI: 10.3389/fpls.2017.00145

Cullis, C. and Kunert, K.J. (2016) Unlocking the potential of orphan legumes. *Journal of Experimental Botany* 68(8), 1895–1903. DOI: 10.1093/jxb/erw437

Egbujie, A.E. and Okoye, J.I. (2019) Quality characteristics of complementary foods formulated from sorghum, African yam bean and crayfish flours. *Science World Journal* 14(2), 16–22.

Eludoyin, O.M., Adelekan, I.O., Webster, R. and Eludoyin, A.O. (2014) Air temperature, relative humidity, climate regionalization and thermal comfort of Nigeria. *International Journal of Climatology* 34(6), 2000–2018. DOI: 10.1002/joc.3817.

Ene-obong, E.E. and Okoye, F.I. (1992) Interelationships between yield and yield component in African yam bean. *Beitrage zur tropischen Landwirtschaft und Veterinarmedizin* 30, 283–290.

Enujiugha, V.N., Talabi, J.Y., Malomo, S.A. and Olagunju, A.I. (2012) DPPH radical scavenging capacity of phenolic extracts from African yam bean (*Sphenostylis stenocarpa*). *Food and Nutrition Sciences* 3, 7–13. DOI: 10.4236/fns.2012.31002

Esan, Y.O. and Fasasi, O.S. (2013) Amino acid composition and antioxidant properties of African yam bean (*Spenostylis stenocarpa*) protein hydrolysates. *African Journal of Food Science and Technology* 4(5), 100–105.

Fowler, C. (1994) *Unnatural Selection: Technology, Politics and Plant Evolution.* Gordon and Breach Science Publishers. Yverdon, Switzerland.

Foyer, C.H., Lam, H., Nguyen, H.T., Siddique, K.H.M., Varshney, R.K. *et al.* (2016) Neglecting legumes has compromised human health and sustainable food production. *Nature Plants* 2, 16112. DOI: 10.1038/NPLANTS.2016.112

Gulisano, A., Alves, S., Martins, J.N. and Trindade, L.M. (2019) Genetics and breeding of *Lupinus mutabilis*: an emerging protein crop. *Frontiers in Plant Science* 10, 1385. DOI: 10.3389/fpls.2019.01385

Ibeawuchi, I.I., Ofoh, M.C., Nwufo, M.I. and Obiefuna, J.C. (2007) Effect of landraces legumes – velvet bean, Lima bean and African yam bean on the performances of yam, cassava based crop mixture. *Journal of Plant Sciences* 2(4), 374–386.

Igbabul, B.D., Iorliam, B.M. and Umana, E.N. (2015) Physicochemical and sensory properties of cookies produced from composite flours of wheat, cocoyam and African yam beans. *Journal of Food Research* 4(2), 150–158.

Ikhajiagbe, B. and Mensah, J.K. (2012) Genetic assessment of three colour variants of African yam bean [*Sphenostylis stenocarpa*] commonly grown in the Midwestern region of Nigeria. *International Journal of Modern Botany* 2(2), 13–18 DOI: 10.5923/j.ijmb.20120202.01

James, S., Nwabueze, T.U., Onwuka, G.I., Ndife, J. and Usman, M.A. (2020) Chemical and nutritional composition of some selected lesser known legumes indigenous to Nigeria. *Heliyon* 6, e05497. DOI: 10.1016/j.heliyon.2020.e05497

Kay, D.E. (1987) *Crop and Product Digest. No. 2-Root Crops*, revised by Gooding, E.G.B. (2nd edn). Tropical Development and Research Institute, London.

Kingsbury, N. (2009) *Hybrid: The History and Science of Plant Breeding.* The University of Chicago Press, Chicago and London.

Klu, G.Y.P., Amoatey, H.M., Bansa, D. and Kumaga, F.K. (2001) Cultivation and use of African yam bean (*Sphenostylis stenocarpa*) in the volta region of Ghana. *Journal of Food Technolnology Africa* 6, 74–77. DOI: 10.4314/jfta.v6i3.19292

Kothari, A. (1997) *Understanding Biodiversity: Life, sustainability and equity.* Orient Longman, New Delhi.

Machuka, J. and Okeola, O.G. (2000). One- and Two-Dimensional gel electrophoresic identification of African yam bean seed proteins. *Journal of Agriculture and Food Chemistry* 48:2296–2299.

Mayes, S., Massawe, F.J., Alderson, P.G., Roberts, J.A., Azam-Ali, S.N. and Hermann, M. (2012) The potential for underutilized crops to improve security of food production. *Journal of Experimental Botany* 63(3), 1075–1079. DOI: 10.1093/jxb/err396.

McCouch, S., Navabi, K., Abberton, M., Anglin, N.L., Barbieri, R.L. *et al.* (2020) Mobilizing crop biodiversity. *Molecular Plant* 13, 1341–1344. DOI: 10.1016/j.molp.2020.08.011

Moe, K.T., Kwon, S.-W. and Park, Y.-J. (2012) Trends in genomics and molecular marker systems for the development of some underutilized crops. *Genes & Genomics* 34, 451–466.

Moteetee, A.N. and Van Wyk, B.E. (2012) Revision of the genus *Sphenostylis* (Fabaceae: Phaseoleae) in South Africa and Swaziland. *Bothalia* 42(1), 1–6.

Moyib, O.K., Gbadegesin, M.A., Aina, O.O. and Odunola, O.A. (2008) Genetic variation within a collection of Nigerian accessions of African yam bean (*Sphenostylis stenocarpa*) revealed by RAPD primers. *African Journal of Biotechnology* 7 (12), 1839–1846. DOI: 10.5897/AJB08.117

National Research Council (1979) *Tropical Legumes: Resources for the Future.* National Academy of Sciences, Washington DC.

National Research Council (2006) *Lost Crops of Africa. Volume II: Vegetables.* The National Academies Press, Washington, DC, 323–343.

Ndidi, U.S., Ndidi, C.U., Muhammad, A.O.A., Billy, F.G. and Okpe, O. (2014) Proximate, antinutrients and mineral composition of raw and processed (boiled and roasted) *Sphenostylis stenocarpa* seeds from Southern Kaduna, Northwest Nigeria. *International Scholarly Research Notices* Article ID 280837. DOI: 10.1155/2014/280837

Ng, N.Q. (1990) Recent development in Cowpea germplasm collection, conservation evaluation and research at the Genetic Resources Unit, IITA. In: Ng, N.Q. and Monti, L.M. (eds) *Cowpea Genetic Resources.* International Institute of Tropical Agriculture (IITA), Ibadan, Nigeria, pp 13–28.

Ngwu, E.K., Aburime, L.C. and Ani, P.N. (2014) Effect of processing methods on the proximate composition of African yam bean (*Sphenostylis stenocarpa*) flours and sensory characteristics of their gruels. *International Journal of Basic and Applied Sciences* 3 (3), 285–290.

Nnamani, C.V., Ajayi, S.A., Oselebe, H.O., Atkinson, C.J., Igboabuchi, A.N. and Ezigbo, E.C. (2017) *Sphenostylis stenocarpa* (ex. A. rich.) Harms. a fading genetic resource in a changing climate: a prerequisite for conservation and sustainability. *Plants* 6, 30. DOI: 10.3390/plants6030030

Nnamani, C.V., Afiukwa, C.A., Oselebe, H.O., Igwe, D.O., Uhuo, C.A. *et al.* (2019) Genetic diversity of some African yam bean accessions in Ebonyi State assessed using inter-simple sequence repeat (ISSR) markers. *Journal of Underutilized Legumes* 1 (1), 20–33.

Nnamani, C.V., Ajayi, S.A., Oselebe, H.O., Atkinson, C.J., Adewale, B.D., Igwe, D.O. and Akinwale, R.O. (2018) Updates on nutritional diversity in *Sphenostylis stenocarpa* (Hoechst ex. A. Rich.) Harms, for food security and conservation. *American Journal of Agricultural and Biological Sciences* 13, 38–49. DOI: 10.3844/ajabssp.2018.38.49

Nwankwo, M.O., Ganyam, M.M. and James, G.O. (2021) Ethnopharmacological use of *Sphenostylis stenocarpa* (Hochst ex. A. Rich.) Harms seed milk as alternative medicine to control diabetes mellitus in Streptozocin-induced rats. *Journal of Research in Pharmaceutical Science* 7(7), 5–13.

Nwofia, G.E., Awaraka, R. and Mbah, E.U. (2014) Yield and yield component assessment of some African yam bean genotypes (*Sphenostylis stenorcarpa* Hochst Ex A. Rich) Harms in lowland humid tropics of South Eastern Nigeria. *American-Eurasian Journal Agriculture and Environmental Science* 14, 923–931. DOI: 10.5829/idosi.aejaes.2014.14.09.12405

Nwokolo, E. (1996) The need to increase consumption of pulses in the developing world. In: Nwokolo, E. and Smart, J. (eds) *Food and Feed From Legumes and Oilseeds.* Chapman and Hall, London, pp. 3–11. DOI: 10.1007/978-1-4613-0433-3_1

Nwosu, J.N. (2013) Evaluation of the proximate composition and anti-nutritional properties of African yam bean (*Sphenostylis sternocarpa*) using malting treatment. *International Journal of Basic and Applied Sciences* 2(4), 157–169.

Nyananyo, B.L. and Nyingifa, A.L. (2011) Phytochemical investigation on the seed of *Sphenostylis stenocarpa* (Hochst ex A. Rich.) Harms (Family Fabaceae). *Journal of Applied Sciences & Environmental Management* 15 (3), 419–423.

Nyananyo, B.L. and Osuji, J.O. (2007) Biosystematic investigation into *Sphenostylis stenocarpa* (Hochst ex A. Rich) Harms (Fabaceae) in Nigeria. *Nigerian Journal of Botany* 20 (2), 411–419.

Nyanayo, B.L. and Nyigifa, A.L. (2011) Phytochemical investigation on the seed of *Sphenostylis stenocarpa* (Hochst Ex. A. Rich.) Harms (family Fabaceae). *Journal of Applied Sciences & Environmental Management* 15 (3), 419–423.

Ogbaji, M.I. and Okeh, J. (2021) Comparative screening of some varieties of African yam bean (*Sphenostylis stenocarpa*) grown in Benue state for growth and yield performance. *Second National Annual Conference Proceedings, Crop Science Society of Nigeria,* pp. 162–164.

Ogunsola, K.E., Ojuederie, O.B. and Emmanuel, B. (2016) In vitro morphogenic responses of African yam bean (*Sphenostylis stenocarpa* (Hochst Ex. A. Rich.) Harms) accessions to plant growth regulators. *Plant Cell Tissues Organ Culture* 127, 613–622.

Ohaegbulam, P.O., Okorie, S.U. and Ojinnaka, M.C. (2018) Evaluation of the engineering properties of two varieties of African yam bean (*Sphenostylis stenocarpa*) seeds. *Journal of Human Nutrition and Food Science* 6(1), 11158.

Ohanmu, E.O., Ikhajiagbe, B. and Edegbai, B.O. (2018) Nitrogen distribution pattern of African yam bean (*Sphenostylis stenocarpa*) exposed to cadmium stress. *Journal of Applied Science and Environmental Management* 22 (7), 1053–1057.

Ojuederie, O.B. and Balogun, M.O. (2017) Genetic variation in nutritional properties of African yam bean (*Sphenostylis stenocarpa* Hochst Ex. A. Rich. Harms) accessions. *Nigerian Journal of Agriculture Food and Environment* 13(1), 180–187.

Ojuederie, O.B. and Balogun, M.O. (2019) African yam bean (*Sphenostylis stenocarpa*) tubers for nutritional security. *Journal of Underutilized Legumes* 1 (1), 56–68.

Ojuederie, B.O., Balogun, M.O., Fawole, I., Igwe, D.O. and Olowolafe, M.O. (2014) Assessment of the genetic diversity of African yam bean (*Sphenostylis stenocarpa* Hochst Ex. A, Rich, Harms) accessions using amplified fragment length polymorphism (AFLP) markers. *African Journal of Biotechnology* 13(18), 1850–1858. DOI: 10.5897/AJB2014.13734

Ojuederie, B.O., Balogun, M.O., Akande, S.R., Korie, S. and Omodele, T. (2015) Intraspecific variability in agro-morphological traits of African yam bean *Sphenostylis stenocarpa* (Hochst ex. A. Rich) Harms. *Journal of Crop Science and Biotechnology* 18 (2), 53–62.

Ojuederie, O.B., Balogun, M.O. and Abberton, M.T. (2016) Mechanism for pollination in African yam bean. *African Crop Science Journal* 24(4), 405–416.

Okereke, I.H., Okereke, C.O. and Akinmutimi, A.H. (2018) Potentials of replacing soybean meal with toasted African yam bean meal on growth performance of broiler finisher birds. *Nigerian Agricultural Journal* 49(1), 138–142.

Okigbo, B.N. (1973) Introducing the yam bean (*Sphenostylis stenocarpa*) (Hochst ex. A. rich.) Harms. In: *Proceedings of the First IITA Grain Legume Improvement Workshop 29 October–2 November 1973* (Ibadan), pp. 224–238.

Okoye, J.I., Ezigbo, V.O. and Animalu, I.L. (2010) Development and quality evaluation of weaning food fortified with African yam bean *(Sphenostylis stenocarpa)* flour. *Continental Journal of Agricultural Science* 4, 1–6.

Okoye, P.C., Anoliefo, G.O., Ikhajiagbe, B., Ohanmu, E.O., Igiebor, F.A. and Aliu, E. (2019) Cadmium toxicity in African yam bean (*Sphenostylis stenocarpa*) (Hochst. Ex. A. Rich.) Harms genotypes. *Acta Agriculturae Slovenica* 114(2), 205–220.

Okpara, D.A. and Omaliko, C.P.E. (1995) Effects of staking, nitrogen and phosphorus fertilizer rates on yield and yield components of African yam bean (*Sphenostylis stenocarpa*). *Ghana Journal Agricultural Science* 28, 23–28. DOI: 10.4314/gjas.v28i1.2004

Okpara, D.A. and Omaliko, C.P.E. (1997) Response of African yam bean (*Sphenostylis stenocarpa*) to sowing date and plant density. *Indian Journal of Agricultural Science* 67, 220–221.

Olatunde, A.O. and Adarabioyo, M.I. (2004) Flowering, pod formation and abscission and seed yield of African yam bean (*Sphenostylis stenocarpa*) under varied NPK application. *International Journal of Current Research* 7, 46–48.

Olayiwola, O.A., Latona, D.F. and Oyeleke, G.O. (2012) Evaluation of the macronutrients composition of soil, leaves and seeds of African yam bean (*Sphenostylis sternocarpa* Harms). *IOSR Journal of Applied Chemistry* 1(1), 13–17.

Oluwole, O.O., Olomitutu, O.E., Paliwal, R., Oyatomi, O.A., Abberton, M.T. and Obembe, O.O. (2020) Preliminary assessment of the association between DArT-SEQ SNP and some nutritional traits in African yam bean. *Tropical Journal of Natural Product Research* 4(11), 877–879.

Oluwole, O.O., Aworunse, O.S., Aina, A.I., Oyesola, O.L., Popoola, J.O. *et al.* (2021) A review of biotechnological approaches towards crop improvement in African yam bean (*Sphenostylis stenocarpa* Hochst. Ex A. Rich.) *Heliyon* 7, e08481. DOI: 10.1016/j.heliyon.2021.e08481

Omitogun, O.G., Jackai, L.E.N. and Thottappilly, G. (1999) Isolation of insecticidal lectin-enrich extracts from African yam bean (*Sphenostylis stenocarpa*) and other legume species. *Entomologia Experimentalis et Applica* 90, 301–311. DOI: 10.1046/j.1570-7458.1999.00450.x

Onoja, U.S. and Obizoba, I.C. (2009) Nutrient composition and organoleptic attributes of gruel based on fermented cereal, legume, tuber and root flour. *Agro-Science Journal of Tropical Agriculture, Food, Environment and Extension* 8(3), 162–168.

Onyeke, C.C. and Akueshi, C.O. (2012) Pathogenicity and reproduction of *Meloidogyne incognita* (Kofoid and White) Chitwood on African yam bean, *Sphenostylis stenocarpa* (Hochst *Ex*. A. Rich) Harms accessions. *African Journal of Biotechnology* 11(7), 1607–1616. DOI: 10.5897/AJB11.3000

Onyeke, C.C., Akueshi, C.O., Ugwuoke, K.I., Onyeonagu, C.C. and Eze, S.C. (2013) Varietal response of African yam bean, *Sphenostylis stenocarpa* (Hochst Ex. A. Rich) Harms to infection with *Meloidogyne incognita* (Kofoid & White) Chitwood under field conditions. *African Journal of Microbiology Research* 7(30), 3968–3975. DOI: 10.5897/AJMR2013.5715

Opiriari, P.P. and Ogwo, E.E. (2014) The effect of soaking time on some engineering properties of brown-speckled African yam bean. *International Journal of Engineering and Technology* 4(12), 700–708.

Oshodi, A.A., Ipinmoroti, K.O., Adeyeye, E.I. and Hall, G.M. (1995) Amino and fatty acids composition of African yam bean (*Sphenostylis stenocarpa*) flour. *Food Chemistry* 53, 1–6.

Osuagwu, A.N. and Nwofia, G.E. (2014) Effect of spent engine oil on the germination ability of eleven accessions of African yam bean seeds (*Sphenostylis sternocarpa* Hochst ex A. Rich) Harms. *IOSR Journal of Agriculture and Veterinary Science* 7(1), 59–62.

Padulosi, S., Thompson, J. and Rudebjer, P. (2013) *Fighting Poverty, Hunger and Malnutrition with Neglected and Underutilized Species (NUS): Needs, Challenges and the Way Forward*. Bioversity International, Rome.

Paliwal, R., Abberton, M., Faloye, B. and Oyatomi, O. (2020) Developing the role of legumes in West Africa under climate change. *Current Opinion in Plant Biology* 13, 1–17.

Poinetti, C. (2005) *Seed Autonomy: Gender and Seed Politics in Semi-arid India*. International Institute for Environment and Development (IIED), Russell Press, Nottingham, London.

Popoola, J.O., Adegbite, A.E. and Obembe, O.O. (2011) Cytological studies on some accessions of African yam bean (AYB) (*Sphenostylis stenocarpa* Hochst. Ex. A. Rich. Harms). *International Research Journal of Plant Science* 2, 249–253.

Popoola, J., Adebayo, B.M., Adegbite, A., Omonhinmin, C. and Adewale, D. (2017) Fruit morphometric and RAPD evaluation of intraspecific variability in some accessions of African yam bean (*Sphenostylis stenocarpa* Hochst. Ex. A. Rich. Harms). *Annual Research & Review in Biology* 14, 1–10.

Potter, D. (1992) Economic botany of *Sphenostylis* (Leguminosae). *Economic Botany* 46, 262–275.

Potter, D. and Doyle, J.J. (1992) Origin of African yam bean (*Sphenostylis stenocarpa, Leguminosae*): evidence from morphology, isozymes, chloroplast DNA and linguistics. *Economic Botany* 46, 276–292. DOI: 10.1007/BF02866626

Rachie, K.O. (1973) Highlight of grain legume improvement at IITA 1970–73. In: *Proceedings of the First IITA Grain Legume Improvement Workshop 29 October–2 November 1973* (Ibadan).

Raji, M.O., Adeleye, O.O., Osuolale, S.A., Ogungbenro, S.D., Ogunbode, A.A. *et al.* (2014) Chemical composition and effect of mechanical processed of African yam bean on carcass characteristics and organs weight of broiler finisher. *Asian Journal of Plant Science and Research* 4(2), 1–6.

Ramprasad, V. and Clements, A. (2016) Lessons for access and benefit sharing from community seed banks in India. *Farming Matters* April 2016, 50–53.

Saka, J.O., Ajibade, S.R., Adeniyan, O.N., Olowoyo, B. and Ogunbodede, B.A. (2004) Survey of grain legume production systems in South-west agricultural zone of Nigeria. *Journal of Agriculture and Food Information* 6, 93–108. DOI: 10.1300/J108v06n02_08

Sale, F.A., Olujobi, O.J. and Abba, S.B. (2016) Influence of watering regime on germination and early growth of African yam bean (*Sphenostylis stenocarpa*) under nursery condition in Anyigba. *Journal of Biology, Agriculture and Healthcare* 6(10), 48–54.

Shitta, N.S., Edemodu, A.C., Abtew, W.G. and Tesfaye, A.A. (2021) A review on the cooking attributes of African yam bean (*Sphenostylis stenocarpa*). In: Jimenez-Lopez, J.C. and A. Clemente A. (eds) *Legumes Research, Vol. 2*. IntechOpen, London.

Shitta, N.S., Unachukwu, N., Edemodu, A.C., Abebe, A.T., Oselebe, H.O. and Abtew, W.G. (2022) Genetic diversity and population structure of an African yam bean (*Sphenostylis stenocarpa*) collection from IITA GenBank. *Scientific Reports* 12, 4437. https://doi.org/10.1038/s41598-022-08271-4

Snapp, S., Rahmanian, M. and Batello, C. (2018) *Pulse Crops for Sustainable Farms in Sub-Saharan Africa*. FAO, Rome. DOI: 10.18356/6795bfaf-en

Stifel, L.D. (1990) The genetic resources. In: Ng, N.Q. and Monti, L.M. (eds) *Cowpea Genetic Resources*. International Institute of Tropical Agriculture (IITA), Ibadan, Nigeria, pp. 3–4.

Talabi, A.O., Vikram, P., Thushar, S., Rahman, H., Ahmadzai, H. *et al.* (2022) Orphan crops: a best fit for dietary enrichment and diversification in highly deteriorated marginal environments. *Frontiers in Plant Science* 13, 839704. DOI: 10.3389/fpls.2022.839704

Thanthriarachchi, A., Green, S. and Wright, J. (2007) Increasing the availability of traditional seeds in Sri Lanka. *LEISA Magazine* 23(2), 22–23.

Uchegbu, N.N. and Amulu, N.F. (2015) Effect of germination on proximate, available phenol and flavonoid content, and antioxidant activities of African yam bean (*Sphenostylis stenocarpa*). *International Journal of Biological, Biomolecular, Agricultural, Food and Biotechnological Engineering* 9(1), 106–109.

Ukom, A.N., Adiegwu, E.C., Ojimelukwe, P.C. and Okwunodulu, I.N. (2019) Quality and sensory acceptability of yellow maize ogi porridge enriched with orange-fleshed sweet potato and African yam bean seed flours for infants. *Scientific African* 6, e00194.

Usoroh, J.I., Eneobong, E.E., Abraham, S.O. and Umoyen, A.J. (2019) Evaluation of the genetic diversity of African yam bean (*Sphenostylis stenocarpa* (Hoechst. ex. A.Rich.) Harms.) using seed protein marker. *Archives of Current Research International* 17(4), 1–10. DOI: 10.9734/ACRI/2019/v17i430117

Villa, T.C., Maxted, N., Scholten, M.A. and Ford-Lloyd, B.V. (2005) Defining and identifying crop landraces. *Plant Genetic Research* 3, 373–384. DOI: 10.1079/PGR200591

Vodouhe, S.R., Atta-Krah, K., Achigan-Dako, G.E., Eyog-Matig, O. and Avohou, H. (2007) Plant genetic resources and food security in West and Central Africa. *Regional Conference, 26–30 April 2004.* Bioversity International, Rome, pp. 11.

Wokoma, E.C. and Aziagba, G.C. (2001) Sensory evaluation of Dawa Dawa produced by the traditional fermentation of African yam bean (*Sphenostylis stenocarpa* Harms) seeds. *Journal of Applied Sciences and Environmental Management* 5(1), 85–91.

3 Adzuki Bean (*Vigna angularis* (Willd.) Ohwi & Ohashi)

Gopal Katna[1]*, Parul Sharma[2] and Kanishka Chandora[3]
[1]*CSK, Himachal Pradesh Krishi Vishvavidyalaya, Palampur, India;* [2]*Punjab Agricultural University, Ludhiana, India;* [3]*ICAR-National Bureau of Plant Genetic Resources, Regional Station, Shimla, India*

Abstract

Today, focusing just on the productivity of existing cereal crops would not address the problem of food security because these crops are frequently selected and produced under high-intensity agriculture, making them more susceptible to future biotic and abiotic pressures. Instead, leveraging the vast pool of minor and underutilized or potential crops to diversify the agricultural system is required to preserve food and nutritional security and stability for the ever-growing population. One such group of crops is potential crops, and among these, adzuki bean (*Vigna angularis*), often known as red bean, red mung bean or red cowpea, is one of East Asia's most significant potential crops. Adzuki bean belongs to the subgenus *Cerototrapis* of the genus *Vigna* and is a self-pollinating diploid legume with 22 chromosomes. The extensively cultivated *Vigna angularis* var. *nipponensis* in East Asia is considered the wild progenitor of the adzuki bean. It has a small genetic base because of its propensity for self-pollination; consequently, the available germplasm can be essential for enlarging the genetic base and developing high-yielding cultivars with desired traits. In this technological era, the adzuki bean genome sequence is a crucial resource for studying the genetic structure and functional components of the crop. While the genetic diversity of the adzuki bean germplasm has been somewhat assessed, many germplasm accessions are still poorly characterized in terms of their genetic features, agronomic performance and stress tolerance. So, characterizing available germplasm and using biotechnological approaches open new avenues for adzuki bean crop improvement programmes. This chapter provides an overview of the adzuki bean by discussing the distribution, crop gene pool, conservation and preservation of germplasm resources, and various aspects of breeding programmes.

Keywords: Adzuki bean, underutilized legume, resilient, gene pools, germplasm collection, characterization

3.1 Introduction

Pulses occupy a prime place in the diet of the vegetarian population because they are an important, unique and cheap source of protein (Marcello, 2006). The unique qualities of pulse crops include atmospheric nitrogen fixation, low water requirement, soil conservation, better yield than cereals in marginal lands and broader adaptability under varied agro-climatic conditions. Although grain legumes have been among the most important food crops along with cereals, and they have a strong connection to the evolution of civilizations, the domestication of Asiatic food legumes remains problematic (Smartt, 1990). Adzuki bean (*Vigna angularis*),

*Corresponding author: gkatna@gmail.com

© CAB International 2024. *Potential Pulses: Genetic and Genomic Resources*
(eds R. Chandora, T. Basavaraja and A. Pratap)
DOI: 10.1079/9781800624658.0003

also known as red bean, red mung bean or red cowpea, is a minor but significant potential crop in East Asia where it is also called an 'Asian Legume Crop' (Isemura *et al.*, 2011). It is mainly grown in China, Japan, Taiwan, South Korea, Bhutan, India, Nepal, the USA, South America, New Zealand and Africa. Outside East Asia, Canada and Australia have recently begun to produce adzuki bean, mainly for export purposes. It has a genome size of about 500 Mb and belongs to the Fabaceae family with chromosome number 2n = 2x = 22 (Kang *et al.*, 2015; Yang *et al.*, 2015). It is a tall, bushy and indeterminate annual vine that matures late with trifoliate leaves, bright yellow flowers and thin, slightly curved pods. Carbohydrates comprise the majority of adzuki bean, between 40% and 65.5% of their total weight (Orsi *et al.*, 2017). Compared to other major cereals, adzuki bean has a high protein content ranging from 15% to 29%. Folate (vitamin B9) is present in adzuki bean seeds, along with several other minerals and little to no fat (Lin *et al.*, 2002). Functional components, such as saponins and general flavone, have been discovered in addition to the regular nutritional content (Liu *et al.*, 2017). Seed coat colour varies from wine red to black, white, yellow, speckled purple and mottled purple. It is a multipurpose legume crop grown primarily for food, fodder and green manure. Seeds of the adzuki bean are an essential source of protein, starch, mineral elements and vitamins (Lin *et al.*, 2002; Xu *et al.*, 2008a).

Due to its low calorie and fat content, digestible protein and high concentration of bioactive substances, the adzuki bean is sometimes referred to as a 'weight loss bean' (Amarowicz *et al.*, 2009; Kitano-Okada *et al.*, 2012). The adzuki bean is related to several health advantages, including a decreased likelihood of type 2 diabetes and heart disease, greater intestinal motility and decreased body weight. It has a long history of usage in China as a diuretic, an antidote and a treatment for dropsy and beriberi (Li, 2010; Pharmacopoeia, 2010). However, it is frequently used in sweets, as a paste in pastries, cake, porridge, adzuki rice, jelly, adzuki milk and ice cream in Japan and China. Although the immature pods and leaves are eaten as vegetables, the entire seed, split grain and flour are the forms in which it is most commonly consumed (Dua *et al.*, 2009).

The adzuki bean plants can be used for forage purposes, soil conservation and as green manure. Owing to its high nutritional value, adzuki bean is a significant food legume planted for human consumption in temperate and subtropical climates. It has a wide range of adaptability and improves soil conditions by fixing nitrogen up to 100 kg per ha. As a result, the adzuki bean has the potential to become a significant agricultural commodity with an enormous future market for pulses. Farmers develop and preserve a variety of landraces. Because of a lack of genetic stock knowledge and a scarcity of information on the current landraces, however, the genetic improvement of this crop is gradual. In addition, being a short-day annual, most adzuki bean cultivars have a short-day need for floral induction, with blooming generally delayed under long-day circumstances. Adzuki bean cultivars raised in higher latitudes must have evolved to long days by losing their sensitivity to blossom under these circumstances. Numerous scientists from all over the globe have revealed the molecular regulation of a flowering date, suggesting the breeding of adzuki bean is possible at higher latitudes (Aoyama and Shimada, 2014; Yamamoto *et al.*, 2016; Imoto *et al.*, 2022). These reports and studies open new doors for adzuki bean crop improvement programmes.

The demand of adzuki bean is increasing day by day in Asian countries such as Japan, China, Korea and others because it is a popular ingredient in these regions. Because of its attractive colour, it is commonly used in both savoury and sweet items. Being a nutritious crop, it can be used as an alternative to other legumes in plant-based diets and recipes. According to the Food and Agriculture Organization of the United Nations (FAO), the global production of adzuki bean has been increasing gradually in recent years and the market is projected to reach a valuation of US$1.2 billion by 2033 (Fact. MR, 2023). It is, therefore, clear that adzuki bean has the capability to be a significant crop for agriculture and has much potential for satisfying the need for pulses in the future. Owing to the crop's limited genetic diversity, breeding programmes must be strengthened by adding new germplasm, gathering local genotypes and implementing interspecific hybridization. As a result, the characterization and evaluation of adzuki bean genotypes are urgently required to ascertain donor(s) for various attributes and to use these identified sources in various breeding schemes. To provide a general overview of the crop, this chapter discusses the origin, distribution and preservation of germplasm resources and technological advancements in several fields. It promotes their integration to understand the genetics of characteristics and use them in crop development programmes.

3.2 Origin, Distribution and Botanical Description

The adzuki bean is a self-pollinating diploid legume with 22 chromosomes that is part of the subgenus *Cerototrapis* of the genus *Vigna*, along with mung bean, rice bean and black gram (Tomooka *et al.*, 2002). The adzuki bean, also known as the Asian legume, is a warm-season pulse crop that is widely produced in East and South-east Asia, notably in China, Japan and Korea (Tomooka *et al.*, 2002; Kramer *et al.*, 2012) (Fig. 3.1). It is believed that the widely distributed *Vigna angularis* var. *nipponensis* in East

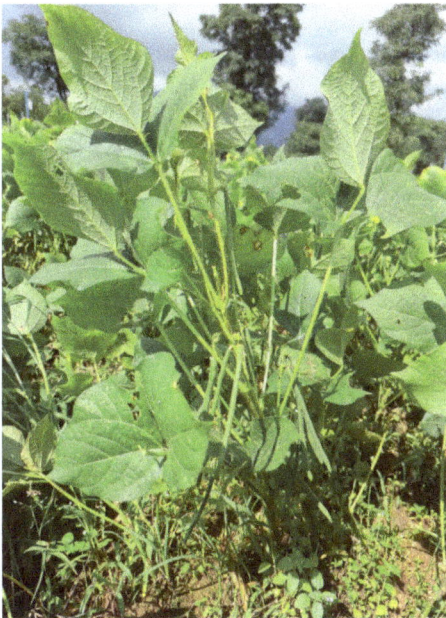

Fig. 3.1. Adzuki bean plant showcasing the leaf morphology. (Author's own figure.)

Asia is the wild ancestor of the domesticated adzuki bean (Ohwi and Ohashi, 1969; Marechal *et al.*, 1978; Tateishi and Ohashi, 1990; Yamaguchi, 1992). According to Tomooka *et al.* (2002), the wild species is spread across an extensive range, starting in Japan and ending in the Korean peninsula, China to Nepal and Bhutan. Archaeologists believe domestication began about 3000 BC and that the evolution from a wild progenitor to *V. angularis* var. *angularis* occurred roughly 50,000 years ago (Lee, 2013; Kang *et al.*, 2015). Adzuki beans were first cultivated in China around 12,000 years ago (Liu *et al.*, 2013), and they are now cultivated in over 30 nations, primarily in East Asia (Tomooka *et al.*, 2002; Kramer *et al.*, 2012). As the precise location of its domestication is unknown, several East Asian origins have been proposed (Kaga *et al.*, 2008). As a result of its long history of cultivation and use in these regions, it is thought to have originated in Asia. The leading producers of adzuki bean include China, Japan, South Korea, Taiwan, etc. The adzuki bean is so significant that it is also cultivated for profit in the USA, South America, India, New Zealand, Congo and Angola. It is now the second-largest pulse crop in Japan and the sixth-largest crop worldwide. According to Schuster *et al.* (1998) and Jansen (2006), Japan is the world's biggest consumer of adzuki bean, importing them from countries including China, Korea, Colombia, Taiwan, the USA, Thailand and Canada.

Vigna comprises about 80 species and occurs throughout the tropics. *V. angularis* belongs to the genus *Vigna*, the family Fabaceae and the subfamily Faboideae. The majority of domesticated landraces are prostrate, although current breeding cultivars are upright, growing to a height of 30–60 cm. The two wild varieties of adzuki bean are wild and weedy. East Asia, South-east Asia and the Himalayan region are among the regions where the wild adzuki bean may be found (Tomooka *et al.*, 2002). With thin stems and an erratic growth pattern, wild forms are twining herbs. The majority of weedy adzuki bean are found in Eastern Nepal and nearby, and they always resemble both wild and domesticated varieties in appearance. The germination of adzuki bean takes place in 7–20 days and is hypogeal. The colour of the stem is usually green and it is sparsely pubescent with long, straight, soft, spreading or erect hairs. The roots of the adzuki

bean have a taproot system that can reach a
depth of 40–50 cm. The leaves are simple, alter-
nate, cordate, trifoliate, pinnate and arranged
alternately along the stem on a long petiole. The
leaflets are lanceolate to ovate and about 5–10
cm long and 5–8 cm wide. Some germplasm possesses anthocyanidins that give the stem and
petiole an overall purple appearance; this characteristic has been utilized to distinguish variants. The inflorescence of the adzuki bean is an
axillary false raceme with 2–20 flowers and a
peduncle that varies in length from long in the
lower nodes to extremely short in the high nodes
(Fig. 3.2). The flower anatomy includes bright
yellow, bisexual, papilionaceous flowers with an
extrafloral nectary at the base; bracteoles that
are longer than the calyx; corollas that are 15 to
18 mm long; standard orbicular; stamens that
are diadelphous and superior ovaries. The pods
are 5–14 cm long, contain 2–14 seeds, have
smooth, thin walls and are cylindrical (Fig. 3.3).
Pods start green and turn white or grey as they
grow. Owing to its indeterminate nature, the
shattering of pods during harvesting is a big
challenge under certain conditions. The seeds
are cylindrical with rounded ends, flattened,

with a length of 5–9.1 mm, width of 4–6.3 mm,
and thickness of 4.1–6 mm. The 100-seed
weight ranged from 5 g to 20 g. Seed colour is
variable, from maroon to blue–black or mottled
(Fig. 3.4). Seeds retain their viability for up to
5 years when stored with 13% moisture content
at 15% relative humidity.

3.3 Crop Gene Pool, Evolutionary Relationships and Systematics

The quantity and accessibility of the genetic
resources available to enhance grain legumes vary
substantially. Available germplasm resources for
certain crops were categorized by Harlan and de
Wet (1971) as primary (GP1), secondary (GP2)
and tertiary (GP3) gene pools. The biological

Fig. 3.2. Flower morphology of adzuki bean.
(Author's own figure.)

Fig. 3.3. Variation in colour of mature pods in
cultivated adzuki bean. (Author's own figure.)

Fig. 3.4. Variability in seed colour and shape in adzuki bean germplasm. (Author's own figure.)

species, or cultigens, as well as any wild forms that are compatible with it, are included in the GP1. All biological species that can exchange genes through interspecific hybrids are included in the GP2. The GP3 includes taxa that can hybridize with one other, but the hybrids are either unviable or sterile, making it impossible for regular hybridization to transfer genes. As a result, in the crop development programme, desirable recombinants might be generated from the domesticated and wild forms in the primary gene pool. In Japan, where the adzuki bean grows in cultivated, wild and weedy forms, Vaughan *et al.* (2004) discovered the adzuki bean crop complex. Additionally, adzuki bean archaeo-botanical remnants from China and Korea (Crawford, 2006) predate the discovery of carbonized seeds at archaeological sites in Japan that date to around 4000 years ago (Maeda, 1987; Yano *et al.*, 2004). These data led researchers to the conclusion that Japan was the first place where the adzuki bean was domesticated. In the natural environment, gene transfers between cultivated and wild gene pools are feasible. Where there are both wild and cultivated adzuki bean, Japan frequently has the highest plant diversity (Kaga *et al.*, 2004). According to Yamamoto *et al.* (2006), variability in the crop is the outcome of 1% outcrossing of the crop's wild and cultivated self-pollinated species.

Genetic diversity available in adzuki bean was studied by various workers with the help of different techniques such as: those based on agronomic traits (Wang *et al.*, 2001), seed protein electrophoresis (Isemura *et al.*, 2001), and those based on different markers i.e. randomly amplified polymorphic DNA (RAPD; Yee *et al.*, 1999; Xu *et al.*, 2000a; Isemura *et al.*, 2002), amplified fragment length polymorphism (AFLP; Yee *et al.*, 1999; Xu *et al.*, 2000b; Zong *et al.*, 2003) and simple sequence repeats (SSR; Wang *et al.*, 2004; Han *et al.*, 2005). Xu *et al.* (2008b) used a sum of 616 accessions selected from more than 2000 accessions conserved in the National Institute of Agrobiological Science, Tsukuba, Japan (NIAS gene bank). These accessions belong to the eight Asian countries that commonly cultivated adzuki bean. With the help of SSR markers, they observed that wild germplasm from Japan and cultivated from China, Korea and Japan were the most diverse and genetically different from each other, which indicates that the cultivation of adzuki bean was very long in isolation in each country. Cultivated accessions from eastern Nepal and Bhutan, however, were similar to each other and distinct from others. Himalayan wild collections showed the highest level of variation and could be used as a desirable source for adzuki bean improvement programmes. Chen *et al.* (2015b) genotyped 261 accessions of adzuki bean and grouped them into 10 clusters, indicating that accessions from North China were well separated from the accessions from South China.

As discussed earlier in the chapter, adzuki bean is mainly cultivated in Japan, Korea, and northern and central China (Lumpkin and Mc-Clary, 1994). The cultivation area extends through southern China as far as Nepal. Wild and cultivated adzuki bean are found in a similar fashion throughout the Asian region. Adzuki bean has been widely cultivated in more than 30 countries worldwide (Kramer *et al.*, 2012; Tomooka *et al.*, 2012; Takahama *et al.*, 2013). Owing to its cultivation in the Asian region, it is also known as the Asian legume crop (Isemura *et al.*, 2011). Xu-xiao *et al.* (2003) used 146 accessions of adzuki bean (*V. angularis* var. *angularis* and *V. angularis* var. *nipponensis*) from six Asian countries to study geographical distribution and

evolutionary relationships between cultivated and wild adzuki bean by AFLP analysis. These accessions are grouped into seven diverse evolutionary groups: Chinese cultivated, Japanese cultivated, Japanese complex–Korean cultivated, Chinese wild, China Taiwan wild, Nepal–Bhutan cultivated and Himalayan wild. They concluded that four originators with, as a minimum, three geographical origins have been included in the domestication of cultivated adzuki bean (Xu-xiao *et al.*, 2003). Table 3.1 illustrates the exploitation of wild species in adzuki bean crop enhancement programmes.

3.4 Germplasm Collection, Characterization, Evaluation and Conservation

Improvement in any crop is based on the genetic variability naturally available in that crop as the cultivated and wild forms and artificial variability created by plant scientists through different means. Moreover, like other crops, knowledge on the genetic variation in adzuki bean is important to understand the genetic variability available and its potential use in the breeding programme (Pusadee *et al.*, 2009). From the overall agricultural genetic diversity that is currently available, humankind has relied on a relatively small number of crops to satisfy the needs of staple foods and a very small number of key non-food crops to satisfy related needs. The future supply of food and rural incomes are at risk owing to the reduction in the number of crops that are essential for global food security and economic expansion. Plant genetic resources (PGRs) are regarded as one of nature's most significant contributions to

Table 3.1. Exploitation of wild species in crop improvement programmes of adzuki bean. (Author's own table.)

Specific trait	Wild species as donor	References
Salt tolerance-related quantitative trait loci (QTLs)	*Vigna riukiuensis, Vigna nakashimae*	Ogiso-Tanaka *et al.* (2023)
Yield related traits	*Vigna angularis* var. *nipponensis*	Hu *et al.* (2022)
Root architectural and morphological traits	Wild adzuki bean accession	Tayade *et al.* (2022a)
Salt tolerance	Wild species	Yoshida *et al.* (2016)
Seed and yield-related traits	*Vigna angularis* var. *nipponensis*	Kaga *et al.* (2008)

humanity. They are the culmination of the diversity that has been cultivated over time through domestication and natural selection. The world's food and nutritional security rest on PGRs, which also hold the key to the fundamental building blocks of agriculture (Singh *et al.*, 2020).

The preservation of genetic stocks, which are crucial to agricultural development efforts, and the choice of suitable plant species for food, fuel and medicine are instances of conservation. PGRs are essential for economic growth, improved food production, poverty reduction and agricultural development (Robinson, 2018). A region inside a natural ecosystem can be designated for *in-situ* or *ex-situ* (in-field gene banks) conservation of genetic resources. In current gene banks, substantial collections of most of the world's essential food crops have been gathered and preserved (Ishaq *et al.*, 2007).

It is important to preserve traditional landraces and cultivars and to assess them for desirable characteristics. In the near future, these features can be exploited for genetic modification to solve new issues in certain crops. The collection, upkeep, assessment and distribution of several significant food crops are all key tasks that international agricultural research institutes perform. Due to its self-pollination tendency, adzuki bean has a limited genetic basis; therefore, the available germplasm can be crucial in extending the genetic base and creating high-yielding cultivars with desirable features. According to reports, several adzuki bean accessions are available in different countries, with China having the most extensive collection of accessions kept

by the Institute of Crop Germplasm Resources (CAAS) (Jansen, 2006; Ning *et al.*, 2009). Major centres that are conserving the germplasm of adzuki bean worldwide include the Institute of Crop Germplasm Resources (CAAS), Beijing (5500), Tokachi Agricultural Experimental Station, Hokkaido Ken, Japan (3600), Genetic Resources Division, Rural Development Administration, Korea (3200), Genetic Resources Center, National Agriculture and Food Research Organization, Japan (1900), National Plant Germplasm System, USA (660), Australia Plant Genetic Resources System, Australia (339), ICAR-NBPGR (National Bureau of Plant Genetic Resources), New Delhi (175) and Asian Vegetable Research and Development Center, Taiwan (150) (Pandiyan *et al.*, 2021).

High-yielding locally adapted cultivars, namely 'Baihong No. 1' in China, 'Erimo' in Japan, 'Chungwonpat' in Korea and 'Kaohsiung No. 3' in Taiwan were developed from the local landraces. A popular landrace available in Himachal Pradesh, India, 'Totru Local' is used as a check in the AICRN network of Potential Crops in India (AICRN, Kharif Annual Report: Kaushik *et al.*, 2020) (Fig. 3.5). *Vigna angularis* var. *nipponensis*, a close wild relative of the adzuki bean, was procured by ICAR-NBPGR from the Lower Subansiri district of Arunachal Pradesh (Pandiyan *et al.*, 2021). Epicotyl explants are frequently used to create *in-vitro* adzuki bean plants. A genetic transformation technique for adzuki bean uses Agrobacterium-mediated transfer. Using morphological and molecular markers, a genetic linkage map has been created.

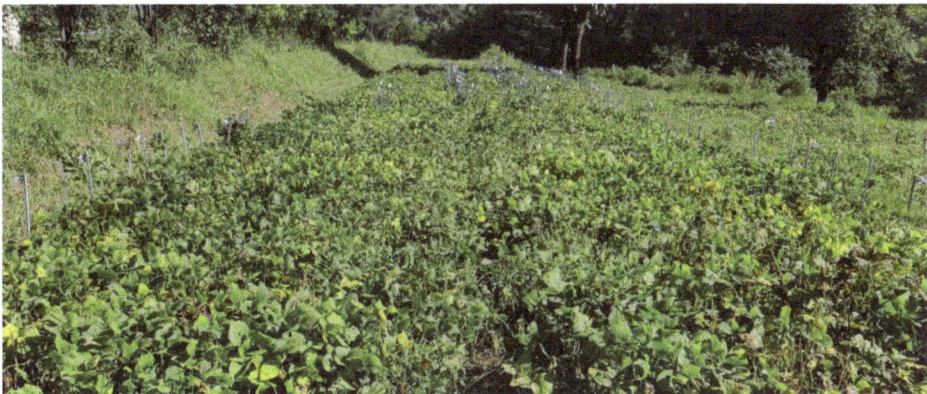

Fig. 3.5. Adzuki bean field view in Himachal Pradesh, India. (Author's own figure.)

Adzuki bean genetic diversity is examined by many researchers utilizing agronomical, molecular and biochemical markers across the world (Wang *et al.*, 2001; Zong *et al.*, 2003; Wang *et al.*, 2004; Xu *et al.*, 2008b; Redden *et al.*, 2009; Zhen-xing *et al.*, 2011; Anjali and Bhardwaj, 2020; Leipe, 2022).

3.5 Use of Germplasm in Crop Improvement by Conventional Approaches

Adzuki bean is a potential crop that is propagated by seeds. Seed yield varies from 500–3500 kg per hectare depending upon climatic conditions and varieties grown in the area. Adzuki bean has broad adaptability, high tolerance to poor soil fertility, and is a high-value rotation crop that improves soil condition through nitrogen fixation (Duan, 1989; Sikkema *et al.*, 2006). Additionally, the adzuki bean can be used as a model species, especially for non-oilseed legumes, because of its short growth period and small genome size (Parida *et al.*, 1990; Yamada *et al.*, 2001). Table 3.2 presents a compilation of globally developed adzuki bean varieties, outlining the improved traits, breeding techniques applied, and the countries/organizations involved. The primary breeding objectives for the adzuki bean are described here.

3.5.1 Yield

Yield is the main objective in any breeding programme. Because adzuki bean is a self-pollinated and neglected crop, the average yield is less than for

Table 3.2. List of recently developed varieties of adzuki bean, trait improvement, breeding methods employed, and country/organization involved at the global level. (Author's own table.)

Trait improvement	Varieties developed	Breeding methods employed	Country/organization	References
High yield	Suhong No. 3	–	China	Zhang *et al.* (2015)
	Erimo syouzu	Hybridization	Japan	Wang *et al.* (2019)
Biotic stress				
Mosaic virus resistance	Yeonkeum	Hybridization	Japan	Moon *et al.* (2006)
Abiotic stress				
Cold weather tolerance	Kita-roman; Erimo syouzu	Hybridization	Japan	Aoyama *et al.* (2009); Wang *et al.* (2019)
Stress resistance	Saegil	Hybridization	Mileage	Han *et al.* (2013)
Nutritional traits				
Antioxidant capacity	Kuro-azuki, Nezumi-azuki, Tamba-dainagon, Hanayome, Ao-azuki, Sanwa-zairai	Selection (landraces)	Kuro-azuki from China and others from Japan	Nagao *et al.* (2023)
Agronomic traits				
Lodging resistance	Honggyeong	Artificial cross	Korea	Song *et al.* (2021)
Soil-borne disease resistance	Syumari	Hybridization	Japan	Fujita *et al.* (2002)
Soil-borne disease resistance	Homare-dainagon	Hybridization	Japan	Tazawa *et al.* (2015)
Large-seeded and early-maturing cultivars	Toyomi dainagonn and Akane dainagonn	Pedigree method	Japan	Wang *et al.* (2019)
Early-maturing and lodging-resistant	Kyungwon	Hybridization	Korea	Wang *et al.* (2019)

other pulse crops. So, breeding for high-yielding stable genotypes is the main objective of the adzuki bean improvement programme (Cheng and Wang, 2009). Various high-yielding varieties have been developed to date by using pedigree selection, hybridization and other breeding methods in Japan, Korea and China (Wang *et al.*, 2019).

3.5.2 Biotic stresses

Resistant varieties are the best alternative to overcome the problem of diseases and insect pests. Breeding resistant varieties adapted to the demands of the location allows them to perform well in adverse situations. According to reports, brown stem rot and mosaic virus are the two main diseases that affect adzuki bean in Japan (Han *et al.*, 1982; Takahashi *et al.*, 1998; Kim *et al.*, 2014). Stem rot will occur more rapidly in areas with high humidity. Additionally, the mosaic virus causes a specific yield loss, resulting in dwarf plants and leaf atrophy. Among pests, bruchids including *Callosobruchus chinensis* are the most damaging crop pests in East Asia (Fernandez and Talekar, 1990).

3.5.3 Varieties suitable for mechanization

In countries such as China, the government has provided specific financial assistance to increase the levels of mechanization in response to significant changes in agricultural structure. New adzuki bean cultivars with traits suited for mechanized activities are therefore urgently required (Wang *et al.*, 2019).

3.5.4 Abiotic stresses

Germplasm available in a particular crop is well adapted to a particular situation. So, desirable traits from the landraces or local germplasm should be used to breed tolerant varieties to tackle adverse climatic conditions. Other traits, such as early maturing, determinate type, seed size etc., can play an essential role in the adzuki bean improvement programme (Tian *et al.*, 2004; Cheng and Wang, 2009).

Progress in crop improvement greatly depends on conserving genetic resources to ensure their effective and long-term use. Wild progenitors and landraces of cultivated plants are valuable sources as donor parents for desirable traits to overcome the defect of a particular plant type. Introducing new germplasm, local landraces and hybridization through conventional approaches can strengthen the adzuki bean breeding programme. The identification of donor(s) for different traits in the available germplasm is the first requirement for any crop improvement programme. So, desirable features for creating superior varieties are provided mainly by the existing germplasm. Characterization entails estimating the heterogeneity that already exists among individuals in a population (Franco, 2003). The conservation and use of germplasm are aided by describing the traits of a crop species using accepted descriptors (Kumari *et al.*, 2017). The production of resistant cultivars can employ resistant genotypes as donor parents. In order to create interspecific hybrids, Kaushal and Singh (1988) crossed the adzuki bean with the urdbean. This introduced adzuki bean resistance to combat the urdbean viral and fungal infections. Zhu *et al.* (2019) screened 80 adzuki bean germplasm lines for drought tolerance and observed three highly tolerant, nine tolerant, 37 moderately tolerant and four low tolerant lines. The identified drought-tolerant germplasms can be employed in adzuki bean breeding programmes focused on drought resistance while also offering opportunities for further investigation into the molecular mechanisms underlying drought tolerance (Table 3.2). In the realm of adzuki bean crop research, Table 3.3 delineates the characterization of germplasm and the identification of donor-specific traits by various research groups.

3.6 Integration of Genomic and Genetic Resources for Crop Improvement

Integrating genomic and genetic resources has revolutionized crop improvement strategies, allowing breeders to accelerate the development of improved varieties with desired traits. Adzuki bean (*V. angularis*) improvement can significantly benefit from utilizing genomic and genetic

Table 3.3. Germplasm characterization and donor-specific traits identified by different research groups for adzuki bean. (Author's own table.)

Specific trait	Donor/gene identified	References
Abiotic stress		
Drought stress	VaPIP2-1, VaPIP2-5 in root and VaPIP1-1, VaPIP1-7 in leaf linked with water uptake	Tayade *et al.* (2022b)
Salt tolerance	JP205833 of *V. riukiuensis* (strain 'Tojinbaka') and JP107879 of *V. nakashimae* (strain 'Ukushima')	Yoshida *et al.* (2016)
Heat tolerance	VrLEA gene	Singh *et al.* (2022)
Nutritional traits		
Flavonoid metabolic pathway	VaSDC1 gene	Chu *et al.* (2021)
Agronomic traits		
Epicotyl length	qECL7.1 region	Kachapila *et al.* (2023)
Flowering time	Major QTL LG03 and two minor QTL LG05	Liu *et al.* (2016)
Black seed coat colour	VaUGT candidate gene	Li *et al.* (2017)

resources. The availability of the adzuki bean genome sequence (Kang *et al.*, 2015; Yang *et al.*, 2015) has provided a fundamental resource for understanding its genetic make-up and functional elements. Genome sequencing allows researchers to identify and characterize genes responsible for important traits, facilitating marker development and genetic mapping. During the past 20 years, SSRs (Gupta and Varshney, 2000) and single nucleotide polymorphisms (SNPs; Varshney *et al.*, 2010) are two molecular markers that have made it easier to analyse complex traits that hinder crop production using genome-wide association mapping and QTL mapping techniques (Varshney *et al.*, 2015). The concept of genomics-assisted breeding (GAB; Varshney *et al.*, 2005) was developed to incorporate genomics into breeding, and it has proven to be highly effective for several traits in both legumes (Varshney, 2016; Pratap *et al.*, 2017) and cereals (Septiningsih *et al.*, 2013; Varshney *et al.*, 2006).

With these advancements during the past decade, genetic and genomics approaches have been integrated into legumes to breed climate-resilient crops (Prince *et al.*, 2017). At present, genomic resources are now readily available for most legume crops. There are several examples where molecular breeding has been successfully used in legume crop improvement programmes (Varshney *et al.*, 2013; Varshney, 2014). However, adzuki beans are underutilized for their genetic resources, and molecular breeding is not as widespread as for other species of the

Vigna genus (such as mung bean). This might be due to several factors, including the difficulty in obtaining additional polymorphic markers owing to low genetic variation (Kaga *et al.*, 1996, 2000) or the fact that adzuki beans are a minor crop and are not as widely grown as other important crops, which means that breeding research on them is not as well funded. Despite the organelle genome sequences (including mitochondria and chloroplast) being finished, the nuclear reference genome sequence of the adzuki bean is still not accessible (Naito *et al.*, 2013). The first set of adzuki bean genomic SSR (gSSR) markers was made using genomic DNA from the Erimoshouzu cultivar for gene flow investigations (Chen *et al.*, 2015a). The genetic linkage map of the adzuki bean was subsequently effectively anchored with 196 SSR markers; however, there were still over 15 cM gaps (Han *et al.*, 2005).

A breakthrough in the molecular breeding of adzuki bean has come through the discovery of 143,113 adzuki bean SSRs and 200,808 mung bean SSRs by whole-genome sequencing (Kang *et al.*, 2014, 2015). In adzuki bean, the assembly of EST-SSR (eSSR) markers has also been reported (Chankaew *et al.*, 2014). Researchers across the world have used molecular markers in adzuki bean to construct genetic linkage maps based on which several genes and QTLs related to stems, leaves, flowering time, pods and seeds as well as bruchid resistance have been identified (Kaga *et al.*, 1996; Han *et al.*, 2005; Somta *et al.*, 2008; Aoyama *et al.*, 2011;

Horiuchi *et al.*, 2015; Yamamoto *et al.*, 2016; Li *et al.*, 2017). Since there has been much investigation on the genetic diversity of cultivated adzuki bean and a relatively low diversity was found based on DNA molecular markers (Xu *et al.*, 2008a; Wang *et al.*, 2009b), even though agronomic characteristics exhibit a significant degree of variation (Xu *et al.*, 2008a; Bai *et al.*, 2014), the majority of the published genetic maps were created using combinations of the cultivated adzuki bean and its wild relatives. Table 3.4 provides information on germplasm, breeding lines, mapping populations, DNA markers and their application in the enhancement of adzuki bean crops.

Compared to other legumes like the common bean, chickpea, pigeon pea and soybean (Choudhary *et al.*, 2009; Dutta *et al.*, 2011; Zhang *et al.*, 2013), adzuki bean has been reported to have many fewer SSR markers, which is a significant restriction in its molecular breeding. Other molecular technologies such as next-generation sequencing (NGS) and transcriptome analysis have also been utilized in adzuki bean. Transcriptome analysis involves studying the complete set of RNA transcripts produced by the adzuki bean genome. It provides insights into gene expression patterns, differential gene regulation under different conditions and the identification of candidate genes associated with specific traits. Utilizing NGS of genomic DNA and RNA, researchers have recently found more gSSRs and eSSR markers (Kang *et al.*, 2014; Chen *et al.*, 2015a, b). In a study conducted by Chen *et al.* (2015b), the creation of eSSR markers utilized part of the 65,950 unique genes that were found from transcriptome sequencing of adzuki bean. In addition, 53 SSR markers were developed, and 261 chosen accessions were genotyped using 110 SSR markers, including 57 that were previously in use.

Table 3.4. Details of germplasm/breeding lines/mapping population, DNA markers and their utilization in adzuki bean crop improvement. (Author's own table.)

Trait	No. of germplasm lines screened	Method used/ molecular markers used	References
Biotic			
Genetic variations	176	SSR markers	Wang *et al.* (2012)
Phylogenetic analysis	34	EST-SSR markers	Chen *et al.* (2015a)
Genetic variability and yield-related traits	475	ANOVA, PCA, correlation and cluster analysis	Hu *et al.* (2022)
Abiotic			
Root morphology and soil water content	1	Fractal analysis	Chun *et al.* (2021)
Saline stress	3	Quadratic regression model	Han *et al.* (2023)
Phosphorus deficiency stress	1	SPSS software	Lian *et al.* (2019)
Nutritional			
Phenolic content	4	Folin–Ciocalteu colorimeter method	Desta *et al.* (2022)
Partitioning of nutritional and phytochemical constituents	9	R Studio	Johnson *et al.* (2022)
Nutritional composition, phytochemicals and antioxidant activity	17	Gas chromatography, Kjeldahl method, RP-HPLC and HPLC	Shi *et al.* (2017)
Agronomic			
Seed coat colour	2	SSR markers	Chu *et al.* (2021)
Seed size	143	SNP markers	Wang *et al.* (2021)
Short-day photoperiod effects on plant growth, flower bud differentiation and yield formation	9	Excel 2003, DPSv2000 and SPSS17.0	Dong *et al.* (2016)

3.7 Erosion of Genetic Diversity from Traditional Areas and Limitations in Germplasm Use

Genetic diversity is a crucial component of crop improvement because it provides the raw material for breeding programmes to develop new and improved varieties. So, understanding the genetic diversity across vast populations of adzuki bean is crucial for the conservation, selection and use of genetic resources during the production of better cultivars (Wang *et al.*, 2019; Li *et al.*, 2020). However, genetic diversity among genotypes of the adzuki bean is believed to be smaller than that of other legume genotypes; hence, successful breeding of improved adzuki bean cultivars is thought to be less prevalent. Likewise, wild relatives of the adzuki bean, which often possess unique disease resistance and stress tolerance genes, are underutilized in breeding programmes. Challenges in hybridization and introgression limit the exploitation of these valuable genetic resources. In addition, the intensification of agriculture, accompanied by the adoption of modern farming practices, has replaced traditional adzuki bean landraces with a limited number of high-yielding varieties. This has resulted in a loss of genetic diversity as farmers increasingly rely on a narrow range of commercial varieties.

Although the germplasm of the adzuki bean has been characterized to some extent for genetic diversity, many germplasm accessions remain poorly characterized in terms of their genetic traits, agronomic performance and stress tolerance. This lack of comprehensive information hinders breeders from effectively utilizing available germplasm resources. Furthermore, there are only a few reports on the genes that have been functionally characterized in adzuki bean (Chu *et al.*, 2021; Imoto *et al.*, 2022). In contrast to the critical legume crops, where these resources have been frequently used in genetic development, the application of these resources is noticeably extremely low in underutilized pulses like adzuki bean (Petereit *et al.*, 2022) imposing difficulties for breeders searching for crop improvement programmes in these crops.

3.8 Cultivation Practices

The adzuki bean is a kharif season crop that thrives well in temperate climatic conditions and, in general, its climatic needs and growth circumstances are comparable to those of soybean (Myers, 1998; Duke, 2012). Adzuki bean can handle some drought. It can be grown in a broader range of soil types, including silt loams and sandy soils, which are loose and rich in organic matter, with proper drainage, preferring slightly acidic to neutral soil pH in the range 5.0–7.5 (Hardman *et al.*, 1989; Duke, 2012). The preferred seed rate is 60–70 kg/ha, and seeds are planted 3–5 cm deep, 10–15 cm spaced out in rows 30–45 cm apart (Kaushik *et al.*, 2020). Inoculation with crop-specific *Rhizobium* culture is advised for seeds not planted in regularly-cropped soils. The optimum sowing time is from the last week of May to mid-June. The congenial soil temperature for germination is 15°C, and for crop cultivation is 15–30°C (Hardman *et al.*, 1989; Duke, 2012). Emergence may take up to 20 days in soils below 13°C. However, emergence in 10–14 days is common in soils with high temperatures. Like other crops, it performs poorly with weed species, so weed management practices, particularly during the early growth stages of the crop, are recommended for the best yield. Being a leguminous crop, the fertilizer requirement (FYM: 10–15 q/ha; NPK: 20:40:20) is less than other cereal crops (Kaushik *et al.*, 2020). The crop may take up to 110–120 days after sowing to reach maturity. For vegetable purposes, the green pods are harvested after 60 days of sowing. For grain purposes, the crop is harvested when pods turn tan and dry. The yield may vary in the range 15–18 q/ha depending on the cultural practices and environmental conditions (Kaushik *et al.*, 2020).

3.9 Conclusion

The adzuki bean is a nutritional powerhouse that has stood the test of time. Its rich history, culinary versatility and health benefits make it a valuable addition to any diet. Breeding for adzuki bean improvement combines traditional breeding methods, molecular techniques and biotechnological tools to develop varieties with enhanced traits and performance. The continuous efforts of breeders contribute to the development of improved adzuki bean varieties that address the challenges of agriculture while meeting the demands of consumers. The future of adzuki bean breeding holds promise in terms of increased yields, improved nutritional quality and resilience to changing environmental conditions, ultimately benefitting farmers and consumers alike.

Integrating genomic and genetic resources has significantly advanced adzuki bean breeding programmes, providing breeders with tools to enhance genetic gain, accelerate variety development and improve important traits. The availability of the adzuki bean genome sequence, along with transcriptome analysis, genetic diversity studies, marker-assisted selection, genomic selection and functional genomics techniques, enables breeders to make informed decisions and develop improved adzuki bean varieties with enhanced yield, disease resistance, abiotic stress tolerance and nutritional quality. The erosion of genetic diversity in the adzuki bean and the limitations in germplasm use pose challenges for breeders seeking to develop improved varieties. Conservation efforts, both *in situ* and *ex situ*, are crucial for maintaining and utilizing the genetic diversity of adzuki bean. Collaboration, germplasm exchange and comprehensive characterization of germplasm resources are key to addressing the limitations and maximizing the potential of available genetic diversity for adzuki bean breeding programmes. By conserving and effectively utilizing genetic resources, breeders can develop resilient and improved adzuki bean varieties that meet the evolving needs of agriculture and society. The future of adzuki bean breeding lies in addressing the challenges of climate change, disease pressures, nutritional demands and market preferences. Integrating genomic tools, molecular breeding techniques, participatory approaches, and collaboration among researchers, farmers and stakeholders will be key to achieving these objectives. By focusing on these areas, adzuki bean breeding programmes can develop improved resilient, high-yielding, nutritious varieties aligned with the needs of both farmers and consumers.

References

Amarowicz, R., Estrella, I., Hernandez, T., Dueñas, M., Troszynska, A., Agnieszka, K. and Pegg, R.B. (2009) Antioxidant activity of a red lentil extract and its fractions. *International Journal of Molecular Sciences* 10, 5513–5527.

Anjali and Bhardwaj, N. (2020) Morphological characterisation and diversity analysis of adzuki bean [*Vigna angularis* (Willd.) Ohwi and Ohashi] germplasm. *International Journal of Current Microbiology and Applied Sciences* 9(10), 603–617.

Aoyama, S. and Shimada, H. (2014) Development of a method for evaluating genetic resources tolerant to flowering and pod setting damage by cool shading in azuki bean. *Japanese Journal of Crop Science* 83(4), 326–332.

Aoyama, S., Shimada, H., Hasegawa, N., Murata, K., Fujita, S. and Matsukawa, I. (2009) A new early maturity adzuki bean variety, "Kita-roman", with cool-weather tolerance and soil-borne disease resistance. *Bulletin of Hokkaido Prefectural Agricultural Experiment Station* 94, 1–16.

Aoyama, S., Onishi, K. and Kato, K. (2011) The genetically unstable dwarf locus in azuki bean (*Vigna angularis* (Willd.) Ohwi&Ohashi). *Journal of Heredity* 102(5), 604–609.

Bai, P., Cheng, X.Z. and Wang, L.X. (2014) Evaluation in agronomic traits of adzuki bean accessions. *Journal of Plant Genetic Resources* 15, 1209–1215.

Chankaew, S., Isemura, T., Isobe, S., Kaga, A., Tomooka, N. *et al.* (2014) Detection of genome donor species of neglected tetraploid crop *Vigna reflexo-pilosa* (creole bean) and genetic structure of diploid species based on newly developed EST-SSR markers from azuki bean (*Vigna angularis*). *PLoS One* 9(8), 104990.

Chen, H., Liu, L., Wang, L., Wang, S., Somta, P. and Cheng, X. (2015a) Development and validation of EST-SSR markers from the transcriptome of adzuki bean (*Vigna angularis*). *PLoS One* 10(7), 0131939.

Chen, H., Liu, L., Wang, L., Wang, S., Wang, M.L. and Cheng, X. (2015b) Development of SSR markers and assessment of genetic diversity of adzuki bean in the Chinese germplasm collection. *Molecular Breeding* 35, 1–14.

Cheng, X.Z. and Wang, S.M. (2009) Records of Chinese food legumes cultivars. *China Agricultural Science and Technology Press (in Chinese)*, 125–197.

Choudhary, S., Sethy, N.K., Shokeen, B. and Bhatia, S. (2009) Development of chickpea EST-SSR markers and analysis of allelic variation across related species. *Theoretical and Applied Genetics* 118, 591–608.

Chu, L., Zhao, P., Wang, K., Zhao, B., Li, Y., Yang, K. and Wan, P. (2021) VaSDC1 is involved in modulation of flavonoid metabolic pathways in black and red seed coats in adzuki bean (*Vigna angularis* L.). *Frontiers in Plant Science* 12, 679892.

Chun, H.C., Sanghun, L.E.E., Choi, Y.D., Gong, D.H. and Jung, K.Y. (2021) Effects of drought stress on root morphology and spatial distribution of soybean and adzuki bean. *Journal of Integrative Agriculture* 20(10), 2639–2651.

Crawford, G.W. (2006) East Asian plant domestication. In: Stark, M.T. (ed.) *Archaeology of Asia*. Blackwell Publishing Ltd, Oxford, UK, pp. 77–95.

Desta, K.T., Yoon, H., Shin, M.J., Lee, S., Wang, X.H., Choi, Y.M. and Yi, J.Y. (2022) Variability of anthocyanin concentrations, total metabolite contents and antioxidant activities in adzuki bean cultivars. *Antioxidants* 11(6), 1134.

Dong, W., Zhang, Y., Zhang, Y., Ren, S., Wei, Y. and Zhang, Y. (2016) Short-day photoperiod effects on plant growth, flower bud differentiation, and yield formation in adzuki bean (I). *International Journal of Agriculture & Biology* 18(2), 337–345.

Dua, R.P., Raiger, H.L., Phogat, B.S. and Sharma, S.K. (2009) *Underutilised Crops: Improved Varieties and Cultivation Practices*. ICAR, The Netherlands.

Duan, H., Long, J. and Lin, L. (1989) Small bean. *Edible Bean Crops*. Science Publishing House, Beijing, pp. 160–171.

Duke, J. (2012) *Handbook of Legumes of World Economic Importance*. Springer Science & Business Media, Berlin, Germany.

Dutta, S., Kumawat, G., Singh, B.P., Gupta, D.K., Singh, S. *et al.* (2011) Development of genic-SSR markers by deep transcriptome sequencing in pigeonpea [*Cajanus cajan* (L.) Millspaugh]. *BMC Plant Biology* 11(1), 1–13.

Fact.MR (2023) Global Adzuki Beans Market Outlook (2023 to 2033) https://www.factmr.com/report/adzuki-beans-market#:~:text=Global%20Adzuki%20Beans%20Market%20Outlook%20(2023%20to%202033)&text=With%203.6%25%20projected%20growth%20from,end%20of%20the%20forecast%20period.https://www.factmr.com/report/adzuki-beans-market#:~:text=Global%20Adzuki%20Beans%20Market%20Outlook%20(2023%20to%202033)&text=With%203.6%25%20projected%20growth%20from,end%20of%20the%20forecast%20period

Fernandez, G.C.J. and Talekar, N.S. (1990) Genetics and breeding for bruchid resistance in Asiatic *Vigna* species. In: Fujii, K., Gatehouse, A.M.R., Johnson C.D., Mitchel, R. and Yoshida, T. (eds) *Bruchids and Legumes: Economics, Ecology and Evolution*. Springer, Dordrecht, The Netherlands, pp. 209–217.

Franco, T.L. (2003) *Analisis Estadistico de Datos de Caracterizacion Morfologica de Recursos Fitogeneticos. BoletinTecnico IPGRI No. 8*. Bioversity International, Rome, Italy.

Fujita, S., Murata, K., Shimada, H., Aoyama, S., Chiba, I. *et al.* (2002) A new adzuki bean variety, "Syumari", with soil-borne disease resistance and excellent processing quality. *Bulletin of Hokkaido Prefectural Agricultural Experiment Station* 82, 31–40.

Gupta, P.K. and Varshney, R.K. (2000) The development and use of microsatellite markers for genetic analysis and plant breeding with emphasis on bread wheat. *Euphytica* 113(3), 163–185.

Han, E.H., Lee, C.J., Sin, C.S. and Lee, E.K. (1982) An investigation on the stem rot red bean caused by *Phytophthora vignae* Purss. *Research Report Office Rural Development* 24, 69–71 (in Korean with English abstract).

Han, O.K., Kaga, A., Isemura, T., Wang, X.W., Tomooka, N. and Vaughan, D.A. (2005) A genetic linkage map for azuki bean [*Vigna angularis* (Willd.) Ohwi&Ohashi]. *Theoretical and Applied Genetics* 111, 1278–1287.

Han, S.I., Han, W.Y., Baek, I.Y., Shin, S.O., Kim, H.T. *et al.* (2013) A new azuki bean cultivar 'Saegil' with red seed coat, had good processing property. *Korean Journal of Breeding Science* 45(3), 258-261.

Han, S., Lu, K., Guo, T., Zhang, Y., Wang, B. and Wu, B. (2023) Effect of saline treatment on seed germination of adzuki beans. *Seed Science and Technology* 51(1), 31–42.

Hardman, L.L., Oplinger, E.S., Doll, J.D., and Combs, S.M. (1989) *Alternative Field Crops Manual: Adzuki Bean*. University of Wisconsin-Cooperative Extension, Madison, Wisconsin.

Harlan, J.R. and de Wet, J.M. (1971) Toward a rational classification of cultivated plants. *Taxon* 20(4), 509–517.

Horiuchi, Y., Yamamoto, H., Ogura, R., Shimoda, N., Sato, H. and Kato, K. (2015) Genetic analysis and molecular mapping of genes controlling seed coat colour in adzuki bean (*Vigna angularis*). *Euphytica* 206, 609–617.

Hu, L., Luo, G., Zhu, X., Wang, S., Wang, L. and Cheng, X. (2022) Genetic diversity and environmental influence on yield and yield-related traits of adzuki bean *(Vigna angularis L.)*. *Plants* 11, 1132.

Imoto, Y., Yoshikawa, S., Horiuchi, Y., Iida, T., Oka, T. *et al.* (2022) Flowering Date1, a major photoperiod sensitivity gene in adzuki bean, is a soybean floral repressor E1 ortholog. *Breeding Science* 72(2), 132–140.

Isemura, T., Noda, C., Mori, S., Yamashita, M., Nakanishi, H., Inoue, M. and Kamijima, O. (2001) Genetic variation and geographical distribution of Azuki bean *(Vigna angularis)* landraces based on the electrophoregram of seed storage proteins. *Breeding Science* 51(4), 225–230.

Isemura, T., Ishii, T., Saito, H., Noda, C., Sanjuo, S. and Ueshima, S. (2002) Genetic diversity of indigenous adzuki bean lines assessed by RAPD analysis. *Breeding Studies* 4(3) 125–135. [In Japanese with English summary.]

Isemura, T., Tomooka, N., Kaga, A. and Vaughan, D.A. (2011) Comparison of the pattern of crop domestication between two Asian beans, azuki bean *(Vigna angularis)* and rice bean *(V. umbellata)*. *Japan Agricultural Research Quarterly* 45(1), 23–30.

Ishaq, M.N., Wada, A.C., Ochigbo, A.A. and Falusi, O.A. (2007) Genetic resources multiplication and utilisation. In: *Plant Genetic Resources and Food Security in West and Central Africa. Regional Conference, 26-30 April 2004*. Bioversity International, Rome, Italy, 472, pp. 129.

Jansen, P.C.M. (2006) *Vigna angularis* (Willd). Ohwi and Ohashi H. In: Brink, M. and Belay, G. (eds) *PROTA (Plant Resources of Tropical Africa) / Ressources végétales de l'Afrique tropicale)*, Wageningen, Netherlands.

Johnson, J.B., Neupane, P., Bhattarai, S.P., Trotter, T. and Naiker, M. (2022) Partitioning of nutritional and phytochemical constituents in nine adzuki bean genotypes from Australia. *Journal of Agriculture and Food Research* 10, 100398.

Kachapila, M., Horiuchi, Y., Nagasawa, H., Michihata, N., Yoshida, T. *et al.* (2023) Fine-mapping of qECL7. 1 is a Quantitative Trait Locus contributing to epicotyl length in adzuki bean *(Vigna angularis)*. *Agriculture* 13(7), 1305.

Kaga, A., Ohnishi, M., Ishii, T. and Kamijima, O. (1996) A genetic linkage map of azuki bean constructed with molecular and morphological markers using an interspecific population *(Vigna angularis x V. nakashimae)*. *Theoretical and Applied Genetics* 93, 658–663.

Kaga, A., Ishii, T., Tsukimoto, K., Tokoro, E. and Kamijima, O. (2000) Comparative molecular mapping in *Ceratotropis* species using an interspecific cross between azuki bean *(Vigna angularis)* and rice bean *(V. umbellata)*. *Theoretical and Applied Genetics* 100, 207–213.

Kaga, A., Han, O.K., Hirashima, S., Saravankumar, P. and Kumari, H.M.P.S. (2004) Collecting and monitoring of the azuki bean *(Vigna angularis)* complex populations in Tottori prefecture, Japan. Annual report on exploration and introduction of plant genetic resources. *National Institute of Agrobiology Science* 20, 61–74.

Kaga, A., Isemura, T., Tomooka, N. and Vaughan, D.A. (2008) The genetics of domestication of the azuki bean *(Vigna angularis)*. *Genetics* 178, 1013–1036.

Kang, Y.J., Kim, S.K., Kim, M.Y., Lestari, P., Kim, K.H. *et al.* (2014) Genome sequence of mungbean and insights into evolution within *Vigna* species. *Nature Communications* 5(1), 5443.

Kang, Y.J., Satyawan, D., Shim, S., Lee, T., Lee, J. *et al.* (2015) Draft genome sequence of adzuki bean, *Vigna angularis*. *Scientific Reports* 5(1), 8069.

Kaushal, R.P. and Singh, B.M. (1988) Interspecific hybridisation between urdbean *(Vigna mungo* (L.) Hepper) and adzuki bean *(Vigna angularis* (Willd.) Ohwi and Ohashi). *Euphytica* 39(1), 53–57.

Kaushik, S.K., Raiger, H.L., Yadav, S.K., Kumar, S., Singh, M.C. *et al.* (2020) *Progress Report Kharif*. All India Coordinated Research Network on Potential Crops, ICAR-NBPGR, New Delhi.

Kim, M.K., Jeong, R.D., Kwak, H.R., Lee, S.-H., Kim, J.-S. *et al.* (2014) First report of cucumber mosaic virus isolated from wild *Vigna angularis* var. *nipponensis* in Korea. *The Plant Pathology Journal* 26, 93–98.

Kitano-Okada, T., Ito, A., Koide, A., Nakamura, Y., Han, K.H. *et al.* (2012) Anti-obesity role of adzuki bean extract containing polyphenols: in vivo and in vitro effects. *Journal of the Science of Food and Agriculture* 92, 2644–2651.

Kramer, C., Soltani, N., Robinson, D.E., Swanton, C.J. and Sikkema, P.H. (2012) Control of volunteer adzuki bean in soybean. *Journal of Agricultural Science and Technology* 3, 501–509.

Kumari, S., Nirala, R.B.P., Rani, N. and Prasad, B.D. (2017) Selection criteria of linseed *(Linum usitatissimum* L.) genotypes for seed yield traits through correlation and path coefficient analysis. *Journal of Oilseeds Research* 34(3), 171–174.

Lee, G.A. (2013) Archaeological perspectives on the origins of azuki (*Vigna angularis*). *The Holocene* 23(3), 453–459.

Leipe, C., Lu, J.C., Chi, K.A., Lee, S.M., Yang, H.C., Wagner, M. and Tarasov, P.E. (2022) Evidence for cultivation and selection of azuki (*Vigna angularis* var. angularis) in prehistoric Taiwan sheds new light on its domestication history. *Quaternary International* 623, 83–93.

Li, J., Qi, C., Gu, J. and Jin, Z. (2020) Effect of sire population on the genetic diversity and fitness of F1 progeny in the endangered Chinese endemic *Sinocalycanthus chinensis*. *Ecology and Evolution* 10(9), 4091–4103.

Li, S. (1590, Ming Dynasty) (2010) *Compendium of Materia Medica*, 255. Yunnan Educ Press, Kunming, China.

Li, Y., Yang, K., Yang, W., Chu, L., Chen, C. *et al.* (2017) Identification of QTL and qualitative trait loci for agronomic traits using SNP markers in the adzuki bean. *Frontiers in Plant Science* 8, 840.

Lian, H., Qin, C., Zhang, L., Zhang, C., Li, H. and Zhang, S. (2019) Lanthanum nitrate improves phosphorus-use efficiency and tolerance to phosphorus-deficiency stress in *Vigna angularis* seedlings. *Protoplasma* 256, 383–392.

Lin, R.F., Chai, Y., Liao, Q., Sun, S.X. and Tian, J. (2002) *Minor Grain Crops in China*, ed Yun L. China Agricultural Science and Technology Press, Beijing, pp. 192–209.

Liu, C., Fan, B., Cao, Z., Su, Q., Wang, Y., Zhang, Z. and Tian, J. (2016) Development of a high-density genetic linkage map and identification of flowering time QTLs in adzuki bean (*Vigna angularis*). *Scientific Reports* 6(1), 39523.

Liu, L., Bestel, S., Shi, J., Song, Y. and Chen, X. (2013) Palaeolithic human exploitation of plant foods during the last glacial maximum in North China. *Proceedings of the National Academy of Sciences* 110(14), 5380–5385.

Liu, R., Zheng, Y., Cai, Z. and Xu, B. (2017) Saponins and flavonoids from adzuki bean (*Vigna angularis* L.) ameliorate high-fat diet-induced obesity in ICR mice. *Frontiers in Pharmacology* 8, 687.

Lumpkin, T.A. and McClary, D.C. (1994) *Azuki Bean: Botany, Production and Uses*. CAB International, Wallingford, UK.

Maeda, K. (1987) *Legumes and Humans: A 10,000 Year History,* Kokonshin, Tokyo *(in Japanese)*.

Marcello, D. (2006) Grain legume proteins and nutraceutical properties. *Fitoterapia* 77, 67–82.

Maréchal, R., Mascherpa, J.M. and Stainier, F. (1978) Etude taxonomique d'un groupe complexe d'espèces des genres *Phaseolus* et *Vigna* (Papilionaceae) sur la base de données morphologiques et polliniquestraitées par l'analyse informatique. *Boissiera* 28, 1–273.

Moon, J.K., Lee, Y.H., Han, W.Y. *et al.* (2006) A new light green seed-coated azuki bean cultivar, "Yeonkeum", with middle seed size and good seed quality. *Korean Journal of Breeding* 38(4), 303–304.

Myers, C. (1998) *Specialty and Minor Crops Handbook*. UCANR Publications, Oakland, California.

Nagao, N., Sakuma, Y., Funakoshi, T. and Itani, T. (2023) Variation in antioxidant capacity of the seven azuki bean (*Vigna angularis*) varieties with different seed coat colour. *Plant Production Science* 26(2), 164–173.

Naito, K., Kaga, A., Tomooka, N. and Kawase, M. (2013) De novo assembly of the complete organelle genome sequences of azuki bean (*Vigna angularis*) using next-generation sequencers. *Breeding Science* 63(2), 176-182.

Ning, X.U., Cheng, X.Z., Li-Xia, W., Su-Hua, W., Chang-You, L., Lei, S. and Li, M. (2009) Screening SSR marker for Adsuki bean and its application in diversity evaluation in chinese Adzuki bean germplasm resources. *Acta Agronomica Sinica* 35(2), 219-227.

Ogiso-Tanaka, E., Chankaew, S., Yoshida, Y., Isemura, T., Marubodee, R. *et al.* (2023) Unique salt-tolerance-related QTLs, evolved in *Vigna riukiuensis* (Na+Includer) and *V. nakashimae* (Na+Excluder), shed light on the development of super-salt-tolerant azuki bean (*V. angularis*) cultivars. *Plants* 12, 1680.

Ohwi, J. and Ohashi, H. (1969) *Vigna umbellata* (Thunb.). *The Journal of Japanese Botany* 44, 31.

Orsi, D.C., Nishi, A.C.F., Carvalho, V.S. and Asquieri, E.R. (2017) Chemical composition, antioxidant activity and development of desserts with azuki beans (*Vigna angularis*). *Brazilian Journal of Food Technology* 20. DOI: 10.1590/1981-6723.17416

Pandiyan, M., Sivakumar, P., Krishnaveni, A., Chinnasamy, S., Radhakrishnan, Mallaian, V. and Tomooka, N. (2021) Adzuki bean. *The Beans and the Peas* 89–103.

Parida, A., Raina, S.N. and Narayan, R.K.J. (1990) Quantitative DNA variation between and within chromosome complements of *Vigna* species (Fabaceae). *Genetica* 82(2), 125–133.

Petereit, J., Marsh, J.I., Bayer, P.E., Danilevicz, M.F., Thomas, W.J., Batley, J. and Edwards, D. (2022) Genetic and genomic resources for soybean breeding research. *Plants* 11(9), 1181.

Pharmacopoeia (2010) *Commission Chinese Pharmacopoeia*. People's Medical Publishing House, Beijing.

Pratap, A., Chaturvedi, S.K., Tomar, R., Rajan, N., Malviya, N. *et al.* (2017) Marker-assisted introgression of resistance to fusarium wilt race 2 in Pusa 256, an elite cultivar of *desi* chickpea. *Molecular Genetics and Genomics* 292(6), 1237–1245.

Prince, S.J., Murphy, M., Mutava, R.N., Durnell, L.A., Valliyodan, B., Shannon, J.G. and Nguyen, H.T. (2017) Root xylem plasticity to improve water use and yield in water-stressed soybean. *Journal of Experimental Botany* 68(8), 2027–2036.

Pusadee, T., Jamjod, S., Chiang, Y.C., Rerkasem, B. and Schaal, B.A. (2009) Genetic structure and isolation by distance in a landrace of Thai rice. *Proceedings of the National Academy of Sciences USA* 106, 13880–13885.

Redden, R.J., Basford, K.E., Kroonenberg, P.M., Islam, F.A., Ellis, R. *et al.* (2009) Variation in adzuki bean (*Vigna angularis*) germplasm grown in China. *Crop Science* 49(3), 771–782.

Robinson, G.M. (2018) New frontiers in agricultural geography: Transformations, food security, land grabs and climate change. *Boletin de la Asociacion de Geografos Espanoles* 78, 1–48.

Schuster, W.H., Alkamper, J., Marquard, R., Stahlin, A. and Stahlin, L. (1998) *Leguminosen zur Kornnutzung*. FordervereinTropen institute Geissen.

Septiningsih, E.M., Collard, B.C., Heuer, S., Bailey-Serres, J., Ismail, A.M. and Mackill, D.J. (2013) Applying genomics tools for breeding submergence tolerance in rice. *Translational Genomics for Crop Breeding: Abiotic Stress, Yield and Quality* 2, 9–30.

Shi, Z., Yao, Y., Zhu, Y. and Ren, G. (2017) Nutritional composition and biological activities of 17 Chinese adzuki bean (*Vigna angularis*) varieties. *Food and Agricultural Immunology* 28(1), 78–89.

Sikkema, P.H., Soltani, N., Shropshire, C. and Robinson, D.E. (2006) Response of adzuki bean to pre-emergence herbicides. *Canadian Journal of Plant Science* 86(2), 601–604.

Singh, C.M., Kumar, M., Pratap, A., Tripathi, A., Singh, S. *et al.* (2022) Genome-wide analysis of late embryogenesis abundant protein gene family in *Vigna* species and expression of VrLEA encoding genes in *Vigna glabrescens* reveals its role in heat tolerance. *Frontiers in Plant Science* 13, 843107.

Singh, K., Gupta, K., Tyagi, V. and Rajkumar, S. (2020) Plant genetic resources in India: management and utilisation. *Vavilov Journal of Genetics and Breeding* 24(3), 306–314.

Smartt, J. (1990) *Grain legumes: Evolution and Genetic Resources*. Cambridge University Press, Cambridge, UK.

Somta, P., Kaga, A., Tomooka, N., Isemura, T., Vaughan, D.A. and Srinives, P. (2008) Mapping of quantitative trait loci for a new source of resistance to bruchids in the wild species *Vigna nepalensis* Tateishi & Maxted (*Vigna* subgenus *Ceratotropis*). *Theoretical and Applied Genetics* 117, 621–628.

Song, S.B., Choe, M.E., Chu, J.H., Kim, J.Y., Lee, B.W. and Han, S. (2021) An adzuki bean (*Vigna angularis* L.) cultivar 'Honggyeong' with bright red seed-coat color and resistance to lodging stress. *Korean Society of Breeding Science* 53 (4), 489–494.

Takahama, U., Yamauchi, R. and Hirota, S. (2013) Isolation and characterization of a cyanidin-catechin pigment from adzuki bean (*Vigna angularis*). *Food Chemistry* 141(1), 282–288.

Takahashi, Y., Tamagata, Y., Matsumura, T., Uyeda, I., Mink, G.I. and Berger, P.H. (1998) Nucleotide sequence of the 3'-terminal region of the genome of the azuki bean mosaic strain of the bean common mosaic virus and its phylogenetic relationship to viruses in the bean common mosaic virus group. *Japanese Journal of Phytopathology* 64, 478–480.

Tateishi, Y. and Ohashi, H. (1990) Systematics of the adzuki bean group in the genus *Vigna*. In: Fujii, K., Gatehouse, A.M.R., Johnson, C.D., Mitchel, R. and Yoshida, T. (eds) *Bruchids and Legumes: Economics, Ecology and Coevolution*. Kluwer Publications, The Netherlands, pp. 189–199.

Tayade, R., Kim, S.H., Tripathi, P., Choi, Y.D., Yoon, J.B. and Kim, Y.H. (2022a) High throughput root imaging analysis reveals wide variation in root morphology of wild adzuki bean (*Vigna angularis*) accessions. *Plants* 11, 405.

Tayade, R., Rana, V., Shafiqul, M., Nabi, R.B.S., Raturi, G. *et al.* (2022b) Genome-wide identification of aquaporin genes in adzuki bean (*Vigna angularis*) and expression analysis under drought stress. *International Journal of Molecular Sciences* 23(24), 16189.

Tazawa, A., Sato, H., Shimada, H., Aoyama, S., Fujita, S. *et al.* (2015) A new Dainagon-brand adzuki bean variety, "Homare-dainagton", with soil-borne disease resistance and high processing adaptability. *Hokkaido Central Agricultural Experimental Station Reports* 99, 1–11.

Tian, J., Fan, B. and Zhang, X. (2004) Breeding of a new adzuki bean variety, Jihong 9218. *Journal of Hebei Agricultural Sciences* 8, 74–76.

Tomooka, N., Vaughan, D.A., Moss, H. and Maxted, N. (2002) *The Asian Vigna: Genus Vigna subgenus Ceratotropis Genetic Resources.* Kluwer Academic Publishers, Dordrecht, The Netherlands.

Tomooka, N., Vaughan, D., Moss, H. and Maxted, N. (2012) *The Asian Vigna: Genus Vigna Subgenus Ceratotropis Genetic Resources.* Springer Science & Business Media, Berlin, Germany.

Varshney, R.K. (2016) Exciting journey of 10 years from genomes to fields and markets: some success stories of genomics-assisted breeding in chickpea, pigeonpea and groundnut. *Plant Science* 242, 98–107.

Varshney, R.K., Graner, A. and Sorrells, M.E. (2005) Genomics-assisted breeding for crop improvement. *Trends in Plant Science* 10(12), 621–630.

Varshney, R.K., Hoisington, D.A. and Tyagi, A.K. (2006) Advances in cereal genomics and applications in crop breeding. *Trends in Biotechnology* 24(11), 490–499.

Varshney, R.K., Glaszmann, J.C., Leung, H. and Ribaut, J.M. (2010) More genomic resources for less-studied crops. *Trends in Biotechnology* 28(9), 452–460.

Varshney, R.K., Gaur, P.M., Chamarthi, S.K., Krishnamurthy, L., Tripathi, S. *et al.* (2013) Fast-track introgression of "QTL-hotspot" for root traits and other drought tolerance traits in JG 11, an elite and leading variety of chickpea. *The Plant Genome* 6(3), 7.

Varshney, R.K., Mohan, S.M., Gaur, P.M., Chamarthi, S.K., Singh, V.K. *et al.* (2014) Marker-assisted backcrossing to introgress resistance to *Fusarium* wilt race 1 and *Ascochyta* blight in C 214, an elite cultivar of chickpea. *The Plant Genome* 7(1), 10.

Varshney, R.K., Kudapa, H., Pazhamala, L., Chitikineni, A., Thudi, M. *et al.* (2015) Translational genomics in agriculture: some examples in grain legumes. *Critical Reviews in Plant Sciences* 34(1-3), 169–194.

Vaughan, D.A., Tomooka, N. and Kaga, A. (2004) Genetic resources, chromosome engineering and crop improvement. In: Singh, R.J. and Jauhar, P. (eds) *Grain Legumes.* CRC Press, Boca Raton, Florida, pp. 341–353.

Wang, L., Wang, J. and Cheng, X. (2019) Adzuki bean (*Vigna angularis* (Willd.) Ohwi & Ohashi) breeding. *Advances in Plant Breeding Strategies Legumes* 7, 1–23.

Wang, L.X., Cheng, X.Z. and Wang, S.H. (2009) Genetic diversity of adzuki bean germplasm resources revealed by SSR markers. *Acta Agronomica Sinica* 35, 1858–1865.

Wang, L.X., Cheng, X.Z., Wang, S.H. and Jing, T.I.A.N. (2012) Analysis of an applied core collection of adzuki bean germplasm by using SSR markers. *Journal of Integrative Agriculture* 11(10), 1601–1609.

Wang, L.X., Wang, J., Luo, G.L., Yuan, X.X., Gong, D. *et al.* (2021) Construction of a high-density adzuki bean genetic map and evaluation of its utility based on a QTL analysis of seed size. *Journal of Integrative Agriculture* 20(7), 1753–1761.

Wang, S.M., Redden, R.J., Jiapeng, J.H., Desborough, P.J., Lawrence, P.L. and Usher, T. (2001) Chinese adzuki bean germplasm: 1. Evaluation of agronomic traits. *Australian Journal of Agricultural Research* 52(6), 671–681.

Wang, X.W., Kaga, A., Tomooka, N. and Vaughan, D.A. (2004) The development of SSR markers by a new method in plants and their application to gene flow studies in azuki bean [*Vigna angularis* (Willd.) Ohwi & Ohashi]. *Theoretical and Applied Genetics* 109, 352–360.

Xu, H.X., Jing, T., Tomooka, N., Kaga, A., Isemura, T. and Vaughan, D.A. (2008a) Genetic diversity of the azuki bean (*Vigna angularis* (Willd.) Ohwi & Ohashi) gene pool as assessed by SSR markers. *Genome* 51(9), 728–738.

Xu, N., Cheng, X.Z., Wang, S.H., Wang, L.X. and Zhao, D. (2008b) Establishment of an adzuki bean (*Vigna angularis*) core collection based on geographical distribution and phenotypic data in China. *Acta Agronomica Sinica* 34(8), 1366–1373.

Xu, R.Q., Tomooka, N., Vaughan, D.A. and Doi, K. (2000a) The *Vigna angularis* complex: genetic variation and relationships revealed by RAPD analysis, and their implications for in situ conservation and domestication. *Genetic Resources and Crop Evolution* 47, 123–134.

Xu, R.Q., Tomooka, N. and Vaughan, D.A. (2000b) AFLP markers for characterising the azuki bean complex. *Crop Science* 40(3), 808–815.

Xu-xiao, Z., Vaughan, D., Tomooka, N., Kaga, A., Xin-wang, W., Jian-ping, G. and Shu-min, W. (2003) Preliminary study on geographical distribution and evolutionary relationships between cultivated and wild adzuki bean (*Vigna angularis* var. *angularis* and var. *nipponensis*) by AFLP analysis. *Plant Genetic Resources* 1(2-3), 175–183.

Yamada, T., Teraishi, M., Hattori, K. and Ishimoto, M. (2001) Transformation of azuki bean by *Agrobacterium tumefaciens. Plant Cell, Tissue and Organ Culture* 64, 47–54.

Yamaguchi, H. (1992) Wild and weed azuki beans in Japan. *Economic Botany* 46, 384–394.

Yamamoto, H., Horiuchi, Y., Ogura, R., Sakai, H., Sato, H. and Kato, K. (2016) Identification and molecular mapping of Flowering Date1 (FD 1), a major photoperiod insensitivity gene in the adzuki bean (*Vigna angularis*). *Plant Breeding* 135(6), 714–720.

Yamamoto, Y., Sano, C.M., Tatsumi, Y. and Sano, H. (2006) Field analyses of horizontal gene flow among *Vigna angularis* complex plants. *Plant Breeding* 125(2), 156–160.

Yang, K., Tian, Z., Chen, C., Luo, L., Zhao, B. *et al.* (2015) Genome sequencing of adzuki bean (*Vigna angularis*) provides insight into high starch and low-fat accumulation and domestication. *Proceedings of the National Academy of Sciences* 112(43), 13213–13218.

Yano, A., Yasuda, K. and Yamaguchi, H. (2004) A test for molecular identification of Japanese archaeological beans and phylogenetic relationship of wild and cultivated species of subgenus *Ceratotropis* (genus *Vigna*, Papilionaceae) using sequence variation in two non-coding regions of the Trnl and trnf genes. *Economic Botany* 58(1), S135–S146.

Yee, E., Kidwell, K.K., Sills, G.R. and Lumpkin, T.A. (1999) Diversity among selected *Vigna angularis* (azuki) accessions on the basis of RAPD and AFLP markers. *Crop Science* 39(1), 268–275.

Yoshida, Y., Marubodee, R., Ogiso-Tanaka, E., Iseki, K., Isemura, T. *et al.* (2016) Salt tolerance in wild relatives of adzuki bean, *Vigna angularis* (Willd.) Ohwi et Ohashi. *Genetic Resources and Crop Evolution* 63, 627–637.

Zhang, G.W., Xu, S.C., Mao, W.H., Hu, Q.Z. and Gong, Y.M. (2013) Determination of the genetic diversity of vegetable soybean [*Glycine max* (L.) Merr.] using EST-SSR markers. *Journal of Zhejiang University Science B* 14, 279–288.

Zhang, H.M., Chen, H.T., Yuan, X.X., Liu, X.Q., Cui, X.Y., Chen, X. and Gu, H.P. (2015) Study on high-yield cultivation techniques of new adzuki bean variety 'Suhong No. 3'. *Acta Agriculturae Shanghai* 6, 108–111.

Zhen-xing, L., Gui-mei, Z. and Jian, C. (2011) The morphological diversity of adzuki bean landraces from Tangshan. *Journal of Hebei Agricultural University* 34, 1–4.

Zhu, Z., Chen, H., Liao, F., Li, L., Liu, C. *et al.* (2019) Evaluation and screening of adzuki bean germplasm resources for drought tolerance during germination stage. *Journal of Southern Agriculture* 50(6), 1183–1190.

Zong, X.X., Kaga, A., Tomooka, N., Wang, X.W., Han, O.K. and Vaughan, D. (2003) The genetic diversity of the *Vigna angularis* complex in Asia. *Genome* 46(4), 647–658.

4 Cowpea (*Vigna unguiculata*)

Martial Nounagnon[1], Gautier Roko[1], Nadège Adoukè Agbodjato[1], Durand Dah-Nouvlessounon[1], Olubukola Oluranti Babalola[2]* and Lamine Baba-Moussa[1]*

[1]*Laboratoire de Biologie et de Typage Moléculaire en Microbiologie (LBTMM), Université d'Abomey-Calavi (UAC), Cotonou, Bénin;* [2]*Food Security and Safety Focus Area, Faculty of Natural and Agricultural Sciences, North-West University, Mmabatho, South Africa*

Abstract

Cowpea (*Vigna unguiculata*) is a significant tropical leguminous crop, comprising a crucial protein source in the nourishment of many people. This chapter comprehensively examines the advancements in cowpea research, offering insights into the diverse aspects of this versatile legume crop. Starting with the taxonomy and botanical description of cowpea, the chapter explores origin and geographical distribution, historical and cultural significance, and the role of cowpea in food security, poverty alleviation and sustainable agriculture across various regions. The genetic variation in cowpea, its importance for breeding improved cultivars with desirable traits, and recent advancements in genomics and biotechnology applications are discussed. The review also addresses the challenges pests and diseases pose in cowpea production and emphasizes integrated pest management strategies. Furthermore, the adaptation of cowpea to drought conditions, its nutritional value and post-harvest management techniques are explored. The chapter concludes by identifying research gaps and proposing future directions, highlighting the need for multidisciplinary approaches, collaboration and innovation to further harness the potential of cowpea. This critical review is valuable for researchers, agronomists, breeders, policy makers and stakeholders involved in cowpea research and production, contributing to advancing sustainable agriculture, food security and rural development.

Keywords: *Vigna unguiculata*, genomics, food security, sustainable agriculture, SDG2

4.1 Introduction

Legumes, also known as pulses, belong to the Leguminosae or Fabaceae family, the third-largest family of flowering plants, which includes species with unique characteristics that contribute to livelihoods, nutrition and sustainable production systems (Guimarães *et al.*, 2023). Throughout the world, leguminous plants help to maintain soil fertility and produce diversified, protein-rich plant products and fodder (Snapp *et al.*, 2018). *Vigna unguiculata* (L.) Walp. is a species of legume known as cowpea, a tropical herbaceous plant native to sub-Saharan Africa, its primary centre of diversity worldwide (Kpatinvoh *et al.*, 2016). It is a leguminous plant crucial to food security and has considerable agricultural and economic importance worldwide (Xu *et al.*, 2016). The yearly worldwide output ranges from 3 to 5.5 million tonnes of dried seeds (FAOSTAT, 2010), of which

*Corresponding author: olubukola.babalola@nwu.ac.za; laminesaid@yahoo.fr

© CAB International 2024. *Potential Pulses: Genetic and Genomic Resources* (eds R. Chandora, T. Basavaraja and A. Pratap)
DOI: 10.1079/9781800624658.0004

more than 64% are produced in Africa (Nkouan-nessi, 2005). The major producing countries are Nigeria and Niger, accounting for almost half the world's production (Langyintuo *et al.*, 2003). Cultivated today on all continents, cowpea originates from Africa and is described as the meat of low-income people (Alzouma, 1995).

In Africa, cowpea, with its high protein content (19–25%), carbohydrates and mineral elements, is crucial in human nutrition and efforts to combat malnutrition (Stoilova and Pereira, 2013). With its remarkable adaptability to diverse environments, resilience to harsh conditions and high nutritional value, cowpea has emerged as a valuable crop for food security, poverty alleviation and sustainable agriculture. The leaves, pods, and green and dry seeds are employed for human nutrition, and the tops are used in animal feed (Bebe *et al.*, 2005; Bello *et al.*, 2016). In addition to its nutritional value, cowpea is known for its adaptability and versatility to different production conditions. It is resistant to water loss, making it suitable for hot arid regions (Silva *et al.*, 2018). It maintains the soil by recycling nutrients by nitrogen fixation and nodulating bacteria (Omomowo and Babalola, 2021). In Benin, cowpea accounts for 7% of the annual crop area, producing 81,152 t (Abadassi, 2014). It is grown throughout the country mainly for its edible seeds (Gbaguidi, 2013). For several years, extensive research efforts have been dedicated to exploring and enhancing various aspects of cowpea cultivation, breeding, physiology and utilization.

In Africa, cowpea is predominantly grown as a bushy, short-podded grain crop, while in Asia, the climbing, long-podded vegetable cowpea is more popular. The diversification of cowpea into these two types and the role of selection in this process are poorly understood (Xu *et al.*, 2016). Cowpea genetic improvement can be achieved by generating selection populations from highly competitive but genetically diverse parents (Edema, 2023). Certain varieties are projected to yield about 3 t/ha in areas with abundant rainfall and approximately 2 t/ha in drier regions (Emmanuel *et al.*, 2021). Unfortunately, these improved varieties did not meet yield expectations in the study environment. Sustainable cowpea production requires substantial inputs of NPK (nitrogen, phosphorus, potassium) and micronutrients (Emmanuel *et al.*, 2021).

Genetic studies have examined the plurality and composition of the cowpea population. In a comprehensive investigation, 768 cultivated cowpea varieties originating from 56 different countries were analysed. This research identified three distinct genetic clusters: Group 1 comprised North America, Latin America, Oceania, East Central Africa, India and South Africa; Group 2 solely consisted of West Africa; and Group 3 encompassed the American cultivar, East Asia, West Central Asia and Europe. This study also traced the migration and domestication history of cowpea, suggesting that it originated from West and East Africa (Xiong *et al.*, 2016).

In terms of cultivation and productivity, different cowpea lines have been evaluated. In South Africa, a comparative study of three main lines showed that Veg1 and Qukawa had significantly better vegetative and reproductive parameters than M 217. These lines were identified as suitable for dual-purpose cultivation (Mfeka *et al.*, 2019). Another study evaluated the productivity and profitability of cowpea varieties in intercropping systems with maize in Nigeria and found that certain cowpea varieties had superior performance in pod production, dry pod weight, and grain yield (Egbe *et al.*, 2010). Despite this high level of production and genetic diversity, biotic and abiotic constraints (pathogenic organisms, water, high temperature) limit cowpea cultivation, resulting in low yields for farmers (Bello and Baco, 2015). Thus, cowpea is susceptible to pests, such as *Callosobruchus maculatus* (cowpea bruchid), which causes significant damage to stored cowpea grains, and developing cowpea lines resistant to bruchid infestation is a practical approach to mitigate losses (Kpoviessi *et al.*, 2019). Also, the varietal selection of prized individuals throughout the chain, from production to storage to the final consumer's plate, are challenges to expanding this legume that has great potential for food and the economy. Given the importance of nutritious food for humans and the difficulties of global supply dependent on the Russian–Ukrainian conflict, it is imperative and urgent that the populations of Africa develop endogenous resources, including legumes and, more specifically cowpea.

This chapter examines the recent advancements in cowpea research, considering several research topics on the plant. By analysing and synthesizing the latest scientific research, this chapter seeks to offer a thorough overview of the existing knowledge by presenting additional data on cowpea concerning origin, nutritional composition, genetic diversity, the diseases that disrupt its proper development and storage, and to identify understudied aspects and propose future directions for further investigation in the field of cowpea research.

4.2   Origin, Distribution and Botanical Description

The exact location of the origin of the species is hard to pin down, but cowpea is believed to have originated in Africa (Ketema *et al.*, 2020). Different authors have suggested various areas in Africa as potential sites of domestication, including North-east Africa, Central Africa, Southern Africa, and West Africa (Ketema *et al.*, 2020). Additional research has demonstrated that cowpea (*Vigna unguiculata*) was first processed in North-east Africa (Pasquet, 1999). It was then domesticated again in West Africa (Pasquet, 1996; Garba and Pasquet, 1998) and India (Yoyo *et al.*, 2022). It is currently grown in tropical and temperate regions.

Regarding geographical distribution, cowpea is grown worldwide, particularly in warm to hot areas and semi-arid regions (Huynh *et al.*, 2013; Ji *et al.*, 2019). It is widely cultivated in Africa, such as in Nigeria, Niger, Benin, Burkina Faso, Ghana, Mali, Senegal and Cameroon; in Asia, including India, China, Indonesia, Bangladesh and Thailand; and in the Americas, such as Brazil, the USA, Haiti and Jamaica, where it serves as both a staple food and an important cash crop (Tan *et al.*, 2016; Chen *et al.*, 2017). It is important to note that in Africa, cowpea is grown in various countries, and each region specializes in different subspecies. For example, western Africa is known for cultivating the *unguiculata* subspecies, while eastern Africa is known for cultivating the *cylindrica* and *sesquipedalis* subspecies (Ketema *et al.*, 2020).

The adaptability of cowpea to diverse environments has facilitated their cultivation in a wide range of agroecological zones, including arid and semi-arid areas where other crops struggle to thrive. Because of its symbiotic relationship with specific bacteria, cowpea is adept at converting atmospheric nitrogen, making it a suitable choice for low-input farming systems and enhancing soil fertility (Huynh *et al.*, 2013). *Vigna unguiculata* is a prominent legume crop product globally, especially in regions characterized by warm to hot climates and semi-arid conditions (Huynh *et al.*, 2013). Its taxonomy is mainly based on morphological attributes (Fatokun *et al.*, 1993). It is a dicotyledon belonging to the order Fabales and the genus *Vigna*. It falls under the classification of *Vigna unguiculata*, a Fabaceae member, specifically the subfamily *Faboitleae*, the tribe *Phaseoleae*, the subtribe *Phaseolinae* and the section Catiang. This family is commonly referred to as the legume or pea family (Marechal *et al.*, 1978; Marubodee *et al.*, 2015). Within the *V. unguiculata* species, there are five recognized subspecies, with three being actively cultivated (*cylindrica*, *unguiculata* and *sesquipedalis*) and two remaining wild (*mensensis* and *dekindtiana*) (Ketema *et al.*, 2020).

Botanically, cowpea is an herbaceous plant that can be either an annual or perennial, depending on the subspecies. It has a climbing or trailing growth habit, with stems that can reach lengths of up to 3 m. The leaves are compound and trifoliate, with oblong to lanceolate leaflets. The flowers are typically whitish, pink or purple, and are arranged in racemes. The cowpea fruit is a pod containing several seeds (Fig. 4.1) (Marubodee *et al.*, 2015).

4.3   Crop Gene Pool, Evolutionary Relationships and Systematics

Cowpea is an important crop for ensuring global food self-sufficiency, being a grain legume native to Africa and Asia and widely planted in the tropics (Elharadallou *et al.*, 2015).

The gene pool of cowpea (*V. unguiculata*) has been studied to understand the genetic architecture of this crop. Huynh *et al.* (2013) conducted a study that genotyped a worldwide collection of cowpea landraces and African ancestral wild cowpea using single nucleotide polymorphism (SNP) markers. They discovered the existence of two primary gene pools within cultivated cowpea in Africa. The first gene pool is predominantly found in western Africa, whereas

Fig. 4.1. The cowpea plant and a variety of seeds of cowpea. (Author's own figure.)

the second gene pool is in eastern Africa. This suggests distinct domestication processes that resulted in the development of these two gene pools. The investigation further unveiled that those landraces from non-African nations displayed a slightly higher level of genetic diversity than their African counterparts. Accessions from Asia and Europe demonstrated a closer genetic affinity with those from western Africa. In contrast, accessions from the Americas showed a more robust genetic association with those from eastern Africa (Huynh *et al.*, 2013).

Another study examined quantitative trait loci (QTL) for symbiotic nitrogen fixation (SNF) in black beans. Although this study focused on black beans, it provides insights into the genetic variation and traits related to nitrogen fixation in legume crops. The study identified 17 unique QTL linked to SNF characteristics, most of which are located in three major clusters on specific chromosomes. This suggests that there is genetic control over SNF traits in legume crops, including cowpea (Heilig *et al.*, 2017).

Regarding genetic diversity, cowpea has a vast gene pool that includes wild species with kinship links likely to fertilize with the cultivated species. These wild relatives represent a primary gene pool readily available for crop improvement

(Brozynska *et al.*, 2015). The gene pool encompasses more distant relatives, requiring increasing effort to access, resulting in secondary and tertiary gene pools. This expanded gene pool constitutes a vital asset for agriculture (Brozynska *et al.*, 2015). Moreover, domesticated cowpea exhibits limited genetic variability, with most genetic diversity preserved in the wild genotype. Therefore, the wild cowpea germplasm is an important resource for breeding programmes to improve yield and resilience to biotic and abiotic stress (Kouam *et al.*, 2012). A combined repository of cowpea genomic information and gene expression data promises to significantly expedite breeding efforts and facilitate the introduction of new genetic attributes to the cowpea crop (Spriggs *et al.*, 2018).

Genomic studies have been conducted on cowpea to understand its genetic make-up. The cowpea genome has been sequenced and analysed, providing insights into its genetic composition (Lonardi *et al.*, 2019). The IT97K-499-35 cowpea genome assembly was developed using advanced sequencing and mapping techniques (Lonardi *et al.*, 2019). This research has contributed to understanding the characteristics linked to the genetics of cowpea and can contribute to crop selection efforts.

Furthermore, studies have been conducted to explore the genetic diversity within cowpea populations. Genetic markers such as microsatellites and random amplified polymorphic DNA (RAPD) have been used to assess the genetic affinities and associations between cowpea breeding lines and cultivars (Li et al., 2001; Ba et al., 2004). This research has produced valuable information on the genetic diversity within cowpea populations, which can be used for breeding programmes and conservation efforts.

It should be noted that the evolutionary relationships of cowpea (V. unguiculata (L.) Walp.) have been studied using various genetic markers and phylogenetic approaches. One study by Boukar et al. (2018) used restriction fragment length polymorphism (RFLP) markers, RAPD, inter-simple sequence repeat (ISSR), amplified fragment length polymorphisms (AFLPs) and simple sequence repeat (SSR) markers to characterize different accessions of cowpea. These markers helped us understand the genetic diversity and relationships among different species of the genus Vigna. In their study, Mohammed et al. (2018) examined the relationship between phylogeny and functional traits of microsymbionts that nodulate cowpea. They employed BOX-PCR amplifications to assess the genetic diversity of Bradyrhizobium species found in cowpea root nodules. Through phylogenetic analysis involving multiple genes, they identified a range of Bradyrhizobium species, some closely related to known species and others entirely novel, within the examined soils. This indicates that the specific environmental conditions of these areas play a significant role in shaping the diversity of microsymbionts associated with cowpea.

Another study by Burstin et al. (2015) investigated genetic diversity and trait genomic prediction in a pea diversity panel. The researchers used high-throughput SNP arrays to analyse the genetic variation in a collection of pea accessions. They then used these markers to predict flowering time, yield and weight phenotypes. Despite the limited number of markers used, the researchers could reliably predict seed weight. This study demonstrates the potential of marker-assisted selection in pea breeding. It provides valuable information for understanding genetic diversity and evolutionary relationships in cowpea.

In addition to these studies, there is also evidence of domestication-related traits in cowpea. Amkul et al. (2020) identified QTLs for domestication-related traits in Zombi pea (Vigna vexillata). Although this study focused on a different species within the Vigna genus, it provides insights into the genetic basis of domestication-related traits that may also be relevant to cowpea.

Cowpea is a crop exhibiting morphological and molecular diversity. In studies by Ghalmi et al. (2009), the genetic variation within cowpea landraces was lower within agro-ecological regions than in the variation among these regions. Correlations were also found between qualitative morphological data and geographical data, indicating that the degree of morphological variation between local breeds is proportional to their geographical distance. However, no intra-landrace variability was found in the analysis of genetic diversity using RAPD markers and ISSR markers. Although there was no significant correlation between morphological and RAPD data, significant correlations were observed between geographical and genetic data.

Bozokalfa et al. (2017) focused their studies on the genetic diversity of cowpea varieties favoured by Turkish farmers. They utilized 36 agro-morphological traits, both qualitative and quantitative, to evaluate the genetic diversity of 32 cowpea genotypes. The study revealed significant genetic variation and phenotypic diversity in these traits. Their principal component analysis demonstrated that properties related to seed dimensions, including weight, eye colour, width, length, diameter, immature pod pigmentation, along with leaf and pod colour characteristics, were the primary factors distinguishing cowpea genotypes. Similarly, bottom-up analysis classified the genotypes into five groups, and no clear correlation between geographical origin and agro-morphological characteristics was observed.

In summary, studying the gene pool, evolutionary relationships and systematics of cowpea is essential for understanding their domestication, genetic variability and potential for breeding improvements. The wild gene pool of cowpea holds significant genetic variation that can be utilized in breeding programmes. Genomic resources and databases have been developed to facilitate genetic research and accelerate breeding efforts for cowpea. A standardized utilization assessment system for wild crop relatives would greatly aid in conserving and utilizing these valuable genetic resources.

4.4 Germplasm Collection, Characterization, Evaluation and Conservation

Germplasm collection, characterization, evaluation, and conservation are crucial aspects of cowpea research and breeding programmes. These processes help understand the genetic diversity of cowpea accessions, identify desirable traits, and develop conservation strategies for this important crop. However, the International Institute of Tropical Agriculture (IITA) has curated an extensive assortment of cowpea germplasm, encompassing upwards of 15,000 cultivated varieties sourced from over 100 countries, along with more than 1500 accessions of wild *Vigna* species. This repository provides a crucial reservoir of foundational strains for crossbreeding, paving the way for the advancement of superior cowpea cultivars (Boukar *et al.*, 2020). The genetic diversity within cowpea germplasm collections has been studied extensively. For example, a study conducted in Ghana used SSR markers to assess the genetic diversity of cowpea accessions in the PGRRI germplasm collection. The study found that there was a high level of genetic diversity among the accessions, which is important for the preservation and utilization of germplasm resources (Asare *et al.*, 2010). Igwe *et al.* (2017) compared the effectiveness of ISSR and start codon targeted (SCoT) markers in evaluating the genetic diversity of cowpea accessions from different regions in Nigeria. They found that both markers were informative in assessing the genetic diversity of *V. unguiculata* accessions. Carvalho *et al.* (2017) conducted a comparative study using high-density SNP markers to analyse the genetic diversity and structure of Iberian Peninsula cowpea compared to worldwide cowpea accessions. This study provided insights into the genetic diversity of southern European cowpea, which had not been extensively studied before.

In addition, the geographic origin of cowpea germplasm collections is also of interest. The cowpea landraces that were examined in this study were incorporated into germplasm repositories more than 30 years ago, and there is minimal possibility of intermingling among landraces across extensive geographical areas. This highlights the importance of maintaining and studying germplasm collections to capture the genetic variation present in different regions (Luigi *et al.*, 1995; Huynh *et al.*, 2018).

Because cowpeas are known for their high protein content and nutritional value, the characterization of germplasm collections can also contribute to developing food products and improving nutritional security (Elharadallou *et al.*, 2015). Thus, the characterization of cowpea (*V. unguiculata* (L.) Walp) germplasm collections involves studying various aspects of the genetic diversity, population structure and traits of different accessions. Several studies have been conducted to explore these aspects and provide valuable insights into the cowpea germplasm. The applicability of molecular markers for germplasm characterization and diversity studies has also been demonstrated. Shitta *et al.* (2016) analysed the genetic diversity of African yam bean (AYB) using SSR markers derived from cowpea. They found that cowpea SSR markers were applicable for germplasm characterization, breeding applications, diversity studies, and comparative mapping between AYB and cowpea. One study focused on the genetic diversity and population structure of cowpea accessions from different origins (Sodedji *et al.*, 2021). The researchers used SNP markers to analyse 274 cowpea accessions. The study revealed the presence of significant genetic diversity and population structure among the accessions, which is crucial for breeding efforts and crop improvement. Another study in Senegal investigated the genetic diversity and relatedness of cowpea germplasm using SSR markers. The study aimed to better manage phylogenetic resources and included analysing important local cowpea varieties and inbred lines. The results provided insights into the diversity and relatedness of cowpea germplasm in Senegal, which can contribute to conserving and utilizing these genetic resources. Then, phenotypic diversity in cowpea germplasm collections was investigated using the UC-Riverside cowpea mini-core collection (Dareus *et al.*, 2021). The researchers characterized various phenological and agronomic traits in 292 accessions, along with three lines released by the USDA and seven cultivars. The study estimated variance components and calculated genetic parameters, providing valuable information for future breeding efforts. Other aspects of cowpea research have also been explored. Carvalho et al. (2017) reviewed different aspects of cowpea, including using molecular markers to assess genetic diversity and evaluate cowpea drought stress tolerance. This integration of molecular and biochemical/transcriptomic data can contribute to the improvement of cowpea in challenging environments.

Furthermore, studies have explored the genetic diversity and population structure of cowpea germplasm in specific regions. Indeed, studies conducted in Senegal (Sarr et al., 2020), Ethiopia (Ketema et al., 2020) and Togo (Gbedevi et al., 2021) analysed the genetic diversity and population structure of cowpea accessions using different molecular markers. These studies highlighted the importance of conserving and characterizing cowpea germplasm to support breeding programmes and ensure food security.

The evaluation and conservation of cowpea germplasm collections are crucial for improving and preserving this important grain legume crop. However, genetic diversity studies have been conducted to assess the population structure and linkage disequilibrium among cowpea accessions. These studies utilize SNPs and microsatellites to analyse the genetic diversity and relationships among cowpea accessions. The results of these studies provide valuable information for the conservation and selection of parental material for further genetic improvement (Gbedevi et al., 2021; Sodedji et al., 2021; Dagnon et al., 2022). Furthermore, cowpea germplasm has been evaluated using agro-morphological traits to characterize the diversity among different accessions. These evaluations help to identify the variation in characteristics such as plant height, leaf shape, pod length and seed size, which are important for crop improvement (Nalawade et al., 2020). It should also be noted that the conservation of cowpea germplasm is essential to maintain the genetic diversity of this crop. Germplasm collections serve as a valuable resource for breeders and researchers to develop improved varieties with desirable traits. Conservation efforts involve collecting, preserving and storing cowpea germplasm to ensure its availability for future use (Gbedevi et al., 2021). In addition to genetic

diversity, studies have also focused on evaluating the resilience of cowpea germplasm to abiotic and biotic stresses. These studies aim to identify sources of stress resistance and enhance the use and conservation of genetic resources for legume breeding (Zonneveld et al., 2020).

Overall, these studies provide valuable insights into cowpea germplasm collection, characterization, evaluation and conservation. They highlight the importance of using molecular markers, such as SNP, ISSR, SCoT and SSR, for assessing genetic diversity and developing conservation strategies. Additionally, integrating molecular data with agro-morphological traits and other biochemical/transcriptomic information can contribute to improving cowpea in challenging environments. As part of a study carried out in Benin, Houindé (2018) used more than 20 markers for the polymerase chain reaction, the data for which are given in Table 4.1. These markers were selected from the cowpea database (http://cowpeagenomics.med.virginia.edu/CGKB/).

4.5 Use of Germplasm in Crop Improvement by Conventional Approaches

The use of germplasm in cowpea crop improvement through conventional approaches has been widely recognized as an effective strategy. The availability of a diverse germplasm collection is essential for long-term crop improvement programmes (Table 4.2). It provides a valuable source of parental strains for hybridization and the development of improved varieties (Boukar et al., 2020). Efforts have been made to identify genetic elements associated with drought stress response in cowpea. Several genes and QTL associated with drought tolerance have been identified and mapped in cowpea (Barrera-Figueroa et al., 2011). This knowledge can be utilized to select and breed cowpea varieties with improved drought tolerance. Other studies have focused on disease resistance, such as bacterial blight caused by Xanthomonas axonopodis pv. vignicola. Resistance to diseases and pests can often be found in landraces or wild relatives of crops, making these resources valuable for breeding programmes (Durojaye et al., 2019). Similarly, knowledge of resistance in germplasm collections

Table 4.1. Microsatellite markers. (Author's own table.)

Marker names		Length (bp)	Primer sequence
[SSR-6169] CP1	CP1F	20	ACCCAAGGACTTCAAGAGCA
	CP1R	20	CGAGTGCAAGAAATGGTTCA
[SSR-6170] CP2	CP2F	20	ACCTGCATTGCCTCATATCC
	CP2R	20	GCTGATTCGGCTTGTTCTTC
[SSR-6171] CP3	CP3F	20	ATTCGATCCAACCCAATGAC
	CP3R	20	AGCGAAGGCATGTTCGTAAG
[SSR-6172] CP4	CP4F	20	GGAAGACACGCGTTATGGTT
	CP4R	23	TTTTCCCACTAAAAGGTTTGTCA
[SSR-6173] CP5	CP5F	20	AGATCCCACGCTGATTATGG
	CP5R	20	ACTTGACGCAGAGCCATCTT
[SSR-6174] CP6	CP6F	20	TCCTTAGAGGTCCAGCCAGA
	CP6R	20	GGAGGAAGAGAGCACACACA
[SSR-6175] CP7	CP7F	20	GCAAGCTTTTGGAAGTTGGA
	CP7R	20	GGCCAGAAAGCATGAATCACT
[SSR-6176] CP8	CP8F	20	GCCACAAGTGCTTGAAGTGA
	CP8R	20	CCACGTAACGAGGATCAACA
[SSR-6179] CP11	CP11F	27	GGATTCAAGAATATTGGTGTTTTCTCC
	CP11R	26	TGCCATCTCTTATCAAGACACTTTAG
[SSR-6181] CP13	CP13F	20	AATGACCCACAAAGCAAAGT
	CP13R	20	TTGGCCCAAAATATCACACA
[SSR-6185] CP17	CP17F	20	CGGAAAAGTAGAGGGCACAG
	CP17R	20	AGAGGTTTGATACGCGCACT
[SSR-6187] CP19	CP19F	20	ACCGCCTAACCCAAGAGTTT
	CP19R	20	TGGGACCACTTCCTTTTCAG
[SSR-6189] CP21	CP21F	20	CTCAATGTCCAACCAGGTCA
	CP21R	20	CAACTCACCAAAGGGAAGGA
[SSR-6190] CP22	CP22F	20	CGAGTTGCGATATCTCCCTG
	CP22R	20	CGAAGACGACAACACAGTGG
[SSR-6191] CP23	CP23F	24	AAACTGCTAACCAGAAACAGAAAA
	CP23R	20	TGTCAATTTGTTGGCCTCA
[SSR-6192]	CP24F	20	AACGGTCCTAAACGAATGA

Source: Houindé (2018).

Table 4.2. Names of 19 cowpea accessions and their geographic distributions.

Sample number	Accession names	Gene bank accession number	District and locality of collection	Latitude	Longitude	Date of collection
1	Rift Valley 040539	GBK-040539	Turkana; Nadoto	2.7333°N	35.11667°E	18.9.1994
2	Australia 016157	GBK-016157	Uasin Gishu	–	–	5.1.1989
3	Coast 032338	GBK-032338	Kwale; Mwachanda	–	–	11.2.1992
4	Coast 032344	GBK-032344	Kwale; Marenje village	4.46167°S	39.12833°E	11.3.1992
5	Coast 032723	GBK-032723	–	–	–	1.1.1976
6	Rift Valley 040472	GBK-040472	Kabarnet	–	–	9.2.1994
7	Coast 031913	GBK-031913	Busia	0.45694°N	34.191389°E	7.10.1992
8	Western 047102	GBK-047102	Kakamega; Bunyala East	0.44172°N	034.68136°E	22.11.2004
9	Western 047111	GBK-047111	Vihiga; Mudete	0.11785°N	034.76527°E	24.11.2004
10	Eastern 046585	GBK-046585	Mwingi; Nzelune-Makilungi	1.284167°N	38.258611°E	29.8.2003
11	Western 044082	GBK-044082	Meru; Nkubu market	0.066667°S	37.666667°E	2.11.1997
12	Eastern 033061	GBK-033066	Embu; Embu research station	3.508889°S	37.454722°E	1.12.1992
13	Eastern 033066	GBK-033061	Embu; Embu research station	3.508889°S	37.454722°E	1.12.1992
14	Ethiopia 015141	GBK-015141	Siaya; Kigilo	–	–	5.1.1989
15	Eastern 033060	GBK-033060	Embu; Embu research station	3.508889°S	37.454722°E	1.12.1992
16	Upper Volta 022436	GBK-022436	Kilifi	–	–	10.1.1975
17	Western 047048	GBK-047048	Busia; Ageng'a	0.22152°N	034.08540°E	19.11.2004
18	Western 047119	GBK-047119	Vihiga; Serem Tiriki East	0.07745°N	034.8548°E	24.11.2004
19	Rift Valley 032108	GBK-032108	Nandi; Kaptumo location	0.067500°N	35.067500°E	27.8.1992
20	Western 047082	GBK-047082	Busia; Bumala	0.30394°N	034.20103°E	19.11.2004

helps identify accessions with genes associated with resistance to diseases such as bacterial blight (Durojaye *et al.*, 2019). Screening cowpea germplasm collections has led to identifying seed traits that contribute to bruchid resistance (Tripathi *et al.*, 2020).

Furthermore, the evaluation of cowpea germplasm collections based on seed yield and yield components can help improve the adaptability and stability of varieties across different environments (Mbuma *et al.*, 2020). This evaluation can also provide insights into genotype and environment interactions, which are important considerations in crop improvement programmes (Mbuma *et al.*, 2020). The availability of large and diverse cowpea germplasm collections, such as those maintained at IITA, allows researchers worldwide to access valuable genes and improve cowpea cultivars (Simion, 2018). In addition to conventional breeding approaches, the use of emerging gene-based tools, such as marker-assisted selection (MAS), can enhance the efficiency and effectiveness of cowpea improvement programmes (Asare *et al.*, 2010). MAS allows for selecting specific genes or traits linked to important diseases and pest resistance, reducing the time required for breeding (Asare *et al.*, 2010).

However, it is important to note that there may be limitations in characterizing and evaluating germplasm collections, which can restrict their effective use. Lack of information on germplasm characterization hinders the exploitation of genetic resources stored in gene banks (Tripathi *et al.*, 2019). Therefore, efforts should be made to characterize cowpea germplasm collections further and evaluate them to fully utilize their potential for genetic improvement.

Ultimately, germplasm collections are invaluable resources for improving the genetic diversity of cowpea and enhancing various traits such as yield, drought tolerance, disease resistance, pest resistance and nutritional content. These collections provide a wide range of genetic resources for plant breeding programmes and contribute to the overall genetic improvement of cowpea varieties.

4.6 Integration of Genomic and Genetic Resources for Crop Improvement

Integrating genomic and genetic resources has become increasingly important in crop improvement, including for cowpea. Cowpea breeding programmes have made progress through conventional breeding methods, but there is potential for further improvement through molecular genetic tools (Boukar *et al.*, 2016). Advances have been made in developing cowpea genetic linkage maps and identifying quantitative trait loci associated with desirable traits, such as resistance to various pathogens and pests (Boukar *et al.*, 2018). These genetic resources can be utilized for genome-wide association studies (GWAS) to identify favourable alleles for simple and complex traits in cowpea germplasm (Muñoz-Amatriaín *et al.*, 2017). The availability of assembled genome sequences and annotated gene models can enhance the precision and speed of cowpea improvement (Pottorff *et al.*, 2012). Genomic tools have facilitated the construction of genetic maps for cowpea, allowing researchers to identify QTLs associated with desirable traits. This enables breeders to select more precise traits (Gbedevi *et al.*, 2021). In the same way, genomic selection employs genomic data to predict the genetic value of individuals, even before they are phenotypically evaluated. This is particularly useful for complex traits and traits that are difficult or costly to measure (Ali *et al.*, 2020).

In addition to genetic resources, genomic resources have also been developed for cowpea. A consensus genetic map containing thousands of SNPs has been created, which can aid in trait discovery (Boukar *et al.*, 2018); anchored to the genetic map using the same SNP markers, a cowpea physical map has also been established (Pottorff *et al.*, 2012). These combined genomic resources and recognized syntenic connections between cowpea and other legume crops can enhance marker-assisted breeding, association mapping and comparative studies (Pottorff *et al.*, 2012). Additionally, published whole-genome sequencing endeavours on related *Vigna* crops such as mung bean and adzuki bean can potentially augment the creation of genetic and genomic resources for cowpea (Iwata-Otsubo *et al.*, 2016).

The availability of genomic and gene expression data resources has the potential to significantly expedite breeding endeavours and facilitate the introduction of innovative genetic traits in cowpea (Spriggs *et al.*, 2018). These resources hold the key to enhancing yield and fortifying resilience against both biotic and abiotic stresses, attributes of paramount importance for ensuring food security in regions characterized by low-input and smallholder farming practices (Spriggs *et al.*, 2018). Cowpea is highlighted as a crop species with extensive genetic variability, underscoring the importance of comprehending its genetic diversity in the pursuit of crop enhancement (Muñoz-Amatriaín *et al.*, 2021). The UCR Minicore, an assemblage of diverse cowpea accessions, is a valuable asset for cowpea research and breeding initiatives (Muñoz-Amatriaín *et al.*, 2021).

Bhandari *et al.* (2017) also explained in their study that the genetic variability present within and between plant species allows plant breeders to select more efficient genotypes. These genotypes can then be used to create a gene pool for hybridization projects or the propagation of new plant varieties (Bhandari *et al.*, 2017). In addition, it enables breeders to integrate genetically distinct parents into their breeding programmes to increase the productivity of agricultural and horticultural crop varieties (Raffard *et al.*, 2019) (Fig. 4.2).

Ultimately, integrating genomic and genetic resources can potentially enhance cowpea breeding programmes. These resources include genetic linkage maps, QTLs, SNP genotyping platforms, assembled genome sequences and gene expression data. They can be utilized for GWAS, marker-assisted breeding and comparative analyses. Understanding the genetic diversity of cowpea is also important for crop improvement. The availability of these resources can accelerate the development of improved cowpea varieties with increased yield and resilience to stresses, contributing to food security in regions where cowpea is a vital crop.

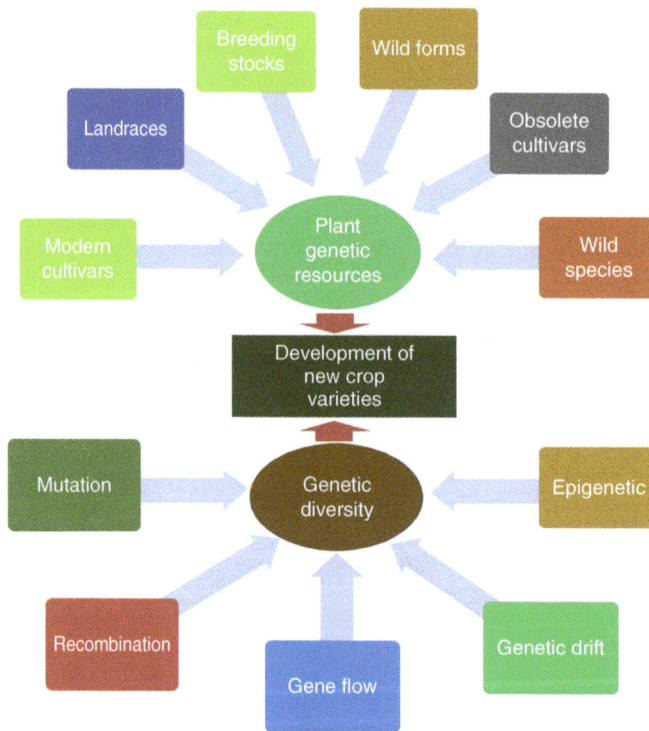

Fig. 4.2. Various origins of genetic diversity and their prospective application in crafting novel crop cultivars. Source: Salgotra and Chauhan (2023). (CC BY 4.0 DEED.)

4.7 Erosion of Genetic Diversity from Traditional Areas and Limitations in Germplasm Use

The erosion of genetic diversity from traditional areas of cowpea is a significant concern in crop domestication areas. Traditional producers in these areas conserve ancestral landraces, which contain the highest genetic diversity and are associated with the knowledge and cultural practices that created this diversity (Martínez-Castillo et al., 2008). However, recent studies have shown that various factors threaten the genetic diversity of cowpea landraces.

The shift towards monoculture farming practices, where a single variety of cowpea is grown extensively, reduces the cultivation of diverse landraces and traditional varieties. This diminishes the genetic diversity within the crop (Martínez-Castillo et al., 2008). The study conducted in Togo found that several accessions of cowpea cultivated in the country are at risk of extinction, leading to genetic erosion (Dagnon et al., 2022). Similarly, adopting high-yielding modern varieties, often bred for specific traits such as uniformity, disease resistance and higher yield, has led to the displacement of diverse, locally adapted landraces. This can result in the loss of unique genetic traits. Indeed, a study in Senegal compared the genetic diversity of improved cowpea varieties to that of traditional cowpea varieties and found lower genetic diversity in the improved varieties (Sarr et al., 2020). This suggests that introducing improved varieties may contribute to the erosion of genetic diversity in cowpea.

Otherwise, market demands for certain cowpea traits, such as uniform pod size, colour or disease resistance, can lead to abandoning traditional varieties in favour of commercially favoured ones. This further reduces the diversity of cultivated cowpea (Osipitan et al., 2021). In addition, introducing exotic, non-local cowpea varieties for commercial purposes can lead to the displacement of traditional varieties. This can result in the loss of unique genetic traits adapted to local environments (Boukar et al., 2020). Likewise, the loss of landraces, which are traditional varieties of crops, is a significant cause of genetic erosion. Landraces are particularly vulnerable to erosion because they concentrate the highest genetic diversity and are cultivated in areas where wild and domesticated varieties coexist, allowing gene flow between them (Martínez-Castillo et al., 2008). In the case of cowpea, the West African sub-region, particularly Nigeria, is the largest producer and contributes significantly to global cowpea production. However, the introduction of improved, higher-yielding modern cultivars has led to the replacement of landraces, resulting in the erosion of crop genetic diversity (Phillip et al., 2018).

Work on cowpea genomic resources is more recent than many other crops. One of the priority objectives has been the molecular characterization of genetic localization. Efforts on genetic richness involve using several marker systems that have evolved over time (Table 4.3) (Boukar et al., 2016). Mahalakshmi et al. (2007) have developed a set of cowpea accessions from germplasm from 89 countries. These accessions have been characterized by 28 agro-morphological descriptors, and 2062 accessions from various groups have been selected to form a genetic base that can be used in breeding programmes (Table 4.4) (Mahalakshmi et al., 2007).

Genetic erosion can also occur owing to changes in agricultural practices. A study on maize genetic erosion in smallholder agriculture found that the adoption of modern varieties has led to a decrease in genetic diversity compared to traditional landraces; however, the distinct features offered by modern varieties have moderated the rate of genetic erosion (Heerwaarden et al., 2009). This suggests that adopting modern agricultural practices can have positive and negative effects on genetic diversity.

In brief, recent studies have highlighted the erosion of genetic diversity from traditional areas of cowpea. Factors such as the introduction of improved varieties, changes in agricultural practices, and the loss of landraces contribute to this erosion. It is crucial to assess and conserve the genetic diversity of cowpea accessions to ensure the preservation of local varieties and the sustainable development of this important legume crop (Carvalho et al., 2017).

The limitations in the germplasm use of cowpea can arise from various factors, including the genetic architecture of domesticated cowpea, genotype and environment interactions, susceptibility to diseases, and lack of resistance

Table 4.3. Cowpea varieties released from 2005 to 2015 in sub-Saharan Africa (IITA).

Year of appearance	Variety	Country
2005	IT93K-452-1, IT90K-277-2	Nigeria
2008	IT97K-499-35	Nigeria
2009	IT89KD-288, IT89KD-391	Nigeria
	IT97K-499-35, IT97K-499-38, IT98K-205-8	Niger
2010	IT97K-499-35, IT93K-876-30	Niger
	IT99K-573-1-1	Mali
2011	IT82E-16, IT00K-1263, IT97K-1069-6	Mozambique
	IT99K-494-6, IT99K-573-1-1, IT99K-573-2-1	Malawi
2012	IT99K-7-21-2-2-1, IT99K-573-1-1	Tanzania
2013	IT99K-573-2-1, IT98K-205-8	Burkina Faso
	IT95K-193-12	Benin
2015	IT00K-1263, IT99K-1122	Tanzania
	IT07K-292-10, IT07K-318-33	Nigeria
	TVx 194801 F, IT05K-321-2, IT97K-390-2,	Swaziland
	IT82E-16, IT82E-18, IT99K-494-4	Sierra Leone
	IT99K-573-1-1, IT99K-573-2-1	

Source: Boukar *et al.* (2016). CC BY 4.0 DEED.

Table 4.4. Composition of global cowpea core developed by IITA, Nigeria. (Author's own table.)

Genetic resources	Total accessions (number)	IITA core set (number)
Landraces	10,227	1,701
Advance lines	1,422	225
Unknown biological status	838	130
Wild and weedy	64	6
Total	12,551	2,062

Source: Tripathi *et al.* (2019).

to certain viruses and nematodes. Thus, cowpea landraces used in germplasm collections may have limited genetic variation owing to the lack of international transfer and admixture across large geographic regions (Huynh *et al.*, 2013). This can restrict the diversity available for breeding programmes and limit the potential for improving adaptability and stability across different environments (Mbuma *et al.*, 2020). Access to diverse and well-documented cowpea germplasm can be challenging, particularly for smaller breeding programmes or farmers in remote areas. This can hinder the incorporation of valuable traits into breeding programmes. In the same context, many traditional or landrace cowpea varieties may not be well-documented or characterized regarding their genetic traits, making it difficult to utilize fully their potential in breeding programmes (Jarvis and Hodgkin, 1999).

Another limitation is the susceptibility of cowpea to diseases and incompatibility with modern agricultural practices. Cowpea is susceptible to fungal, bacterial and viral diseases, which can have devastating effects on production. Seed-borne viruses, particularly, can be a significant constraint to cowpea production. The presence of these diseases in cowpea germplasm can limit its use and require careful screening and selection of disease-resistant accessions (Ogundiwin *et al.*, 2002; Ojuederie *et al.*, 2010). Moreover, some traditional cowpea varieties may not align with modern agricultural practices, such as mechanized farming or high-input systems, which can limit their adoption and use in mainstream agriculture (Jarvis *et al.*, 2017).

Furthermore, while diversity is valuable, some traditional cowpea varieties may also carry undesirable traits, such as lower yield potential or susceptibility to specific pests or diseases, which can limit their direct use (Dempewolf and Eastwood, 2017). Cowpea germplasm may lack resistance to certain pests such as root-knot nematodes, which can cause significant damage to cowpea crops, leading to yield losses (Dareus *et al.*, 2021). The absence of resistance in cultivated cowpea lines necessitates the exploration of wild *Vigna* species, such as *Vigna vexillata*, for resistance traits (Ogundiwin *et al.*, 2002).

Additionally, the evaluation and characterization of cowpea germplasm collections can be limited by the availability of resources and the need for extensive field experiments (Mbuma *et al.*, 2020; Nalawade *et al.*, 2020). Assessing the genetic diversity of cowpea germplasm requires using molecular markers, such as microsatellites, which can be time-consuming and costly (Li *et al.*, 2001; Sarr *et al.*, 2020). However, other limitations such as introducing wild relatives, biological barriers to hybridization, complex genetic backgrounds, legal and policy constraints, cost and time constraints and the adaptation to local environments can also be mentioned. Thus, comprehending the genetic diversity of indigenous germplasm is pivotal for enhancing crops and preserving genetic integrity, particularly in light of climate change (Sarr *et al.*, 2020).

4.8 Cultivation Practices

Cultivation practices of cowpea involve various aspects such as nitrogen fixation, response to abiotic stresses, genetic diversity and domestication history (El-Taher *et al.*, 2021).

Cowpea is a versatile crop that can be grown in a wide range of agroecological zones, from arid and semi-arid regions to more humid tropical climates (Singh *et al.*, 2003). One aspect of cowpea cultivation is symbiotic nitrogen fixation (SNF), which is the process by which legumes convert atmospheric nitrogen into a form that can be used by plants. Various legume crops exhibit significant disparities in the proportion of nitrogen they fix from the atmosphere, with cowpea, for instance, demonstrating a nitrogen

fixation rate ranging from approximately 54% to 58%. (Heilig *et al.*, 2017). This ability to fix nitrogen makes cowpea an important crop for sustainable agriculture. Cowpea is also known for its ability to tolerate and respond to abiotic stresses such as salt stress, water deficit and hypoxia (Jayawardhane *et al.*, 2022). These stresses can have detrimental effects on plant growth and productivity, but cowpea has mechanisms in place to mitigate these effects. For example, the use of plant growth-promoting rhizobacteria and arbuscular mycorrhizal fungi has demonstrated the ability to enhance the resilience of cowpea plants, mitigating the adverse impacts of drought (Rocha *et al.*, 2019). Cowpea thrives in well-drained soils with a pH range of 6.0–7.5. It can tolerate a wide range of soil types, from sandy to clayey, but performs best in loamy soils (Ndiiri *et al.*, 2009). In addition, the land must be well prepared before planting cowpea. Proper land preparation involves ploughing or tilling the soil to a fine tilth. This helps to create a suitable seedbed for planting cowpea seeds (Muchero *et al.*, 2009). The recommended seed rate for cowpea is also around 20–30 kg/ha, with row spacing in the range of 60–75 cm. Intra-row spacing should be about 10–15 cm for bush-type varieties and 30–45 cm for spreading types (Omoigui *et al.*, 2020). Cowpea seeds are generally sown at a depth of 3–5 cm. The optimal sowing time varies depending on the location, but it is typically during the warm season when soil temperatures exceed 20°C (Nieuwenhuis and Nieuwelink, 2002). Cowpea is relatively drought-tolerant, but adequate moisture is essential, especially during flowering and pod formation stages. Proper irrigation practices should be employed to ensure consistent moisture (Singh *et al.*, 2003). Concerning fertilization, cowpea has low to medium nutrient requirements. Generally, 20–40 kg/ha of nitrogen, 30–60 kg/ha of phosphorus and 30–60 kg/ha of potassium are recommended, but this can vary based on soil nutrient levels (Ogunbodede *et al.*, 2006). Weed competition can significantly reduce cowpea yields. Pre-emergence and post-emergence herbicides, as well as mechanical weeding, are commonly used for weed control (Ojulong and Saxena, 2010). Likewise, several integrated strategies are used to control pests (Ezeaku *et al.*, 2017). Cowpea pods are typically ready for harvest about 60–90 days after planting, depending on the variety and growing conditions. Harvesting should be done when pods are mature but before they become overripe (Nelson *et al.*, 2008). It should be added that proper post-harvest practices, such as drying and storage in well-ventilated conditions, are crucial for maintaining seed quality and preventing post-harvest losses (Njonjo *et al.*, 2019).

The genetic diversity of cowpea is another crucial aspect of its cultivation. Cowpea is a diploid plant with a diverse genetic pool. The wild gene pool comprises perennial and annual types, while the forms produced are all annuals (Herniter *et al.*, 2020). Knowledge of the genetic richness of cowpea is essential for breeding programmes and the emergence of improved varieties.

The domestication history of cowpea is also of interest. Cowpea was domesticated in sub-Saharan Africa; however, it is now produced on every continent except Antarctica. The spread of cultivated cowpea has been reconstructed using archaeological, textual and genetic resources (Lo *et al.*, 2017). This information provides insights into the historical global spread of cowpea and its importance as a crop.

4.9 Conclusion

This critical review chapter comprehensively examined the advancements in cowpea (*Vigna unguiculata*) research, shedding light on the diverse aspects of this remarkable crop. From its origin and geographical distribution to its genetic diversity, nutritional value, pests, diseases and adaptation to drought, this chapter provided a holistic overview of the current knowledge and research gaps in cowpea. The review also highlighted the advancements in genomic studies, biotechnology applications and agronomic practices that have contributed to a deeper understanding of cowpea and their potential for sustainable agriculture.

The chapter, therefore, contributes to the existing knowledge of cowpea research, highlighting its significance as a versatile and resilient crop. It calls for further investigations in underexplored areas, such as genomics, abiotic stress tolerance, nutritional enhancement, climate change adaptation, post-harvest management and value chain development. Thus, continued research on cowpea will ensure this vital crop for addressing global agricultural challenges.

References

Abadassi, J. (2014) Agronomic traits of cowpea (*Vigna unguiculata* (L.) Walp.) populations cultivated in Benin. *International Journal of Science and Advanced Technology* 4, 1–4.

Ali, M., Zhang, Y., Rasheed, A., Wang, J. and Zhang, L. (2020) Genomic prediction for grain yield and yield-related traits in Chinese winter wheat. *International Journal of Molecular Science* 21, 1342. DOI: 10.3390%2Fijms21041342

Alzouma, I. (1995) Connaissance et contrôle des coléoptères Bruchidae ravageurs des légumineuses alimentaires au Sahel: Sahel Integrated Pest Management (IPM)/Gestion Phytosanitaire Intégrée. *Revue Institut CILSS du Sahel.*

Amkul, K., Somta, P., Laosatit, K. and Wang, L. (2020) Identification of QTLs for domestication-related traits in zombi pea [*Vigna vexilla* (l.) A. Rich], a lost crop of Africa. *Frontiers in Genetique* 11, 00803. DOI: 10.3389/fgene.2020.00803

Asare, A.T., Gowda, B.S., Galyuon, I.K.A., Aboagye, L.L., Takrama, J.F. *et al.* (2010) Assessment of the genetic diversity in cowpea (*Vigna unguiculata* L. Walp.) germplasm from Ghana using simple sequence repeat markers. *Plant Genetic Resources* 8, 142–150. DOI: 10.1017/s1479262110000092

Ba, F., Pasquet, R. and Gepts, P. (2004) Genetic diversity in cowpea [*Vigna unguiculata* (l.) Walp.] as revealed by rapd markers. *Genetic Resources and Crop Evolution* 51, 539–550. DOI: 10.1023/b:gres.0000024158.83190.4e

Barrera-Figueroa, B., Gao, L., Diop, N., Wu, Z., Ehlers, J. *et al.* (2011) Identification and comparative analysis of drought-associated micrornas in two cowpea genotypes. *BMC Plant Biology* 11, 127. DOI: 10.1186/1471-2229-11-127

Bebe, A.G., Assefa, T., Harrun, H., Mesfine, T. and Al-Tawaha, A.R.M. (2005) Participatory selection of drought tolerant maise varieties using mother and baby methodology: a case study in the semi arid zone of the central rift valley of Ethiopia. *World Journal of Agricultural Sciences* 1, 22–27.

Bello, S. and Baco, M.N. (2015) Importance, typologie des détenteurs et taxonomie locale de la diversité variétale du niébé au Nord-Est du Bénin. *Annales des Sciences Agronomiques* 19, 337–366.

Bello, S., Affokpon, A., Djihinto, C.A. and Idrissou-Touré, M. (2016) Pest susceptibility, grain production, and agropastoral interests of cowpea variety IT 95K-193-12 in Southern Benin. *Document Technique et d'informations*. Bibliothèque Nationale du Bénin, Benin.

Bhandari, H.R., Bhanu, A.N., Srivastava, K., Singh, M.N. and Shreya, H.A. (2017) Assessment of genetic diversity in crop plants—An overview. *Advances in Plants & Agriculture Research* 7, 279–282.

Boukar, O., Fatokun, C., Huynh, B., Roberts, P.A. and Close, T.J. (2016) Genomic tools in cowpea breeding programs: status and perspectives. *Frontiers in Plant Science* 7. DOI: 10.3389/fpls.2016.00757

Boukar, O., Belko, N., Chamarthi, S., Togola, A., Batieno, J. *et al.* (2018) Cowpea (*Vigna unguiculata*): genetics, genomics and breeding. *Plant Breeding* 138, 415–424. DOI: 10.1111/pbr.12589

Boukar, O., Abberton, M., Oyatomi, O., Togola, A., Tripathi, L. and Fatokun, C. (2020) Introgression breeding in cowpea [*Vigna unguiculata* (l.) Walp.]. *Frontiers in Plant Science* 11. DOI: 10.3389/fpls.2020.567425

Bozokalfa, M., Aşçioğul, T. and Eşiyok, D. (2017) Genetic diversity of farmer-preferred cowpea (*Vigna unguiculata* L. Walp) landraces in Turkey and evaluation of their relationships based on agromorphological traits. *Genetika* 49, 935–957. DOI: 10.2298/gensr1703935b

Brozynska, M., Furtado, A. and Henry, R. (2015) Genomics of crop wild relatives: expanding the gene pool for crop improvement. *Plant Biotechnology Journal* 14, 1070–1085. DOI: 10.1111/pbi.12454

Burstin, J., Salloignon, P., Chabert-Martinello, M., Magnin-Robert, J., Siol, M. *et al.* (2015) Genetic diversity and trait genomic prediction in a pea diversity panel. *BMC Genomics* 16, 105. DOI: 10.1186/s12864-015-1266-1

Carvalho, M., Muñoz-Amatriaín, M., Castro, I., Lino-Neto, T. and Matos, M. (2017) Genetic diversity and structure of Iberian peninsula cowpeas compared to worldwide cowpea accessions using high-density SNP markers. *BMC Genomics* 18, 891. DOI: 10.1186/s12864-017-4295-0

Chen, H., Wang, L., Liu, X., Hu, L., Wang, S. and Cheng, X. (2017) De novo transcriptomic analysis of cowpea (*Vigna unguiculata* L. Walp.) for genic SSR marker development. *BMC Genomics* 18, 65. DOI:10.1186/s12863-017-0531-5

Dagnon, Y.D., Palanga, K.K., Bammite, D., Bodian, A., Akabassi G.C., Fonseka, D. and Tozo, K. (2022) Genetic diversity and population structure of cowpea [*Vigna unguiculata* (l.) Walp.] accessions from Togo using SSR markers. *PLoS One* 17,e0252362 DOI:10.1371/journal.pone.0252362

Dareus, R., Acharya, J., Paudel, D., Souza, C. and Gouveia, B. (2021) Phenotypic diversity for phenological and agronomic traits in the uc-riverside cowpea (*Vigna unguiculata* l. Walp) mini-core collection. *Crop Science* 5, 3551–3563. DOI: 10.1002/csc2.20544

Dempewolf, H. and Eastwood, R.J. (2017) Plant genetic resources conservation in the Anthropocene. *Conservation Letters* 10, 540–549.

Durojaye, H., Moukoumbi, Y., Dania, V., Boukar, O. and Bandyopadhyay, R. (2019) Evaluation of cowpea (*Vigna unguiculata* (l.) Walp.) landraces to bacterial blight caused by *Xanthomonas axonopodis* pv. *vignicola*. *Crop Protection* 116, 77–81. DOI: 10.1016/j.cropro.2018.10.013

Edema, R., Adjei, E.A., Ozimati, A.E., Tusiime, S.B. and Badji, A. (2023) Genetic diversity of cowpea parental lines assembled for breeding in Uganda. *Plant Molecular Biology Reporter* 41, 713–725. DOI: 10.1007/s11105-023-01394-6

Egbe, O., Alibo, S. and Nwueze, I. (2010) Evaluation of some extra-early- and early-maturing cowpea varieties for intercropping with maize in Southern Guinea Savanna of Nigeria. *Agriculture and Biology Journal of North America* 5, 845–858.

Elharadallou, S., Khalid, I., Gobouri, A. and Abdel-Hafez, S. (2015) Amino acid composition of cowpea *Vigna unguiculata* l. Walp) flour and its protein isolates. *Food and Nutrition Sciences* 6, 790–787. DOI: 10.4236/fns.2015.69082

El-Taher, A., El-Raouf, H., Osman, N., Azoz, S. and Omar, M. (2021) Effect of salt stress and foliar application of salicylic acid on morphological, biochemical, anatomical, and productivity characteristics of cowpea (*Vigna unguiculata* L.) plants. *Plants* 11, 115. DOI: 10.3390/plants11010115

Emmanuel, O.C., Akintola, O.A., Tetteh, F.M. and Babalola, O.O. (2021) Combined application of inoculant, phosphorus and potassium enhances cowpea yield in savanna soils. *Agronomy* 11, 15. DOI:10.3390/agronomy11010015

Ezeaku, I.E., Mbah, B.N. and Baiyeri, K.P. (2017) Response of cowpea (*Vigna unguiculata* (L.) Walp) genotypes to sowing dates and insecticide spray in southeastern Nigeria. *The Journal of Animal & Plant Sciences* 27, 239–345.

FAOSTAT (2010) Agricultural production, crop primary database. Food and Agricultural Organization of the United Nations, Rome. Available at: http://faostat.fao.org/faostat (accessed 27 November 2023).

Fatokun, C.A., Danesh, D., Young, N.D. and Stewart, E.L. (1993) Molecular taxonomic relationships in the genus *Vigna* based on RFLP analysis. *Theoretical and Applied Genetics* 86, 97–104. DOI: 10.1007/BF00223813

Garba, M. and Pasquet, R.S. (1998) The *Vigna vexillata* (L.) A. Rich. gene pool. In: *Proceedings of 2nd International Symposium on Tuberous Legumes, Celaya, Guanajuato, Mexico, 5–8 August 1996*, pp. 61–71.

Gbaguidi, A.A., Dansi, A., Loko, L.Y., Dansi, M. and Sanni, A. (2013) Diversity and agronomic performances of the cowpea (*Vigna unguiculata* Walp.) landraces in southern Benin. *International Research Journal of Agricultural Science and Soil Science* 3, 121–133.

Gbedevi, K., Boukar, O., Abe, A., Ongom, P. and Unachukwu, N. (2021) Genetic diversity and population structure of cowpea [*Vigna unguiculata* (l.) Walp.] germplasm collected from Togo based on dart markers. *Genes* 12, 1451. DOI: 10.3390/genes12091451

Ghalmi, M., Malice, M., Jacquemin, J., Ounane, S., Mekliche, L. and Baudoin, J.-P. (2009) Morphological and molecular diversity within algerian cowpea (*Vigna unguiculata* (l.) Walp.) landraces. *Genetic Resources and Crop Evolution* 57, 371–386. DOI: 10.1007/s10722-009-9476-5

Guimarães, J.B., Nunes, C., Pereira, G., Gomes, A., Nhantumbo, N. *et al.* (2023) Genetic diversity and population structure of cowpea (*Vigna unguiculata* (L.) Walp.) landraces from Portugal and Mozambique. *Plants* 12, 846. DOI: 10.3390/plants12040846

Heerwaarden, J., Hellin, J., Visser, R. and Eeuwijk, F. (2009) Estimating maize genetic erosion in modernized smallholder agriculture. *Theoretical and Applied Genetics* 119, 875–888. DOI: 10.1007/s00122-009-1096-0

Heilig, J., Beaver, J., Wright, E., Song, Q. and Kelly, J. (2017) QTL analysis of symbiotic nitrogen fixation in a black bean population. *Crop Science* 57, 118–129. DOI: 10.2135/cropsci2016.05.0348

Herniter, I., Muñoz-Amatriaín, M. and Close, T. (2020) Genetic, textual, and archeological evidence of the historical global spread of cowpea (*Vigna unguiculata* [l.] Walp.). *Legume Science* 2, e57. DOI: 10.1002/leg3.57

Houindé, P.J. (2018) Caracterisation genetique moleculaire des accessions de niebe (*Vigna unguiculata* L. Walp) utilisant des marqueurs microsatellites (SSRs). Mémoire de fin de formation de master. Université Nationale d'Agriculture de Ketou, Benin.

Huynh, B., Close, T., Roberts, P., Hu, Z., Wanamaker, S. *et al.* (2013) Gene pools and the genetic architecture of domesticated cowpea. *The Plant Genome* 6. DOI: 10.3835/plantgenome2013.03.0005

Huynh, B., Ehlers, J., Huang, B., Muñoz-Amatriaín, M., Lonardi, S. (2018) A multi-parent advanced generation inter-cross (magic) population for genetic analysis and improvement of cowpea (*Vigna unguiculata* L. Walp). *Plant Journal* 6, 1129–1142.

Igwe, D., Afiukwa, C., Ubi, B., Ogbu, K., Ojuederie, O. and Ude, G. (2017) Assessment of genetic diversity in *Vigna unguiculata* l. (Walp) accessions using inter-simple sequence repeat (ISSR) and start codon targeted (scot) polymorphic markers. *BMC Genetics* 18, 98. DOI: 10.1186/s12863-017-0567-6

Iwata-Otsubo, A., Lin, J., Gill, N. and Jackson, S. (2016) Highly distinct chromosomal structures in cowpea (*Vigna unguiculata*), as revealed by molecular cytogenetic analysis. *Chromosome Research* 24, 197–216. DOI: 10.1007/s10577-015-9515-3

Jarvis, A. and Hodgkin, T. (1999) Wild relatives and crop cultivars: detecting natural introgression and farmer selection of new genetic combinations in agroecosystems. *Molecular Ecology* 8, 159–173. DOI: 10.1046/j.1365-294X.1999.00799.x

Jarvis, D.I., Padoch, C. and Cooper, H.D. (2007) *Managing Biodiversity in Agricultural Ecosystems*. Columbia University Press, Columbia.

Jayawardhane, J., Goyali, J., Zafari, S. and Igamberdiev, A. (2022) The response of cowpea (*Vigna unguiculata*) plants to three abiotic stresses applied with increasing intensity: hypoxia, salinity, and water deficit. *Metabolites* 12, 38. DOI: 10.3390/metabo12010038

Ji, J., Zhang, C., Sun, Z., Wang, L., Duanmu, D. *et al.* (2019) Genome editing in cowpea *Vigna unguiculata* using Crispr-cas9. *International Journal of Molecular Sciences* 10, 2471.

Ketema, S., Tesfaye, B., Keneni, G., Fenta, B., Assefa, E. *et al.* (2020) Dartseq SNP-based markers revealed high genetic diversity and structured population in Ethiopian cowpea [*Vigna unguiculata* (L.) Walp] germplasms. *PLoS One* 15, p.e0239122. DOI: 10.1371/journal.pone.0239122

Kouam, E., Pasquet, R., Campagne, P., Tignegre, J. and Thoen, K. (2012) Genetic structure and mating system of wild cowpea populations in West Africa. *BMC Plant Biology* 12, 1–14. DOI: 10.1186/1471-2229-12-113

Kpatinvoh, B., Adjou, E.S., Dahouenon, A.E. and Konfo, T.R.C. (2016) Problématique de la conservation du niébé (*Vigna unguiculata* (L), Walp) en Afrique de l'Ouest : étude d'impact et approche de solution. *Journal of Agricultural and Food Chemistry* 3, 4831–4842.

Kpoviessi, D., Agbahoungba, S., Agoyi, E., Chougourou, D. and Assogbadjo, A. (2019) Resistance of cow-pea-to-cowpea bruchid (*Callosobruchus maculatus* Fab.): knowledge level on the genetic advances. *Journal of Plant Breeding and Crop Science* 8,185–195. DOI: 10.5897/jpbcs2019.0818

Langyintuo, A., Lowenberg-DeBoer, J., Faye, M., Lambert, D., Ibro, G. *et al.* (2003) Cowpea supply and demand in West Africa. *Field Crops Research* 82, 215–231.

Li, C., Fatokun, C., Ubi, B., Singh, B. and Scoles, G. (2001) Determining genetic similarities and relation-ships among cowpea breeding lines and cultivars by microsatellite markers. *Crop Science* 41, 189–197. DOI:10.2135/cropsci2001.411189x

Lo, S., Muñoz-Amatriaín, M., Boukar, O., Herniter, I. and Cisse, N. (2017) Identification of genetic factors controlling domestication-related traits in cowpea (*Vigna unguiculata* l. Walp). *bioRxiv* p.202044. DOI: 10.1101/202044

Lonardi, S., Muñoz-Amatriaín, M., Liang, Q., Shu, S. and Wanamaker, S. (2019) The genome of cowpea (*Vigna unguiculata* [l.] Walp.). *The Plant Journal* 98, 767–782. DOI: 10.1111/tpj.14349

Luigi, G., Rao, R.V. and Goldberg, E. (1995) *Collecting Plant Genetic Diversity Technical Guidelines.* CAB International, Wallingford, UK.

Mahalakshmi, V., Ng, Q., Lawson, M. and Ortiz, R. (2007) Cowpea [*Vigna unguiculata* (L.) Walp.] core col-lection defined by geographical, agronomical and botanical descriptors. *Plant Genetic Resources: Characterization and Utilization* 5, 1479–2621. DOI: 10.1017/S1479262107837166.

Maréchal, R., Mascherpa, J.M. and Stainier, F. (1978) Etude taxonomique d'un groupe d'espèces des genres *Phaseolus* et *Vigna* (Papilionaceae) sur la base des données morphologiques et polliniques, traitées pour l'analyse informatique. *Boissiera* 28, 1–273.

Martínez-Castillo, J., Colunga-GarcíaMarín, P., Zizumbo-Villarreal, D. (2008) Genetic erosion and in situ conservation of lima bean (*Phaseolus lunatus* l.) landraces in its mesoamerican diversity center. *Genetic Resources and Crop Evolution* 55,1065–1077. DOI: 10.1007/s10722-008-9314-1

Marubodee, R., Ogiso-Tanaka, E., Isemura, T., Chankaew, S. and Kaga, A. (2015) Construction of an SSR and Rad-marker based molecular linkage map of *Vigna vexillata* (L.) A. Rich. *PLoS One* 10, p.e0138942. DOI:10.1371/journal.pone.0138942

Mbuma, N., Gerrano, A., Lebaka, N., Mofokeng, A. and Labuschagne, M. (2020) The evaluation of a south-ern african cowpea germplasm collection for seed yield and yield components. *Crop Science* 61, 466–489. DOI: 10.1002/csc2.20336

Mfeka, N., Mulidzi, R. and Lewu, F. (2019) Growth and yield parameters of three cowpea (*Vigna unguicula-ta* L. Walp) lines as affected by planting date and zinc application rate. *South African Journal of Sci-ence* 115, 1–9. DOI: 10.17159/sajs.2019/4474

Mohammed, M., Jaiswal, S. and Dakora, F. (2018) Distribution and correlation between phylogeny and functional traits of cowpea (*Vigna unguiculata* l. Walp.)-nodulating microsymbionts from Ghana and South Africa. *Scientific Reports* 8, p.18006. DOI: 10.1038/s41598-018-36324-0

Muñoz-Amatriaín, M., Mirebrahim, H., Xu, P., Wanamaker, S., Luo, M. *et al.* (2017) Genome resources for climate-resilient cowpea, an essential crop for food security. *The Plant Journal* 89, 1042–1054. DOI: 10.1111/tpj.13404

Muñoz-Amatriaín, M., Lo, S., Herniter, I., Boukar, O., Fatokun, C., Carvalho, M. *et al.* (2021) The ucr mini-core: a resource for cowpea research and breeding. *Legume Science* 3, p.e95. DOI: 10.1002/leg3.95

Nalawade, A., Patil, S., Rajwade, P. and Kauthale, V. (2020) Evaluation of cowpea germplasm by using agro-morphological characters. *Indian Journal of Agricultural Research* 55, 364–368. DOI:10.18805/ijare.a-5490

Ndiiri, J.A., Zingore, S. and Herrero, M. (2009) Suitability of different bean and cowpea varieties for the major agro-ecological zones of Kenya. *African Crop Science Journal* 17, 115–130.

Nelson, M., Dempster, W.F., Allen, J.P., Silverstone, S., Alling, A. *et al.* (2008) Cowpeas and pinto beans: Performance and yields of candidate space crops in the laboratory biosphere closed ecological system. *Advances in Space Research* 41, 748–753. DOI: 10.1016/j.asr.2007.03.001

Nieuwenhuis, R. and Nieuwelink, J. (2002) *La Culture du Soja et D'autres Légumineuses.* Digigrafi, Wageningen, Pays-Bas.

Njonjo, M.W., Muthomi, J.WJ. and Mwangombe, A.W. (2019) Production practices, postharvest handling, and quality of cowpea seed used by farmers in Makueni and Taita Taveta counties in Kenya. *International Journal of Agronomy* 1607535. DOI: 10.1155/2019/1607535

Nkouannessi, M. (2005) The genetic, morphological and physiological evaluation of African cowpea geno-types. MSc Thesis, University of the Free State Bloemfontein, South Africa.

Ogunbodede, B.A., Afolayan, R.A. and Olakojo, S.A. (2006) Soil nutrient status and cowpea performance as influenced by organic and inorganic fertilizer amendments in a tropical Alfisol. *International Journal of Agriculture and Biology* 8, 633–638.

Ogundiwin, E., Thottappilly, G., Aken'ova, M., Ekpo, E. and Fatokun, C. (2002) Resistance to cowpea mottle carmovirus in *Vigna vexillata*. *Plant Breeding* 121, 517–520. DOI: 10.1046/j.1439-0523.2002.00769.x

Ojuederie, O., Odu, B. and Ilor, C. (2010) Serological detection of seed borne viruses in cowpea regenerated germplasm using protein a sandwich enzyme linked immunorsorbent assay. *African Crop Science Journal* 17. DOI: 10.4314/acsj.v17i3.54212

Ojulong, H.F. and Saxena, K.B. (2010) Fast-track breeding of extra-early pigeonpea (*Cajanus cajan*) for improved yield and *Striga* resistance. *Field Crops Research* 118, 140–145.

Omoigui, L.O., Kamara, A.Y., Kamai, N., Ekeleme, F. and Aliyu, K.T. (2020) *Guide to Cowpea Production in Northern Nigeria*. International Institute of Tropical Agriculture (IITA) Ibadan, Nigeria.

Omomowo, O.I. and Babalola, O.O. (2021) Constraints and prospects of improving cowpea productivity to ensure food, nutritional security and environmental sustainability. *Frontiers in Plant Science* 12, 731–751.

Osipitan, O.A., Fields, J.S., Lo, S. and Cuvaca, I. (2021) Production systems and prospects of cowpea (*Vigna unguiculata* (L.) Walp.) in the United States. *Agronomy* 11, 2312. DOI: 10.3390/agronomy11112312

Pasquet, R.S. (1996) Cultivated cowpea (*Vigna unguiculata*): genetic organisation and domestication. In: B. Pickersgill and J.M. Lock (ed.) *Advances in Legume Systematics 8: Legumes of Economic Importance*, Royal Botanic Gardens, Kew, pp. 101–108.

Pasquet, R.S. (1999) Genetic relationships among subspecies of *Vigna unguiculata* (L.) Walp. based on allozyme variation. *Theoretical and Applied Genetics* 98, 1104–1119.

Phillip, J., Udensi, O., Phillip, I. and Abimiku, S. (2019) Protein-based fingerprinting in cowpea accessions [*Vigna uniguiculata* (I.) Walp] for deciphering genetic relatedness. *International Journal of Plant Breeding and Genetics* 13, 1–11. DOI: 10.3923/ijpbg.2019.1.11

Pottorff, M., Wanamaker, S., Yaqin, Q., Ehlers, J., Roberts, P. and Close, T. (2012) Genetic and physical mapping of candidate genes for resistance to *Fusarium oxysporum* f.sp. *tracheiphilum* race 3 in cowpea [*Vigna unguiculata* (I.) Walp]. *Plos.One* 7, e41600. DOI: 10.1371/journal.pone.0041600

Raffard, A., Santoul, F., Cucherousset, J. and Blanchet, S. (2019) The community and ecosystem consequences of intraspecific diversity: A meta-analysis. *Biological Review* 94, 648–661.

Rocha, I., Ma, Y., Vosátka, M., Freitas, H., Oliveira, R. (2019) Growth and nutrition of cowpea (*Vigna unguiculata*) under water deficit as influenced by microbial inoculation via seed coating. *Journal of Agronomy and Crop Science* 205, 447–459. DOI: 10.1111/jac.12335

Salgotra, R.K. and Chauhan, B.S. (2023) Genetic diversity, conservation, and utilization of plant genetic resources. *Genes* 14, 174. DOI: 10.3390/genes14010174

Sarr, A., Bodian, A., Gbedevi, K., Ndir, K., Ajewole, O. *et al.* (2020) Genetic diversity and population structure analyses of wild relatives and cultivated cowpea (*Vigna unguiculata* (I.) Walp.) from Senegal using simple sequence repeat markers. *Plant Molecular Biology Reporter* 39, 112–124. DOI: 10.1007/s11105-020-01232-z

Shitta, N.S., Aberton, M.T., Adesoye, A.I., Adewale, B.D. and Oyatomi, O. (2016) Analysis of genetic diversity of African yam bean using SSR markers derived from cowpea. *Plant Genetic Resources* 14, 50–56.

Silva, V., Boleta, E., Lanza, M., Rabêlo, F., Martins J. *et al.* (2018) Physiological, biochemical, and ultrastructural characterization of selenium toxicity in cowpea plants. *Environmental and Experimental Botany* 150, 172–182.

Simion, T. (2018) Breeding cowpea *Vigna unguiculata* I. Walp for quality traits. *Annals of Reviews & Research* 3(2), 555609. DOI: 10.19080/arr.2018.03.555609

Singh, B.B., Ajeigbe, H.A., Tarawali, S.A., Fernandez-Rivera, S. and Abubakar, M. (2003) Improving the production and utilization of cowpea as food and fodder. *Field Crops Research* 84, 169–177.

Snapp, S., Rahmanian, M. and Batello, C. (2018) *Légumes Secs et Exploitations Durables en Afrique Subsaharienne, sous la Direction de T. Calles*. FAO, Rome.

Sodedji, F., Agbahoungba, S., Agoyi, E., Kafoutchoni, K., Choi, J. *et al.* (2021) Diversity, population structure, and linkage disequilibrium among cowpea accessions. *The Plant Genome* 14, e20113. DOI: 10.1002/tpg2.20113

Spriggs, A., Henderson, S., Hand, M., Johnson, S., Taylor, J. and Koltunow, A. (2018) Assembled genomic and tissue-specific transcriptomic data resources for two genetically distinct lines of cowpea (*Vigna unguiculata* (I.) Walp). *Gates Open Research* 2, 7. DOI: 10.12688/gatesopenres.12777.2

Stoilova, T. and Pereira, G. (2013) Assessment of the genetic diversity in a germplasm collection of cowpea (*Vigna unguiculata* (L.) Walp.) using morphological traits. *African Journal of Agricultural Research* 8, 208–215.

Tan, H., Huang, H., Tie, M., Tang, Y., Yunsong, L. *et al.* (2016) Transcriptome profiling of two asparagus bean (*Vigna unguiculata* subsp. *sesquipedalis*) cultivars differing in chilling tolerance under cold stress. *PLoSOne* 11, e0151105. DOI: 10.1371/journal.pone.0151105

Tripathi, K., Gore, P., Ahlawat, S., Tyagi, V. and Semwal, D., Gautam, J.C. and Kumar, R.A. (2019) A cowpea genetic resources and its utilization: Indian perspective – a review. *Legume Research* 42, 437–446. DOI: 10.18805/lr-4146

Tripathi, K., Prasad, T., Bhardwaj, R., Jha, S., Semwal, D. *et al.* (2020) Evaluation of diverse germplasm of cowpea [*Vigna unguiculata* (l.) Walp.] against bruchid [*Callosobruchus maculatus* (Fab.)] and correlation with physical and biochemical parameters of seed. *Plant Genetic Resources* 18, 120–129. DOI: 10.1017/s1479262120000180

Yoyo, F.D.M., Gonne, S., Kosma, P., Pierre, T.F., Liliane, I. *et al.* (2022) Morphological and biometric diversity of *Colletotrichum capsici* isolates, causal agent of cowpea brown blotch disease (*Vigna unguiculata* (L.) Walp) in the Sudano-Sahelian zone of Cameroon. *Open Journal of Applied Sciences* 12, 1837–1855. DOI: 10.4236/ojapps.2022.1211127

Wamalwa, E.N., Muoma, J., Wekesa, C. (2016) Genetic diversity of cowpea (*Vigna unguiculata* (L.) Walp.) accession in Kenya gene bank based on simple sequence repeat markers. *International Journal of Genomics* 8956412. DOI: 10.1155/2016/8956412

Xiong, H., Shi, A., Mou, B., Qin, J., Motes, D. *et al.* (2016) Genetic diversity and population structure of cowpea (*Vigna unguiculata* L. Walp). *PLoSOne* 11, p.e0160941. DOI: 10.1371/journal.pone.0160941

Xu, P., Wu, X., Muñoz-Amatriaín, M., Wang, B., Wu, X. *et al.* (2016) Genomic regions, cellular components and gene regulatory basis underlying pod length variations in cowpea (*V. unguiculata* L. Walp). *Plant Biotechnology Journal* 5, 547–557.

Zonneveld, M., Rakha, M., Tan, S., Chou, Y., Chang, C. *et al.* (2020) Mapping patterns of abiotic and biotic stress resilience uncovers conservation gaps and breeding potential of *Vigna* wild relatives. *Scientific Reports* 10, 2111. DOI: 10.1038/s41598-020-58646-8

5 Cluster Bean (*Cyamopsis tetragonoloba*)

Mithlesh Kumar[1]*, Kirti Rani[2], Shailesh Kumar Jain[3] and Moti Lal Mehriya[1]

[1]*Agricultural Research Station, Mandor, Agriculture University, Jodhpur, Rajasthan, India;* [2]*Indian Council of Agricultural Research – National Bureau of Plant Genetic Resources, Regional Station, Jodhpur, Rajasthan, India;* [3]*Rajasthan Agricultural Research Institute – Sri Karan Narendra Agriculture University, Durgapura, Jaipur, Rajasthan, India*

Abstract

Cluster bean is an essential annual leguminous crop in north-west parts of India, primarily in Rajasthan and sections of Punjab, Gujarat and Haryana, which account for 80% of world production. It is a natural hydrocolloid source (galactomannan / 'guar gum') with unique qualities and broad applicability. The gum has versatile applications in various industries, including textiles, paper, printing, cosmetics, pharmaceuticals, mining, petroleum, well drilling, natural gas and oil. Green pods are nutritious and eaten as vegetables in north-western India. Intercropping cluster bean has also been shown to be beneficial. Cluster bean meal, which is rich in protein and derived from the seed coat and germ cell, has significant potential as a valuable feed ingredient for monogastric animals. The progress of intense breeding endeavours in cluster bean has been impeded by the limited genetic diversity found among farmed types and the reduction in crop production resulting from both biotic and abiotic stressors. Traditional breeding techniques such as distant hybridization, induced mutations and induced male sterility have been used to improve the genetic diversity of cluster bean. Hybridization has been unsuccessful owing to the delicate flower, and mutation breeding has facilitated the creation of novel genetic variability for a few essential characteristics but has had little effect on cluster bean breeding. Research using biotechnological interventions has also been initiated to boost the varietal improvement of cluster bean because of their crucial role in rainfed agriculture. However, the development of these tools is still in its infancy. Molecular markers that are based on DNA are being utilized in genetic diversity analysis and phylogenetic research. In addition, efforts are being made to develop genetic and genomic resources that could be utilized in breeding cluster bean. Recent reports also include some preliminary findings from limited transcriptomic and mi-RNA research. It is possible, given the availability of sophisticated molecular techniques, to build resources that would assist in the development of cluster bean varieties that have improved yield potential and high gum content. This would position these beans as a desirable export item.

Keywords: Cluster bean, hybridization, molecular markers, gum content

5.1 Introduction

Guar, also known as a cluster bean or *Cyamopsis tetragonoloba* (L.) Taub, is a self-pollinating Leguminosae plant (2n = 14). It is a significant legume cultivated primarily in the Indian subcontinent under arid and semi-arid conditions. Because of its deep tap roots, it can withstand droughts and even extreme high temperatures (Kumar and Rodge, 2012). In the arid regions of India (Rajasthan, Haryana, Gujarat and Punjab), cluster bean is explicitly planted for adhesive purposes, whereas it

*Corresponding author: mithleshgenetix@gmail.com

© CAB International 2024. *Potential Pulses: Genetic and Genomic Resources* (eds R. Chandora, T. Basavaraja and A. Pratap)
DOI: 10.1079/9781800624658.0005

is cultivated as a vegetable in other regions (Rai and Dharmatti, 2013). The production of cluster bean globally is approximately 3 million tons, with India accounting for more than 2.5 million tons. Cluster bean legumes are an excellent source of nutrition for both humans and animals, being consumed as a vegetable and fed to cattle, and are also utilized as green manure. Each 100 g of fruits used as vegetables contains 10.8 g carbohydrates, 3.2 g protein, 1.4 g of minerals, 316 international units of vitamin A, and 47 mg vitamin C. Moreover, it provides inexpensive access to such essential nutrients. Its seed contains three essential components: husk or hull (15–17%), which forms the outer seed coat, endosperm (43–47%), and germ cell or embryo (35–42%), an interior delicate part. Thus, an average of 38% of the total seed weight is recovered as endosperm, which contains between 68 and 70% pure gum. After dehulling the seed, the gum is produced primarily from pulverized endosperm (Sabahelkheir et al., 2012). The endosperm of cluster bean seeds contains approximately 18% protein, 32% fibre and 30–33% pectin. Cluster bean gum is a galactomannan, a naturally occurring hydrocolloid ($C_{18}H_{32}O_{16}$, consisting of a linear backbone of mannose with galactose side chains. The material is utilized in the paper, textile, mining and explosive industries. Additionally, it is utilized as a stabilizer in various culinary products. A by-product of the gum business, cluster bean meal comprises the external seed coat and embryo section of the cluster bean. It has around 35–47.5% crude protein, approximately 1.5 times greater than the protein found in other leguminous seeds. Animals, including poultry, domesticated fish and livestock, are administered this substance. Cluster bean seeds have a higher concentration of minerals and charcoal than other pulses (Pathak, 2015). Cluster bean green forage is an excellent source of nutrition, containing 60% digestible dry matter, 46% total digestible protein and 16% unrefined protein. Therefore, cluster legume is an ideal crop for both human and animal consumption and industrial use, making it an appealing commodity.

5.2 Origin, Distribution and Botanical Description

There are four different species of *Cyamopsis*, all of which belong to the Leguminosae family and the Papilionoideae subfamily. These are *Cyamopsis dentata* (N.E.Br.) Torre, *Cyamopsis senegalensis* Guill. & Perr., and *Cyamopsis serrata* Schinz as a wild species (Guill. & Perr., Schinzas); *Cyamopsis tetragonoloba* is a cultivated species (Chevalier, 1939). The karyotypes of *C. senegalensis*, *C. serrata*, and *C. tetragonoloba* were analysed, and the results showed that all three species had the same number of chromosomes (2n = 14) and a similar chromosome structure (Arora et al., 1985). The precise history on where cluster bean first appeared is still hotly debated. Researchers have found that the dry regions of West Africa and India are the most likely places of origin and diversity, respectively (Gillett, 1958). Cluster bean can be traced back to *Cyamopsis senegalensis*, an African species with similar gum and viscosity properties (Strickland and Ford, 1984; Mudgil et al., 2014). In contrast to popular belief, *C. senegalensis* is more related to *C. serrata* than *C. tetragonoloba*, as evidenced by allozymes, restriction fragment length polymorphism (RFLP), and random amplified polymorphic DNA (RAPD) profiling (Hiremath et al., 1996; Weixin et al., 2009).

In addition to being a slow-growing crop, *C. serrata* has narrow, trifoliate leaves and pink seeds. It is a wild species that matures quickly (40–50 days) and has useful agronomic and economic properties, such as resistance to drought (Menon, 1973), insensitivity to light and heat (Ahlawat et al., 2013a), and lack of susceptibility to disease (Orellana, 1966). The seeds are tiny, weighing 1.0–1.4 g per 100 seeds. *Cyamopsis dentata* is a 30–35 cm annual herb with upright branches. It has biramous (equally or unequally branched) hairs, glabrous and erect pods, and mauve–purple flowers on either entire or dentate leaflets. This species is found in similar environments to both *C. serrata* and *C. senegalensis* (Pathak, 2015), and is thus regarded as a transitional form between the two. *Cyamopsis tetragonoloba* is a slow-growing annual herb that can reach 50–150 cm in height and features four to ten rigid, upright branches. Trifoliate leaves and an angled stem describe this plant. Axillary racemes bear tiny white or purple flowers. Fruits are upright, compressed and 4–10 cm in length. Five to twelve white, grey or black seeds can be found in each fruit (Undersander et al., 2006). Plants that are either upright or branched or have basal branching are seen. The erect types either do not branch at all or have only a couple of branches at most. Plants of the branched variety have additional stems and leaves growing out from their main stem, while those of the basal branching type have three or more branches at the plant's base. Cluster bean varieties that do not branch out produce pod clusters all along a single stem (Fig. 5.1). The trifoliate leaves are medium in size and can be either pubescent or glabrous. It blooms with slender flowers in shades of pink and purple. The cluster bean is cleistogamous, with very low rates of outcrossing (0.3–7.9%), according to various sources, such as Saini *et al.* (1981) and Ahlawat *et al.* (2012), and others. Raceme-shaped inflorescences are typical of cluster bean, with lengths in the range 9–13 cm

in branched types to 15–20 cm in erect (non-branched or sparsely branched) types. Menon (1973) found that branched types had 40–60 flowers, while erect types with sparse branching had 50–70. The flower changes from white to deep blue as it grows from the bud to the petal drop stage. Pink and purple tones predominate among flowers. However, there are also varieties that feature white flowers. A fully developed bud will change from a creamy white to a pale pink or white colour. Small, grey/white, and minutely tuberculate seeds are housed in the pods, which are flattened, elongated (2.5–13 cm) and slightly beaked (Singh *et al.*, 2009). The seeds have a greyish hue and a compressed, square shape (Fig. 5.2). Because the pods at the crown of the plant remain green for an extended period of time after harvest, the quality of the seeds is diminished. Brief descriptions of different species of cluster bean are presented in Table 5.1.

5.3 Genetic Resources in Cluster Bean and their Utilization

Crop improvement programmes require genetic diversity collection, maintenance, use and conservation (Poehlman and Sleper, 1995). The genetic variability in the germplasm of cluster bean has been assessed using agro-morphological and biochemical traits. The goal of agriculture is to develop superior cultivars with higher yields and better adaptation. The diverse germplasm gives plant breeders all crop plant genetic variations. The genetic diversity of *Cyamopsis* spp. can be used to create the desired plant type. Trait-specific germplasm changes plant architecture to maximize and stabilize yield. Short duration (early flowering and early maturing), plant height, pattern in branching, length of pods, internode length, peduncle length and determinate growth habit have helped develop new varieties. In cluster bean most varieties have similar plant architecture, maturity group and yield potential, limiting improvement.

Cluster bean seed types have a narrow range of economically important traits such as seed weight (~3.5 g), length of pods (~6.0 cm), number of seeds per pod (~6) and crop duration (~95 days). On realizing the lack of trait diversity for yield improvement and suitable plant type in guar,

Fig. 5.1. Pods on cluster bean plants. (Author's own figure.)

Fig. 5.2. Variability in colour and shape of cluster bean seeds. (Author's own figure.)

trait-specific genotypes were identified as donors. Genotypes with short lifespans, longer seedlings, longer pods and high seed weight have been reported. About 4313 cluster bean accessions from two wild species have been conserved at −18°C in the National Bureau of Plant Genetic Resources (NBPGR) gene bank in New Delhi, India. Various researchers have reported the wide range of agro-morphological and biochemical traits in cluster bean germplasm. Plant height (46.75–239 cm), number of clusters per plant (1.75–64.5), number of pods per plant (2.25–262.35), length of pods (1.85–19.3 cm), number of seeds per pod (4.15–13.0), maturity days (128–185 days), seed yield (0.95–59.7 g/plant) and 100-seed weight (1.9–4.75 g) showed high diversity. The NBPGR, New Delhi and the Central Arid Zone Research Institute (CAZRI) in Jodhpur have evaluated 4869 germplasm resources and reported early maturing and high-yielding genotypes. The three mutants, namely CAZG-20-17 (early maturing), CAZG-109 (high seed weight) and CAZG-110 (slender with long grains), have been identified as potential donors at ICAR-CAZRI (Mahla *et al.*, 2022). Several promising lines with enhanced disease resistance against bacterial leaf blight (BLB), *Alternaria* blight and root rot have been identified in S K Nagar, Gujarat (Kumar, 2008). Morris (2010) characterized 73 accessions from India, Pakistan and the USA and found greater genetic variability in 100-seed weight (2.3–4.8 g), pod length (32–110 mm), and days to 50% maturity (96–185 days). Pathak *et al.* (2009) found significant variation in the endosperm (30.4–46.3%), gum content (23.9–34.2%), crude fibre content (4.1–8.0%), oil content (1.8–5.2%), crude protein content (28.3–35%), carbohydrate (38.8–59.1%) and minerals (3.5–6.0%) in cluster bean germplasm. Kays *et al.* (2006) found a narrow range for dietary fibre (52.4–57.7%), fat (2.88–3.45%), protein (22.9–30.6%) and ash (3.04–3.53%).

Table 5.1. Description of four different species of the genus *Cyamopsis*.

Species	Native	Plant height	Pod features	Seed features	Special features
C. senegalensis	Dry tropical Africa to Arabian Peninsula	0.5 m	Erect, curved, narrowly oblong, 4–6 cm long, including a beak of 5–6 mm with 7–9 seeds	The 100-seed weight range is1.0–1.4 g. Seeds grey, square, elliptic in cross-section, and minutely tuberculate	Wild species
C. serrata	Botswana, Cape Provinces, Namibia, Northern Provinces	Up to 20 cm	Erect, 4-angled pod	Small seeds with 100-seed weight of 1.0–1.4 g	Wild species. It is an early maturing species, i.e. matures in 40–50 days and possesses various essential traits such as drought resistance, insensitivity to photoperiod and temperature and resistance to disease
C. dentata	Angola to Botsawana	30–35 cm	Glabrous and erect pods	Seeds cubic or nearly so, surface slightly tubercular bulged	Wild species
C. tetragonoloba	West Africa	50–150 cm	Flattened, 2.5–13 cm oblongate, slightly beaked, containing 5–12 seeds	The seeds have a compact form, a square outline, a minute tuberculate surface, and a drab-coloured appearance	Domesticated species

Source: Whistler and Hymowitz (1979); Duke (1981). Table used with permission.

Suvidha, Sona, Naven, PLG-85, PLG-850, RGC-471, Pusa Sadabahar, Pusa Navbahar, and Sharad Bahar are grain and vegetable cluster bean lines (Kumar and Rodge, 2012). Cluster bean maturing in 80–90 days is also reported, for lines HG-365, RGC-936 and RGC-563 (Dabas *et al.*, 1982). There is a need to identify lines with >32% gum content and high viscosity profiles (4000 cP) (Kumar, 2008). NBPGR reported lines with high gum content, including 34 lines with >40%, 98 lines with 30–39.9% and 31 lines with 19.2–29.9% gum content. HGS 870 had 31.78% gum and 5116 cP viscosity (Gandhi *et al.*, 1978; Kumar, 2008). It was found that 'GAUG 63' is tolerant to powdery mildew, while HG75, G85, G102 and G225 are resistant to BLB. The promising resistance lines against BLB, GG-1, GAUG-9406, RGC-1027, *Alternaria* leaf blight, GG-1, GAUG-9005, GAUG-9003, and root rot, GG-1, GAUG-9406, HGS-844, were reported by evaluating over 375 accessions (Kumar, 2008). At present, the National Gene Bank located at the NBPGR in India houses a collection of 5000 accessions of cluster bean. Additionally, the gene bank also preserves two notable wild relatives of guar, namely *C. serrata* and *C. senegalensis.* Genetic improvement of cluster bean in Durgapura (Rajasthan), Hisar (Haryana) and CAZRI (Rajasthan) is focused on germplasm collection, their evaluation and characterization, and selecting/identifying several high-yielding and trait specific genotypes. According to Mishra *et al.* (2009),

NBPGR, India, conserves 4878 indigenously collected accessions in medium-term storage and conserves 3714 accessions as *ex-situ* collections. India has released nearly 40 vegetable and seed varieties, mostly using mass selection and pedigree methods (Pathak, 2015). The breeding of cluster bean is at a relatively slow pace and necessitates an immediate expansion of their genetic diversity through the use of germplasm, mutagenesis and distant hybridization. This is imperative owing to the insufficient morphological variability observed in the crop, which hinders the development of economically desirable features (Jukanti *et al.*, 2015). Similar to NBPGR in India, the United States Department of Agriculture (USDA) and Agricultural Research Service (ARS) via the Plant Genetic Resources Conservation Unit (PGRCU) are also engaged in the preservation of 1298 accessions originating from Pakistan, India and the USA (Morris, 2010). Figure 5.3 shows a comprehensive view of a cluster bean field, with a vast expanse of lush green foliage, characteristic of the crop.

5.4 Approaches for Genetic Improvement in Cluster Bean

Globally, the three primary breeding objectives for cluster bean improvement are high yield, stress tolerance and seed gum content. A limited

Fig. 5.3. Cultivation of cluster bean in India. (Author's own figure.)

breeding programme, a restricted genetic base and a limited use of cutting-edge genomics tools are significant obstacles to the genetic improvement of cluster bean. Listed below are the genetic advancement techniques currently utilized in cluster bean improvement programmes. Crossbreeding between different species or between genetically distinct members of the same species generates genetic diversity in crops.

5.4.1 Hybridization

Because of its small cleistogamous blooms, cluster bean is primarily a self-pollinating crop (Saini *et al.*, 1981). The name 'cluster bean' comes from the fact that this primarily autogamous plant produces pods or beans in clusters. Therefore, cluster bean hybridization presents significant challenges. Cluster bean, like many other self-pollinating crop species, is often improved using pedigree, bulk pedigree and backcross techniques (Knauft and Ozias-Akins, 1995). Despite the advancements achieved by utilizing high-yielding varieties and hybrids, a persistent requirement remains for the continued creation of novel and enhanced cultivars with superior yield potential. Using hybridization, desirable characteristics from wild species (such as *C. serrata* and *C. senegalensis*) can be introduced to domesticated varieties. Cluster bean landraces and germplasm accessions can also be an excellent source of variation for several economically relevant characteristics. Hybridization programmes have made good use of cluster bean genotypes such as FS 277 and HG 884 (Buttar *et al.*, 2008; Jitender *et al.*, 2014) owing to their high yield, gum, protein and proline content. Stigmatic incompatibility and poor pollen viability prevent interspecific hybridization in cluster bean (Sandhu, 1988). Hybridization between *C. tetragonoloba* and *C. serrata*, and *C. tetragonoloba* and *C. senegalensis* was attempted through both traditional and non-traditional breeding techniques (Ahlawat *et al.*, 2013a). However, only an interspecific cross of *C. tetragonoloba* × *C. serrata* resulted in a successful hybrid. *C. tetragonoloba* × *C. serrata* had a 10.47% pod set, with only 83 pods recovered from 792 crosses. When using the traditional breeding method, however, neither

cross showed any signs of pod setting. Only the *C. tetragonoloba* × *C. serrata* cross produced viable offspring when pollination was attempted by smearing the stub of the female parent *(C. tetragonoloba)* with a solidified pollen germination medium. Attempts to circumvent stigmatic incompatibility through bud pollination or stigma/style amputation were unsuccessful. Additional potential barriers to achieving effective interspecific hybridization may arise from variances in the nutritional needs of the pollen across different *Cyamopsis* species, a prolonged period of dormancy for pollen germination in natural environments, and the extended duration of stigma receptivity and the elongated style in *C. tetragonoloba*.

Cluster bean has been studied for its extensive heterosis, which has been documented by multiple researchers (Chaudhary *et al.*, 1981; Saini *et al.*, 1990; Arora *et al.*, 1998). Heterosis has commercial potential, but only if male sterility can be maintained and fertility fully restored. Few studies on heterosis breeding in cluster bean have been reported because of the lack of a reliable source of male sterility. Unless methods are developed for the efficient mass production of hybrid seed, the potential for heterosis breeding will remain low (Pathak, 2015). Thus, other methods, such as tissue culture and recombinant DNA technology, may also serve as hybridization alternatives. Important methods for improving crops through parental selection include heterosis and combining ability (Vasal, 1998). Although it is possible to create hybrids of cluster bean, doing so may prove difficult for crop breeders in the absence of a reliable male sterility system. Different types of cluster bean have been bred for different uses, including as food, animal fodder and for use in weaponry. South Indian cultivars are used for vegetables, while north-west Indian cultivars are grown for seeds. Cluster bean come in both large and small sizes. Most vegetable varieties are small and featureless. Most varieties used for livestock feed are hairy. Cluster bean varieties that resulted from traditional breeding techniques are listed in Table 5.2. The potential donors of cluster bean to be used in the pre-breeding programme are given in Table 5.3. Figure 5.4 shows the flowering and pod-bearing cluster bean plants.

Table 5.2. Potential donors identified for cluster bean.

Genetic resources/cultivars	Important traits
GC-1	Donor for tall plant (80–100 cm)
RGM112	Branched-type genetic stock, moderately resistant to bacterial leaf blight and root rot
Esser	Medium to late maturing; resistant to disease
Lewis	Medium to late maturing; high yield potential
Kinman	Early maturing; resistance to bacterial leaf blight and *Alternaria* leaf spot
Sant Cruz	Sparsely branching, indeterminate growth habit and glabrous cultivar
BR 99	Early maturing variety, single-stemmed and high-yielding potential
BR 90	Single-stemmed variety, glabrous and erect growth habit
CAZG-20-1	Early maturing genotype (79 days)
CAZG-109	Long pod length (12.3 cm) and bold seed (test weight 46.8 g)
CAZG-110	Mutant line having thin stems and longer seedlings

Source: Ministry of Agriculture and Farmers Welfare, India (http://farmer.gov.in); Seednet India Portal (https://seednet.gov.in); Indian Council of Agricultural Research (ICAR), Central Arid Zone Research Institute (CAZRI), Jodhpur, India (http://mkisan.gov.in); Alternative Field Crop Manual, New Crop Resource Online, Purdue University, West Lafayette, Indiana, (https://hort.purdue.edu/ newcrop/afcm/guar.html); Saleem *et al.* (2002). Table used with permission.

Table 5.3. Salient features of cluster bean varieties released and notified for grain and forage purposes in India (1976–2023).

Name of the variety	Breeding method/source	Recommended areas	Special features
Fodder and grain type			
FS-277	Selection from local material (Hisar)	Entire guar growing areas	–
HFG-119	Selection	Entire guar growing areas	It matures in 130–135 days, yielding 25–30 t/ha of green and 5–6 t/ha of dry fodder. It is highly tolerant to drought; pods are non-shattering and also resistant to *Alternaria* leaf spot
Guara-80	Pedigree selection in the intervarietal cross (FS 277 × Strain No. 119)	The northwestern zone (NWZ) of the country	It yields 26.8 t/ha of green fodder and 15 q/ha of seed
HG-182	An individual plant selection from the genetic stock (accession HFC-182)		It matures in 110–125 days
Maru Guar (2470/12)	From the germplasm material supplied by NBPGR, New Delhi	Suitable for western Rajasthan	It yields 22.5 t/ha of green fodder and 9.5 q/ha of seed
HFG-156	Selection	Haryana	It yields 35 t/ha of green fodder
Bundel Guar-1	Single plant selection	Entire guar growing areas	The crop yields 35.0 metric tonnes per hectare of green fodder and 6.5 metric tonnes per hectare of dry fodder. The plant exhibits resistance to lodging, responsiveness to nutrients, tolerance to drought, and a non-shattering trait

Continued

Table 5.3. Continued.

Name of the variety	Breeding method/source	Recommended areas	Special features
Bundel Guar- 2 (IGFRI-2395-2)	Selection	Entire guar growing areas of the country	This variety's green fodder, dry fodder, and crude protein yield range from 25.0–30.0, 5.0–6.0, and 0.12–0.15 t/ha, respectively. It is favourable to bovine animals, with a digestibility rate of approximately 74% on a dry matter basis. Additionally, it exhibits superiority in terms of grain and gum production. The cultivar exhibits high fertilizer responsiveness and is moderately resistant to bacterial leaf blight. It is also tolerant to lodging, drought and pod shattering
Bundel Guar- 3 (IGFRI 1019-1)	Selection from indigenous material, Rajasthan (RGC-19-1) collected from Durgapura	Entire guar growing area of India	The cultivar exhibits a moderate resistance level to bacterial blight and powdery mildew, displays a positive response to fertilizers, demonstrates a high tolerance to shattering, and exhibits a reasonable level of resistance to drought conditions. The maturity period of the crop ranges from 50 to 55 days, resulting in a yield of 35 to 40 metric tonnes per hectare of green fodder
Durga Ajay	Individual plant selection from the material of Nagaur, Rajasthan	Rajasthan	It yields 27 t/ha of green fodder and 12.6 q/ha of seed
Durgapura Safed	Individual plant selection from local material of Rajasthan	Rajasthan	It yields green fodder of 25 t/ha
Agaita Guara-111	The inter-varietal cross of G325 and FS277 was followed by the pedigree method of selection	Entire guar growing areas of Punjab	It yields green fodder 23 t/ha and 4.4 t/ha of dry fodder
Agaita Guara-112	Hybridization (326 × FS 277) × 315 followed by pedigree selection	Entire guar growing areas of Punjab	It matures at an early stage. The plants exhibit trichomes and lack branching, resulting in a greater abundance of pods. The variety yields a total of 30 metric tonnes per hectare of green fodder and 6.4 metric tonnes per hectare of dry fodder
HG-75	Selection	Entire guar cultivating areas of the country	The variety yields green fodder of 25 t/ha and 20 q/ha seed

Continued

Table 5.3. Continued.

Name of the variety	Breeding method/source	Recommended areas	Special features
Grain type/vegetable type			
Pusa Mausami	Selection from a local collection of Jaipur, Rajasthan		Densely branching, first picking in 65–80 days after sowing
Pusa Sadabahar	Selection from a local collection of Jaipur, Rajasthan	Suitable for sowing both in summer and rainy season	Non-branching type pods green and 12–13 cm long, tender and fibreless. First picking in 45–55 days after sowing. It is used for vegetable purpose
Pusa Navbahar	A cross between Pusa Mausami and Pusa Sadabahar	Suitable for both summer and rainy season	Non-branching type used for vegetable purpose
Sharad Bahar	Developed through selection		Profused branching, a single plant produces about 133 pods
P-28-1-1		Suitable for both summer and rainy season	Photo insensitive variety
P281	Cross between Pusa Navbahar × IC11521	Suitable for both summer and rainy season	Photo insensitive
Goma Manjari			Resistance to powdery mildew (PM), bacterial leaf blight (BLB) and leaf spot (LS)
Durga Bahaar			Used for vegetable purposes
RGC 197	Selection from local collection of Nagaur district, Rajasthan	Suitable for the rainy season in Rajasthan	The gum content is 32–33%. It matures in 110 to 120 days and has an average yield of 10–18 q/ha
RGC 936	E.C. 248 × RGC 401	Most popular variety in Rajasthan. Adopted in Gujarat, Haryana, Rajasthan, Tamil Nadu, UP and West Bengal	It is a branched type of early maturing (85–90 days). It is a drought-tolerant variety. Grains are medium-sized and light pink in colour. This variety is resistant to many diseases. It yields 8–10 q/ha
RGC 1002	NC 46/P2-1 × RGC 516	Suitable for the rainy season in Rajasthan	It is suitable for rainfed and irrigated conditions. 1000 seed weight varies from 3.2 to 3.6 grams
RGC 1003	NAVEEN/HG75	Adopted in Gujarat, Rajasthan, MP and UP	It is suitable for rainfed conditions. The seeds contain about 30% gum and mature in 110 to 115 days
Haryana Guar 365 (HG 365)	DURGAJAY × HISSAR LOCAL	This is suitable for the rainfed conditions of Rajasthan and Haryana	It is early maturing. It has brisk podding behaviour. It can be grown during both the summer season and in irrigated conditions

Continued

Table 5.3. Continued.

Name of the variety	Breeding method/source	Recommended areas	Special features
HG 2-20	HG 365 × PS 277	This is suitable for the rainfed and irrigated conditions of Rajasthan and Haryana	It is a branched, bold-seeded variety. It is moderately resistant to wilt, Alternaria leaf blight, BLB and root rot (RR). It can also be grown during the summer season
Gujarat Guar 1 (GAUG-34)	Mutant of Kutch-8 local variety of Guar (10 K.R. of Gama rays)	This is suitable for the guar growing states of India like Gujarat, Haryana, and Rajasthan	Dark green foliage, the serrated margin of leaves, branched type, matured in 91 to 122 days and gave 10–13 q/ha seed yield
RGC 1017	NAVEEN × HG 75	It is suitable for both rainfed and irrigated conditions. Suitable for Rajasthan, MP, UP and Punjab	It is moderately resistant to BLB, PM, and RR. It grows to a height of 56.7 to 74.4 cm and matures in 90–100 days, with an average yield of 12–14 q/ha
HG 563	DURGAJAY × HISSAR LOCAL	Suitable for all the guar growing areas	It contains 33% gum and has brisk podding behaviour. The plants grow to a height of about 60 to 100 cm. They mature in 85–90 days and have an average yield of 12–13 q/ha
Surya (RGM 112)		Suitable for kharif rain-fed areas of North India	Leaves are pubescent and tripinnate with a smooth margin and grey seeds. This variety is moderately resistant to bacterial leaf blight and root rot disease. Matures in 92–95 days with an average yield of 12–14 q/ha
HG 884	HG 75 × HG 296	Suitable for all the guar growing areas of Haryana, Gujarat, MP and UP	High-yielding, medium to early maturity, 30-32% gum content with 2000-3500 cP values. Matures in 100-110 days with an average yield of 10-11 q/ha
RGC 1031(Guar Kranti)	RGC936 × RGC 986/P-10	Suitable for all the guar-growing areas of Rajasthan	It is tolerant to major diseases; gum content is 28.1 to 30%. Matures in 90–100 days with an average yield of 15–16 q/ha
RGC 1055 (Guar Uday)	RGC 1013 × RGC - 986/P6-1	Suitable for all the guar growing areas of Rajasthan, both kharif and zaid season	It is semi photo insensitive; it has profuse podding and high yield potential
RCG 1038	RGC 516 × HG 75	It is suitable for summer and kharif seasons in Rajasthan	This is a branched type variety, and it is somewhat photo-insensitive. It is heavily pod-bearing. Matures in 100–105 days with an average yield of 12–15 q/ha

Continued

Table 5.3. Continued.

Name of the variety	Breeding method/source	Recommended areas	Special features
RGC 1066 (Guar Lathi)	RGC936 × RGC 197 / P 20-1	It is suitable for cultivation in kharif and during the zaid season in Rajasthan	This variety is resistant to BLB and RR. It is early and high-yielding, suitable for intercropping and mixed cropping. It matures in 97–105 days and yields an average 10–14 q/ha yield
Guar Kunjal (RGC 1033)	RGC 1013 × RGC 986/P-6	It is suitable for cultivation in the kharif season in North India	It is moderately tolerant to ALB, BLB and root rot. Gum content is 29.9 to 31.5%. Matures in 95–100 days with an average yield of 15–20 q/ha
Karan Guar-1 (RGr 12-1)	RGC 197 × RGC 1017	It is suitable for cultivation in the kharif season in Rajasthan	Moderately resistant to ALB, BLB and RR. Gum content is 29.9 to 31.5%. Matures in 90–95 days with an average yield of 15–20 q/ha
Gujarat Guar-3		It is suitable for cultivation in kharif season in Gujarat	Resistant to bacterial leaf blight, whitefly and leaf hopper. 29.40% gum content. Matures in 98–100 days with an average yield of 13–14 q/ha
Karan Guar 14 (RGr 18-1)	RGC 1038 × RGC 1017	It is suitable for cultivation in the kharif season in North India	Moderately resistant to Alternaria leaf blight, tolerant to BLB and root rot, 29.9 to 31.5% gum content. Matures in 90–100 days with an average yield of 15–18 q/ha
X 10	Selection from the cross X 6 × HG 563	Suitable for all the guar growing areas	Medium maturity, suitable for rainfed under both high and low fertility conditions, resistant to root rot and bacterial blight diseases

Source: Roy *et al.* (2018); Jindal *et al.* (2021); Directorate of Research, SKNAU, Jobner (2023); Seednet (2023). Table used with permission.

5.4.2 Tissue culture and genetic transformation

Tissue culture techniques, such as micropropagation and genetic transformation/manipulation, have the potential to serve as supplementary methodologies to the enhancement of cluster bean through conventional breeding methods. Micropropagation methods offer several advantages. For example, they allow the circumvention of post-fertilization barriers, and enable the large-scale production of true-to-type plants with a single gender (Kumar *et al.*, 2010a, b). According to a study conducted by Modi *et al.* (2012), there is evidence for the cultivation of plants that exhibit a diminished germination rate. According to Karupussamy (2009), two key strategies for enhancing the production of secondary chemicals are increasing their production and establishing regeneration protocols for genetic transformation. The success rate of the *in-vitro* regeneration of explants is influenced by various factors, such as the nature of explants, medium composition, hormones and culture conditions (Kalia *et al.*, 2014). Researchers have conducted several studies to develop standardized procedures for the micropropagation of cluster bean (Prem *et al.*, 2003, 2005; Ahmad and Anis, 2007; Gargi *et al.*, 2012; Ahlawat *et al.*, 2013b; Mathiyazhagan *et al.*, 2013; Verma *et al.*, 2013;

Fig. 5.4. A close-up view of cluster bean plants. (Author's own figure.)

Sheikh *et al.*, 2015). The investigation of genetic transformation in cluster bean was commenced by Prem (1999) through the utilization of *Agrobacterium tumefaciens*. Notably, the expression of β-glucuronidase (GUS) was observed to be favourable in explants that were 14 days old. Stable transgenic plants of cluster bean were generated by introducing β–glucuronidase and neomycin phosphotransferase genes. A previous endeavour to modify cluster bean involved the introduction of an α-galactosidase gene from *Senna occidentalis* (coffee senna). While achieving success, this attempt led to a decrease in the galactomannan content of 30% within the transformed plants (Joersbo *et al.*, 2001).

5.4.3   Mutation breeding

Mutation breeding is an effective technique employed to deliberately induce genetic variability that is crucial for enhancing crop improvement. According to Bhosle and Kothekar (2010), mutations can be induced in both seed and vegetatively propagated crops. The mutagenesis process encompasses the comprehensive assessment and identification of plants exhibiting favourable characteristics that hold potential for breeding endeavours. Previous studies have documented various natural mutations in cluster

bean, including partial to complete male sterility (Mittal *et al.*, 1968). Additionally, declined fertility in a rosette-type inflorescence has also been observed (Stafford, 1988, 1989). While there have been reports of these mutants in cluster bean, their practical application in enhancing crop quality has not yet been proven (Arora and Pahuja, 2008). Various physical mutagens and chemical mutagens have been identified and effectively employed in crop improvement to augment the mutation frequency rate (Mullainathan *et al.*, 2014; Pathak and Roy, 2015).

The utilization of gamma rays, specifically from a ^{60}Co source, was the initial application of physical mutagens that proved to be effective in inducing mutations in cluster bean, as demonstrated by Vig in 1965. According to Singh and Agarwal (1986), plants that were exposed to gamma-ray irradiation demonstrated reduced fertility and displayed a trisomic phenotype. According to Lather and Chowdhury (1972), the application of gamma-ray treatment led to a decrease in the percentage of germination, pollen fertility and seedling survival as the dosage increased within the range of 10–200 kR. Furthermore, it led to a range of chromosomal abnormalities. In contrast, Chaudhary *et al.* (1973) reported an augmentation in crop yield, gum production and grain protein content by applying low levels of gamma-ray irradiation, ranging from 2 to 20 kR. In their study, Chowdhury *et al.* (1975) observed significant variability among various traits within irradiated populations of two distinct cluster bean varieties. Singh *et al.* (1981) reported that the gamma-ray irradiation of cluster bean resulted in the emergence of a particularly intriguing mutant characterized by early flowering and determinate growth. Plants of the determinate type exhibited several notable characteristics, including diminished plant height, a non-branching growth pattern, enlarged cluster size, early and synchronized maturation, and, notably, the primary shoot terminated in either a leaf or an inflorescence. In addition, it was observed that subjecting Pusa Navbahar seeds to a 10 kR X-ray treatment resulted in the emergence of a stable, homozygous mutant characterized by early flowering and an enhanced pod count (Rao and Rao, 1982). Singh and Aggarwal (1986) conducted a study in which they employed gamma-rays in a dose of 80 kR/800 Gray, ethyl methanesulfonate (EMS) of 0.1%, and N-nitrosoN-methylurea (NMU) of 0.01%, either individually or in combination. The objective of their study was to generate mutants of PLG143 genotypes. The mutants exhibited desirable traits such as early flowering, long pods, improved yields and a gum percentage of 17%.

Lingakumar and Kalandaivelu (1998) conducted intriguing experiments in which Pusa Navbahar seedlings were subjected to UV-B, UV-A and white light. Applying UV-B treatment in isolation led to a decrease in pigment content, photosynthetic activity and overall plant growth. The addition of UV-A supplementation counteracted certain consequences of UV-B exposure because it facilitated the overall growth of seedlings and enhanced the synthesis of chlorophyll and carotenoids. The activation of UV-B damage was not observed, however, when UV-B irradiation was followed by white light. In addition, a study conducted by Joshi et al. (2007) found that the decrease in photosynthetic pigments and oxygen production caused by UV-B exposure was partially mitigated when combined with UV-A radiation.

Cluster bean plants derived from seeds treated with EMS had a notable insufficiency of chlorophyll, accompanied by vigorous vegetative growth (Gohal et al., 1970). Additionally, alterations in leaf morphology and texture, growth patterns and pod dimensions were observed in these plants. Several pod mutants displayed pleiotropic phenotypes, encompassing characteristics such as significant branching, delayed flowering and alteration of seed colour from the typical violet to a light grey/brown shade. In a study conducted by Rao et al. (1982), the effects of treating seeds of Pusa Navbahar with different concentrations of kitazin (600, 400, 200 ppm) or Saturn (3000, 2000, 1000 ppm) for durations of 12 and 24 h were investigated. The treatment resulted in developing determinate growth habits and spreading seed variants. The investigation of mutation induction in cluster bean has encompassed the utilization of physical rays and chemical mutagens, resulting in the development of a diverse range of varieties through the combined application of these mutagens (Chopra, 2005). A study conducted by Singh and Agarwal (1986) examined the effects of a combination treatment involving physical mutagens (gamma rays at doses of 80 and 100 kR) and chemical mutagens (aqueous solution of 20.1–0.3% EMS and 0.01–0.03% of N-methyl-N-nitrosourea) on the seeds of two

cultivars namely, Suvidha and PLG 143. The results revealed that this combination treatment led to the development of mutants exhibiting increased traits such as the number of pods, pod length and early maturity. Moreover, these mutant varieties had enhanced productivity, characterized by elevated gum content in both the endosperm and seeds. Heterophylly was observed in cluster bean resulting from gamma-ray and sodium 4 azide treatment, as documented by Badami and Bhalla (1992). In a study conducted by Babariya et al. (2008), the effects of gamma rays and EMS treatment on two cultivars, Pusa Navbahar and VRS-culture, were investigated. The researchers discovered that, in addition to creating variability, mutagens also influenced the correlation between the traits under study. As a result, it became feasible to breed desired traits selectively by manipulating the direction of selection. In their studies, Mahla et al. (2005, 2010) and Velu et al. (2012) conducted physical (gamma-rays) and chemical (EMS) mutagenesis experiments. They observed a progressive reduction in plant height and various agronomic characteristics in cluster bean cultivars as the mutagen dose increased. An increase in the mutation frequency was observed with increased doses of gamma rays and chemical mutagens, such as EMS and NaN_3. Vig (1965) and Bewal et al. (2009) investigated the potential of polyploidization to enhance variability, but no noteworthy advancements were documented. To date, a restricted quantity of cluster bean mutants possesses advantageous characteristics. Therefore, it is crucial to carry out extensive testing that includes a wide range of lower doses of different mutagens, such as physical, chemical or combined, with the aim of promoting desirable genetic variety. The variety that is produced has the potential to be harnessed in the selective breeding of cultivars that demonstrate favourable characteristics, including but not limited to increased yield, accelerated maturation, enhanced resistance to diseases and elevated seed gum content.

5.5 Biotechnological Approaches

The success of any breeding strategy relies heavily on the precise estimation of the available genetic diversity. In the context of traditional breeding, plant breeders commonly rely on agro-morphological characteristics as outlined by Henry and Mathur (2005) and Kumar and Rodge (2012). The success of any breeding strategy is contingent upon the precise estimation of both accessible and heritable genetic diversity. Nevertheless, assessing variability using physical attributes often leads to limited genetic progress in selection owing to their inadequate representation of genetic variation and susceptibility to environmental factors (Pathak *et al.*, 2011c). Consequently, molecular markers have been employed in the process of genotype selection for heterosis, as well as for conducting purity tests on cultivars/hybrids, identifying varieties/genotypes and analysing genetic diversity. Furthermore, utilizing molecular markers presents a viable approach for monitoring a noteworthy gene before its manifestation in terms of phenotype, particularly during specific developmental stages (Lübberstedt and Varshney, 2013). Nevertheless, the available data regarding the extent and nature of genetic variation in cluster bean, specifically at the DNA/genome level, are insufficient (Sharma *et al.*, 2014b). The investigation of isozyme variation in guar germplasm for the purpose of genetic diversity analysis has been limited despite the existence of substantial variability at the agro-morphological level. Electrophoretic technology has not been extensively utilized for this purpose. In Mauria's (2000) study, an initial effort was made to investigate the domestication of cluster bean by analysing the diversity of isozymes. This analysis focused on primitive landraces of India, as well as varieties that were released from the USA. Additionally, two wild relatives of guar, namely *C. serrata* and *C. senegalensis*, were included in the study. Subsequently, Brahmi *et al.* (2004) further endeavoured to employ allozyme markers to investigate diversity in cluster bean. Their findings indicated that the level of interpopulation diversity surpassed the overall genetic diversity observed in guar. Due to the heightened environmental sensitivity, lack of consistency, reduced quantity and limited polymorphism exhibited by protein-based markers such as isozymes and allozymes, DNA-based markers are considered more preferable (Shah *et al.*, 2015). During the past decade, there has been a limited number of studies conducted to investigate the genetic diversity and phylogeny of cluster bean through the utilization of DNA-based

markers. RAPD has emerged as a prominent DNA marker system for various applications in cluster bean, including the assessment of genetic diversity, characterization of germplasm, identification of cultivars, determination of genetic purity and gene tagging (Weixin *et al.*, 2009; Pathak *et al.*, 2011c; Rodge *et al.*, 2012; Kuravadi *et al.*, 2013; Kumar *et al.*, 2013, 2017). However, it has been observed that RAPD markers exhibit limited reproducibility and lack consistency in their results. According to Sharma *et al.* (2014b), the utilization of inter-simple sequence repeat (ISSR) markers has been found to be more effective in identifying polymorphisms with higher precision and the ability to assess diversity at both intra- and inter-genomic levels (Kuravadi *et al.*, 2013; Sharma *et al.*, 2014a, b). Moreover, both RAPD and ISSR techniques are characterized by their simplicity, efficiency and independence from prior sequence information. It is worth noting, however, that these methods exhibit limitations because of poor reproducibility and low stability. In order to overcome these issues, Paran and Michelmore (1993) employed molecular cloning techniques to convert the RAPD markers into the sequence-characterized amplified region (SCAR) marker, as documented by Cheng *et al.* (2015). The SCAR marker exhibits co-dominance, locus specificity, stability and a high level of reproducibility, as reported by Dhawan *et al.* (2013) and Shah *et al.* (2015). In their seminal work, Sharma *et al.* (2014a) were the first to successfully develop three polymorphic SCAR markers and one region-specific SCAR marker using RAPD techniques. Additionally, they also developed a genotype-specific SCAR marker using ISSR techniques, specifically targeting the RGC-1031 genotype. The variation in the number of DNA bases and arrangement of nucleotides in the internal transcribed spacer (ITS) regions of ribosomal DNA (rDNA) may occur as a consequence of selection pressure during the process of evolution (Sharma *et al.*, 2002). The versatility of ITS spacers as genetic markers for phylogenetic and diversification studies is widely recognized, primarily attributed to their ease of amplification and high level of polymorphism (Powers *et al.*, 1997; Beltrame-Botelho *et al.*, 2005). The genetic diversity in cluster bean (Pathak *et al.*, 2011c) was evaluated using primers targeting the nuclear ribosomal ITS region, including ITS-1, ITS-2, and 5.8S rDNA, as previously described by White *et al.* (1990). The analysis of the rDNA

ITS DNA sequence unveiled a closely related lineage consisting of different genotypes. This analysis demonstrated notable occurrences of single nucleotide polymorphisms (SNPs) at seven specific sites within an amplified conserved DNA region. In contrast to other legumes, it has been observed that sequence-based DNA markers, particularly microsatellites or simple sequence repeats (SSRs), are insufficient for cluster bean (Kuravadi *et al.*, 2013; Kumar *et al.*, 2015, 2017). Despite the high industrial demand for cluster bean as a crop, adopting various 'omics' approaches such as transcriptomics and genomics has been slow. As a result, there are currently only 16,476 expressed sequences tags (ESTs) available in the National Center for Biotechnology Information (NCBI) database, as submitted by Naoumkina *et al.* in 2007. The transcriptomic study conducted by Rawal *et al.* (2017) revealed the presence of 127,706 transcripts and 48,007 non-redundant high-quality unigenes in cluster bean leaf, shoot and flower tissue. In addition, a total of 8687 potential simple sequence repeats (SSRs) have been identified, exhibiting an average frequency of one SSR per 8.75 kilobases (kb). A database named Cluster gene DB was developed to facilitate the efficient retrieval of unigenes and microsatellite markers. The study's findings, including identifying tissue-specific genes and molecular markers, will contribute to advancing genetic enhancement in cluster bean. The transcriptome analysis from cluster bean leaf tissues of two widely cultivated varieties, namely M-83 and RGC-1066, was conducted using the RNASeq methodology (Tanwar *et al.*, 2017). The present study successfully identified a comprehensive set of 62,146 non-redundant unigenes, which were further annotated with a total of 175,882 Gene Ontology (GO) annotations. In addition, a total of 11,308 unigenes were subjected to annotation and characterization, resulting in their classification into 6 clusters and 55 subclasses. A cumulative count of 5773 potential SSRs and 3594 high-quality SNPs were documented. The ESTs were examined and analysed by Kuravadi *et al.* (2014) and Kumar *et al.* (2015), identifying 187 and 100 SSR markers, respectively. The diversity analysis studies of 32 cluster bean genotypes using EST-SSR markers revealed a limited genetic diversity, as indicated by the low polymorphism information content (PIC) of 0.13 and average dissimilarity coefficient of 0.09, as

reported by Kumar *et al.* (2015). The level of similarity observed in the study conducted by Kumar *et al.* (2015) differed from previous findings reported by Pathak *et al.* (2011a, b), Kumar *et al.* (2013) and Sharma *et al.* (2014b). This discrepancy could be attributed to using distinct marker systems in these studies. Therefore, considering the findings obtained from SSR analysis, it is recommended that the development of genomic SSR markers or the utilization of more advanced marker systems such as SNPs or GBS (genotyping-by-sequencing) (Elshire *et al.*, 2011) is crucial for the successful implementation of markers in the enhancement of cluster bean. The researchers conducted a tissue-specific investigation on the cluster bean variety RGC-936 to gain insights into the regulatory function of mi-RNAs in galactomannan biosynthesis (Tyagi *et al.*, 2018). The present study successfully identified a total of 187 known and 171 novel miRNAs that exhibited differential expression. Subsequently, a subset of 10 miRNAs was subjected to validation procedures. Two novel unigenes, namely ManS (mannan synthase/mannosyl transferase) and UGE (UDP-D-glucose 4-epimerase), were identified and confirmed as targets for three newly discovered miRNAs (Ct-miR3135, Ct-miR3130 and Ct-miR3157). The aforementioned findings have potential utility in elucidating the mechanisms governing galactomannan biosynthesis regulation. Furthermore, these findings may be harnessed in forthcoming breeding endeavours aimed at augmenting the gum content in guar.

5.6 Cultivation Practices of Cluster Bean

5.6.1 Selection of field/land preparation, seed treatments, sowing time and seed rate/sowing method

The field should be well levelled, weed-free, well drained and fertile, and should be prepared by one ploughing with a soil turning plough and two harrowings followed by planking. Seed should be treated with carbendazim at 2 g/kg of seed followed by *Rhizobium* culture, PSB and PGPR each at a rate of 600 g/ha seed. Cluster bean sowing should be performed in kharif after the onset of the monsoon or the first week of July. Line sowing is done with a row-to-row distance of 45–30 cm and a plant-to-plant distance of 10 cm. Sowing with a seed drill or desi plough is generally recommended in cluster bean cultivation. In cluster bean, 15–20 kg/ha seed rate is sufficient to raise a good crop.

5.6.2 Irrigation, fertilizer, weed management, disease and pest control

A rainfed crop of cluster bean does not require extra irrigation in kharif crop. If irrigation facilities are available, life-saving irrigation during times of moisture stress or critical stages, like the flowering and pod-filling stages, increases the seed yield. Fertilizer at the rate of 20 kg N/ha and 40 kg P_2O_5/ha as basal application at the time of sowing through drill application is sufficient at the initial stage for normal growth and development of cluster bean. Applying zinc sulphate at 15–20 kg/ha is also recommended to improve the seed yield. The first hand weeding should be performed at 20–25 days after sowing and subsequently at 20 days after the first weeding. Alternatively, pre-emergence application of pendimethalin at 0.75–1.0 kg a.i./ha accompanied with one hand weeding at 35–40 days after sowing or post-emergence application of imazethapyr at 40g/ha at 15–20 days after sowing can be carried out. A spray of dimethoate 30 EC at 0.1% and repeated at intervals of 15 days is recommended to control insect pests during cluster bean cultivation.

5.6.3 Harvesting and seed yield

Harvesting should be done at the physiological maturity stage for a good seed yield. A managed crop of cluster bean would provide an average seed yield of about 10–12 q/ha.

5.7 Future Prospects and Conclusion

The cluster bean (*C. tetragonoloba*) serves as a significant option for farmers residing in arid and semi-arid regions of India and other parts of the world. Hybridization, mutation breeding and germplasm/varieties selection/evaluation are among the prevailing techniques employed for germplasm enhancement and the creation of novel cultivars. In order to acquire cultivars that are not affected by light, can withstand drought, are resistant to diseases and pests, and have higher yields, the utilization of genomics, namely marker-assisted selection, along with advanced sequencing and molecular techniques, may prove to be the most effective approach. The crop has lower-quality seed owing to its uncertain growth pattern and the presence of elevated amounts of blackened seed during the harvest. Hence, the creation of cultivars with synchronized ripening is of paramount importance. Molecular methodologies, such as tissue culture techniques, can be employed to address the fertility challenge at hand, as evidenced by the favourable outcomes observed in interspecific crossings involving wild species, in contrast to the unsuccessful outcomes observed in crosses between wild and cultivated species. In conclusion, cluster bean is a crucial leguminous plant that possesses substantial economic value. Hence, the principal objectives of breeding projects should prioritize cultivating types well-suited for seed, vegetable and gum utilization.

References

Ahlawat, A., Dhingra, H.R. and Pahuja, S.K. (2012) Biochemical composition of stigma and style in *Cyamopsis* spp. *Forage Research* 38(1), 53–55.

Ahlawat, A., Dhingra, H.R. and Dhankar, J.S. (2013a) *In vitro* regeneration of wild species of guar (*Cyamopsis serrata* and *Cyamopsis senegalensis*). *Journal of Krishi Vigyan* 1(2), 48–55.

Ahlawat, A., Pahuja, S.K. and Dhingra, H.R. (2013b) Overcoming interspecific hybridisation barriers in *Cyamopsis* species. *International Journal of Biotechnology and Bioengineering Research* 4(3), 181–190.

Ahmad, N. and Anis, M. (2007) Rapid plant regeneration protocol for cluster bean (*Cyamopsis tetragonoloba* L. Taub.). *The Journal of Horticultural Science and Biotechnology* 82(4), 585–589.

Arora, R.N. and Pahuja, S.K. (2008) Mutagenesis in guar (*Cyamopsis tetragonoloba* (L.) Taub.). *Plant Mutation Reports*, 2(1), 7–9.

Arora, R.N., Saini, M.L., Singh, J.V., Sareen, P.K. and Paroda, R.S. (1985) Karyotype analysis in three species of genus *Cyamopsis*. *Indian Journal of Genetics and Plant Breeding* 45(2), 302–309.

Arora, R.N., Lodhi, G.P. and Singh, J.V. (1998) Variety cross diallel analysis for grain yield and its components in clusterbean (*Cyamopsis tetragonoloba* (L.) Taub.). *Forage Research* 24(1), 1–1.

Babariya, H.M., Vaddoria, M.A., Mehta, D.R., Madariya, R.D. and Monpara, B.A. (2008) Effect of mutagens on characters association in cluster bean (*Cyamopsis tetragonaloba* L. Taub). *International Journal of Bioscience Reporter* 6, 135–140.

Badami, P.S. and Bhalla, J.K. (1992) Mutagenic effectiveness and efficiency of gamma rays, magnetic fields and sodium azide in cluster bean. *Advances in Plant Science* 5, 534–541.

Beltrame-Botelho, I.T., Gaspar-Silva, D., Steindel, M., Dávila, A.M.R. and Grisard, E.C. (2005) Internal transcribed spacers (ITS) of *Trypanosoma rangeli* ribosomal DNA (rDNA): a valuable marker for inter-specific differentiation. *Infection, Genetics and Evolution* 5(1), 17–28.

Bewal, S., Purohit, J., Kumar, A., Khedasana, R. and Rao, S.R., (2009) Cytogenetical investigations in colchicine-induced tetraploids of *Cyamopsis tetragonoloba* L. *Czech Journal of Genetics and Plant Breeding* 45(4), 143–154.

Bhosle, S.S. and Kothekar, V.S. (2010) Mutagenic efficiency and effectiveness in cluster bean (*Cyamopsis tetragonoloba* (L.) Taub.). *Journal of Phytology* 2(6), 21–27.

Brahmi, P., Bhat, K.V. and Bhatnagar, A.K. (2004) Study of allozyme diversity in guar [*Cyamopsis tetragonoloba* (L.) Taub.] germplasm. *Genetic Resources and Crop Evolution* 51, 735–746.

Buttar, G.S., Brar, K.S. and Singh, S. (2008) Genetic architecture of seed yield and its attributing traits in cluster bean (*Cyamopsis tetragonoloba*) grown under the semi-arid region of Punjab. *Indian Journal of Agricultural Science* 78(9), 795–797.

Chaudhary, B.S., Lodhi, G.P. and Arora, N.D. (1981) Heterosis for grain yield and quality characters in cluster bean. *Indian Journal of Agricultural Science* 51(9), 638–642.

Chaudhary, M.S., Ram, H., Hooda, R.S. and Dhindsa, K.S. (1973) Effect of gamma irradiation on yield and quality of guar (*Cyamopsis tetragonoloba* L. Taub). *Annals of Arid Zone* 12, 19–22.

Cheng, J., Long, Y., Khan, A., Wei, C., Fu, S. and Fu, J. (2015) Development and significance of RAPD-SCAR markers for the identification of *Litchi chinensis* Sonn: by improved RAPD amplification and molecular cloning. *Electronic Journal of Biotechnology* 18(1), 35–39.

Chevalier, A. (1939) Recherches sur les espèces du genre *Cyamopsis*, plantes fourragères pour les pays tropicaux et semi-arides. *Journal d'Agriculture Traditionnelle et de Botanique Appliquée* 19(212), 242–249.

Chopra, V.L. (2005) Mutagenesis: Investigating the process and processing the outcome for crop improvement. *Current Science* 89, 353–359.

Chowdhury, R.K., Chawdhury, J.B. and Singh, R.K. (1975) Induced polygenic variability in cluster bean. *Crop Improvement* 2, 17–24.

Dabas, B.S., Mital, S.P. and Arunachalam, V. (1982) An evaluation of germplasm accessions in Guar (*Cyamopsis tetragonoloba* L.). *Indian Journal of Genetics and Plant Breeding* 42, 56-59.

Dhawan, C., Kharb, P., Sharma, R., Uppal, S. and Aggarwal, R.K. (2013) Development of male-specific SCAR marker in date palm (*Phoenix dactylifera* L.). *Tree Genetics & Genomes* 9, 1143–1150.

Duke, J.A. (1981) *Handbook of Legumes of World Economic Importance*. Plenum Press, New York, pp. 70–73.

Elshire, R.J., Glaubitz, J.C., Sun, Q., Poland, J.A., Kawamoto, K., Buckler, E.S. and Mitchell, S.E. (2011) A robust, simple genotyping-by-sequencing (GBS) approach for high-diversity species. *PloS One* 6(5), e19379.

Gandhi, S.K., Saini, M.L. and Jhorar, B.S. (1978) Screening of cluster bean genotypes for resistance to leaf spot caused by *Alternaria cyamopsidis*. *Forage Research* 4, 169–171.

Gargi, T., Acharya, S., Patel, J.B. and Sharma, S.C. (2012) Callus induction and multiple shoot regeneration from cotyledonary nodes in cluster bean [*Cyamopsis tetragonoloba* L Taub]. *Agressologie* 1, 1–7.

Gillett, J.B. (1958) *Indigofera* (*Microcharis*) in tropical Africa with the related genera *Cyamopsis* and *Rhyncotropis*. *Kew Bulletin Additional Series* 1, 1–66

Gohal, M.S., Kalia, H.R., Dhillon, H.S. and Nagi, K.S. (1970) Effect of ethylmethane sulphonate on the mutation spectrum in guar. *Indian Journal of Heredity* 2(1), 51–54.

Henry, A. and Mathur, B.K. (2005) Genetic diversity and performance of cluster bean varieties for quality and quantitative characters in an arid region. *Journal of Arid Legumes* 2, 145–148.

Hiremath, S.C., Ramamoorthy, J., Cai, Q. and Chinnappa, C.C. (1996) Analysis of genetic relationships in the genus *Cyamopsis* (Fabaceae) using allozymes, RFLP and RAPD markers. *American Journal of Botany* 83, 207.

Jindal, Y., Sehrawat, S.K., Chhabra, A.K., Kumar, N., Kumar, S. *et al.* (2021) Varieties of CCS HAU: Continued efforts towards food security (CCSHAU/PUB #21-058). Directorate of Research, CCS Haryana Agricultural University, Hisar, Haryana, p 152.

Jitender, P.S.K., Verma, N. and Bhusal, N. (2014) Genetic variability and heritability for seed yield and water use efficiency related characters in cluster bean [*Cyamopsis tetragonoloba* (L.) Taub.]. *Forage Research* 39(4), 170–174.

Joersbo, M., Marcussen, J. and Brunstedt, J. (2001) *In vivo* modification of the cell wall polysaccharide galactomannan of guar transformed with a α-galactosidase gene cloned from senna. *Molecular Breeding* 7(3), 211–219.

Joshi, P.N., Ramaswamy, N.K., Iyer, R.K., Nair, J.S., Pradhan, M.K. *et al.* (2007) Partial protection of photosynthetic apparatus from UV-B-induced damage by UV-A radiation. *Environmental and Experimental Botany* 59(2), 166–172.

Jukanti, A.K., Bhatt, R., Sharma, R. and Kalia, R.K. (2015) Morphological, agronomic, and yield characterisation of cluster bean (*Cyamopsis tetragonoloba* L.) germplasm accessions. *Journal of Crop Science and Biotechnology* 18, 83–88.

Kalia, R.K., Rai, M.K., Sharma, R. and Bhatt, R.K. (2014) Understanding *Tecomella undulata*: an endangered pharmaceutically important timber species of hot arid regions. *Genetic Resources and Crop Evolution* 61, 1397–1421.

Karupussamy, S. (2009) A review on trends in production of secondary metabolites from higher plants by *in vitro* tissue, organ and cell cultures. *Journal of Medicinal Plants Research* 3, 1222–1239.

Kays, S.E., Morris, J.B. and Kim, Y. (2006) Total and soluble dietary fibre variation in *Cyamopsis tetragonoloba* (L.) Taub. (guar) genotypes. *Journal of Food Quality* 29(4), 383–391.

Knauft, D.A. and Ozias-Akins, P. (1995) Recent methodologies for germplasm enhancement and breeding. In: Pattee, H.E., Stalker, H.T. (eds) *Advances in Peanut Science.* American Peanut Research and Education Society, Stillwater, Oklahoma, pp 54–94.

Kumar, D. (2008) Arid legumes - an introduction. In: Kumar, D. and Henry, A. (eds) A Souvenir, 3rd National Symposium on Arid Legumes. Indian Arid Legumes Society, CAZRI, Jodhpur, India, pp. 25–44.

Kumar, D. and Rodge, A.B. (2012) Status, scope and strategies of arid legumes research in India. A review. *Journal of Food Legume* 25, 255–272.

Kumar, N., Modi, A.R., Singh, A.S., Gajera, B.B., Patel, A.R. *et al.* (2010a) Assessment of genetic fidelity of micropropagated date palm (*Phoenix dactylifera* L.) plants by RAPD and ISSR markers assay. *Physiology and Molecular Biology of Plants* 16, 207–213.

Kumar, N., Singh, A.S., Modi, A.R., Patel, A.R., Gajera, B.B. and Subhash, N. (2010b) Genetic stability studies in micropropagated date palm (*Phoenix dactylifera* L.) plants using microsatellite marker. *Journal for Environment Science* 26(1), 31-36.

Kumar, S., Joshi, U.N., Singh, V., Singh, J.V. and Saini, M.L. (2013) Characterisation of released and elite genotypes of guar [*Cyamopsis tetragonoloba* (L.) Taub.] from India proves unrelated to geographical origin. *Genetic Resources and Crop Evolution* 60, 2017–2032.

Kumar, S., Parekh, M.J., Patel, C.B., Zala, H.N., Sharma, R. *et al.* (2015) Development and validation of EST-derived SSR markers and diversity analysis in cluster bean (*Cyamopsis tetragonoloba*). *Journal of Plant Biochemistry and Biotechnology* 25, 263–269.

Kumar, S., Modi, A.R., Parekh, M.J., Mahla, H.R., Sharma, R. *et al.* (2017) Role of conventional and biotechnological approaches for genetic improvement of cluster bean. *Industrial Crops and Products* 97, 639–648.

Kuravadi, A.N., Tiwari, P.B., Choudhary, M., Tripathi, S.K., Dhugga, K.S., Gill, K.S. and Randhawa, G.S. (2013) Genetic diversity study of cluster bean (*Cyamopsis tetragonoloba* (L.) Taub.) landraces using RAPD and ISSR markers. *International Journal of Advanced Biotechnology and Research* 4(4), 460–471.

Lather, B.P.S. and Chowdhury, J.B. (1972) Studies on irradiated guar. *Nucleus* 15, 16–22.

Lingakumar, K. and Kalandaivelu, G. (1998) Differential responses of growth and photosynthesis in *Cyamopsis tetragonoloba* L. grown under ultraviolet-B and supplemental long-wavelength radiations. *Photosynthetica* 35, 335–343.

Lübberstedt, T. and Varshney, R. (2013) *Diagnostics in Plant Breeding.* Springer, The Netherlands.

Mahla, H.R., Shekhawat, A., Kumar, D. and Bhati, P.S. (2005) Induced variability in cluster bean (*Cyamopsis tetragonoloba*). *Journal of Arid Legumes* 2, 282–286.

Mahla, H.R., Kumar, D. and Shekhawat, A. (2010) Effectiveness and efficiency of mutagens and induced variability in cluster bean (*Cyamopsis tetragonoloba*). *Indian Journal of Agricultural Sciences* 80, 1033–1037.

Mahla, H.R., Sharma, R., Rathore, V.S., Meena, S.C. and Swami, S. (2022) Trait specific genetic stocks for cluster bean [*Cyamopsis tetragonoloba* (L.) Taub.] improvement. *Annals of Arid Zone* 61(3&4), 187–192.

Mathiyazhagan, S., Pahuja, S.K., Ahlawat, A. (2013) Regeneration in cultivated (*Cyamopsis tetragonoloba* L.) and wild species (*C. serrata*) of guar. *Legume Research* 36, 180–187.

Mauria, S. (2000) Isozyme diversity in relation to domestication of guar [*Cyamopsis tetragonoloba* (L.) Taub.]. *Indian Journal of Plant Genetic Resources* 13, 1–10.

Menon, U. (1973) A comparative review on crop improvement and utilisation of cluster bean (*Cyamopsis tetragonoloba* (L.) Taub). Department of Agriculture (Rajasthan), India. *Monograph Series* 2, 1–5.

Mishra, S.K., Singh, N. and Sharma, S.K. (2009) Status and utilisation of genetic resources of arid legumes in India. In: Kumar, D., Henry, A. (eds) *Perspective Research Activities of Arid Legumes in India*. Indian Arid Legumes Society, CAZRI, Jodhpur, India, pp. 23–30.

Mittal, S.P., Dabas, B.S. and Thomas, T.A. (1968) Male sterility in guar (*Cyamopsis tetragonoloba* (L.) Taub.). *Current Science* 37, 357.

Modi, A.R., Patil, G., Kumar, N., Singh, A.S. and Subhash, N. (2012) A simple and efficient *in vitro* mass multiplication procedure for *Stevia rebaudiana* Bertoni and analysis of genetic fidelity of *in vitro* raised plants through RAPD. *Sugar Tech* 14, 391–397.

Morris, J.B. (2010) Morphological and reproductive characterisation of guar (*Cyamopsis tetragonoloba* L. Taub.) genetic resources regenerated in Georgia, USA. *Genetic Resources and Crop Evolution* 57, 985–993.

Mudgil, D., Barak, S. and Khatkar, B.S. (2014) Guar gum: processing, properties and food applications-a review. *Journal of Food Science and Technology* 51, 409–418.

Mullainathan, L., Aruldoss, T. and Velu, S. (2014) Cytological studies in cluster bean by the application of physical and chemical mutagens (*Cyamopsis tetragonoloba* L.). *International Letters of Natural Sciences* 16, 35–40.

Naoumkina, M., Torres-Jerez, I., Allen, S., He, J., Zhao, P.X., Dixon, R.A. and May, G.D. (2007) Analysis of cDNA libraries from developing seeds of guar (*Cyamopsis tetragonoloba* (L.) Taub). *BMC Plant Biology* 7(1), 1–12.

Orellana, R.G. (1966) A new occurrence of tobacco ring spot of guar in the United States. *Plant Dissertation* 50, 7–10.

Paran, I. and Michelmore, R.W. (1993) Development of reliable PCR-based markers linked to downy mildew resistance genes in lettuce. *Theoretical and Applied Genetics* 85, 985–993.

Paroda, R.S. and Saini, M.L. (1978) Guar breeding. In: Paroda and Arora (eds) Guar: its improvement and management. *Forage Research* 4, 9–39.

Pathak, R. (2015) *Cluster Bean: Physiology, Genetics and Cultivation*. Springer Singapore. pp. 125–143.

Pathak, R. and Roy, M.M. (2015) Climatic responses, environmental indices and interrelationships between qualitative and quantitative traits in clusterbean *Cyamopsis tetragonoloba* (L) Taub. Under arid conditions. *Proceedings of the National Academy of Sciences, India Section B: Biological Sciences* 85, 147–154.

Pathak, R., Singh, M., Henry, A. (2009) Genetic divergence in cluster bean (*Cyamopsis tetragonoloba* (L.) Taub.) for seed yield and gum content under rainfed conditions. *Indian Journal of Agricultural Sciences* 79, 559–561.

Pathak R., Singh, M. and Henry, A. (2011a) Genetic diversity and interrelationship among cluster bean genotypes for qualitative traits. *Indian Journal of Agricultural Sciences* 81(5), 402–406.

Pathak, R., Singh, M., Singh, S.K. and Henry, A. (2011b) Genetic variation in morphological traits and their interrelationships with gum content and yield in cluster bean. *Annals of Arid Zone* 50(1), 77–79.

Pathak, R., Singh, S.K. and Singh, M. (2011c) Assessment of genetic diversity in cluster bean based on nuclear rDNA and RAPD markers. *Journal of Food Legumes* 24, 180–183.

Poehlman, J.M. and Sleper, D.A. (1995) *Breeding Field Crops*. Iowa State University Press, Ames.

Powers, T.O., Todd, T.C., Burnell, A.M., Murray, P.C.B., Fleming C.C. *et al*. (1997) The rDNA internal transcribed spacer region as a taxonomic marker for nematodes. *Journal of Nematology* 29, 441–450.

Prem, D. (1999) Callus induction, regeneration and *Agrobacterium* mediated transformation of clusterbean (*Cyampsis tetragonologba* L. Taub.) MSc Thesis, CCS Haryana Agricultural University, Hisar, India.

Prem, D., Singh, S., Gupta, P.P., Singh, J. and Yadav, G. (2003) High-frequency multiple shoot regeneration from cotyledonary nodes of guar (*Cyamopsis tetragonoloba* L. Taub). *In Vitro Cellular & Developmental Biology-Plant* 39, 384–387.

Prem, D., Singh, S. and Gupta, P.P. (2005) Callus induction and *de novo* regeneration from callus in guar (*Cyamopsis tetragonoloba*). *Plant Cell Tissue Organ Culture* 80, 209–214.

Rai, P.S. and Dharmatti, P.R. (2013) Genetic divergence studies in cluster bean (*Cyamopsis tetragonoloba* (L.) Taub.). *Global Journal of Science Frontier Research* 13, 1–5.

Rao, S. and Rao, D. (1982) Studies on the effect of X-ray-irradiation on *Cyamopsis tetragonoloba* (L.) Taub. *Proceedings of the National Academy of Sciences, India Section B: Biological Sciences* 48, 410–415.

Rao, S.R.M., Murthy, P.K. and Rao, D. (1982) Note on determinate and spreading variants in cluster bean. *Current Science* 51(19), 945–956.

Rawal, H.C., Kumar, S., Mithra, S.V.A., Solanke, A.U., Nigam, D. *et al.* (2017) High-quality unigenes and microsatellite markers from tissue-specific transcriptome and development of a database in cluster bean (*Cyamopsis tetragonoloba* L. Taub). *Genes* 8(11), 313.

Rodge, A.B., Sonkamble, S.M., Salve, R.V. and Hashmi, S.I. (2012) Effect of hydrocolloid (guar gum) incorporation on the quality characteristics of bread. *Journal of Food Processing and Technology* 3(2), 1–7.

Roy, A.K., Agrawal, R.K., Ahmad, S., Kumar, R.V., Mall, A.K. *et al.* (2018) Database of forage crop varieties: 2018. AICRP on Forage Crops and Utilization, ICAR-IGFRI, Jhansi, Uttar Pradesh, India.

Sabahelkheir Murwan, K., Abdalla Abdelwahab, H. and Nouri Sulafa, H. (2012) Quality assessment of guar gum (endosperm) of guar (*Cyamopsis tetragonoloba*). *ISCA Journal of Biological Sciences* 1(1), 67–70.

Saleem, M., Shah, S.A.H. and Akhtar, L.H. (2002) BR-99 a new guar cultivar released for general cultivation in Punjab Province. *Asian Journal of Plant Sciences* 1(3), 266–268.

Saini, M.L., Arora, R.N. and Paroda, R.S. (1981) Morphology of three species of genus *Cyamopsis*. *Guar Newsletter* 2, 7–11.

Saini, M.L., Singh, J.V. and Jhorar, B.S. (1990) Guar. *Agricultural Science Digest* 10, 113–116.

Sandhu, H.S. (1988) Interspecific hybridisation studies in genus *Cyamopsis*. PhD thesis, CCS Haryana Agricultural University, Hisar, India.

Seednet (2023) Seednet India Portal. Available at: https://seednet.gov.in (accessed 15–18 September 2023).

Shah, K.P., Kathiria, K.B. and Kumar, S. (2015) Development of SCAR marker linked to sex determination locus in *Trichosanthes dioica*. *Molecular Plant Breeding* 6(15), 1–6.

Sharma, P., Kumar, V., Raman, K.V. and Tiwari, K. (2014a) A set of SCAR markers in cluster bean (*Cyamopsis tetragonoloba* L. Taub) genotypes. *Advances in Bioscience and Biotechnology* 5(2), 131–141.

Sharma, P., Sharma, V. and Kumar, V. (2014b) Genetic diversity analysis of cluster bean [*Cyamopsis tetragonoloba* (L.) Taub] genotypes using RAPD and ISSR markers. *Journal of Agricultural Science and Technology* 16(2), 433–443.

Sharma, S., Rustgi, S., Balyan, H.S. and Gupta, P.K. (2002) Internal transcribed spacer (ITS) sequences of ribosomal DNA of wild barley and their comparison with ITS sequences in common wheat. *Barley Genetics Newsletter* 32, 38–45.

Sheikh, W.A., Dedhrotiya, A.T., Khan, N., Gargi, T., Patel, J.B. and Acharya, S. (2015) An efficient *in vitro* regeneration protocol from cotyledon and cotyledonary node of cluster bean (*Cyamopsis tetragonoloba* L. Taub). *Current Trends in Biotechnology and Pharmacy* 9(2), 175–181.

Singh, C., Singh, P. and Singh, R. (2009) Cluster bean [Guar] (*Cyamopsis tetragonoloba* L.). In: *Modern Techniques of Raising Field Crops* (2nd edn). Oxford & IBH Publishing Co. Pvt Ltd, New Delhi, pp. 452–457.

Singh, V.P. and Agarwal, S. (1986) Induced high-yielding mutants in cluster bean. *Indian Journal of Agricultural Science* 56(10), 695–700.

Singh, V.P., Yadav, R.K. and Chowdhury, R.K. (1981) Note on the determinate mutant of cluster bean. *Indian Journal of Agricultural Science* 51, 682–683.

Stafford, R.E. (1988) Inheritance of rosette-raceme in guar. *Crop Science* 28, 609–610.

Stafford, R.E. (1989) Inheritance of partial male sterility in guar. *Plant Breeding* 103, 43–46.

Strickland, R.W. and Ford, C.W. (1984) *Cyamopsis senegalensis*: potential new crop source of guaran. *Journal of the Australian Institute of Agricultural Science* 50, 47–49.

Tanwar, U.K., Pruthi, V. and Randhawa, G.S. (2017) RNA-Seq of guar (*Cyamopsis tetragonoloba*, L. Taub.) leaves: *de novo* transcriptome assembly, functional annotation and development of genomic resources. *Frontiers in Plant Science* 8, 91.

Tyagi, A., Nigam, D., Mithra, S.V.A., Solanke, A.U., Singh, N.K., Sharma, T.R. and Gaikwad, K. (2018) Genome-wide discovery of tissue-specific miRNAs in cluster bean (*Cyamopsis tetragonoloba*) indicates their association with galactomannan biosynthesis. *Plant Biotechnology Journal* 16(6), 1241–1257.

Undersander, D.J., Putnam, D.H., Kaminski, A.R. *et al.* (2006) *Alternative Field Crops Manual.* University of Wisconsin-Madison, USA, pp 34–38.

Vasal, S.K. (1998) Hybrid maize technology: Challenges and expanding possibilities for research in the next century. In: Vasal, S.K., Gonzalez, C.F. and Xingming, F. (eds) *Proceedings of the 7th Asian Region Maize Workshop.* 23–27 February, Los Banos, Philippines, pp. 58–62.

Velu, S., Mullainathan, L. and Arulbalachandran, D. (2012) Induced morphological variations in clusterbean (*Cyamopsis tetragonoloba* (L.) Taub). *International Journal of Current Trends in Research* 1(1), 48–55.

Verma, S., Gill, K.S., Pruthi, V. *et al.* (2013) Callus induction and multiple shoot regeneration from cotyledonary nodes in cluster bean [*Cyamopsis tetragonoloba* L Taub]. *Indian Journal of Experimental Biology* 51, 1120–1124.

Vig, B.K. (1965) Effect of a reciprocal translocation on cytomorphology of guar. *Science as Culture* 31, 531–533.

Weixin, L., Anfu, H., Peffley, E.B. and Auld, D.L. (2009) Genetic relationship of guar commercial cultivars. *Chinese Agricultural Science Bulletin* 25(2), 133–138.

Whistler, R.L. and Hymowitz, T. (1979) *Guar: Agronomy, Production, Industrial Use and Nutrition.* Purdue University Press, West Lafayette, Indiana.

White, T.J., Bruns, S.L., Lee, S.B. and Taylor, J.W. (1990) Amplification and direct sequencing of fungal ribosomal RNA genes for phylogenetics. In: Innis, M.A., Geltand, D.H. and Sninsky, T.T. (eds) *PCR Protocols: A Guide to Methods and Applications.* Academic Press, San Diego, p. 315.

6 Rice Bean (*Vigna umbellata* (Thunb.) Ohwi & Ohashi)

Gayacharan[1]*, Swarup K. Parida[2], Amit Kumar Singh[1], Debasis Chattopadhyay[2], D.C. Joshi[3] and Gopal Katna[4]

[1]*ICAR-National Bureau of Plant Genetic Resources, New Delhi, India;* [2]*DBT-National Institute of Plant Genome Research, New Delhi, India;* [3]*ICAR-Vivekananda Parvatiya Krishi Anusandhan Sansthan, Almora, India;* [4]*CSK, Himachal Pradesh Agriculture University, Palampur, India*

Abstract

Rice bean (*Vigna umbellata* (Thunb.) Ohwi & Ohashi) is a legume crop widely distributed throughout South and South-east Asia, and other parts of the world. It is a nutritionally rich legume crop and plays a crucial role in securing food and nutritional requirements in traditional farming systems. Its grains are a rich source of quality protein (18–32%) and minerals such as Ca (68–230 mg/100 g), P (209–370 mg/100 g), Mg (9–16 mg/100 g), K (8–1122 mg/100 g) and Fe (2.61–6.4 mg/100 g). Rice bean grain contains vitamins such as thiamine (0.5–1.09 mg), riboflavin (0.18–0.5 mg) and niacin (2.0–3.6 mg). The crop is grown in diverse agro-climatic conditions by diverse ethnic groups for food, fodder, cover crop, living hedges, etc. More importantly, the species is relatively free from pests and diseases. Therefore, it is being utilized as a donor species in pre-breeding programmes for trait introgression and genetic base broadening of *V. radiata, V. mungo* and *V. angularis.* Even though the crop has several beneficial traits and plays a crucial role in local nutritional and food security, rice bean improvement has been relatively inadequate. Therefore, the crop remains an orphan legume, and the area under crop production has continued to decline amid competition from more profitable similar crops such as mung bean, urd bean and cowpea. Nevertheless, a substantial amount of rice bean crop diversity has been collected and conserved to avoid any risk of losing it. National and international project-based initiatives are reviving, improving and introducing crop cultivation in its original habitats and new areas. Recently, significant genetic and genomic resources have been generated, which will help in crop improvement programmes.

Keywords: Adaptation, antinutritional factors, biochemical composition, breeding programmes, diversity, *ex-situ* collections, germplasm enhancement, pre-breeding

6.1 Introduction

Rice bean (*Vigna umbellata* (Thunb.) Ohwi & Ohashi) is also known as *Azukia umbellata* (Thunb.) Ohwi, *Vigna calcarata* (Roxb) Kurz, *Phaseolus calcaratus* (Roxb), *Dolichos umbellatus* (Thunb), *Phaseolus pubescens* (Blume), *Phaseolus riccardianus* (Ten.), *Phaseolus ricciardus* (Ten.), *Phaseolus torosus* (Roxb.) and *Vigna papuana* (Baker f.). It has several local vernacular names such as climbing mountain bean, mambi bean, oriental bean, chakhawai, masyang, jhilinge, gurous, siltung, naurangi dal, red bean, lazy-man pea, haricot de riz, frijol arroz, judia de arroz,

*Corresponding author: gayacharan@icar.gov.in

© CAB International 2024. *Potential Pulses: Genetic and Genomic Resources* (eds R. Chandora, T. Basavaraja and A. Pratap)
DOI: 10.1079/9781800624658.0006

mambi bean, anipay and bamboo bean. It is consumed as a pulse or vegetable and considered a future crop amid climate change owing to its high range of adaptability and high level of resistance to biotic stresses. Rice bean grains contain 18–32% protein, which is about 60% digestible and are rich in methionine and lysine compared to mung bean and black gram (Sharma *et al.*, 2023). Antinutritional factors such as trypsin inhibitors are lowest, and calcium content is highest when compared with other pulses (Andersen and Chandyo, 2010). Therefore, rice bean is also gaining worldwide attention for its nutritional value, particularly in addressing protein deficiency-related issues. The area under crop cultivation has reduced significantly. The gradual dis-adoption of the crop by farmers is primarily linked to disadvantageous traits of existing rice bean landraces and enhanced competition from crops with similar uses that are agronomically superior (Andersen, 2012). As rice bean cultivation has remained confined to traditional growing areas as part of subsistence farming systems, its produce is primarily used by the producer or sold in the local market. However, very little of the rice bean grain production enters the organized market. Therefore, production statistics are not available. Japan is the leading importer, while Thailand, Myanmar, China and Madagascar are the few exporting countries. Its average annual export from 1998 to 2000 was only around 1100 tons. In India, the area under cultivation is roughly around 15,000 ha (Dua *et al.*, 2009). Another comparatively recent study indicates that in 2019 rice bean was grown on 20,000 ha of land in India (Singh *et al.*, 2020).

6.2 Origin and Distribution of Rice Bean

Rice bean mainly grows in India, Central China and the Indochinese Peninsula. The centre of origin and diversity of the species is the South-east Asia region, and it was probably first domesticated in Thailand (Tomooka *et al.*, 1991; Tomooka and Akimiti, 2009; Tomooka *et al.*, 2011). The wild forms of rice bean in India are widespread from Kerala through the Western and Eastern Ghats to the North-Eastern Himalayas and the north-eastern States (Jain and Mehra, 1978; Bisht *et al.*, 2005;). A significant amount of variability in wild forms is documented in the north-east region and sporadically in the Western Himalayas and the Eastern and Western Ghats (Arora, 1991). *V. umbellata* var *gracilis*, the immediate progenitor species of rice bean, is found from India to Malaysia, the Philippines and the Central China region. Furthermore, it is distributed in diverse geographical areas, and the cultivated landraces are adapted to diverse climatic conditions. Major rice bean-growing countries are India, Nepal, Bhutan, Burma, Thailand, China, Vietnam, Indonesia and Japan. The crop was first introduced into Egypt by the Arabs, along the Eastern Coast of Africa and to the Indian Ocean Islands (Fuller and Boivin, 2009). Recently, the crop has spread to other continents as well, like the USA, Australia, Fiji, south-western Asia and tropical Africa, but it is grown on a very small scale. In India, this crop is primarily grown in the north-eastern Himalayan region, including Assam, Meghalaya, Mizoram, Manipur, Himachal Pradesh, Uttarakhand, and Jammu and Kashmir. In the Eastern Himalayan region, where the climate is moist and subtropical, rice bean has a high degree of variability and is grown in shifting cultivation. Meanwhile, in the Western Himalayas, the weather conditions are dry and temperate, and rice bean has relatively less variability and is grown under terrace farming. To a small extent, it is grown in tribal areas of Odisha, Madhya Pradesh, Chhattisgarh, Kerala, Maharashtra and Gujarat (Fig. 6.1).

6.3 Crop Gene Pool, Evolutionary Relationships and Systematics

Rice bean is a diploid species (2n = 2x = 22) among the five cultivated Asiatic *Vigna* species, mung bean (*V. radiata* (L.) Wilczek), urd bean (*V. mungo* (L.) Hepper), adzuki bean (*V. angularis* (Willd.) Ohwi & Ohashi) and moth bean (*V. aconitifolia* (Jacq.)). The species belongs to the order Fabales (Bromhead), family Fabaceae (Lindl.), sub-family Papillionoideae, genus *Vigna* (Savi) and subgenus *Ceratotropis* (Piper) Verdc. *Vigna umbellata* var. *gracilis* is the immediate progenitor species and is classified in the primary gene pool of the rice bean. *V. umbellata* var. *gracilis* is characterized by small, narrow leaflets, long peduncles, thin stems, freely branching, photoperiod sensitivity, indeterminate growth habits,

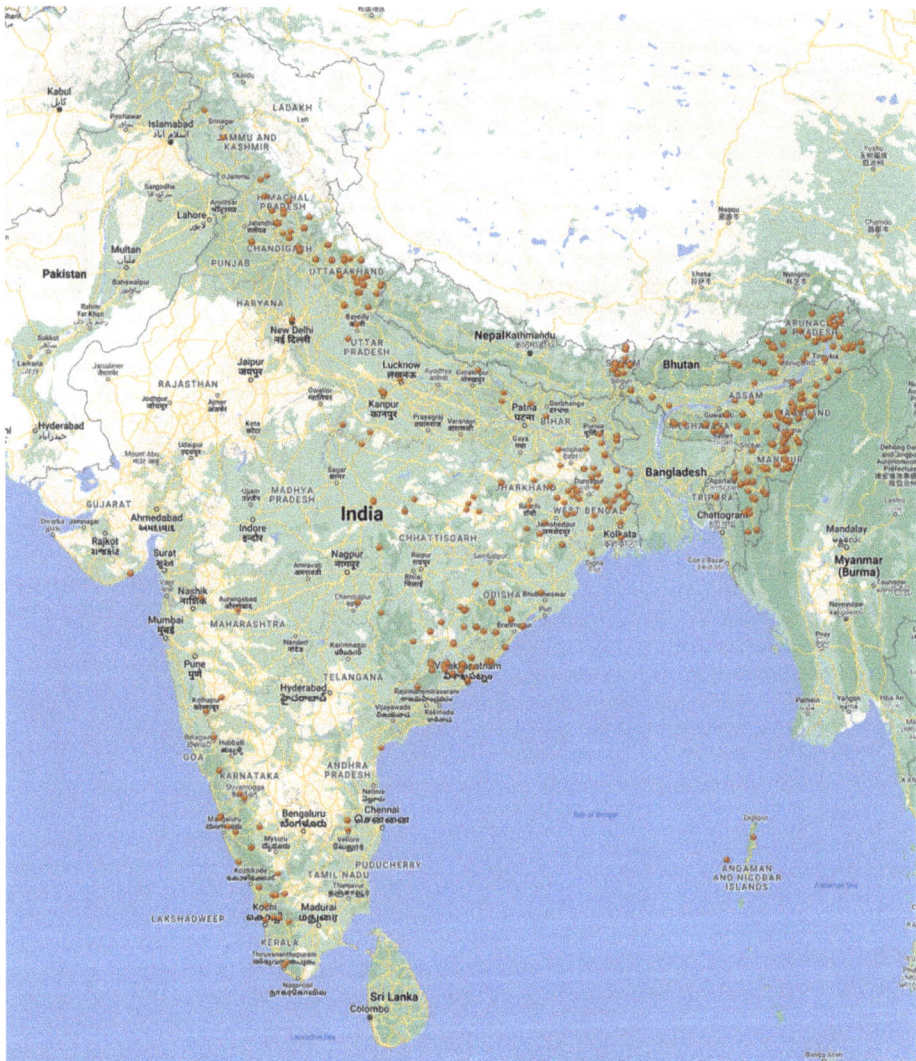

Fig. 6.1. Rice bean collections on map of India. (Author's own figure.)

asynchronous flowering, pod shattering, and small and hard seeds. This taxon resembles *V. minima*. *V. angularis* is another close relative of *V. umbellata*, which can be crossed via embryo culture using *V. umbellata* as the female parent. *V. minima*, a wild species, is even more closely related to *V. umbellata*. The species *V. angularis*, *V. minima*, *V. dalzelliana*, *V. exelis*, *V. hirtella*, *V. nakashimae*, *V. nepalensis*, *V. reflexopilosa*, *V. riukiuensis*, *V. tenuicaulis* and *V. trinervia* make the secondary gene pool of rice bean. *Vigna* species in the tertiary gene pool are *V. mungo*, *V. radiata*, *V. aconitifolia*, *V. trilobata*, *V. aridicola*, *V. grandiflora*, *V. khandalensis*, *V. stipulacea* and *V. subramaniana* (Crop Wild Relatives Portal: https://www.cwrdiversity.org/). The species grows naturally as a perennial herb but is typically grown as an annual crop. Plant growth behaviour is erect, semi-erect or twining (Fig. 6.2). the plant height can range from 0.30 to 3 m. Its tap root has an extensive root system:up to 100–150 cm deep in the soil. Stem, branches, leaves and peduncles

Fig. 6.2. Typical morphological features of rice bean plant at different growth stages. (Author's own figure.)

are covered with fine hair. The stem is branched, and it bears alternate and trifoliate leaflets. The leaflets are broadly ovate to ovate-lanceolate with an area of 5–10 cm × 1.5–6 cm, entire or 2–3 lobed; the lateral leaflets are unequal-sided, membranous and almost glabrous. The stipules are lanceolate, ca. 1.5 cm long, petioles are 5–10 cm long and stipels (the stipule of a leaflet) are linear-lanceolate with ca. 0.5 cm length. Rice bean bears typical papilionaceous flowers. The species requires a short-day length to produce flowers and pods. The floral formula is · | ·or ↑ ♂ $K_{(5)}$ $C_{1+2+(2)}A_{(9)+1}G_1$, Flowers are zygomorphic, bisexual, borne on 5–10 cm long axillary racemes. Each raceme contains 5–20 flowers with a peduncle up to 20 cm long and a pedicel ca. 5 cm long. The plant germination is hypogeal in nature.

6.4 Common Uses

Underutilized or orphan crops are often linked to food and nutritional security, especially because of their richness in micronutrient content (Fassil *et al.*, 2000; Johns and Sthapit, 2004). Rice bean is considered as one of the best cost-effective potential options for reducing malnourishment while ensuring food security and improving livelihood in subsistence farming systems (Khadka

and Acharya, 2009). The crop is an integral part of cultural and religious ceremonies in traditional rice bean growing areas. In local growing areas of Nepal, its grains are offered to deities in several traditional rituals and used in marriage ceremonies. All parts of the rice bean plant are edible and used in food preparations. Its grains, which are the primary product of the crop, are used for making various local dishes. Its grains are used in making almost all dishes for which mung bean and urd bean are used owing to a similar taste. In South India, rice bean grains are used as a substitute for urd bean for making South Indian dishes such as Idli, Vada, Dosa, and so on. The matured grains are used for making soup or dal (split grains are boiled in water to make a soup-like dish), which is generally taken with rice. Unlike using other pulses, however, making dal from grains is difficult due to their fibrous mucilage that prevents hulling and separation of the cotyledons. Grain from young pods and sprouted seeds are used as vegetables. There are a variety of dishes made from rice bean. The nutritional significance of rice bean is well recognized in Madagascar. In the country, rice bean seed flour is used to make food and provide a nutritious diet to children. Similarly, immature pods and sprouted seeds are recommended as a nutritious diet in the Philippines. Its seeds are also considered a perfect broiler ration (Martens *et al.*, 2012). As per indigenous knowledge, rice bean genotypes adapted to high altitudes and with yellow-brownish seeds are reported to be more nutritious. As a leguminous crop, it improves soil fertility by adding atmospheric carbon and nitrogen. It reduces soil erosion, particularly in highland areas. Therefore, the rice bean is often grown as a cover crop.

6.5 Germplasm *Ex-situ* Collections

The major *ex-situ* rice bean germplasm collections holder gene banks are ICAR-National Bureau of Plant Genetic Resources, New Delhi, India, which holds over 2027 accessions of rice bean (Gayacharan *et al.*, 2023), the Institute of Crop Germplasm Resources (CAAS, China) with around 1400 accessions, and the Plant Genetic Resources Unit of the NARC, Nepal, maintains around 300 accessions. Genesys, a global gene bank portal (available at: https://www.genesys-pgr.org; accessed 13 May 2024) indicates that more than 1020 accessions are conserved in other gene banks such as the World Vegetable Center, Taiwan, Centro Internacional de Agricultura Tropical (CIAT), Cali Columbia, Institute of Plant Breeding-National Plant Genetic Resources Laboratory, Philippines, Australian Grains Gene bank, Australia, and Embrapa Recursos Genéticose Biotecnologia, Brazil.

6.6 Climatic and Soil Adaptation

Rice bean is adapted to a diverse range of climatic conditions. It grows in humid tropical to subtropical to cool temperate climatic areas in South and South-east Asia. The rice bean genotypes are adapted to enormous altitudes ranging from plains to mid-hills (400–1700 masl) to even higher altitudes (3000 masl). The crop requires an average temperature of 18–30°C for better growth. It thrives well in 25–35°C temperatures, and well-distributed rainfall of 80–150 cm is required during the growing period. As most rice bean genotypes are photo period sensitive, the vegetative phase is prolonged and flower initiation is delayed if grown in the high-temperature conditions of the sub-tropics. Well-drained

loamy to sandy loam and red soils with moderate fertility are considered the best. The optimum pH is 6.8–7.5. Rice bean is highly sensitive to salinity and water logging. Therefore, well-drained field conditions are required, particularly in the seedling stage of the crop. The crop also has a moderate level of tolerance to drought and acidic soils (Chatterjee and Mukherjee, 1979; Mukherjee *et al.*, 1980; Dwivedi, 1996). The current trend of climate change indicates that the crop may not be highly impacted soon because the crop gene pool comprises genotypes from a diverse range of climatic conditions.

6.7 Nutritional Significance

The nutritional composition of rice bean grains closely resembles that of other legumes in the *Vigna* genus across almost all criteria. These grains are recognized for their high levels of protein, minerals and vitamins. Further, protein content varies from 18 to 32%, total mineral (ash) ranges from 3.5 to 4.9% and crude fibre from 3.6 to 5.5% (Khadka and Acharya, 2009). The biological value of rice bean protein is good because it is rich in limiting amino acids, i.e. methionine and tryptophan (Carvalho and Vieira, 1996). Proteins mainly comprise albumins (6.13–7.47%) and globulins (13.11–15.56%). Important minerals are Ca (68–230 mg/100 g), P (209–370 mg/100 g), Mg (9–16 mg/100 g), K (8–1122 mg/100 g) and Fe (2.61–6.4 mg/100 g). Fat content ranges between 1.2% and 2.1% and contains a relatively higher proportion of unsaturated fatty acids, namely linoleic and linolenic acid, which are nutritionally desirable in the diet (Katoch, 2013). Its grain also contains vitamins such as thiamine (0.5–1.09 mg), riboflavin (0.18–0.5 mg), niacin (2.0–3.6 mg) and ascorbic acid in lesser quantities (Institute of Nutrition, Mahidol University, 2014).

A comprehensive biochemical study using a diverse set of 32 rice bean accessions revealed 23.23–27.33% protein and 2.27–16.69% dietary fibre (Sharma *et al.*, 2023). Sucrose is observed to be most abundant (up to 370 mg/100 g) among free sugars. The concentration of essential micronutrients such as iron (3.49–7.46 mg/100 g), zinc (1.90–3.72 mg/100 g) and selenium (0.28–4.48 μg/100 g) varied very significantly among the rice bean germplasm lines.

There have been more biochemical analyses of rice bean indicating similar nutrient compositions (Saikia *et al.*, 1999). A study on rice bean indicated that treatment of rice bean seeds with 100 μmol methyl jasmonate (MeJA) significantly increased the production of secondary metabolites contributing to antioxidant activity in sprouts between the fourth and sixth days of germination (Li *et al.*, 2018).

6.7.1 Antinutritional factors

In rice bean, antinutritional compounds include total phenolics (1.63–1.82%), tannins (1.37–1.55%), condensed tannins (0.75–0.80%), hydrolysable tannins (0.56–0.79%), trypsin inhibitor (245–372 mg/100 g), phytic acid (732–817 mg/100 g) and lipoxygenase activity (703–950 units/mg) (Katoch, 2013). Another study by Sharma *et al.* (2023) on 32 diverse rice bean germplasms revealed a similar range of variation for phytic acid (303–760 mg/100 g), saponin (19–46 mg/g), trypsin inhibitor (309–1076 mg/100 g), and oxalate (219–431 mg/100 g). Antinutritional factors are considered as the major drawback for rice bean acceptability as food. However, these so-called antinutritional factors also play some critical roles, as phenolics act as antioxidants, trypsin inhibitor helps in controlling obesity and metabolic disorders, and stimulate the pancreas for releasing digestive enzymes (Melmed and Bouchier, 1969; Cristina Oliveira de Lima *et al.*, 2019). Phytate is the major (up to 80%) storage form of phosphorus, which helps in the accumulation of phosphorus in grains and chelates the minerals. However, it is also known to reduce/suppress colon cancer (Graf and Eaton, 1993). Saponins, which range from 1.2 to 3.1 mg/100 g of rice bean seed, are known to improve lipid profile, improve capacity to respond to oxidative stress and lower cardiovascular disease incidences (Johns *et al.*, 1999; Simnadis *et al.*, 2015). Another common problem of eating pulses and even rice bean is flatulence owing to the presence of high amounts of oligosaccharides (fructans and non-fructans). In rice bean, oligosaccharides such as raffinose, stachyose and verbascose range from 1.66 to 2.58%, 0.94 to 1.88%

and 0.85 to 1.23%, respectively (Katoch, 2013). Sharma *et al.* (2023) reported oligosaccharides such as raffinose and stachyose in the range of 47–186 mg/100 g and 117–5765 mg/100 g, respectively; however, verbascose was not detected. Flatulence-causing oligosaccharides also help human intestinal immunity by acting as prebiotics. They have nutritional properties owing to their resistance to hydrolysis in the upper part of the gastrointestinal tract, which leads to extensive fermentation in the large bowel. This results in health-beneficial changes in the composition of the colonic microbiome (Ziemer and Gibson, 1998; Rastall *et al.*, 2005). Probably, indigenous knowledge about the nutritional significance of rice bean is well recognized, which is reflected in diverse traditional dishes based on this crop.

6.8 Poor Adoption Attributes of the Species

Rice bean cultivation is not spreading; rather, its cultivation area is steadily decreasing (Andersen, 2012). Past research and crop improvement status indicate that the crop has not received the due attention of breeders and researchers. The poor phenotypic characteristics of rice bean and competition from crops with similar uses are the two primary reasons that have sidelined the rice bean crop from mainstream research and crop improvement programmes. Nevertheless, recently, modern scientific interventions have been applied to explore the genome and genetic resources of rice bean to accelerate rice bean crop domestication (Kaul *et al.*, 2019; Guan *et al.*, 2022; Verma *et al.*, 2022; Francis *et al.*, 2023; Sahu *et al.*, 2023; Sharma *et al.*, 2023). The crop suffers drawbacks of pod shattering, asynchronous maturity, indeterminate growth, late maturity and high levels of antinutritional compounds (phenolic compounds, phytic acids, saponins, oligosaccharides, enzyme inhibitors and lipoxygenase), which are the major disadvantageous factors for its poor adoption in new areas and dis-adoption in its traditional growing areas. Despite the rice bean having nutritionally rich seeds, high yield, less or no fertilizer requirement, longer shelf life, resistance to devastating insects and diseases, better tolerance to storage pests (bruchids), and broader adaptability, the crop remains restricted to a small production area compared to other cultivated *Vigna* taxa. The major hindrance to its slow domestication process and acceptance among farmers and consumers is its plant type and palatability. Almost all the collections are of an asynchronous type and have a seed-shattering problem, making the crop unsuitable for commercial cultivation and even growing crops on a small scale is challenging. Another major problem with the rice bean is it is less palatable than other *Vigna* spp. Poor palatability and comparatively higher amounts of antinutritional factors are the main drawbacks of the crop. Breeding high-yielding varieties utilizing the diversity of rice bean landraces has not made much difference to the crop area under production and yield compared to other cultivated *Vigna* crops. Almost all the rice bean varieties released to date are single-plant selection-based varieties. Probably, the lack of sufficient morphological variability in rice bean collections is another constraint in selecting contrasting parents for bi-parental crosses. Further, rice bean is reported to have a moderately high level of outcrossing (Tian *et al.*, 2013; Guan *et al.*, 2022). The improvement of rice bean and its genetic resource management has become a challenging task.

6.9   Crop Improvement Programmes

Rice bean breeding programmes have not had an encouraging impact on crop improvement, and the crop gene pool remains underutilized. Farmers growing rice bean rely on their landraces because of the unavailability of high-yielding varieties (Joshi *et al.*, 2008). Traits such as early maturity, synchronous flowering, high yielding, and non-shattering are available in rice bean germplasm, and the current breeding programmes should be based on combining these traits in modern cultivars. However, a major drawback of the rice bean crop is that the rice bean gene pool lacks the genotype, having traits such as determinate growth habits and dwarf plant types. This hinders breeders from developing rice bean genotypes suitable for commercial cultivation. Major developments in rice bean are discussed below.

6.9.1   Characterization and evaluation

More than 2000 accessions of rice bean have been evaluated from 1976 to 2009 at ICAR-National Bureau of Plant Genetic Resources under the All India Coordinated Research Network (AICRN) on Underutilized Crops Reports. Further detailed characterization and evaluation of entire rice bean *ex-situ* collections at the Indian National Gene Bank has been initiated under a DBT-funded project at ICAR-NBPGR. Most economic traits, such as plant height, yield/plant, pod length, flowering period, biomass, etc., are highly influenced by environmental variation, which discourages crop breeding activities. As a result, a large G×E (genetic × environment) interaction for these traits makes breeding activities difficult (Dobhal and Gautam, 1994; Singh *et al.*, 1998; Shukla *et al.*, 2003). Much variability exists, however, in the cultivated gene pool of rice bean (Fig. 6.2). As per the AICRN trials undertaken during 2006 and 2007, a range of variability was observed for traits such as plant height (57–144 cm), days to flowering (58.5–83.0), days to maturity (77.5–138.2), 100-seed weight (4.42–9.25 g) and grain yield/ha (279–1204 kg) (Gautam *et al.*, 2007). The most significant variation in rice bean is reported for seed coat colour and size (Tomooka and Akimiti, 2009). Heritability for seed weight and seed yield/plant is relatively high (Gupta *et al.*, 2009).

6.9.2   Breeding programmes

Systematic breeding activities in rice bean lag far behind any cultivated *Vigna* species despite the crop having beneficial traits and great yield potential. Breeding programmes in rice bean are only taking place in India, and some progress has been made regarding improved cultivar release. The AICRN on Potential Crops played a critical role in rice bean germplasm characterization, evaluation, multi-location trials and varietal release of grain type. Some crucial varieties released in India suited for hills are PRR-1 (PRR 8801), PRR-2 (PRR 8901), RBL-1, RBL-6, RBL-35, RBL-50 and VRB-3. Few important landraces/ elite lines are BRS-1, BRS-2, Naini, Megha Rumbaija 1 (RCRB 1-6), MNPL-1, MNPL-2, Bidhan-1, KHRB-1, KHRB-3, KRB 1 to 18, LRB

31-5, PDRB-1, JRB JO 5-4, JRB-2 and RBL series. Crop improvement programmes on fodder-type rice bean are part of AICRP on Forage Crops. Bidhan Rice bean-2 (KRB-4), Bidhan Rice bean-3, Bidhan-1 (BC-15/K-1), Jawahar Rice bean-1, RBL-6, Shymalima and Surabhi are the rice bean varieties released mainly for fodder purpose. A mutational breeding approach (using gamma radiation) on three cultivars from Manipur, namely RBM-6, RBM-13 and RBM-31, added some significant variability to the rice bean gene pool. Two dwarf mutants from RBM-6 (50% reduction in plant height relative to parent) and an early maturing mutant of RBM-13 (120 from 154 days of the parent) are in the M_2 generation (Devi and Singh, 2006). Some of the important varieties and elite cultivars of rice bean, along with their yield and specific characteristics, are listed in Table 6.1.

6.9.3 Pre-breeding and genetic enhancement

Although significant progress is lacking in the rice bean improvement programmes, the species is being used for interspecific crossing programmes for the introgression of biotic stress resistance traits such as mungbean yellow mosaic virus (MYMV) resistance. Interspecific hybridization utilizing rice bean and its wild relative *V. minima* was undertaken for the first time to understand the fertility–sterility relationship (Gopinathan and Babu, 1985; Gopinathan *et al.*, 1986). There had been many other such attempts at interspecific hybridization utilizing *V. umbellata* for understanding the status of crossability, broadening the genetic base of other cultivated *Vigna* species and developing an interspecific linkage map (Chowdhury and Chowdhury, 1977; Chen *et al.*, 1983; Somta *et al.*, 2006; Ujianto *et al.*, 2019). *V. umbellata* is known to have high-level resistance to important biotic stresses such as MYMV disease and bruchid storage pests (Pandiyan *et al.*, 2008; Sehrawat *et al.*, 2016). Therefore, the species is being utilized to transfer these traits to susceptible cultivated *Vigna* species such as *Vigna radiata* and *Vigna mungo* (Mittal *et al.*, 2008; Pandiyan *et al.*, 2008; Thiyagu *et al.*, 2008; Singh *et al.*, 2013; Singh, 2014; Kaur *et al.*, 2017). However, despite several efforts for trait introgression from

rice bean to mung bean and urd bean, minimal success could be achieved due to the presence of pre-fertilization or post-fertilization barriers. Only a single variety of urd bean, i.e. Mash 114, is reported to have been developed utilzing rice bean as a donor species (Singh *et al.*, 2013).

This indicates that the species is not only challenging to shape for commercial cultivation but also that its genome is not compatible with other cultivated species for use in pre-breeding programmes.

6.8.4 Interventions of modern tools for crop improvement

As the potential crops are highly adapted to traditional growing areas, expanding their production in non-conventional areas becomes challenging. Marred by other disadvantageous factors, as discussed above, the potential crops, such as rice bean, could not gain attention from breeders and researchers who are currently involved in developing climate resilience in crops to sustain and achieve food and nutritional security. Modern tools, particularly genomics, play a crucial role in accelerating crop improvement and attaining higher genetic gain in various legume crops (Bohra *et al.*, 2020; Sinha *et al.*, 2021; Varshney *et al.*, 2021). Genomics has unravelled whole genome information in various legume crops (Gayacharan *et al.*, 2023). Rice bean has been lagging in the area of genomics until recently, with the first *de novo* rice bean genome assembly at the scaffold level (Kaul *et al.*, 2022). The study identified 31,276 genes, including genes associated with some important traits such as late-flowering, palatability, and abiotic and biotic stresses. Guan *et al.* (2022) presented a reference grade rice bean genome assembly of rice bean landrace FF25, along with sequencing 440 collections of rice bean landraces to assess the phylogenetic position and speciation time of rice bean in relation to other *Vigna* species. The assembly covered 90.49% (475.64 Mb) of the estimated total rice bean genome size in 351 contigs, with an N50 of 18.26 Mb. Annotation of the FF25 genome resulted in the identification of 26,736 protein-coding genes, with an average gene density of one gene per 17.79 Kb. Further, genome-wide association studies (GWAS) revealed loci associated with flowering,

Table 6.1. Rice bean varieties released in India for grain and fodder purposes. (Author's own table.)

Variety	Pedigree	Release year	Area recommended	Maturity period (days)	Grain yield (Q/ha)	Green fodder yield (Q/ha)	Specific trait
RBL-1	Pure line selection from Nagour, Rajasthan, during 1976	1986	Punjab region	125–135	16	150–175	Pod length is 10–12 cm, grains are oblong, seeds are large (5 g/100), smooth and light green coloured with good culinary properties.
PRR-1 (PRR 8801)	Pure line selection from Jagdhar (Tehri) collection	1995	Hills of Uttarakhand	111–165	15	–	It is a bluish-black seeded variety. Seeds are medium in size, and 100-seed weight is ~7 g.
PRR-2 (PRR-8901)	Selection from the Dargi collection	1997	Western and Eastern Himalayas	130–140	12	150–175	It is suitable for mid- to high-altitude hilly areas and can be sown in timely conditions. The seed is relatively bold (100 seeds weigh 10 g).
RBL-6	Pure line selection from a landrace of Nagour, Rajasthan	2000	Plain regions of India	115–122	13.3	270	Seed colour is light green. Mature pods are brown to black.
Bidhan rice bean–1 (BC–15/K-1)	Selection from a landrace in Haringhata area, West Bengal	2000	North-eastern region	170–180	20–25	350–400	Seeds are slightly ovate, bold, dirty green in colour and 7–9 seeds/pod. Tolerant to drought and cold.
Konkan Rice bean-1 (RB-10)	Single plant selection from germplasm through multi-environment testing at Dapoli, Maharashtra	1997	Konkan region of Maharashtra	–	–	200–220	It is suitable for green fodder purposes. Protein content 18.37%.
RBL 35	Pure line selection of a germplasm collection from Nagaur, Rajasthan	2003	Plain regions of India	92	15	–	Early maturing type, greenish-brown and 6.2 g per 100-seed weight.
RBL 50	Pure line selection of a germplasm collection from Nagaur, Rajasthan	2003	Plain regions of India	101	15.5	–	Dark brown mature pods, dark green seed, 100-seed weight around 6 g.
Bidhan Rice bean-2 (KRB-4)	Pure line selection of a landrace from Kalimpong, West Bengal	2003	Assam, Bengal, Bihar and Odisha, subtropical to tropical regions	130–150	16–18	290–360	Released for forage purposes. Its pod colour is deep brown to black colour at maturity.

Variety	Development	Year	Region				Remarks
BRS-1	Pure line selection of local germplasm developed by NBPGR, Bhowali	2003	North-west hilly region	140	14.5	–	Early maturing and high seed yield. Released for fodder purposes. The seeds are black in colour, like urd bean.
Naini	Pure line selection from IC26973, a collection from Imphal	–	Central Himalayan region	150	8	–	Bold seed (100-seed weight 10.3 g).
Jawahar Rice bean-1 (JRBJ 05-2)	Selection of a landrace from Dindori, Madhya Pradesh	2011	Specific areas in Madhya Pradesh and Chhattisgarh	–	5–6	280–310	Seed colour is brown and mottled. It is a semi-erect variety for green fodder purposes.
Shymalima	Selection of a landrace from Manipur	2011	Plain and hill zone of Assam	–	14–15	300–310	It is also drought and cold-tolerant. Released for fodder purposes.
VRB-3	Single plant selection of a collection (IC538080) from Nainital, Uttarakhand	2012	North-west and North-east hill regions	130	17.08	170	Seeds are bold (7.56 g/100) and light green in colour, and farmers like them because they resemble mung bean.
Surabhi	Selection of KRB 6	2013	Uplands and homesteads of three southern districts of Kerala (Trivandrum, Kollam and Pathanamthitta)	–	–	380	Released for fodder purposes. It has 18.9% crude protein content and 20% crude fibre.
RBHP-43	Selection from the local germplasm	2015	Himachal Pradesh	–	–	–	It is a high-yielding variety with medium duration suitable for low and mid hills.
Bidhan Rice bean 3 (BRB-3/ KRB-19)	Selection from germplasm (IC545622) collected from Tengnaupal, Manipur	2016	North-east region, Jharkhand, West Bengal, Odisha, Assam, Manipur and Kerala	–	9–11	360	Seed colour is brown with black spot mosaic.

yield and adaptation-associated traits (Guan *et al.*, 2022). Moreover, this study identified several significant findings, e.g. copy-number-variation for *VumCYP78A6* is found to regulate seed yield traits, a recently acquired InDel in *TFL1* (*TERMINAL FLOWER1*) affecting stem determinacy. The phylogenetic analysis indicated that the rice bean is the sister species of the adzuki bean (*Vigna angularis*), and they diverged 1.75 million years ago (mya) (Guan *et al.*, 2022).

Recently, a high-quality reference grade *de novo* genome assembly of VRB-3, a rice bean cultivar, has been published (Francis *et al.*, 2023). This is the largest chromosome-scale assembly among the *Vigna* genus. The assembly covered approximately 99.5% (619.01 Mb) of the estimated rice bean genome size of 630.5 Mb. This assembly identified 33,004 transcripts using the IsoSeq method, 37,489 protein-coding genes, 669 duplicated syntenic blocks containing 12,912 genes, and several other pieces of information that will be helpful in rice bean crop improvement. This study has also constructed a rice bean pan-genome of 679.32 Mb using the whole genome sequence data (8.31x coverage) of 353 rice bean accessions (Francis *et al.*, 2023). Based on a GWAS using 2,145,937 SNPs on 353 accessions, 241 and 64 genomic loci were linked to 11 agro-morphological and nine nutritional quality traits (Francis *et al.*, 2023). The study also revealed whole genome duplication (WGD) 51.23 mya, which led to the origin of Papilionoideae and divergence between rice bean and adzuki bean 2.54 mya.

The crop is now progressing well in terms of genetic and genomic resources. A transcriptome study using stage-specific tissues from contrasting genotypes identified candidate genes (*PHO1*, *cytokinin dehydrogenase*, A-type cytokinin, and *ARR* response negative regulator) for seed weight and seed size (Verma *et al.*, 2022). A downregulation of this gene activity was observed in small-seeded genotypes of rice bean. Another study on associative transcriptomics involving various GWAS models identified 22 markers on 21 transcripts for seed weight, 26 markers on 22 transcripts for maturity, and 82 markers on 48 transcripts for flowering (Sahu *et al.*, 2023). A study revealed that secretion of citrate from the root apex of rice bean is delayed by several hours in response to high Al concentration in soils through regulation of VuMATE1. This transporter transports

secondary metabolite into the vacuoles (Fan *et al.*, 2014). The study also identified several genes that responded to the mechanism involved. Further, another study revealed that the expression of rice bean multidrug and toxic compound extrusion (VuMATE) protein is controlled by Al, not by Fe, in acid soils, which helps in the adaptation of rice bean in acid soils (Liu *et al.*, 2016). There have been some more molecular studies in rice bean such as understanding heterochromatin distribution (Shamurailatpam *et al.*, 2015), generation of interspecific F_2 population with adzuki bean and linkage mapping using molecular markers (Kaga *et al.*, 2000).

6.9   Cultivation Practices

Rice bean is grown for multiple purposes such as vegetables, pulses, fodder, cover crops and living fences of biological barriers (Joshi *et al.*, 2008). Therefore, cultivation practices vary accordingly. However, some of the most common practices are highlighted here.

6.9.1   Soil and fertilizer requirement

Rice bean is mainly grown in subsistence farming systems by resource-poor farmers on marginal lands. The crop is highly sensitive to saline soils and waterlogging conditions, particularly during the early stage of the plant. The crop can withstand moderate levels of drought and acidic soils. Sandy loam and red soils with moderate fertility and a pH range of 6.8–7.5 are considered the best. Generally, fertilizers are not applied to the crop to avoid heavy vegetative growth and grain yield loss. In the case of rice bean as an intercrop with maize, fertilizers used for maize are enough. For the sole crop, inorganic fertilizers at the rate of 20:40:20 kg NPK/ha or five tons of compost application are advised for better yield.

6.9.2   Sowing time

Because rice bean is a highly sensitive crop for photo period length and atmospheric temperature, crop sowing time is an important factor for desired growth and yield. The optimum sowing time for rice bean crops in the foothills of the Indian Himalayas is the 15–30 May. In the climatic conditions of the north Indian plains, if the purpose of cultivation is green manuring or fodder, early sowing during the middle to end of July is the best time. For seed production of rice bean in central plain zones, the optimum sowing time is the last week of July to the first week of August. However, staking rice bean plants becomes a demanding and challenging task. Therefore, to avoid plant staking, crop sowing can be further delayed by 15–20 days to reduce excessive vegetative growth. Dense planting can compensate for the yield. In this case, the rice bean plant escapes heavy vegetative growth and enters the reproductive phase earlier due to its short-day photo period requirement. This results in less vegetative growth, less twining and high yield.

6.9.3   Seed rate and spacing

Because hard seed is common in most rice bean genotypes, seed soaking for 10–12 h before sowing is recommended. In addition, chemical treatment of concentrated H_2SO_4 for 120 seconds is also effective, promoting better seed germination. Seed rate varies in the range 30–37 kg/ha if the broadcasting method is followed, and 15–22 kg/ha in case of the dibbling method of seed sowing. Row-to-row spacing is kept at 45–60 cm for vining and vigorous types and 30–40 cm for erect types. Plant-to-plant distance of 10–15 cm can be maintained. To avoid high vegetative growth and increase seed yield, cutting off the tip (de-topping) is practised to enhance lateral branching. In sole cropping, vine trailing avoids seed loss through rodents, fungal infestation, proper pod bearing and plant growth.

6.9.4   Inter-culture operations

Pre-emergent application of pendimethalin is advised to control weeds. Subsequently, 30–45 days of crop hoeing and weeding are crucial for crop growth and yield. As rice bean plants are tall (sometimes more than 3 m) and have indeterminate growth habits, plant staking is required. Rice bean plants may rot without staking due to attacks by saprophytic fungi and other fungal diseases. Rodent attacks also cause significant yield loss. In the case of mixed crops or intercropping with maize, sorghum, pigeon pea and so on staking is not required. When rice bean is grown with maize or sorghum, sometimes the tops of rice bean vines are removed so that the maize/sorghum stalk is tall and strong enough to support the rice bean plants. This also results in comparatively higher grain yield.

6.9.5   Harvesting

The crop maturity period varies depending on the climatic conditions, e.g. in temperate high altitudes, it matures in 120–150 days, while in the plains where the temperature is a bit higher throughout the growing period, the crop matures within 75–100 days. Most of the crop gene pool has indeterminate growth habits, resulting in asynchronous maturity and pod shattering during maturity; two to three pod pickings are advised to obtain a better yield. Generally, early-type genotypes are ready along with maize for harvest, but medium to late types require another 3–4 weeks. Plants can be harvested when ~75% of pods turn brown. Harvesting during early morning hours helps in reducing seed loss due to shattering. Productivity is around 1–2 tons/ha depending upon the farmers' management practices and growing climatic conditions.

6.10   Cropping Patterns and Crop Rotations

In Nepal and North and North-east India, the crop is grown as mixed or intercropped with maize, in home gardens, margins of maize fields, rice field bunds, etc. Seeds are generally sown by broadcasting and, for proper intercropping, dibbling is used between maize or sorghum plants. It is reported that intercropping rice bean with maize in a 2:1 ratio results in the best equivalent yield in hills. In peninsular India, growing pigeon pea and rice bean in a 1:2 ratio gives the highest rice bean equivalent yield, with a land equivalent ratio of 1.34 and a benefit–cost ratio of 2.1. The crop is also grown as the sole crop in home gardens or on small pieces of land. Major crop rotations followed in the hills are rice bean–wheat, rice bean–mustard, rice bean–pea, rice bean + finger millet–wheat, rice bean + grain amaranth–wheat, rice bean + other pulses–wheat/barley.

6.11 Future Prospects

The rice bean crop has two major advantageous traits: adaptability to a diverse range of agroclimatic conditions and a high level of biotic stress resistance. These crop qualities are highly relevant to the changing climate. Therefore, it is expected that the crop will play a critical role in food security as well as nutritional security. Further, the rice bean gene pool will play a significant role in the improvement of other important legume crops such as mung bean and urd bean with which rice bean genome is easily crossable, utilizing embryo rescue or growth hormone treatments. Advances in genomics technology, the availability of the whole genome sequence in the near future, and the delineation of marker regions with traits of economic importance such as seed shattering, indeterminate growth, asynchronous maturity, photo period sensitivity and antinutritional compounds will help in the development of ideal rice bean genotypes suitable for commercial cultivation.

References

Andersen, P. (2012) Challenges for under-utilized crops illustrated by rice bean (*Vigna umbellata*) in India and Nepal. *International Journal of Agricultural Sustainability* 10(2), 164–174.

Andersen, P. and Chandyo, R.K. (2010) Orphan crops and nutrition: the potential of rice bean (*Vigna umbellata*) to reduce 'hidden hunger' among rural women in India and Nepal. In: *Conference on International Research on Food Security, Natural Resource Management and Rural Development. ETH, Zurich, 14–16 September.*

Arora, R.K. (1991) Plant diversity in the Indian gene centre. In: Paroda, R.S. and Arora, R.K. (eds) *Plant Genetic Resources Conservation and Management Concepts and Approaches.* IBPGR, New Delhi, pp. 25–54.

Bisht, I.S., Bhat, K.V., Lakhanpaul, S., Latha, M., Jayan, P.K., Biswas, B.K. and Singh, A.K. (2005) Diversity and genetic resources of wild Vigna species in India. *Genetic Resources and Crop Evolution* 52(1), 53–68.

Bohra, A., Chand Jha, U., Godwin, I.D. and Kumar Varshney, R. (2020) Genomic interventions for sustainable agriculture. *Plant Biotechnology Journal* 18(12), 2388–2405.

Carvalho, N.M. de and Vieria, R.D. (1996) Rice bean [*Vigna umbellata* (Thunb.) Ohwi and Ohashi]. In: Nkowolo, E. and Smartt, J. (eds) *Legumes and Oilseeds in Nutrition.* Springer, New York, pp. 222–228.

Chatterjee, B.N. and Mukherjee, A.J. (1979) A new rice bean for fodder and grain pulse. *Indian Farming* 29(5), 29–31.

Chen, N.C., Baker, L.R. and Honma, S. (1983) Interspecific crossability among four species of Vigna food legumes. *Euphytica* 32(3), 925–937.

Chowdhury, R.K. and Chowdhury, J.B. (1977) Intergeneric hybridisation between *Vigna mungo* (L.) Hepper and *Phaseolus calcaratus* Roxb. [bean (Phaseolus), India]. *Indian Journal of Agricultural Sciences* 47, 117–121.

Cristina Oliveira de Lima, V., Piuvezam, G., Leal Lima Maciel, B. and Heloneida de Araújo Morais, A. (2019) Trypsin inhibitors: promising candidate satietogenic proteins as complementary treatment for obesity and metabolic disorders? *Journal of Enzyme Inhibition and Medicinal Chemistry* 34(1), 405–419.

Devi, T.R. and Singh, N.B. (2006) Mutagenic induction of variability and selection in M2 generation of selected rice bean {*Vigna umbellata* (Thunb.) Ohwi and Ohashi} cultivars of Manipur. *Legume Research-An International Journal* 29(2), 150–153.

Dobhal, V.K. and Gautam, N.K. (1994) Stability analysis for yield and component characters in rice bean (*Vigna umbellata*). *Indian Journal of Agricultural Sciences* 64, 237–239.

Dua, R.P., Raiger, H.L., Phogot, B.S. and Sharma, S.K. (2009) Underutilized crops: improved varieties and cultivation practices. All India Coordinated Research Network (Underutilized Crops), National Bureau of Plant Genetic Resources, New Delhi.

Dwivedi, G.K. (1996) Tolerance of some crops to soil acidity and response to liming. *Journal of the Indian Society of Soil Science* 44(4), 736–741.

Fan, W., Lou, H.Q., Gong, Y.L., Liu, M.Y., Wang, Z.Q., Yang, J.L. and Zheng, S.J. (2014) Identification of early Al-responsive genes in rice bean (*Vigna umbellata*) roots provides new clues to molecular mechanisms of Al toxicity and tolerance. *Plant, Cell & Environment* 37(7), 1586–1597.

Fassil, H., Guarino, L., Sharrock, S., BhagMal, Hodgkin, T. and Iwanaga, M. (2000) Diversity for food security: improving human nutrition through better evaluation, management, and use of plant genetic resources. *Food and Nutrition Bulletin* 21(4), 497–502.

Francis, A., Singh, N.P., Singh, M., Sharma, P., Gayacharan *et al.* (2023) The rice bean genome provides insight into Vigna genome evolution and facilitates genetic enhancement. *Plant Biotechnology Journal* 21(8), 1522.

Fuller, D.Q. and Boivin, N. (2009) Crops, cattle and commensals across the Indian Ocean. Current and potential archaeobiological evidence. *Études Océan Indien* 42–43, 13–46.

Gautam, R., Kumar, N., Yadavendra, J.P., Neog, S.B., Thakur, S., Khanal, A. *et al.* (2007) Food Security through Rice Bean Research in India and Nepal (FOSRIN). Report 1. Distribution of rice bean in India and Nepal. FOSRIN.

Gayacharan, Parida, S.K., Mondal, N., Yadav, R., Vishwakarma, H. and Rana, J.C. (2023) Mining legume germplasm for genetic gains: An Indian perspective. *Frontiers in Genetics* 14, 996828.

Gopinathan, M.C. and Babu, C.R. (1986) Meiotic studies of the F1 hybrid between rice bean (*Vigna umbellata*) and its wild relative *V. minima. Genetica* 71(2), 115–117.

Gopinathan, M.C., Babu, C.R. and Shivanna, K.R. (1986) Interspecific hybridisation between rice bean (*Vigna umbellata*) and its wild relative (*V. minima*): fertility-sterility relationships. *Euphytica* 35(3), 1017–1022.

Graf, E. and Eaton, J.W. (1993) Suppression of colonic cancer by dietary phytic acid. *Nutrition and Cancer* 19(1), 11–19.

Guan, J., Zhang, J., Gong, D., Zhang, Z., Yu, Y. *et al.* (2022) Genomic analyses of rice bean landraces reveal adaptation and yield-related loci to accelerate breeding. *Nature Communications* 13(1), 5707.

Gupta, S., Kozak, M., Sahay, G., Durrai, A.A., Mitra, J. *et al.* (2009) Genetic parameters of selection and stability and identification of divergent parents for hybridisation in rice bean (*Vigna et al.*. (Ohwi and Ohashi)) in India. *The Journal of Agricultural Science* 147(5), 581–588.

Jain, H.K. and Mehra. K.L. (1980) Evolution, adaptation, relationships, and uses of the species of Vigna cultivated in India. In: Summerfield, R.J. and Bunting, A.H. (eds) *Advances in Legume Science.* Kew Botanical Gardens, London, pp. 459–468.

Johns, T. and Sthapit, B.R. (2004) Biocultural diversity in the sustainability of developing-country food systems. *Food and Nutrition Bulletin* 25 (2), 143–155.

Johns, T., Mahunnah, R.L., Sanaya, P., Chapman, L. and Ticktin, T. (1999) Saponins and phenolic content in plant dietary additives of a traditional subsistence community, the Batemi of Ngorongoro District, Tanzania. *Journal of Ethnopharmacology* 66(1), 1–10.

Joshi, K.D., Bhandari, B., Gautam, R., Bajracharya, J. and Hollington, P.B. (2008) Rice bean: a multi-purpose, underutilised legume. In: Smartt, J. and Haq, N. (eds) *New Crops and Uses: Their Role in a Rapidly Changing World.* CUC, UK, pp. 234–248.

Kaga, A., Ishii, T., Tsukimoto, K., Tokoro, E. and Kamijima, O. (2000) Comparative molecular mapping in *Ceratotropis* species using an interspecific cross between azuki bean (*Vigna angularis*) and rice bean (*V. umbellata*). *Theoretical and Applied Genetics* 100, 207–213.

Katoch, R. (2013) Nutritional potential of rice bean (*Vigna umbellata*): an underutilised legume. *Journal of Food Science* 78(1), C8–C16.

Kaul, T., Eswaran, M., Thangaraj, A., Meyyazhagan, A., Nehra, M. *et al.* (2019) Rice bean (*Vigna umbellata*) draft genome sequence: unravelling the late flowering and unpalatability related genomic resources for efficient domestication of this underutilized crop. BioRxiv. DOI: 10.1101/816595

Kaul, T., Easwaran, M., Thangaraj, A., Meyyazhagan, A., Nehra, M. *et al.* (2022) De novo genome assembly of rice bean (*Vigna umbellata*)–A nominated nutritionally rich future crop reveals novel insights into flowering potential, habit, and palatability-centric–traits for efficient domestication. *Frontiers in Plant Science* 13, 739654.

Kaur, S., Bains, T.S. and Singh, P. (2017) Creating variability through interspecific hybridisation and its utilisation for genetic improvement in mungbean [*Vigna radiata* (L.) Wilczek]. *Journal of Applied and Natural Science* 9 (2), 1101–1106.

Khadka, K. and Acharya, B.D. (2009) *Cultivation practices of rice bean.* Research and Development (LI-BIRD), Local Initiatives for Biodiversity, Pokhara.

Li, L., Gong, X., Ren, H., Wang, X., He, Y. and Dong, Y. (2018) Increased polyphenols and antioxidant activity of rice bean (*Vigna umbellata* L.) sprouts induced by methyl jasmonate: the promotion effect of methyl jasmonate on rice bean sprouts. *Food Science and Technology* 39, 98–104.

Liu, M., Xu, J., Lou, H., Fan, W., Yang, J. and Zheng, S. (2016) Characterization of VuMATE1 expression in response to iron nutrition and aluminum stress reveals adaptation of rice bean (*Vigna umbellata*) to acid soils through cis regulation. *Frontiers in Plant Science* 7, 511.

Martens, S.D., Tiemann, T.T., Bindelle, J., Peters, M. and Lascano, C.E. (2012) Alternative plant protein sources for pigs and chickens in the tropics–nutritional value and constraints: a review. *Journal of Agriculture and Rural Development in the Tropics and Subtropics* 113(2), 101–123.

Melmed, R.N. and Bouchier, I.A.D. (1969) A further physiological role for naturally occurring trypsin inhibitors: the evidence for a trophic stimulant of the pancreatic acinar cell. *Gut* 10(12), 973–979.

Mittal R.K., Sood, B.C., Sharma, R. and Katna, G. (2008) Interspecific hybridisation and DNA based polymorphism in urdbean (*Vigna mungo*), rice bean (*V. umbellata*) and adzuki bean (*V. angularis*). 3rd Asian Chromosome Colloquium, 1–4 December 2008, Osaka, Japan.

Mukherjee, A.K., Roquib, M.A. and Chatterjee, B.N. (1980) Rice bean for the scarcity period. *Indian Farming* 30, 26–8.

Pandiyan, M., Ramamoorthi, N., Ganesh, S. K., Jebaraj, S., Pagarajan, P. and Balasubramanian, P. (2008) Broadening the genetic base and introgression of MYMV resistance and yield improvement through unexplored genes from wild relatives in mungbean. *Plant Mutation Reports* 2, 33–38.

Rastall, R.A., Gibson, G.R., Gill, H.S., Guarner, F., Klaenhammer, T.R. *et al.* (2005) Modulation of the microbial ecology of the human colon by probiotics, prebiotics and synbiotics to enhance human health: an overview of enabling science and potential applications. *FEMS Microbiology Ecology* 52(2), 145–152.

Sahu, T.K., Verma, S.K., Gayacharan, Singh, N.P., Joshi, D.C. *et al.* (2023) An associative transcriptomics study on rice bean (*Vigna umbellata*) provides new insights into genetic basis and candidate genes governing flowering, maturity and seed weight. *bioRxiv* 2023-02.

Saikia, P., Sarkar, C.R. and Borua, I. (1999) Chemical composition, antinutritional factors and effect of cooking on nutritional quality of rice bean [*Vigna umbellata* (Thunb; Ohwi and Ohashi)]. *Food Chemistry* 67(4), 347–352.

Sehrawat, N., Yadav, M., Bhat, K.V., Sairam, R.K. and Jaiwal, P.K. (2016) Introgression of mungbean yellow mosaic virus resistance in Vigna mungo (L.) Hepper and purity testing of F1 hybrids using SSRs. *Turkish Journal of Agriculture and Forestry* 40(1), 95–100.

Shamurailatpam, A., Madhavan, L., Yadav, S.R., Bhat, K.V. and Rao, S.R. (2015) Heterochromatin distribution and comparative karyo-morphological studies in *Vigna umbellata* Thunberg, 1969 and *V. aconitifolia* Jacquin, 1969 (Fabaceae) accessions. *Comparative Cytogenetics* 9(1), 119.

Sharma, P., Goudar, G., Chandragiri, A.K., Ananthan, R., Subhash, K. *et al.* (2023) Assessment of diversity in anti-nutrient profile, resistant starch, minerals and carbohydrate components in different rice bean (*Vigna umbellata*) accessions. *Food Chemistry* 405, 134835.

Shukla, G.P., Melkania, N.P., Singh, K., Rajpali, S.K. and Arya, O.N. (2003) Genotype x location environment interaction in rice bean [*Vigna umbellata* (Thunb.) Ohwi and Ohashi] with special reference to acidic soil. *Forage Research* 28,194–200.

Simnadis, T.G., Tapsell, L.C. and Beck, E.J. (2015) Physiological effects associated with quinoa consumption and implications for research involving humans: a review. *Plant Foods for Human Nutrition* 70(3), 238–249.

Singh, B. (2014) Exploitation of interspecific hybridization for genetic improvement in mungbean [*Vigna radiata* (L.) Wilczek]. MSc thesis, Punjab Agricultural University, Ludhiana, Punjab.

Singh, G., Chaudhary, B.S. and Singh, S.P. (1998) Stability analysis of some agro-morphological characters in rice bean. *Annals of Agricultural Research* 19, 411–414.

Singh, I., Sandhu, J.S., Gupta, S.K. and Singh, S (2013) Introgression of productivity and other desirable traits from rice bean (*Vigna umbellata*) into black gram (*Vigna mungo*). *Plant Breeding* 132(4), 401–406.

Singh, M., Rundan, V. and Onte, S. (2020) Rice bean: High valued fodder crop. *Indian Farming* 70(6).

Sinha, P., Singh, V.K., Bohra, A., Kumar, A., Reif, J.C. and Varshney, R.K. (2021) Genomics and breeding innovations for enhancing genetic gain for climate resilience and nutrition traits. *Theoretical and Applied Genetics* 134(6), 1829–1843.

Somta, P., Kaga, A., Tomooka, N., Kashiwaba, K., Isemura, T. (2006) Development of an interspecific Vigna linkage map between *Vigna umbellata* (Thunb.) Ohwi & Ohashi and *V. nakashimae* (Ohwi) Ohwi& Ohashi and its use in analysis of bruchid resistance and comparative genomics. *Plant Breeding* 125(1), 77–84.

Thiyagu, K., Jayamani, P. and Nadarajan, N. (2008) Pollen pistil interaction in inter-specific crosses of *Vigna* sp. *Cytologia*. 73(3), 251–257.

Tian, J., Isemura, T., Kaga, A., Vaughan, D.A. and Tomooka, N. (2013) Genetic diversity of the rice bean (*Vigna umbellata*) gene pool as assessed by SSR markers. *Genome* 56(12), 717–727.

Tomooka, N. and Akimiti, T. (2009) The origin of rice bean (*Vigna umbellata*) and azuki bean (*V. angularis*): the evolution of two lesser-known Asian beans. In: *An Illustrated Eco-history of the Mekong River Basin, Bangkok*. White Lotus Co. Ltd, Bangkok, Thailand, pp. 33–35.

Tomooka, N., Lairungreang, C., Nakeeraks, P., Egawa, Y. and Thavarasook, C. (1991) *Mungbean and the Genetic Resources, the Subgenus* Ceratotropis. Tropical Agricultural Research Center, Japan.

Tomooka, N., Kaga, A., Isemura T. and Vaughan, D. (2011) Vigna. In: *Wild Crop Relatives: Genomic and Breeding Resources*. Springer, Berlin, pp. 291–311.

Ujianto, L., Basuki, N. and Kasno, A. (2019) Successful Interspecific hybridization between mungbean [*Vigna radiata* (L.) Wilczek] and rice bean [*V. umbellata* (Thunb.) Ohwi & Ohashi]. *Legume Research - An International Journal* 42(1), 55–59.

Varshney, R.K., Bohra, A., Yu, J., Graner, A., Zhang, Q. and Sorrells, M.E. (2021) Designing future crops: genomics-assisted breeding comes of age. *Trends in Plant Science* 26(6), 631–649.

Verma, S.K., Mittal, S., Wankhede, D.P., Parida, S.K., Chattopadhyay, D. *et al*. (2022) Transcriptome analysis reveals key pathways and candidate genes controlling seed development and size in rice bean (*Vigna umbellata*). *Frontiers in Genetics* 12, 791355.

Ziemer, C.J. and Gibson, G.R. (1998) An overview of probiotics, prebiotics and synbiotics in the functional food concept: perspectives and future strategies. *International Dairy Journal* 8(5-6), 473–479.

7 Grass Pea (*Lathyrus sativus*)

P.S. Basavaraj[1]*, Ramya Rathod[2], Krishna Kumar Jangid[1], K.M. Boraiah[1], C.B. Harisha[1], H.M. Halli[1], Kuldeep Tripathi[3] and K. Sammi Reddy[1]
[1]ICAR-National Institute of Abiotic Stress Management, Baramati, India; [2]Professor Jayashankar Telangana State Agriculture University, Rajendranagar, Hyderabad, Telangana, India; [3]ICAR – National Bureau of Plant Genetic Resources, New Delhi, India

Abstract

Grass pea is an important dual-purpose legume crop. It is one of the hardiest legume crops that can survive and yield under extreme conditions such as drought, problematic soils, floods, and resistance to several pests and diseases. Despite its climate resilience and high protein content, the crop has not attained commercial status mainly due to the problematic existence of oxalyldiaminopropionic acid (ODAP), a neurotoxin, in the seeds and fodder that causes neurolathyrism. Several thousands of grass pea collections are available at national and international gene banks. These have been evaluated and characterized for many agronomically important traits and low ODAP content. Recent advances in next-generation sequencing approaches have made it possible to assemble the large-scale genome of grass pea successfully, offering an array of opportunities to develop suitable marker systems for accelerated breeding for desired traits. Globally, efforts are underway for the genetic enhancement of grass pea using modern genomic tools such as molecular markers and functional genomic approaches, and the most recent genome editing tools promise to provide potential solutions to lower the ODAP content and enhance the other economically essential features.

Keywords: Grass pea, *Lathyrus*, orphan legume, ODAP, Khesari

7.1 Introduction

Grass pea (*Lathyrus* spp.) is an underutilized legume crop cultivated for its edible seeds and animal feed. It is also popularly known as kesar in India. In other parts of the world, it is popularly known as Indian vetch, chickling vetch (UK and the USA), Almitra (Italy), guaya (Ethiopia), and san li dow (China). The crop is one of the most neglected but hardiest legumes that can thrive well even under harsher ecosystems due to its adaptation to climatic extremities such as high- or low-water stress and high-temperature stress (Lambein *et al.*, 2019). The crop thrives well during the winter season, having economic benefits such as food and fodder, and also provides ecological benefits through atmospheric N fixation and soil conservation. The crop is widespread in South Asia and sub-Saharan Africa; however, it is endemic in the rest of the world. The crop is cultivated for seed purposes in India, Pakistan and Bangladesh, and for feed and fodder purposes in other countries (Gonçalves *et al.*, 2022). Furthermore, the crop is an abundant protein source, containing about 24–31%.

*Corresponding author: bassuptl@gmail.com

© CAB International 2024. *Potential Pulses: Genetic and Genomic Resources*
(eds R. Chandora, T. Basavaraja and A. Pratap)
DOI: 10.1079/9781800624658.0007

South Asian nations often cultivate the crop for food and livestock feed, where *Lathyrus* grains are consumed either as a whole or processed split dal after boiling (Deshpande and Campbell, 1992). Because of its high-value nutritional make-up, there is a greater demand for it as fodder than for human consumption in most of the world (Campbell, 1997). Grass pea can survive and grow under extreme conditions, and it can grow in a region where annual rainfall ranges from 300 mm to 1500 mm (Dixit *et al.*, 2016). Further, by virtue of its root system architecture, it has genetic drought (Talukdar, 2009) and flood tolerance ability (Girma and Korbu, 2012). In addition, it can grow well in diverse types of soils (Dixit *et al.*, 2016).

Grass pea is rich in dense nutrients and, when combined with its ability to withstand harsher climatic conditions, nitrogen fixation capacity and low water requirement, has the potential to significantly reduce global malnutrition. Nevertheless, *Lathyrus* has received very little attention to date, mainly becaue it contains an oxalyldiaminopropionic acid (ODAP); continued ingestion of ODAP causes 'lathyrism', a neurological condition in people and livestock (Hanbury *et al.*, 2000).

The genus *Lathyrus* is very rich in its species diversity, and it includes about 187 species with a cosmopolitan distribution. Genetically, most of the *Lathyrus* species are diploid ($2n = 2x = 14$) in nature, and they exhibit comparable chromosome morphology to *Lathyrus sativus* and *Lathyrus cicer* (Kupicha, 1983). The genus *Lathyrus* was separated into 13 sections based on morphological and other classifications. These sections were confirmed using sequencing data for the internal transcribed spacer region of the chloroplastic DNA (Aci *et al.*, 2020).

The only species of *Lathyrus* cultivated widely is *L. sativus*, with two other species, *L. cicera and L. ochrus*, cultivated to a limited extent (Shehadeh *et al.*, 2013). The national and international gene banks worldwide housed more than 20,379 germplasm accessions of *Lathyrus* species. Their utilization is limited, however, to only a set of accessions exploited for the genetic improvement of *Lathyrus* for agronomically important traits. This indicates there is an opportunity to improve grass pea by exploring gene bank diversity using conventional and modern breeding tools. Details of species diversity, origin, germplasm collection, and characterization using conventional and modern breeding and biotechnological tools are described in the following sections.

7.2   Origin, Distribution and Botanical Description

Grass pea is a primaeval legume crop cultivated for thousands of years. The exact origin of grass pea is uncertain; however, it is thought that the Mediterranean region, which includes parts of Greece and Turkey, is where the grass pea originated. It later spread to other continents (Sarker *et al.*, 2001). The crop has been domesticated for a long time, and trade and migration probably helped its spread to different parts of the world. Grass pea is also thought to have been an essential crop in ancient civilizations such as Egypt and Rome. In addition to being cultivated in China and India in the past, it has been mentioned in historical records dating back to the 5th century BC. Today, grass pea is grown in several countries, including Asia, Africa, Europe and the Americas. The crop is crucial in regions with poor soil or harsh climatic conditions, where it can provide a reliable source of food and nutrition for local populations (Lambein *et al.*, 2019).

Grass pea is widely distributed, especially in hot and dry conditions. The crop can thrive in various soil types and is well suited to drought-prone areas. Furthermore, crops are essential in regions where others may not thrive due to poor soil or unfavourable climatic conditions. The crop is widely grown in Asia, particularly India, Bangladesh and Nepal. It is also cultivated in parts of Africa, including Ethiopia, Sudan and Uganda, as well as in the Mediterranean region, including Turkey, Syria and Greece. Grass pea is also grown in some parts of Europe, such as Spain and Italy. The distribution of grass pea is closely linked to its historical use as a food source in many cultures. It is a traditional crop in many parts of the world and is often consumed as soups, stews and curries. However, the consumption of grass pea is limited in some regions owing to the presence of the toxin ODAP.

According to Vavilov (1951), Central Asia and Abyssinia were the origin centres of grass pea; however, archaeological data show it was first cultivated in the Balkan Peninsula during the early Neolithic era. According to Kislev (1989), it originated in South-west and Central Asia before moving to the Eastern Mediterranean. During an excavation in Israel, charred seeds of a species of *Lathyrus* were discovered; it is thought that these seeds were brought to the Levant from the centre of origin or from Philistine immigrants (Mahler-Slasky and Kislev, 2010). However, it is challenging to pinpoint the origin of *L. sativus* precisely because cultivation has completely masked its natural range. The lack of physical distinctions between wild and domesticated populations is thought to have resulted from the simultaneous use of *L. sativus* as forage and grain in its native habitats. Small-seeded accessions and sub-successions are primitive types with hard seeds, but selection for forage use has developed landraces with large leaves, pods and seeds but limited seed yield in the Mediterranean region. *Lathyrus* species are diverse, with populations in Europe, Asia, North America, South America and East Africa; however, the Mediterranean and Irano-Turanian regions remain the primary centres of diversity of *Lathyrus* species. Although *Lathyrus* is acclimated to moderate climates, it can also be found in tropical Africa at high altitudes. With the exception of occurring in Australia and Antarctica, the genus comprises many restricted endemic species (van de Wouw *et al.*, 2001).

The most extensively grown species is *L. sativus*, mainly for human consumption. Other species, such as *L. cicera*, *L. ochrus*, *L. clymenum*, *L. tingitanus*, *L. latifolius* and *L. sylvestris*, are cultivated for forage and grain (Gurung *et al.*, 2010). In contrast to *L. ochrus*, which is found in Cyprus, Greece, Syria and Turkey, *L. cicera* is extensively dispersed and grown mainly in the region of Greece, Cyprus, Iran, Iraq, Jordan, Spain and Syria. In Greece and the southern USA, species such as *L. hirsutus* and *L. clymenum* are planted for forage or fodder (Sarker *et al.*, 2001). Some species, including *L. latifolius*, *L. odoratus* and *L. sylvestris*, are grown for their decorative qualities.

The genus *Lathyrus* comprises more than 187 species (Chatterjee *et al.*, 2019). Kupicha (1983) taxonomically classified these species into 13 sections (Table 7.1). However, further research using molecular, morphological, biochemical and cytogenetic markers is necessary to fully understand the phylogenetic relationships between the sections and the individual species. Five groups were constituted based on morphology and taxonomy, and they are *Clymenum*, which includes annual species, whereas *Aphaca*, *Nissolia*, *Cicerula* and *Lathyrus* include perennials (Kupicha, 1983).

Lathyrus sativus is an annual legume crop that can reach 30 to 100 cm in height. It has a sprawling or climbing growth habit, with slender, hairy and slightly angled stems. The grass pea has pinnate leaves with 2–4 pairs of leaflets and a terminal tendril. The, smooth leaflets have a lance-like form. The flowers are arranged in racemes, with 2–8 blooms per stem. The flowers are usually purple, but also white, pink or blue. The fruit is a pod that is up to 5–10 cm long and contains 1–6 seeds. Grass pea seeds are typically rounded or flattened and can be brown, yellow, green or black. The seeds have a distinctive shape, with slightly flattened and rounded sides. The seed coat is thick and stiff, making the seeds challenging to cook and digest.

7.3 Crop Gene Pool, Evolutionary Relationships and Systematics

Based on crossability and ease of gene transfer, Harlan and de Wet (1971) presented the idea of a gene pool to classify crops and their ancestral counterparts. The crop gene pool of grass pea includes a diverse range of wild and cultivated varieties that exhibit a wide range of genetic variability (Wang *et al.*, 2015). For breeding programmes aimed at enhancing crop yields, resistance to pests and diseases, and lowering the toxicity of the seeds, the genetic diversity in the gene pool of grass pea is crucial. The *L. sativus* secondary gene pool was recently extended further to include *L. chrysanthus*, *L. gorgoni*, *L. marmoratus* and *L. pseudocicera* by Heywood *et al.* (2007). Members of the tertiary gene pool, including the remaining species in the genus, are discussed in detail by Heywood *et al.* (2007).

Grass pea belongs to the family Fabaceae, one of the largest flowering plants. Within the family, grass pea is classified in the subfamily Papilionoideae, which is the largest subfamily of Fabaceae and includes many economically important

Table 7.1. Sections of the genus *Lathyrus* (Kupicha, 1983; Kenicer *et al.*, 2005, 2009). (Author's own table.)

1. *Notolathyrus*	6. *Neurolobus*	11. *Viciopsis*
2. *Orobus*	7. *Orobon*	12. *Nissolia*
3. *Pratensis*	8. *Lathyrus*	13. *Clymenum*
4. *Aphaca*	9. *Orobastrum*	
5. *Lathyrostylis*	10. *Linearicarpus*	

crop species such as beans, peas and soybeans. The genus *Lathyrus*, to which grass pea belongs, has about 160 species, which are distributed throughout the world. Within the genus, grass pea is classified in the section *Lathyrus*, with other annual or biennial *Lathyrus* species cultivated for food or forage. Phylogenetic analyses based on molecular data have shown that *Lathyrus* is a monophyletic group, meaning that all species within the genus share a common ancestor. Grass pea is believed to have diverged from other *Lathyrus* species about 3 million years ago. Within *Lathyrus*, grass pea is closely related to several other species cultivated for food, including *L. cicera*, *L. hirsutus* and *L. ochrus*. These species are sometimes called 'minor pulses' and are essential sources of protein and other nutrients in many parts of the world. Most *Lathyrus* species are diploid (2n = 14), while others are naturally auto- or polyploid or contain both types. Despite having nuclear DNA contents ranging from 6.9 to 29.2 pg/2C (10.6 and 13.4 pg/2C for *L. sativus* and *L. cicera*, respectively), several species exhibit a comparable chromosome architecture (Ali and Osman, 2020).

7.4 Germplasm Collection, Characterization, Evaluation and Conservation

Germplasm collections are a critical component of agricultural research because they provide a repository of genetic diversity that can be used to develop new crop varieties with desirable traits. Germplasm collections typically consist of seeds or other plant material from different varieties or populations of a given crop species. In the case of grass pea, several germplasm collections have been established to preserve and study the genetic diversity of this vital food

crop. For instance, the International Centre for Agricultural Research in the Dry Areas (ICARDA) maintains a grass pea germplasm collection that includes more than 2000 accessions from different parts of the world. Other institutions, such as the United States Department of Agriculture (USDA), also maintain germplasm collections of grass pea. The genetic diversity of the crop can be studied using these collections to identify traits of interest and develop new grass pea varieties that are better adapted to specific environments or have improved nutritional qualities. For instance, two germplasm accessions, IG64782 and IG65197, were identified as resistant to the *Orobanche crenata* weed after screening more than 285 accessions belonging to more than 13 species of *Lathyrus*. Therefore, these accessions can be used in an introgression programme to develop improved *L. sativus* varieties with *O. crenata* resistance. In addition to their scientific value, germplasm collections have essential cultural and historical significance. Many accessions in grass pea germplasm collections have been collected from traditional farming communities and represent a rich diversity of local landraces cultivated for generations. By preserving these accessions, researchers can also help to conserve traditional farming practices and promote cultural heritage. Different institutes worldwide assembled and maintained *ex-situ* both wild and cultivated *Lathyrus* species (Abdallah *et al.*, 2021). Nearly 50% of *L. sativus* accessions are at CBNPMP in France (4477 accessions) and ICARDA (3600 accessions) (Table 7.2).

In-situ preservation is equally as crucial as the previously discussed *ex-situ* collection. *In-situ* conservation, as defined by Maxted *et al.* (1997), is the site-specific management and monitoring of genetic diversity in naturally existing wild populations within bounds set aside for active, long-term conservation. The native *Lathyrus* populations are susceptible to genetic deterioration or

Table 7.2. *Lathyrus* germplasm collections at various institutes worldwide (Vaz Patto and Rubiales, 2014). (Author's own table.)

Institute	Location	No. of accessions
Conservatoire Botanique Nationaldes Pyrénées et de Midi-Pyrénées (CBNPMP)	France	4477
International Center for Agricultural Research in Dry Areas (ICARDA)	Syria	3600
National Bureau of Plant Genetic Resources (NBPGR)	India	2720
Plant Genetic Resource Centre (PGRC), Bangladesh Agricultural Research Institute (BARI)	Bangladesh	1841
Instituto de Investigacion Agraria (INIA)	Chile	1424
Ustymivka Experimental Station of Plant Production	Ukraine	1215
N.I. Vavilov All-Russian Scientific Research Institute of Plant Industry	Russian Federation	1207
Australian Grains Genebank	Australia	1020
Plant Gene Resources of Canada (PGRC)	Canada	840
Institute of Biodiversity Conservation (IBC)	Ethiopia	586
Germplasm Resource Information Network (GRIN) United States Department of Agriculture (USDA)	USA	505
Leibniz Institute of Plant Genetics and Crop Plant Research (IPK)	Germany	515
Centrode Recursos Fitogenéticos (CRF) Institutonacional de Investigación Tecnologia Agraria Alimentaria (INIA)	Spain	429

even extinction because there have not been many efforts to protect the genetic diversity of the species *in situ* (Maxted and Bennett, 2001).

7.4.1 Characterization

There is a need to characterize the diversity already present in *Lathyrus* to increase the effectiveness of germplasm use. Understanding the various levels of variability present in the *Lathyrus* germplasm will help locate sources for expanding better breeding pools and searching for genes and alleles that have not been utilized in contemporary breeding. Some accessions identified for agronomically important traits are described in Table 7.3.

7.4.2 Morphological diversity

Several authors described that *Lathyrus* species show profound genetic variation at morphological levels in terms of leaf shape and size, flower colour, seed shape and size (Jackson and Yunus, 1984; Hanbury *et al.*, 1999; Tay *et al.*, 2000; Benková and Záková, 2001; Kumari, 2001; Table 7.4).

7.4.3 Diversity at the biochemical and molecular level

It is possible to trace the genetic variety among potential parental materials for future breeding programmes using biochemical and molecular markers (Table 7.5). At the biochemical and molecular levels, *L. sativus* germplasm exhibits a great degree of agronomic and morphological variation. *L. sativus* has significant genetic variation worldwide, as shown by isozymes and molecular markers. These markers are typically quite effective at differentiating between various *L. sativus* genotypes. For instance, markers for the ODAP trait exist among several *Lathyrus* species (Hanbury *et al.*, 2000; Abd El Moneim *et al.*, 2001; Granati *et al.*, 2003; Tadesse and Bekele, 2003; Kumar *et al.*, 2011; Grela *et al.*, 2010, 2012). Several isozyme markers were identified and deployed to distinguish species of *Lathyrus* (Yunus *et al.*, 1991; Tadesse and Bekele, 2001; Gutierrez-Marcos *et al.*, 2006). In addition, seed protein (Przybylska *et al.*, 1998, 2000; Lioi *et al.*, 2011) based and other molecular markers such as random amplified polymorphic DNA (RAPD), inter-simple sequence repeats (ISSRs) and expressed sequence tag (EST)-derived simple sequence repeats (SSRs) were

Table 7.3. Promising accessions for agronomically superior traits. (Modified from Dixit *et al.*, 2016; Das *et al.*, 2021.)

Traits	Accessions and varieties identified
Early maturity	IC-120438, IC-120446, IC-120447, IC-120448, IC-120596, IC-1205O7, IC-120420, EC-200325, EC-209076, EC-208952, RLK-10012, RLK-10013, RLK-10031, RLK-10037, RLK-10050, RLK-10048, RLK-266, RLK-287, BANG-267, BANG-310, IFLA-2475
Less than 0.2% ODAP concentration	BioL-212, BioR-202, BioR-231, Biol-222, BioL-208, LS 157-12, BARI Khesari-1, BARI Khesari-2, BARI Khesari-3, BARI, Khesari-4, BARI Khesari-5 BINA Khesari-1, Ratan, Prateek, P-24, Mahateora, Nirmal, Bidhan, Khesari-1, Wasie
Higher pods per plant (>50)	BioL-212, BioL-239, BioR-234, DL-265, RLK-430, IC-12507, IC-120497, IC-120537, IC-120422, NIC-18768, NIC-18849, NIC-18851, NIC-18890
Higher grain yield per plant (10 g/plant)	JRL-47, Pusa 534, RJK-49, RLK-204, RLK-393, RLK-658, RLK-1009, RLK-1081, IC-120479, IC-120512, IC-120530, IC-120531, IC-120535, NIC-18768, NIC-18890, S-270, P-72, P-176, G2178 in BANG1 and ILG1721, ILG1632, ILG1624, ILG1540 and ILG150
Higher 100-seed weight (12 g/100 seed)	EC-209017, EC-209026, EC-209044, EC-200322, RLK-143, RLK-148, RLK-158, BioL-208, BioR-227, LS-8246, Sel. 505
Resistant to the *O. crenata* weed	IG64782 and IG65197
Nematode resistance (root-knot nematode)	PI 236481, UT2921, I 358879, PI 440462
Cyst	IFLA 347 (partially resistant)
Aschochyta blight resistance	ATC 80878
Thrips resistant	RLK-1, RLK-281, RLK-617, RPL-26, RLK-273-1, RLK-273-3, JRL-6 and JRL-41
Rust resistance	BG-15744 and BG-23505
Downy mildew resistance	RLS-1, RLS-2, JRS-115, JRL-43, and JRL-16

Table 7.4. Phenotypic diversity among *Lathyrus* species for various traits. (Author's own table.)

Traits	Species and distribution	Reference
Flower colour, seed and leaf size	*L. sativus*, wild sp. (worldwide)	Jackson and Yunus (1984)
Plant vigour, time of flowering to end of flowering and podding, physiological maturity, seed weight and yield	*L. sativus*, *L. cicera* (worldwide)	Hanbury *et al.* (1999)
Seed size, shape, colour and days to flowering	*L. sativus* (Chile)	Tay *et al.* (2000)
Time to maturity, plant height, first pod height offsetting, plant dry weight, pods/plant, seeds/plant, seed weight and yield, lodging resistance	*L. sativus* (Slovakia)	Benková and Záková (2001)
Time to flowering, podding and maturity, pods/plant seeds/pod, seed weight and yield	*L. sativus* (India)	Kumari (2001)

Table 7.5. Genetic diversity in *Lathyrus* species at biochemical and molecular markers. (Author's own table.)

Marker type	Species and distribution	References
ODAP (oxalyldiaminopropionic acid)	*L. sativus*, *L. cicera* (worldwide)	Hanbury *et al.* (2000)
	L. sativus (worldwide)	Abd El Moneim *et al.* (2001)
	L. sativus, *L. cicera* (worldwide)	Granati *et al.* (2003)
	L. sativus (Ethiopia)	Tadesse and Bekele (2003)
	L. sativus (worldwide)	Kumar *et al.* (2011)
	L. sativus, *L. cicera* (European)	Grela *et al.* (2010, 2012)
Isozymes	*L. sativus* (worldwide)	Yunus *et al.* (1991)
	L. sativus (Ethiopia)	Tadesse and Bekele (2001)
	L. sativus (worldwide)	Gutierrez-Marcos *et al.* (2006)
Seed storage proteins	*L. sativus*, *L. amphicarpos*, *L. blepharicarpus*, *L. cicera*, *L. gorgoni*, *L. marmoratus* (worldwide)	Przybylska *et al.* (1998, 2000)
	L. sativus (southern Italy)	Lioi *et al.* (2011)
RAPD (random amplified polymorphic DNA)	*L. sativus*, *L. cicera*, *L. latifolius*, *L. ochrus* (worldwide)	Chtourou-Ghorbel *et al.* (2001)
	L. sativus (worldwide)	Barik *et al.* (2007)
ISSRs (inter simple sequence repeats)	*L. sativus*, *L. cicera* (worldwide)	Belaid *et al.* (2006)
	L. sativus (Italy)	Lioi *et al.* (2011)
	L. sativus (Iberian Peninsula)	VazPatto *et al.* (2011)
L. sativus- and *Lotus japonicus*-derived EST-SSR	*L. sativus* (Italy)	Lioi *et al.* (2011)
Medicago truncatula- and *L. sativus*-derived EST-SSR	*L. sativus* (Ethiopia)	Shiferaw *et al.* (2012)

developed and employed in genetic diversity studies among and between *Lathyrus* species (Chtourou-Ghorbel *et al.*, 2001; Belaid *et al.*, 2006; Barik *et al.*, 2007; Lioi *et al.*, 2011; VazPatto *et al.*, 2011; Shiferaw *et al.*, 2012).

7.5   Use of Germplasm in Crop Improvement by Conventional Approaches

L. sativus has a floral biology that favours self-pollination. There have been signs, however, that the species does engage in some outcrossing, which is influenced by genetic or environmental variables. During the past ten years, several plant breeders have expressed concern about the potential natural outcrossing in *L. sativus*.

The critical goals of grass pea breeding are creating high-yielding varieties combined with reduced ODAP (<0.2%), enhancing crop management techniques and improving agronomic performance. These goals have been the rallying cry for all grass pea genetic improvement research worldwide. Many research strategies have been suggested to achieve these goals. The following research directions were proposed by Wuletaw *et al.* (1995): (i) hybridization programmes aimed at toxin-free or low toxin content varieties; (ii) the introduction of low toxin or toxin-free source material from other countries; and (iii) the selection, analysis and assessment of landraces grown by local farmers for their low toxin content and other desirable agronomic features, etc.

Grass pea germplasm can be used in crop improvement programmes to develop new varieties with desirable traits through conventional breeding approaches. Conventional breeding methods rely on natural genetic variation, crossing two or more grass pea varieties to produce offspring with new recombinant traits. Conventional breeding primarily focuses on the hybridization between the best parents and evaluating the offspring for the desired traits. In most breeding programmes, a selection criterion has been the high yield potential. Conversely, some factors that affect yield, such as double podding or more seeds per pod, have not often been used. Therefore, trait-based selections allow ample opportunity to improve the grass pea. To develop low OADP varieties, crossing low ODAP accessions with high-yielding varieties was found to be promising (Campbell, 1997). In India, considerable progress has been made through conventional breeding to develop high-yielding grass pea varieties. Notable examples include Pusa 24, Prateek, Mahateora and Nirmal, which are high-yielders with low ODAP content. Likewise, Bangladesh also made some progress through conventional breeding and released BARI Khesari 1 and BARI Khesari 2 for commercial cultivation. Some examples of how grass pea germplasm can be utilized in crop improvement programmes through conventional breeding approaches follow here.

7.5.1 Hybridization

Although grass pea is predominantly self-pollinated, a certain amount of cross-pollination occurs through bees (Hanbury et al., 1999). About 2–16% outcrossing per generation was reported in the earlier studies (Chowdhury and Slinkard, 1997), indicating that, while hybridizing, care should be taken to avoid outcrossing by other unintended pollens. Hybridization involves crossing two or more genetically distinct lines to produce progeny with a recombination of desirable traits. For example, a variety with high yield potential can be crossed with a variety resistant to disease to develop a new variety that combines features of both parents, such as high yield and disease resistance. Although most efforts are directed towards improving yield and quality, some progress has also been made towards resistance breeding. Extensive screening identified species such as L. ochrus and L. clymenum, L. aphaca and hybrid sweet peas (L. odoratus × L. belinensis) that possess powdery mildew resistance. Similarly, Aschochyta blight resistance has been reported in L. ochrus, L. clymenum and L. sativus.

Selection involves identifying and propagating individual grass pea plants or lines with desired features such as high yield, pests, disease resistance and enhanced nutritional qualities. In the past, significant achievements have been made through simple selection approaches

(Table 7.6). The best example is 'Pusa 24', a landmark variety of grass pea developed from selection in 1966. This variety is high-yielding coupled with low ODAP. Later research led to the development of cultivars with low ODAP concentration (0.2%) that best-fit upland (LSD1, LSD2) and rice fallow (LSD3, LSD6, Pusa-305, and Selection 1276) environments (Gautam et al., 1998). Other notable examples include 'Quilablanco' from Chile, developed in 1983 through selection (Campbell et al., 1993). However, due to ODAP's polygenic inheritance, numerous attempts have been unsuccessful in establishing a correlation between ODAP and easily visible traits for ease of selection (Hanbury et al., 1999).

Backcrossing involves crossing a desired grass pea variety with a related but genetically different variety and then repeatedly crossing the resulting progeny back to the original desired variety. This method is intended to transfer specific traits from one variety to another while maintaining the desired genetic background.

Mutagenesis involves exposing grass pea seeds to irradiation or chemicals to induce random mutations in the DNA. Mutations that produce desirable traits, such as improved yield or drought tolerance, can then be selected and propagated to develop new grass pea varieties. It has been used to increase genetic diversity in

Table 7.6. Improved varieties of grass pea released for cultivation in different countries (modified from Kumar et al., 2013.)

Sl. No	Country	Varieties
1	India	Pusa 24, Prateek, Ratan, Mahateora, Nirmal
2	Bulgaria	Strandja
3	Bangladesh	Bari Khesari 1, Bari Khesari 2, Bari Khesari 3, Bina Khesari 1
4	Australia	Ceora, Chalus
5	Canada	LS 82046
6	Russia	Poltavskaya
7	Kazakhstan	Ali Bar
8	Ethiopia	Wasie
9	Chile	Luanco-INIA
10	Turkey	Gurbuz 2001
11	Nepal	CLIMA 2 pink, 19A, 20B, Bari Khesari 2
12	Poland	Derek, Krab
13	Pakistan	Italian

grass pea to produce zero/low ODAP variants (Talukdar, 2009). Two varieties – Poltavskaya in the former USSR and Bina Khesari 1 in Bangladesh – were created through mutation breeding with gamma rays (250 Gy) and ethyl methane sulphonate (EMS; 0.01%), respectively.

The genetic variability and diversity available in the primary gene pool of *L. sativus* were previously explored through conventional means (Chowdhury and Slinkard, 2000). Pre-breeding and distant hybridization are necessary to introduce beneficial alleles from beyond the primary gene pool and expand the genetic basis of the crop. Successful interspecific crosses have been reported between grass pea and other *Lathyrus* species. The number of species in successful inter-specific crosses has expanded owing to embryo rescue (Addis and Narayan, 2000). For several years, the efforts of ICARDA led to collecting and conservating more than 1555 wild accessions (Kumar *et al.*, 2011). These wild species of *Lathyrus* serve as a treasure trove for many of the agronomically important traits. For instance, the toxin-free gene has been identified in *L. tingitanus* and this species would serve to create toxin-free *Lathyrus* varieties (Dixit *et al.*, 2016). In addition, *L. ochrus*, *L. clymenum* (Sillero *et al.*, 2005) and *L. cicera* harbour gene(s) for tolerance to *Orobanche* and cold tolerance (Fernández and Rubiales, 2010) which can also be utilized in the breeding programme. The outcomes of inter-specific hybridization in grass pea suggest there is potential for *Lathyrus* improvement, particularly for crossable species like *L. cicera* and *L. amphicarpus*, by identifying and transferring desirable features from exotic and wild genotypes.

7.6 Integration of Genomic and Genetic Resources for Crop Improvement

With the development of molecular biology and next-generation sequencing technology in recent years, genomic resources in grass pea have significantly improved. However, limited progress has been made in developing genomic resources owing to large genome sizes and poorly characterized genetic resources (Shiferaw *et al.*, 2012; Kumar *et al.*, 2013). Molecular markers have

been generated and utilized to investigate the genetic diversity and evolutionary relationships within the genus *Lathyrus*. These markers include ISSRs, RAPDs, sequence tagged sites (STSs), restriction fragment length polymorphism (RFLP) and amplified fragment length polymorphism (AFLP) (Croft *et al.*, 1999; Chtourou-Ghorbel *et al.*, 2001; Badr *et al.*, 2002; Skiba *et al.*, 2003; Belaid *et al.*, 2006; Barik *et al.*, 2007;). Only 15 SSR primers were previously reported in *Lathyrus* (Lioi *et al.*, 2011; Shiferaw *et al.*, 2012). Twenty-four grass pea accessions were used to identify more than 300 EST-SSR primer pairs and characterize the locus for size polymorphism (Sun *et al.*, 2012). Only 44 SSR loci were polymorphic, whereas 117 were monomorphic. The study of gene mapping and molecular breeding in grass pea will benefit from and be made easier by this new marker system. Various mapping populations, such as recombinant inbred lines, near isogenic lines and targeting induced local lesions in genomes populations, are crucially needed for trait-marker association and gene inactivation/deletion studies when it comes to plant resources for functional genomic studies. More recently, 42 single nucleotide polymorphisms (SNP) markers have been developed and validated using kompetitive allele-specific PCR (KASP) markers for 43 grass pea accessions (Hao *et al.*, 2017). Emmrich *et al.* (2020) made the draft genome assembly available at a 6.3 Gb size. With recent advances in molecular biology and next-generation sequencing technologies, however, the genome of the Pusa-24 variety of *L. sativus* was sequenced and assembled by Rajarammohan *et al.* in 2023. According to the study, the assembled genome measured 3.80 Gb and had a scaffold N50 of 421.39 Mb. According to BUSCO analysis, 98.3% of the highly conserved Viridiplantae genes were found in the assembly. The *Lathyrus* assembly had 50,106 protein-coding genes and 3.17 Gb (83.31%) of repetitive sequences. Thus, the *Lathyrus* genome assembly described here offers a crucial and reliable platform for numerous genetic and genomic studies in this important legume crop. Several marker systems have been developed and discovered in grass pea.

Researchers started searching for other improvement methods after the traditional breeding strategy failed to produce grass pea with zero or low ODAP concentration. Trends suggest that exploring contemporary technologies such as gene editing, the use of antisense RNA technology, and food processing that involves combining grass pea with other pulses and grains would be worthwhile. The amount of protein in grass pea seeds could be increased because of genomics and genetic engineering breakthroughs. Integrating genomic and genetic resources has revolutionized crop improvement programmes, including those focused on grass pea. By combining these approaches, researchers can identify genes and genetic markers associated with desirable traits and use this information to develop new varieties more efficiently. Some of the ways in which genomic and genetic resources can be integrated for crop improvement in grass pea are described here.

7.6.1 Association mapping

This approach involves identifying associations between genetic markers and specific traits in a population of plants. This can help identify genes or genetic markers associated with specific traits, which can be used in breeding programmes. Because the genome assemblies are in place, developing genome-wide markers and using significant gene bank collection marker-trait association is possible, allowing markers associated with traits of interest to be identified rapidly.

7.6.2 Marker-assisted selection (MAS)

With the availability of genome sequences, it is possible to develop genome-wide markers; these markers help fast-track breeding progress and involve using genetic markers to select plants with desirable traits. By identifying markers associated with the desired trait, researchers can screen many plants more efficiently and select only those with the desired trait.

7.6.3 Genomic selection

This approach involves using genome-wide markers to predict the performance of a plant for a specific trait. Researchers can develop new varieties with desired traits more efficiently by selecting plants with the highest predicted performance. There are ample opportunities to employ genomic selection in *Lathyrus* breeding, especially for traits such as yield and reduced ODAP content.

7.6.4 Gene editing

Gene editing technologies, such as CRISPR-Cas9, can enable precise changes to specific genes in the grass pea genome. This can introduce or eliminate specific traits, such as disease resistance or improved nutritional content. This technology may be the future for developing low ODAP grass pea varieties.

7.7 Erosion of Genetic Diversity from Traditional Areas and Limitations in Germplasm Use

Genetic erosion is a serious concern in many orphan grain legumes, including *Lathyrus*. Loss of genetic diversity is mainly caused by urbanization, industrialization replacement of landraces by modern elite varieties, intensive agricultural practices, overgrazing of livestock, decrease in grasslands, and loss of vegetation in the Mediterranean region, notably sclerophyll evergreen trees and garrigue shrub (Maxted and Bisby, 1986). Although the crop has been traditionally grown in many regions of the world, the erosion of genetic diversity from traditional areas and limitations in germplasm use pose significant challenges to the sustainable development and use of *Lathyrus*.

Serious concern for the loss of diversity is associated with the extensive cultivation of high-yielding elite grass pea varieties by replacing traditional landraces that are adopted locally. This caused the loss of several inimitable and treasured varieties, reducing genetic variability and diversity. Further, other factors, such as deforestation and desertification, are significantly responsible for the rapid decrease in genetic diversity. Moreover, climate change factors also negatively impact many crops, including grass pea, by accelerating the loss of valuable biodiversity.

7.8   Cultivation Practices

Grass pea is a hardy crop grown in various environmental conditions and soil types. It can thrive in areas receiving 380–650 mm of precipitation (Kosev and Vaisileva, 2022).

7.8.1   Time of sowing

The ideal time for sowing grass pea depends on the climate, soil conditions and intended use of the crop. It can be sown in the early spring in areas with cold winters. In warmer regions, the crop can be sown in the autumn, about 6–8 weeks before the first frost date, to allow the plants to establish before winter (Campbell, 1997). The time of sowing for grass pea in India typically depends on the specific region and climate. In general, it can be sown during the winter season, from October to November, in northern and central India, and during the summer season, from February to March, in southern India (Dixit *et al.*, 2016).

7.8.2   Soil preparation

Although grass pea can be grown in various soils, they favour a pH of 6.0 to 7.0, and specifically, loam to clay-loam soils are ideal for growing grass pea (Adal, 2009). The soil should be ploughed and tilled to a fine texture before planting.

7.8.3   Sowing

Seeds should be sown at a depth of 3–5 cm at a seed rate of about 70–80 kg per acre at a depth of around 2–3 cm. Seeds should be sown with a spacing of 10–15 cm between plants and 30–40 cm between rows.

7.8.4   Fertilization

Being a legume crop, grass pea is capable of atmospheric nitrogen fixation through a symbiotic relationship with rhizobia bacteria. However, in areas with low soil fertility, it may be beneficial to apply phosphorus and potassium fertilizers to promote plant growth. Grass pea requires moderate nitrogen, phosphorus and potassium for optimum growth (Brunet *et al.*, 2008). A basal dose of 25 kg N, 50 kg P_2O_5, and 50 kg K_2O per hectare is recommended for optimum yield.

7.8.5   Irrigation

Grass pea requires moderate watering, with regular irrigation during dry periods. It is a drought-tolerant crop, but it requires moisture during the critical stages of growth, such as seedling emergence and pod development. Depending on the rainfall, irrigation should be provided for 15–20 days. It is essential, however, to avoid overwatering, which can lead to waterlogging and root rot.

7.8.6   Weed control

The crop is prone to weed infestation, which can reduce the yield. Manual weeding or the application of herbicides is recommended for effective management. To control weeds, farmers can use mechanical methods such as hoeing or hand-weeding, or they can apply herbicides (Lazányi, 2000).

7.8.7   Pest and disease management

Pests such as aphids, spider mites and pod borers, as well as diseases such as *Ascochyta* blight and *Fusarium* wilt, are the predominant ones affecting grass pea, reducing the economic yield significantly. Integrated pest management strategies, including resistant varieties, crop rotation and biological controls, can help minimize the impact of these pests and diseases (Kumar *et al.*, 2021). Regular monitoring and timely control measures, such as fungicides and insecticides, are essential to minimize yield losses.

7.8.8   Harvesting

Grass pea is typically harvested 90–120 days after planting when the pods have turned brown and dry (Biswas *et al.*, 2019). The pods should be harvested carefully to avoid damaging the seeds, which can be threshed and cleaned for storage or sale.

7.9 Conclusions

Grass pea is an underutilized orphan legume crop important to rural and resource-poor farmers around several parts of sub-Saharan Africa and South Asia. It plays a crucial role in food security in these regions owing to its immense hardiness to drought and flood conditions. Despite this potential, the crop has not gained much significance due to the associated problem of neurolathyrism upon consumption of grains of grass pea for a long period. In the past, moderate success has been achieved in the genetic improvement of *Lathyrus* for agronomically essential traits through conventional breeding approaches. However, recent innovations in molecular biology and advanced breeding tools and techniques open avenues for the rapid improvement of complex traits such as yield and other quality traits. An extensive collection of germplasms is available at different gene banks, and coordinated efforts of interdisciplinary scientists are necessary to characterize and utilize them to their fullest potential.

Furthermore, rapid developments in sensors and high-throughput phenotyping tools and platforms offer the opportunity to screen thousands of lines rapidly and can phenotype challenging traits non-destructively. Recently, with the advent of molecular marker technology, the characterization of gene bank collections using molecular markers led to significant improvements in germplasm enhancement, characterization evaluation and utilization in plant breeding. Using molecular markers also considerably helps in the effective management of plant genetic resources and provides deeper insights into the pattern of variability and diversity between and within accessions; evolution history can be traced easily. In addition, integrating genomics with conventional breeding strategies, particularly comparative genomics, helps to understand the nature/mechanism of resistance to pests and disease, and agronomically important traits. This leads to the identification of genes/alleles and quantitative trait loci (QTLs).

Further identified genes/QTLs can be deployed in the main breeding programme through marker-assisted and genomic selection. It is anticipated that breeding processes can be accelerated by using these genomic approaches to improve the agronomically important traits and low ODAP traits. Increased emphasis should be placed on next-generation breeding and phenotyping tools to improve this crop. In addition, a novel tool such as genome editing offers a potential solution to reduce OADP content in grass pea. Thus, significant scientific investment, increased stakeholder cooperation, and developing national, regional and international synergies are necessary to transform this orphan crop into a staple pulse legume worldwide.

7.10 Future Opportunities and Research Needs

Because there are scale collections of germplasm available in several gene banks, characterization using modern high-throughput phenotyping tools and techniques can save time and provide reliable data for future breeding strategies. Nevertheless, the trait of low ODAP content is a prominent target; therefore, searching for germplasm/genes/alleles should be a research priority. In addition, the novel breeding tool genome editing offers a potential solution to reduce OADP content in grass pea by targeting gene(s) associated with ODAP content.

Acknowledgements

The authors thank ICAR-NIASM for supporting the 'Genetic Garden Project'.

References

Abd El Moneim, A.M., Van Dorrestein, B., Baum, M., Ryan, J. and Bejiga, G. (2001) Role of ICARDA in improving the nutritional quality and yield potential of grass pea (*Lathyrus sativus* L.) for subsistence farmers in dry areas. *Lathyrus Lathyrism Newsletter* 2(2), 55–58.

Abdallah, F., Kumar, S., Amri, A., Mentag, R., Kehel, Z., Mejri, R.K. and Amri, M. (2021) Wild *Lathyrus* species as a great source of resistance for introgression into cultivated grass pea (*Lathyrus sativus* L.) against broomrape weeds (*Orobanche crenata* Forsk. and *Orobanche* foetida Poir.). *Crop Science* 61(1), 263–276.

Aci, M.M., Lupini, A., Badagliacca, G., Mauceri, A., Lo Presti, E. and Preiti, G. (2020) Genetic diversity among *Lathyrus* ssp. based on agronomic traits and molecular markers. *Agronomy* 10(8), 1182. DOI:10.3390/agronomy10081182.

Adal, M. (2009) Phenotypic and symbiotic characterization of grass pea (*Lathyrus sativus*) rhizobial isolates from some major growing areas of South Wollo and West Shoa, Ethiopia. Doctoral dissertation, MSc Thesis, Addis Ababa University, Ethopia.

Addis, G. and Narayan, R.K.J. (2000) Interspecific hybridisation of *Lathyrus sativus* (Guaya) with wild *Lathyrus* species and embryo rescue. *African Crop Science Journal* 8(2), 129–136.

Ali, H.B. and Osman, S.A. (2020) Ribosomal DNA localization on *Lathyrus* species chromosomes by FISH. *Journal of Genetic Engineering and Biotechnology* 18, 1–9.

Arora, R.K., Mathur, P.N., Riley, K.W. and Adham, Y. (1996) *Lathyrus Genetic Resources in Asia: Proceedings of a Regional Workshop, 27–29 December 1995, Indira Gandhi Agricultural University, Raipur, India*. IPGRI Office for South Asia, New Delhi.

Badr, A., Shazly, H.E., Rabey, H.E. and Watson, L.E. (2002) Systematic relationships in *Lathyrus* sect. *Lathyrus* (Fabaceae) based on amplified fragment length polymorphism (AFLP) data. *Canadian Journal of Botany* 80(9), 962–969.

Barik, D.P., Acharya, L., Mukherjee, A.K. and Chand, P.K. (2007) Analysis of genetic diversity among selected grass pea (*Lathyrus sativus* L.) genotypes using RAPD markers. *Zeitschrift für Naturforschung C* 62(11–12), 869–874.

Belaid, Y., Chtourou-Ghorbel, N., Marrakchi, M. and Trifi-Farah, N. (2006) Genetic diversity within and between populations of *Lathyrus* genus (Fabaceae) revealed by ISSR markers. *Genetic Resources and Crop Evolution* 53, 1413–1418.

Benková, M. and Záková, M. (2001) Evaluation of selected traits in grass pea (*Lathyrus sativus* L.) genetic resources. *Lathyrus Lathyrism Newsletter* 2, 27–30.

Biswas, S., Jana, K., Khan, R., Agrawal, R.K. and Puste, A.M. (2019) Periodic dry matter accumulation and crop growth rate of oat and lathyrus as influenced by integrated nutrient management in intercropping systems. *International Journal of Current Microbiology and Applied Sciences* 8, 2675–2686.

Brunet, J., Repellin, A., Varrault, G., Terryn, N. and Zuily-Fodil, Y. (2008) Lead accumulation in the roots of grass pea (*Lathyrus sativus* L.): a novel plant for phytoremediation systems?. *Comptes Rendus Biologies* 331(11), 859–864.

Campbell, C.G. (1997) *Grass Pea,* Lathyrus sativus *L. Promoting the Conservation and Use of Underutilized and Neglected Crops* (Vol. 18). Bioversity International, Rome.

Campbell, C.G., Mehra, R.B., Agrawal, S.K., Chen, Y.Z., Abd El Moneim, A.M. and Khawaja, H.I.T. (1993) Current status and future strategy in breeding grass pea (*Lathyrus sativus*). *Euphytica* 73, 167–175. DOI: 10.1007/BF00027192

Chatterjee, C., Debnath, M., Karmakar, N. and Sadhukhan, R. (2019) Stability of grass pea (*Lathyrus sativus* L.) genotypes in different agroclimatic zone in eastern part of India with special reference to West Bengal. *Genetic Resources and Crop Evolution* 66, 1515–1531.

Chowdhury, M.A. and Slinkard, A.E. (1997) Natural outcrossing in grass pea. *Journal of Heredity* 88, 154–156.

Chtourou-Ghorbel, N., Lauga, B., Combes, D. and Marrakchi, M. (2001) Comparative genetic diversity studies in the genus *Lathyrus* using RFLP and RAPD markers. *Lathyrus Lathyrism Newsletter* 2(2), 62–68.

Croft, A.M., Pang, E.C.K. and Taylor, P.W.J. (1999) Molecular analysis of *Lathyrus sativus* L. (grasspea) and related *Lathyrus* species. *Euphytica* 107, 167–176.

Das, A., Parihar, A.K., Barpete, S., Kumar, S. and Gupta, S. (2021) Current perspectives on reducing the β-ODAP content and improving potential agronomic traits in grass pea (*Lathyrus sativus* L.). *Frontiers in Plant Science* 12, 703275. DOI: 10.3389/fpls.2021.703275

Deshpande, S.S. and Campbell, C.G. (1992) Genotype variation in BOAA, condensed tannins, phenolics and enzyme inhibitors of grass pea (*Lathyrus sativus*). *Canadian Journal of Plant Science* 72(4), 1037–1047.

Dixit, G.P., Parihar, A.K., Bohra, A. and Singh, N.P. (2016) Achievements and prospects of grass pea (*Lathyrus sativus* L.) improvement for sustainable food production. *The Crop Journal* 4(5), 407–416.

Emmrich, P.M., Sarkar, A., Njaci, I., Kaithakottil, G.G., Ellis, N., Moore, C. and Wang, T.L. (2020) A draft genome of grass pea (*Lathyrus sativus*), a resilient diploid legume. *BioRxiv* 2020-04.

FAO (2009) International treaty on plant genetic resources for food and agriculture. FAO, Rome.

Fernández-Aparicio, M, and Rubiales, D. (2010) Characterisation of resistance to crenate broomrape (*Orobanche crenata* Forsk.) in *Lathyrus cicera* L. *Euphytica* 173, 77–84.

Gautam, P.L. Singh, I.P. and Karihaloo, J.L. (1998) Need for a crop network on *Lathyrus* genetic resources for conservation and use. In: Mathur, P.N., Rao, V.R. and Arora, R.K. (eds) *Lathyrus. Proceedings of a IPGRI-ICARDA-ICAR Regional Working Group Meeting, 8–10 December 1998*. Genetic Resources Network, National Bureau of Plant Genetic Resources, New Delhi, pp. 15–21.

Girma, D. and Korbu, L. (2012) Genetic improvement of grass pea (*Lathyrus sativus*) in Ethiopia: an unfulfilled promise. *Plant Breeding* 131(2), 231–236.

Gonçalves, L., Rubiales, D., Bronze, M.R. and Vaz Patto, M.C. (2022) Grass pea (*Lathyrus sativus* L.)—A sustainable and resilient answer to climate challenges. *Agronomy* 12(6), 1324.

Granati, E., Bisignano, V., Chiaretti, D., Crinò, P. and Polignano, G.B. (2003) Characterization of Italian and exotic Lathyrus germplasm for quality traits. *Genetic Resources and Crop Evolution* 50, 273–280.

Grela, E.R., Rybiński, W., Klebaniuk, R. and Matras, J. (2010) Morphological characteristics of some accessions of grass pea (*Lathyrus sativus* L.) grown in Europe and nutritional traits of their seeds. *Genetic Resources and Crop Evolution* 57, 693–701.

Grela, E.R., Rybiński, W., Matras, J. and Sobolewska, S. (2012) Variability of phenotypic and morphological characteristics of some *Lathyrus sativus* L. and *Lathyrus cicera* L. accessions and nutritional traits of their seeds. *Genetic Resources and Crop Evolution* 59, 1687–1703.

Gurung, A.M. and Pang, E.C. (2010) Lathyrus. In: *Wild Crop Relatives: Genomic and Breeding Resources: Legume Crops and Forage*. Springer, Berlin, pp. 117–126.

Gutiérrez-Marcos, J.F., Vaquero, F., de Miera, L.E.S. and Vences, F.J. (2006) High genetic diversity in a world-wide collection of *Lathyrus sativus* L. revealed by isozymatic analysis. *Plant Genetic Resources* 4(3), 159–171.

Hanbury, C.D., Siddique, K.H.M., Galwey, N.W. and Cocks, P.S. (1999) Genotype-environment interaction for seed yield and ODAP concentration of *Lathyrus sativus* L. and *L. cicera* L. in Mediterranean-type environments. *Euphytica* 110, 45–60.

Hanbury, C.D., White, C.L., Mullan, B.P. and Siddique, K.H.M. (2000) A review of the potential of *Lathyrus sativus* L. and *L. cicera* L. grain for use as animal feed. *Animal Feed Science and Technology* 87(1–2), 1–27.

Hao, X., Yang, T., Liu, R., Hu, J., Yao, Y., Burlyaeva, M. and Zong, X. (2017) An RNA sequencing transcriptome analysis of grasspea (*Lathyrus sativus* L.) and development of SSR and KASP markers. *Frontiers in Plant Science* 8, 1873.

Harlan, J.R. and de Wet, J.M. (1971) Toward a rational classification of cultivated plants. *Taxon* 20(4), 509–517.

Heywood, V., Casas, A., Ford-Lloyd, B., Kell, S. and Maxted, N. (2007) Conservation and sustainable use of crop wild relatives. *Agriculture, Ecosystems and Environment* 121(3), 245–255.

Jackson, M.T. and Yunus, A.G. (1984) Variation in the grass pea (*Lathyrus sativus* L.) and wild species. *Euphytica* 33, 549–559.

Kenicer, G.J., Kajita, T., Pennington, R.T. and Murata, J. (2005) Systematics and biogeography of *Lathyrus* (Leguminosae) based on internal transcribed spacer and cpDNA sequence data. *American Journal of Botany* 92, 1199–1209.

Kenicer, G., Nieto-Blásquez, E.M., Mikić, A. and Smýkal, P. (2009) *Lathyrus* – diversity and phylogeny in the genus. *Grain Legumes* 54, 16–18.

Kislev, M.E. (1989) Pre-domesticated cereals in the Pre-Pottery Neolithic A period. *BAR. International Series* (508), 147–151.

Kosev, V. and Vasileva, V. (2022) Assessment of the genetic diversity of a collection of grass pea (*Lathyrus sativus* L.) genotypes. *Agriculturae Conspectus Scientificus* 87(3), 191–199.

Kumar, S., Bejiga, G., Ahmed, S., Nakkoul, H. and Sarker, A. (2011) Genetic improvement of grass pea for low neurotoxin (β-ODAP) content. *Food and Chemical Toxicology* 49(3), 589–600.

Kumar, S., Gupta, P., Barpete, S., Sarker, A., Amri, A., Mathur, P.N. and Baum, M. (2013) Grass pea. In: Singh, M. and Upadhya, H. (eds) *Genetic and Genomic Resources of Grain Legume Improvement*. Elsevier, The Netherlands, pp. 269–305.

Kumar, S., Gupta, P., Barpete, S., Choukri, H., Maalouf, F. and Sarkar, A. (2021) Grass pea. In: *The Beans and the Peas*. Woodhead Publishing, Elsevier, The Netherlands, pp. 273–287.

Kumari, V. (2001) Field evaluation of grasspea (*Lathyrus sativus* L.) germplasm for its toxicity in the northwestern hills of India. *Lathyrus Lathyrism Newsletter* 2, 82–84.

Kupicha, F.K. (1983) The infrageneric structure of *Lathyrus*. *Notes from the Royal Botanic Garden Edinburgh* 41, 209–244.

Lambein, F., Travella, S., Kuo, Y.H., Van Montagu, M. and Heijde, M. (2019) Grass pea (*Lathyrus sativus* L.): orphan crop, nutraceutical or just plain food? *Planta* 250, 821–838.

Lazányi, J. (2000) Grass pea and green manure effects in the Great Hungarian Plain. *Lathyrus Lathyrism Newsletter* 1, 28–30.

Lioi, L., Sparvoli, F., Sonnante, G., Laghetti, G., Lupo, F. and Zaccardelli, M. (2011) Characterization of Italian grasspea (*Lathyrus sativus* L.) germplasm using agronomic traits, biochemical and molecular markers. *Genetic Resources and Crop Evolution* 58, 425–437.

Mahler-Slasky, Y. and Kislev, M.E. (2010) Lathyrus consumption in late Bronze and iron age sites in Israel: an Aegean affinity. *Journal of Archaeological Science* 37(10), 2477–2485.

Maxted, N. and Bisby, F.A. (1986) *IBPGR Final Report - Wild Forage Legume Collection in Syria*. IBPGR, Rome, Italy.

Maxted, N. and Bennett, S. (2001) *Plant Genetic Resources of Legumes in the Mediterranean*. Kluwer Academic Publishers, Dordrecht, The Netherlands.

Maxted, N., Ford-Lloyd, B.V. and Hawkes, J.G. (1997) Complementary conservation strategies. In: *Plant Genetic Conservation: The In Situ Approach*. Chapman and Hall, London, pp. 20–55.

Przybylska, J., Zimniak-Przybylska, Z. and Krajewski, P. (1998) Diversity of seed albumins in the grasspea (*Lathyrus sativus* L.): an electrophoretic study. *Genetic Resources and Crop Evolution* 45, 423–430.

Przybylska, J., Zimniak-Przybylska, Z. and Krajewski, P. (2000) Diversity of seed globulins in *Lathyrus sativus* L. and some related species. *Genetic Resources and Crop Evolution* 47, 239–246.

Rahman, M.M., Kumar, J., Rahman, M.A. and Ali Afzal, M. (1995) Natural outcrossing in *Lathyrus sativus* L. *Indian Journal of Genetics* 13, 204–207.

Rajarammohan, S., Kaur, L., Verma, A., Singh, D., Mantri, S., Roy, J.K. and Kandoth, P.K. (2023) Genome sequencing and assembly of *Lathyrus sativus*-a nutrient-rich hardy legume crop. *Scientific Data* 10(1), 32.

Sarker, A., El Moneim, A.A. and Maxted, N. (2001) Grasspea and chicklings (*Lathyrus* L.). In: Maxted, N. and Bennett, S.J. (eds) *Plant Genetic Resources of Legumes in the Mediterranean*. Springer, Dordrecht, The Netherlands, pp. 159–180.

Shehadeh, A., Amri, A. and Maxted, N. (2013) Ecogeographic survey and gap analysis of *Lathyrus* L. species. *Genetic Resources and Crop Evolution* 60, 2101–2113.

Shiferaw, E.M, Porceddu, E. and Ponnaiah, M (2012) Exploring the genetic diversity of Ethiopian grass pea (*Lathyrus sativus* L.) using EST-SSR markers. *Molecular Breeding* 30, 789–797.

Sillero, J.C., Cubero, J.I., Fernández-Aparicio, M. and Rubiales, D. (2005) Search for resistance to crenate broomrape (*Orobanche crenata*) in *Lathyrus*. *Lathyrus Lathyrism Newsletter* 4, 7–9.

Skiba, B., Ford, R. and Pang, E.C. (2003) Amplification and detection of polymorphic sequence-tagged sites in *Lathyrus sativus*. *Plant Molecular Biology Reporter* 21(4).

Sun, X.L., Yang, T., Guan, J.P., Ma, Y., Jiang, J.Y., Cao, R. and Zong, X.X. (2012) Development of 161 novel EST-SSR markers from *Lathyrus sativus* (Fabaceae). *American Journal of Botany* 99(10), e379–e390.

Tadesse, W. and Bekele, E. (2001) Isozymes variability of grasspea (*Lathyrus sativus* L.) in Ethiopia. *Lathyrus Lathyrism Newsletter* 2, 43–46.

Tadesse, W. and Bekele, E. (2003) Variation and association of morphological and biochemical characters in grass pea (*Lathyrus sativus* L.). *Euphytica* 130, 315–324.

Talukdar, D. (2009) Dwarf mutations in grass pea (*Lathyrus sativus* L.): origin, morphology, inheritance and linkage studies. *Journal of Genetics* 88(2), 165.

Tay, J., Valenzuela, A. and Venegas, F. (2000) Collecting and evaluating Chilean germplasm of grasspea (*Lathyrus sativus* L.). *Lathyrus Lathyrism Newsletter* 1, 21.

van de Wouw, M., Enneking, D., Robertson, L.D. and Maxted, N. (2001) Vetches (*Vicia* L.). In: Maxted, N. and Bennett, S.J. (eds) *Plant Genetic Resources of Legumes in the Mediterranean*. Springer, Dordrecht, The Netherlands, pp. 134–158.

Vavilov, N.I. (1951) Phytogeographic basis of plant breeding - the origin, variation, immunity and breeding of cultivated plants. *Chronica Botanica* 13, 1–366.

Vaz Patto, M.C. and Rubiales, D. (2014) Lathyrus diversity: available resources with relevance to crop improvement–*L. sativus* and *L. cicera* as case studies. *Annals of Botany* 113(6), 895–908.

Vaz Patto, M.C., Fernández-Aparicio, M., Moral, A. and Rubiales, D. (2006) Characterisation of resistance to powdery mildew (*Erysiphe pisi*) in a germplasm collection of *Lathyrus sativus*. *Plant Breeding* 125, 308–310.

Vaz Patto, M.C., Hanbury, C., Van Moorhem, M., Lambein, F., Ochatt, S. and Rubiales, D. (2011) Grass pea (*Lathyrus sativus* L.). In: Perez de la Vega, M., Torres, A.M., Cubero, J.I. and Kole, C. (eds) *Genetics,*

Genomics and Breeding of Cool Season Grain Legumes. Series on Genetics, Genomics and Breeding of Crop Plants. CRC Press, Taylor & Francis, Boca Raton, Florida, pp. 151–204.

Wang, F., Yang, T., Burlyaeva, M., Li, L., Jiang, J., Fang, L. and Zong, X. (2015) Genetic diversity of grass pea and its relative species revealed by SSR markers. *PloS One* 10(3), e0118542.

Wuletaw, T., Melesse, W. and Degago, Y. (1995) Identification of alternatives for grass pea production in Fogera plains of Gondar region. In: Haymanot, R.T. (ed.) *Proceedings of International Conference on Lathyrus and Lathyrism: A Decade of Progress.* Addis Ababa, Ethiopia, pp. 74–75.

Yunus, A.G., Jackson, M.T. and Janet, P.C. (1991) Phenotypic polymorphism of six enzymes in the grass pea (*Lathyrus sativus* L.). *Euphytica* 55, 33–42.

8 Lima Bean (*Phaseolus lunatus* L.)

Carine Nono Temegne[1]*, Isabel Milagros Gavilan Figari[2], Esaïe Tsoata[1], Atabong Paul Agendia[1] and Francis Ajebesone Ngome[3]

[1]*Department of Plant Biology, Faculty of Science, University of Yaoundé I, Yaoundé, Cameroon;* [2]*Technological Institute of Production, Lima, Peru;* [3]*Institute of Agricultural Research for Development, Yaoundé, Cameroon*

Abstract

Lima bean is an orphan legume that is indigenous to South America. Despite its global distribution to hot regions and specific temperate locations, wild crop varieties have been found exclusively on the American continent. This study examines the origins, distribution patterns and botanical characteristics of lima bean to explore the potential influence of evolution on their geographic spread. Our focus relies on the study of plant gene pools, evolutionary relationships and classification systems. Additionally, we prioritize the collection, characterization, evaluation and conservation of germplasm. This analysis emphasizes the utilization of germplasm in plant improvement through conventional methodologies. This study presents a comprehensive analysis of the integration of genomic and genetic resources for the purpose of enhancing crop improvement in lima bean. It also explores the utility of genomic tools in identifying introgression events. This study further looks at the reduction of genetic diversity in traditional areas and the constraints in utilizing germplasm to assess the potential disruption of evolutionary dynamics in ecological adaptability and the widespread utilization of cultivars. It is a clear imperative to prioritize the conservation of wild species and thoroughly document the genetic diversity they possess, particularly in relation to the comprehensive impact of introgressions on the whole genome and their significance for improvement programmes. Furthermore, recent advancements in the scientific understanding of plants, utilizing principles of genomics and agronomy, have resulted in significant modifications to production practices.

Keywords: Breeding scheme, centre of diversity, crop distribution, gene integration, genetics, *Phaseolus* bean

8.1 Introduction

Lima bean (*Phaseolus lunatus* L.), also called Cape pea, sieva bean, Madagascar bean, and Java bean, is innate to South America and distributed in regions of the world with hot climates. It is an herbaceous, climbing legume grown in hot countries for its seeds, which are eaten as a vegetable like the common bean.

There are two distinct forms, with large or small seeds. They are twining, but there are also bushy varieties. They are adapted to hot and humid climates, where they are grown mainly as food crops. When ripe, the seeds contain hydrocyanic acid. They must, therefore, be soaked and refreshed with water to cook them. The pods of colourless-seeded landraces can be fed to goats and sheep. Lima bean is well known

---

*Corresponding author: carine.temegne@facsciences-uy1.cm

in South America for its richness in protein, fibre, vitamins, slow sugars and mineral salts, providing highly nutritious properties (Temegne, 2011; Dohle, 2017; Jackson, 2021). This plant is attractive as fodder and hay when well established, and it can endure austere water deficit (Tsoata *et al.*, 2016, 2017a, b). Growing in arid and often insect-infested land, this plant has naturally developed a stabilizing substance capable of protecting its proteins from desiccation and insectivorous attack. The seeds are astringent and are used as a fever remedy. A decoction of the seeds, green pods and stalks is utilized for the treatment of diabetes. The cosmetic industries utilize it as an equilibrating constituent for proteins and amino acids in the skin (Temegne, 2011).

P. lunatus is a legume with an interesting history and wide distribution. It is thought to have arisen in the neotropics of Central and South America, particularly in the Peruvian and Ecuadorian regions (Gutiérrez *et al.*, 1995; Chacón-Sánchez *et al.*, 2017). *P. lunatus* has been farmed for almost 7000 years, making it one of the most ancient domesticated species in the Americas, according to archaeological data (Smith, 1968). Lima beans were an essential part of the diets and cultures in pre-Columbian civilizations, such as the Inca and Moche (Jackson, 2021).

Due to human migration and trade, the *P. lunatus* crop moved beyond its original location. Lima beans were introduced to many regions around the world, namely Africa, Asia, Europe and the Americas, where they flourished under a variety of climatic conditions and became essential crops in many areas. The USA, Mexico, Guatemala, Brazil, China and Indonesia now dominate in lima bean production.

Due to the nutritional, medicinal and ecological importance and wide distribution of lima bean, this work aims to review: the origin, distribution and botanical description of *P. lunatus*; its gene pool, evolutionary relationships and systematics; *P. lunatus* germplasm collection, characterization, evaluation and conservation and its use in crop improvement; the incorporation of genomic and genetic materials for *P. lunatus* enhancement; and the erosion of *P. lunatus* genetic variability in traditional regions and limitations in germplasm use and its cultivation practices.

8.2 Origin, Distribution and Botanical Description

8.2.1 Origin

The crop commonly referred to as lima bean (*P. lunatus* L.) holds immense economic and sustainable significance on a global scale. This crop belongs to the esteemed Fabaceae family, which also comprises other noteworthy legumes such as common bean and soybean. Owing to its exceptional nutritional value and human consumption, the lima bean is acknowledged as the second most crucial species of the *Phaseolus* genus (Kaplan and Lynch, 1999; Temegne *et al.*, 2020; da Silva *et al.*, 2022).

Their cultivation extends over different areas, and it is fair to say that lima bean is virtually cultivated worldwide. The Latin American region now cultivates 8100 hectares of lima bean annually, with a focus on white large-seeded and white small-seeded types with both determinate and indeterminate growth habits (Adebo, 2023). The evolutionary domestication of lima bean has been recorded in Mesoamerica and in the Andean mountains, occurring during their reproductive isolation caused by the geographical obstacles amongst these genetic reservoirs (Bitocchi *et al.*, 2017).

Phaseolus has been domesticated at least twice, with the first domestication episode occurring in Ecuador and Peru and the second episode in central–west Mexico (Andueza-Noh *et al.*, 2013). The origins of the bean can be traced to archaeological findings in Mexico City, where it is thought to have been cultivated around 5000 years ago (Kaplan, 1965; Salgado *et al.*, 1995). *Phaseolus* is a genus with at least 70 species that naturally develop in tropical and subtropical areas of the Americas, from Sinaloa, Mexico, to Salta, Argentina (Ezeagu and Ibegbu, 2010; Bitocchi *et al.*, 2017; Alcázar-Valle *et al.*, 2021; García *et al.*, 2021; Adebo, 2023). It is an autogamous diploid species (2n = 2x = 22) with a relatively small genome (Bitocchi *et al.*, 2017). The domestication of the plant has been credited to five species that are narrowly related: *Phaseolus vulgaris*, *P. lunatus*, *P. coccineus*, *P. dumosus* (previously known as *P. polyanthus*) and *P. acutifolius*. However, it is noteworthy that, unlike other cultivable species, both *P. vulgaris* and

P. lunatus have undergone two separate domestications (Smartt, 1985; Aragão *et al.*, 2011; Porch *et al.*, 2013).

Since then, it has been cultivated in many parts of the world, encompassing Mexico, Guatemala, Honduras, El Salvador, Nicaragua, Costa Rica, Panama, Colombia, Ecuador, Peru, Bolivia, Brazil, Venezuela, the Caribbean, the USA, and some regions of Africa and Asia (Melgar, 2021). The geographical distribution of lima bean is vast owing to its adaptability to different climatic and edaphic conditions (Martínez-Nieto *et al.*, 2020). The traces of *P. lunatus* have been discovered in Peru, where it is believed they were cultivated more than 4000 years ago (López-Alcocer *et al.*, 2016). In both instances, wild lima bean varieties have been identified that are analogous to currently domesticated cultivated varieties. Owing to its ability to tolerate unfavourable soil and climatic conditions, it has been able to survive and domesticate easily and is regarded as a promising crop (Heredia-Pech *et al.*, 2022).

After the arrival of the Spanish in the Americas throughout the conquest era, *P. lunatus* was taken to Europe. Since then, it has spread throughout the world and has come to be an essential food in many states, especially in Latin America and the Caribbean (Baudoin, 1993). Technically, the scientific name *Phaseolus lunatus* means 'crescent', which corresponds to the shape of the bean. Lima bean represents both war and eternal life (Cueva, 2018). The most crucial assortment of bean germplasm is located at the International Center for Tropical Agriculture (CIAT) in Cali, Colombia, with approximately 26,000 entries (Plucknett *et al.*, 1983).

8.2.2 Distribution

Undoubtedly, the pre-Columbian cultures that flourished in South America showed a preference for large fruits and grains. In Peru, for example, the floury corn of the Urubamba Valley is characterized by the large size of its grains, just like the 'Pallares' (lima bean) from the Ica Valley (Hacquenghem, 1984; Cueva, 2018; Pires *et al.*, 2021). It has undoubtedly been a selection work,

not only because of the size but also because of the shape and colour of the seed (Espinoza *et al.*, 2022).

During the expansion period of the Inca Empire, its territory spanned from Pasto (Colombia) in the north to the Maule River (Chile) in the south. The Argentine regions were part of Tahuantinsuyo, so it is unsurprising that Quechua phonemes, such as 'poroto', were widely used throughout the Southern Cone region of America. Additionally, there is a similarity in the shape, size, and colour of beans in much of South America, clearly distinguished from beans in Central America and Mexico (Kaplan, 1965; Kaplan and Kaplan, 1988).

The small bean grown in the Centre of America, coasts of Mexico, Antilles, Venezuela and Brazil probably has a common origin. The black bean from Guatemala might have been dispersed towards the Isthmus of Tehuantepec and the Yucatan Peninsula and then carried by Carib tribes to Cuba, the Caribbean islands and Venezuela, eventually spreading to Brazil. In pre-Columbian times, these trade routes were considered avenues for the dispersion of lima bean (*P. lunatus*), and the same could be said of common bean. The small red beans were also dispersed through this same route, as evidenced by the Rosinha and Roxao types in Brazil, which are more like the small red beans of Central America than any other type of bean from Mexico or South America (Matsubara and Zúñiga-Dávila, 2015; Martínez-Nieto *et al.*, 2022).

In Mexico, Colombia, Ecuador, Peru and Chile, unlike the situation in Central America, Venezuela and Brazil, farmers and consumers prefer large beans (Reynoso Camacho *et al.*, 2007). In the highland regions of Colombia, people prefer long-shaped red beans, either whole or speckled with cream. However, moving further into the southern parts of Colombia and in the Andean region of Ecuador, the beans become more oval-shaped and lighter tones such as cream, yellow or white predominate over red beans, which, although present, lose importance compared to the lighter tones (Ferrer-Vilca and Valverde-Rodríguez, 2020).

Further south, in the Andean highlands of Peru, the predominance of light-coloured beans over red beans is more evident. Around the 7th degree of latitude south, where the Mochica and Chimú cultures thrived, beans began to be cultivated at sea level. They spread southward through the valleys that intermittently emerge along Peru's desert coasts, and their cultivation reached Chiloé, Chile (Kaplan, 1965). In all these cases, apart from export types, the preferred beans are large or medium grain and light coloured (Espinoza *et al.*, 2022).

The selection of varieties for these conditions has produced beans that are highly restricted in adaptation, maintaining a remarkable resemblance to Andean beans and none to Central American and Caribbean types. In Mexico, small beans are preferred in tropical areas, while medium- and large-sized beans are preferred in the rest of the country. In the high valleys and temperate coasts of the north, large or medium-sized beans are cultivated with light-coloured yellow or cream tones or speckled with shades of light and dark brown (López-Alcocer *et al.*, 2016; Dadther-Huaman *et al.*, 2022).

Throughout the process of *P. lunatus* domestication, there has been an enormous phenotypic variability of shapes, sizes and colours of seeds (Chacón-Sánchez *et al.*, 2021; García *et al.*, 2021). In Latin America, lima bean has developed very definite consumption patterns that seem to derive from the dispersion of certain types of beans that occurred in very remote times (López-Alcocer *et al.*, 2016).

8.2.3 Botanical description

Regarding domesticated species, only *P. lunatus* (Fig. 8.1) belongs to a diversified lineage in South America, different from the rest of the species. In the same group are *P. augusti*, *P. bolivianus*, *P. mollis* and *P. pachyrrhizoides*, mainly found in inter-Andean valleys (Delgado-Paredes *et al.*, 2021). According to Baudoin (1993), temperatures below 14°C have been identified as a limiting factor for wild populations. This study examines various cultigroups within the species *P. lunatus* L., including Cv-gr: huge lime; Cv-gr: Sieva; and Cv-gr: potato (Baudet, 1977).

The systematics of lima bean are:

Kingdom: Plantae
Clade: Tracheophytes
Clade: Angiosperms
Clade: Eudicots
Clade: Rosids
Order: Fabales
Family: Fabaceae
Subfamily: Faboideae
Genus: *Phaseolus*
Species: *P. lunatus*
Binomial name: *Phaseolus lunatus*

Morphologically, it is an herbaceous or shrubby crop that can grow as a bushy or climbing vine, depending on the cultivar and evolving circumstances. It has compound leaves with three leaflets (Smýkal *et al.*, 2015). Leaf shape ranges from lanceolate to ovate, with a medium to dark green tint. The leaves are essential for photosynthesis and contribute to the plant's general growth and development (Smith *et al.*, 1997). The flowers of *P. lunatus* are generally white or pale yellow and are arranged in racemes or clusters. They exhibit bilateral symmetry, which

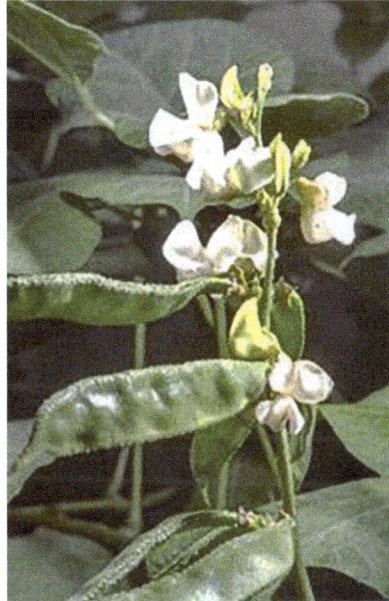

Fig. 8.1. The *Phaseolus lunatus* plant. (Author's own figure.)

is characteristic of legume flowers. The flowers comprise five petals, including a sizeable upper petal known as the standard, two lateral petals called wings, and two lower petals bonded to give the keel. The keel encloses the flower's reproductive organs (Sood and Gupta, 2017). Their blooms produce fruit in the form of pods after pollination. Like the bean, with its pods and seeds larger, kidney-shaped white, cream, with soft brown or greenish spots, it is the second most significant economically important species (López-Alcocer *et al.*, 2016; García *et al.*, 2021). Depending on the cultivar, the pods might be flat or cylindrical (Myers *et al.*, 2022). Regarding the seeds of *P. lunatus*, they are highly variable in size, shape and colour. They are subdivided into Microspermus – sieva or small limes, 1 cm long – and Macruspermus – lima or large limes, 2.5 cm long. According to Sood and Gupta (2017), the length of the pod varies from a few millimetres to several inches. Each pod contains two to four kidney-shaped seeds, constituting the edible part of the *P. lunatus* plant. The shape of the seed varies from flat to round (potato-shaped) and is usually white or cream-coloured, although there are also red, purple, brown, black, green and mottled ones (Matilla *et al.*, 2005; Melgar, 2021). The size of the seeds can also vary with the cultivar, with some generating smaller seeds than others. Variations in size and colour are due to insufficient information about the gene pool, as stated by Cuny *et al.* (2017). Numerous factors (such as genetic, physiological, source-sink and environmental interactions) affect seed development, and the changes that occur cannot be solely linked to the time elapsed after pollination (Ormeño-Orrillo *et al.*, 2006).

8.3   Crop Gene Pool, Evolutionary Relationships and Systematics

The classification and systematic study of lima bean has been influenced by several factors, such as the study of morphological characteristics, the evaluation of genetic information, and the usage of phylogenetic methods (Delgado-Salinas *et al.*, 1999). *P. lunatus* is classified by taxonomists according to its comprehensive visual characteristics, anatomical structure and reproductive traits. Lima bean underwent a process of domestication

and subsequent diversification within the geographical boundaries of the Americas. This domestication process occurred in two distinct sites, resulting in the establishment of two separate gene pools. Mesoamerican (mainly in Mexico) and Andean (southern Peru, Bolivia and north-eastern Argentina) (Andueza-Noh *et al.*, 2013; Bitocchi *et al.*, 2017; Chacón-Sánchez *et al.*, 2021; García *et al.*, 2021; Adebo, 2023). The genetic heritage of *P. lunatus* is a primary source of genetic variability aimed at improving the species. *P. lunatus* is a legume that is extensively grown in Latin America, Africa and Asia, and is a significant origin of proteins and other nutrients for millions of people around the world (Baudoin, 1993).

The genetic heritage of *P. lunatus* includes an extensive assortment of morphological and agronomic characteristics, like seed measurements and form, disease and pest resistance, drought tolerance, and the capacity to fix nitrogen in association with soils bacteria (Martínez-Nieto *et al.*, 2022). One of the most interesting aspects of the genetic heritage of *P. lunatus* is the presence of wild varieties collected in various regions of Latin America (Matsubara and Zúñiga-Dávila, 2015; Oliveira-Silva *et al.*, 2017; Wilker *et al.*, 2020). These wild varieties are vital for the enhancement of the species, because they possess an extensive assortment of agronomically vital characteristics and can be crossed with cultivated varieties to introduce new traits into crops. In addition, the genetic heritage of *P. lunatus* also includes traditional varieties that have been cultivated for centuries in diverse areas of Latin America (Nienhuis *et al.*, 1995; da Silva *et al.*, 2015; Martínez-Nieto *et al.*, 2022; Renzi *et al.*, 2022). These traditional varieties possess great genetic diversity and have been selected by farmers to adapt to different environmental conditions and meet local food needs. Upkeep of the genetic heritage of *P. lunatus* is vital to ensure the availability of genetic resources to enhance this species (Jarvis *et al.*, 2004). *In-situ* conservation, i.e. maintaining cultivars at their places of origin, is essential to preserve genetic variability and maintain relationships amid wild and domesticated varieties. Furthermore, *ex-situ* upkeep and the conservation of seeds in germplasm banks are crucial to ensure the disposal of genetic materials for the improvement of the species in the future.

In researching the domestication history of various present-day cultivated plants, it has been concluded that some originated from singular domestication processes, such as coffee, barley, peas, lentils, oats and others. Other crops, however, arose through multiple domestication processes, either at different times or places, as with chilli peppers, mustard, squash and beans (Saini *et al.*, 2020). Thus, broad knowledge about the geographical origin, wild progenitor, current genetic diversity in both wild and cultivated varieties, and possible number of domestication events is of vital importance for germplasm management and genetic improvement programmes (Gepts, 2000, 2006). Due to centuries of genetic isolation, most cultivars from one centre have characteristics that are not found in those from the other. Mesoamerican varieties of common bean are much more productive, and the frequency of disease and pest tolerance genes and abiotic stresses is higher than in the Andean types, which have a larger bean size and quality (Nienhuis *et al.*, 1995; Melgar, 2021).

The beans are innate to America, but they have been disseminated to other parts of the world over centuries (Kaplan, 1965). During this expansion, its germplasm was introduced to the Iberian Peninsula, and the plant adapted to new environments (Lioi and Piergiovanni, 2013). Unlike in America, where genetic isolation prevailed owing to the great distance amid diverse centres of origin and the slow diffusion process of the plant, in the Iberian Peninsula, the Mesoamerican and Andean genetic stocks of the bean were grown in proximity, which favoured gene flow between them (Piergiovanni and Lioi, 2010). This mixing of different genetic stocks led to the appearance of natural recombinant forms with intermediate genotypic and phenotypic characteristics. The existence of these new genetic forms has been documented by the Legume group of the Galicia Biological Mission-Spanish National Research Council (MBG-CSIC), leading to the Iberian Peninsula being called a secondary centre of diversification for the species (Aragão *et al.*, 2011). This process of genetic mixing and adaptation in the Iberian Peninsula has been of great interest to researchers. It has allowed a greater understanding of the intricacy of the genetic organization of the common bean, which was initially thought to have two main genetic stocks: the

Mesoamerican and the Andean, each with wild and cultivated populations. However, recent studies have shown that the genetic organization of the bean is even more complex, particularly in the Mesoamerican genetic stock, where two distinct groups have been identified, both genetically and geographically: Mesoamerican I and Mesoamerican II (MI and MII, respectively). This discovery has opened new lines of research on the evolution and genetic diversity of this important agricultural species (Bitocchi *et al.*, 2017). In other previous investigations by the Legumes group of the MBG-CSIC, the existence of these new genetic forms has been documented, which is why the Iberian Peninsula is considered a secondary centre of diversification of the species (Casquero *et al.*, 2006).

Indeed, by studying DNA sequence data from various nuclear and chloroplast markers, researchers showed that *P. lunatus* is linked to other domesticated *Phaseolus* species such as *P. vulgaris* and *P. coccineus* (Delgado-Salinas *et al.*, 2006). These species are members of the same subgenus and have a recent common ancestor. In addition, current research has focused on employing high-throughput sequencing technology to investigate the evolutionary links and genetic diversity within the *Phaseolus* genus. Tian *et al.* (2021) sequenced the chloroplast genome of *P. lunatus* and compared it to other legumes to shed light on the genetic links between different *Phaseolus* species and the domestication history and genetic diversity of *P. lunatus*. It is worth noting that the evolutionary relationships within the *Phaseolus* genus are complex and have resulted in reticulate evolution through hybridization events and introgression.

Phylogenetic investigations, which include building evolutionary trees or cladograms from genomic data, have made significant contributions to understanding the relationships of *P. lunatus* with other species. These methods examine DNA sequences or other molecular markers to determine shared ancestry and divergence patterns. Taxonomists can infer evolutionary relationships and position *P. lunatus* into the larger framework of the plant kingdom by evaluating genetic similarities and differences (Delgado-Salinas *et al.*, 1999). Evolutionary relationships reveal information about the evolutionary history and relatedness of organisms. Understanding the evolutionary relationships of *P. lunatus* gives essential information for taxonomy, breeding and conservation initiatives (Lioi, 1994). It is vital to remember that taxonomy and systematics are dynamic fields that are constantly changing as new knowledge becomes available. Advances in genetic sequencing technologies and computational methodologies have brought more profound insights into the links between organisms, resulting in categorization revisions and adjustments, critical in breeding programmes, conservation efforts and agricultural development in general. Understanding the genetic make-up and variety of the *P. lunatus* gene pool is essential for generating novel cultivars with desired features such as disease resistance, yield potential and nutritional quality (Baudoin, 1993; Dohle, 2017; García *et al.*, 2021). Several researchers have examined the genetic variability and structure of *P. lunatus* populations, shedding light on the composition and dispersion of the crop gene pool. One such study, conducted by Pires *et al.* (2022), revealed substantial genetic variation among *P. lunatus* accessions in Brazil. They assessed 61 accessions of lima bean obtained from numerous locations in Brazil. The combination of agro-morphological indicators and single-sequence repeat (SSR) markers allowed a thorough examination of the genetic structure and variability of the *P. lunatus* gene pool. Across the accessions, the researchers

discovered a total of 90 distinct SSR alleles. The findings demonstrated great genetic variety inside the crop gene pool, indicating the presence of various gene variants that can be used in breeding, implying a high degree of genetic variation. Martínez-Castillo *et al.* (2014) also utilized SSR markers to assess the genetic structure and variability of wild *P. lunatus* populations in Mexico. The findings demonstrated remarkable genetic variability within the Mesoamerican *P. lunatus* gene pool, showing many gene variations. The study also discovered discrete genomic clusters, which indicate patterns of population differentiation among wild populations. Geographic considerations were found to have a role in this differentiation, implying limited gene flow and isolation. These findings have significant implications for conserving wild *P. lunatus* populations, highlighting the need to preserve unique genetic resources. The study also shed light on the domestication process of *P. lunatus*, as wild populations showed varying degrees of genetic resemblance to farmed cultivars. This indicates that wild populations may have contributed genetically during domestication. The considerable genetic variety revealed in the landrace germplasm emphasizes its potential for application in breeding programmes, allowing *P. lunatus* to be improved by adding diverse genes associated with desirable features (Fig. 8.2). A substantial

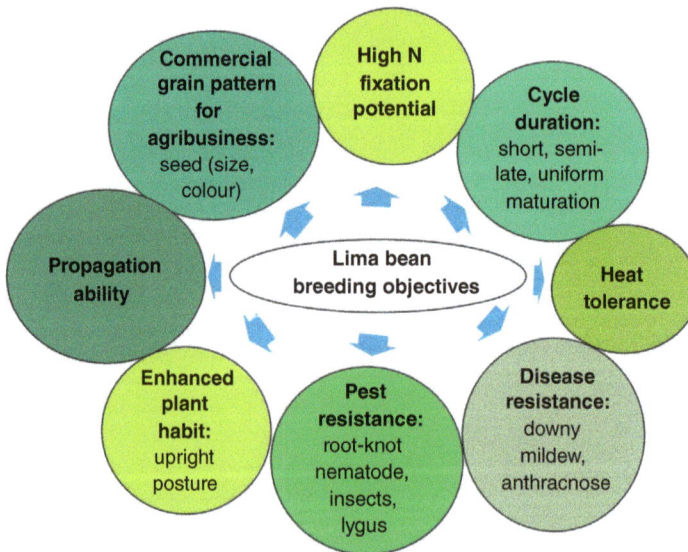

Fig. 8.2. Breeding objectives of *Phaseolus lunatus*. (Author's own figure.)

genetic variety was found among *P. lunatus* landraces gathered from different locations in Ethiopia using the ISSR marker (Nasir *et al.*, 2021). Out of 70 polymorphic ISSR markers, 157 polymorphic bands were found showing a significant amount of genetic variation within the examined landraces. The observed polymorphic loci percentage was 97.46%. Shannon's index values showed the presence of significant genetic diversity. The principal coordinate analysis indicated significant clustering patterns among the *P. lunatus* landraces, demonstrating genetic differentiation and population structure (Nasir *et al.*, 2021).

Montero-Rojas *et al.* (2013) demonstrated the occurrence of various genetic variants within the Caribbean *P. lunatus*, which provides the opportunity for breeders to choose and generate superior varieties with desirable features. However, they discovered variances in cyanogen levels among the different landraces, indicating potential nutritional and toxicological concerns. The cyanogen content variation emphasizes the significance of assessing the safety and nutritional features of these landraces.

Lustosa-Silva *et al.* (2022) found significant genetic variety in *P. lunatus* populations in north-east Brazil, indicating the presence of many gene variations and highlighting the possibility for genetic improvement and breeding programmes. Furthermore, discrete genetic clusters within the populations were detected, demonstrating genetic divergence and regional subgroups. However, other groups showed symptoms of genetic erosion, indicating a decrease in genetic diversity over time. This highlights the significance of conservation efforts to prevent further loss of genetic resources. The study also revealed distinctive *P. lunatus* accessions with distinct genetic traits and high levels of variability, which can be used for breeding and conservation. Camacho-Pérez *et al.* (2018) observed in the Mayan area that *P. lunatus* landraces have a significant level of genetic variability, indicating the presence of a diversified gene pool within the local populations. The genetic study revealed multiple population clusters, implying the existence of genetically diverse groupings within the *P. lunatus* gene pool. Geographic considerations also played a crucial influence in determining the genetic structure since *P. lunatus* populations from different places within the Mayan area had distinct genetic signatures, indicating local adaptation and geographical divergence. Additionally, evidence of gene flow between *P. lunatus* populations was discovered, demonstrating the interchange of genetic material among distinct landraces and contributing to total genetic diversity. Chacón-Sánchez and Martínez-Castillo (2017) underlined the genetic variability of lima bean in Mesoamerica, indicating the presence of a diversified gene pool. The study discovered a genetic bottleneck during domestication, resulting in lower genetic diversity in cultivated types, by comparing the genetic profiles of wild and cultivated *P. lunatus*. They also detected various genetic groupings within Mesoamerican *P. lunatus* populations, which could indicate separate domestication episodes or migration patterns. Also, reports of gene flow amid wild and cultivated populations suggested that genetic variability from wild relatives could be introduced in the cultivated gene pool. *P. lunatus* populations were domesticated independently in various Mesoamerican locations rather than in a single area.

Phaseolus lunatus holds significant importance in global plant genetic resource conservation efforts (Table 8.1). They are actively conserved and exchanged through various international organizations and gene banks. CIAT plays a pivotal role by actively conserving lima bean germplasm in their gene bank, ensuring its availability for research and breeding (Plucknett and Smith, 2014). Additionally, lima bean is recognized by the Food and Agriculture Organisation (FAO) for its contribution to food security and sustainable agriculture, advocating for conservation and responsible utilization (Calles, 2016). In the USA, the National Plant Germplasm System (NPGS) comprehensively conserves lima bean genetic resources, further contributing to the global effort (Greene *et al.*, 2018). These initiatives align with the objectives of the Convention on Biological Diversity (CBD), which encourages member nations to safeguard and sustainably manage plant genetic resources, including lima bean (Ferreira-Miani, 2004). As part of a broader strategy, lima bean is integral to maintaining global crop diversity and ensuring future agricultural resilience (Debouck and Morales de Borja, 2012).

So, systematics has been critical in determining the evolutionary relationships and classification of *P. lunatus*. This knowledge is essential for

Table 8.1. World germplasm bank for the conservation of lima bean seeds. (Author's own table.)

No.	Gene bank	Location	Number of accessions	Type of bank	Website
1	International Center for Tropical Agriculture	Colombia	37,938	Active	CIAT, 2022
2	Agricultural Research Service United States Department of Agriculture (USDA-ARS NPGS (National Plant Germplasm System), 2023) Washington	USA	2,255	Active	USDA-ARS NPGS (National Plant Germplasm System), 2023
3	Germplasm Bank	México	7,376	Active	INIFAP, 2018
4	Germplasm Bank National Institute for Agrarian Innovation	Perú	4,125	Active	INIA, 2023a
5	Agricultural Innovation Institute	Chile	989	Active	INIA, 2023b
6	Germplasm Bank	Brazil	363	Active	CNPAF-EMBRAPA, 2017
7	Leibniz Institute of Plant Genetics and Crop Plant Research	Germania	9,002	Active	IPK, 2023
8	Institute of Crop Germplasm Resources	China	310,000	Active	ICGR-CAAS, 2023
9	All-Russian Institute of Plant Genetic Resources named after N.I. Vavilov (VIR) Ministry of Science and Higher Education (VIR)	Russia	42,000	Active	Federal Research Center, 2023
10	International Institute of Tropical Agriculture	Nigeria	323	Active	IITA–CGIAR, 2023

the appropriate classification, conservation and exploitation of *P. lunatus* in a variety of disciplines, including agriculture, breeding programmes and genetic resource management.

8.4 Germplasm Collection, Characterization, Evaluation and Conservation

Studies of the genetic make-up and evolutionary history of lima bean have revealed interesting insights into the plant's relationships with other members of the *Phaseolus* genus. For example, genetic analysis has shown that the lima bean is narrowly linked to the wild bean species *P. lunatus*, which is found in the wild in Mexico and Central America (Serrano-Serrano *et al.*, 2010; Chacón-Sánchez *et al.*, 2021). In addition, studies have shown that the lima bean has a complex genetic history that involves hybridization between different ancestral populations. It is thought that the lima bean came from the hybridization of two wild bean species, *P. lunatus* and *P. acutifolius*, which are found in Mexico and the south-western USA (Serrano-Serrano *et al.*, 2010). The hybridization event that offered an upsurge to *P. lunatus* is thought to have occurred

around 2000 years ago. Since then, the plant has undergone further genetic changes through artificial selection by humans. The selection process has resulted in the development of many different lima bean cultivars, which are adapted to diverse environments and have different nutritional qualities (Smýkal *et al.*, 2018).

Recent studies of the lima bean genome have provided new insights into evolution and genetic diversity. For example, research revealed that *P. lunatus* has a large and complex genome, about twice the size of the human genome. This complexity is the consequence of the plant's evolutionary past, which has involved multiple rounds of genome duplication and reshuffling of genetic material (Gross and Olsen, 2010).

Overall, the evolutionary relationships of lima bean are complex and multifaceted, reflecting the plant's long history of hybridization and artificial selection by humans. Despite these complexities, however, lima bean remains a vital crop plant that provides essential nutrition to millions of people worldwide. Ongoing research into the genetics and evolution of lima bean will continue to shed new light on its relationship with other leguminous crops. It may ultimately lead to the expansion of novel and improved cultivars of lima bean, which are better adapted to changing environmental conditions (Table 8.2).

Table 8.2. Main characteristics of the evolutionary history of lima bean. (Author's own table.)

Characteristics	Lima bean
Genetic composition	The sequencing of the lima bean genome has been accomplished, revealing its chromosomal constitution of 26 distinct units, with a total size of approximately 470 Mb. Furthermore, it has been ascertained that the lima bean species is an amphiploid organism, possessing two discernible sets of chromosomes, one of which originates from the Mesoamerican region, while the other is of Andean provenance.
Evolutionary history	Lima bean originated in South and Central America and is believed to have been domesticated about 7000 years ago. During its dispersal throughout the region, genetic diversification occurred, resulting in the emergence of different varieties. The lima bean was also introduced to Europe in the 16th century, where it changed to different climates and indigenous varieties were developed. Today, it is cultivated in many countries worldwide for human food and animal feed and as a cover crop.
Relationship with other bean species	Lima bean is deemed to be closely related to other species of beans, specifically the common bean and the ayocote bean. It is widely believed that these species have undergone evolutionary changes from a mutual forebear about 2.5 million years ago. Even though they possess common genetic and phenotypic characteristics, each species exhibits unique peculiarities and has been subjected to natural and artificial selection in diverse regions across the globe.
Economic and cultural importance	Lima bean is an important agricultural species in several areas of Latin America, Africa and Asia. It is an important source of protein, carbohydrates and other essential nutrients, and is grown on small family plots and large commercial farms. In addition, lima bean is of great cultural and culinary importance in many communities, where they are used in traditional dishes, and a wide variety of recipes and preparations have been developed.

8.5 Germplasm in Plant Breeding by Conventional Approaches

Breeding is directed by enhancement concerning resistance to biotic and abiotic limitations. So, the enhancement of cultivars by means of healthier resistance is the primary purpose of most *P. lunatus* breeding programmes. Breeding utilizing the worldwide germplasm assortments between 1973 and 1980 was completed at IITA (Nigeria) to raise harvest, biotic tolerance and grain nutritive value. Then, other works were implemented at CIAT (Colombia) from 1980 to 1992. Breeding programmes utilizing indigenous assortments are being conducted in many African countries.

Screening of pure lines and mass screening of populations were accepted as preferred approaches, and the targeted culture schemes included pure and related crops. The erect form and tolerance to *Rhizoctonia* were key benchmarks to enhance the pseudo-determined bush forms. Initial growth, indifference to photoperiod, resistance to *P. lunatus* golden mosaic virus and adaptability of the related plant for multiple cropping are needed in uphill forms (Baudoin, 2002). In Brazil, a breeding programme for lima bean to breed enhanced varieties with determinate growth form, straight growth, unvarying maturity, short cycle, tolerance to anthracnose and a marketable seed configuration for agroindustry has been conducted by the Federal University of Piauí. It is also useful to develop varieties with an indeterminate growth habit that can be used as an alternative in intercropping with corn. This investigation began in 2007, with the selection of the most vigorous-growing cultivar in Piauí (Martínez-Castillo *et al.*, 2022). Numerous approaches have been utilized for breeding lima bean, such as lineage, backcrossing, single seed descent, screening in gamete, bulk method, repeated screening, and participatory crop breeding (Kelly, 2010). *P. lunatus* improvement was initiated in

2000 at Delaware, with principal outcomes reported from 2006 to 2008. Several scientists at the University of Delaware conducted a breeding program between 2004 and 2018 intending to develop novel cultivars that are fine-tuned to the Delaware environment and cultivation, resistant towards downy mildew of *P. lunatus* (Santamaria *et al.*, 2018), tolerant to stresses (Ernest *et al.*, 2006), and resistant to nematodes (Roberts *et al.*, 2008). Conjoint crosses were completed amongst UC 92 and UC Haskell, utilizing the stigma hook technique to create enhanced large and baby bushes, as well as vine-form cultivars with lygus and nematode resistance. Breeding tools such as genetic maps and molecular markers were generated to aid improvement. Crosses were confirmed using SSR size polymorphic markers. Generations were evolved, grains were augmented and disposable recombinant inbred lines were completed and accessible to other scientists through CIAT (Dohle, 2017).

While the traditional breeding procedure takes many years to choose and develop accurate cultivars, various genetic enhancements on *P. lunatus* have been conducted using these approaches with pure lines and mass screening. Most gene enhancement efforts have focused on evolving erect and bushy forms, with robust sprouts and erect habit capacity (David, 1984), increasing grain harvest, and enhancing pest and disease tolerance and grain nutritive value (Baudoin, 1991).

8.6 Incorporation of Genomic and Genetic Assets for Plant Enhancement

In response to numerous limitations in *P. lunatus* cultivation, novel breeding methods have been applied to yield varieties with excellent resistance to biotic and abiotic constraints. Much of the improvement effort on *P. lunatus* has been conducted using marker-assisted selection (Table 8.3), where screening for the gene is founded on molecular indicators linked to a gene of attention, and a marker is utilized to control the integration of the desired gene. For example, DNA markers are related to essential agronomical features such as resistance towards abiotic

constraints (Ernest *et al.*, 2006), infectious agents (Mhora *et al.*, 2016), pests and nematodes (Dohle, 2017; Temegne *et al.*, 2020). As an alternative to choosing a trait, a breeder may pick an indicator that can be effortlessly separated as a screening strategy. Molecular marker-assisted backcrossing (MABC; Table 8.3) using foreground and background screening with genome-wide SSR indicators for recovery of recurring parental genome is an environment-self-determining, accurate and fast methodology for improvement of cultivars for a trait of interest (Pratap *et al.*, 2017). The goal was to generate 1000 potentially polymorphic DNA fragments swiftly. This involved simplifying the DNA preparation process and, ultimately, employing genetic stocks to differentiate these 1000 DNA segments. Several methods have been tested to sift efficiently through many unmapped random molecular markers within a short timeframe. These approaches aim to identify selectively only the rare markers that are located in close proximity to the target gene. Selected indicator tools have been utilized for *P. lunatus*. Dohle (2017) used a population of *P. lunatus* to create a genetic linkage map containing 515 single nucleotide polymorphisms (SNPs) spanning 1622 cM across 13 linkage clusters. This map was instrumental in identifying 27 quantitative trait loci (QTL) intervals associated with various agronomic traits, including sprouting percentage, shoot height, seed mass, yield, blossoming period, inflorescence site, growth arrangement and hydrogen cyanide concentration in *P. lunatus*. It was found that some QTLs related to these traits in *P. lunatus* were located at similar positions to those found in other beans. Although there has been limited research on the genetic modification of *P. lunatus*, it is often used as a model crop for specific genetic modification techniques. One such technique involves transient modification, which leads to temporary changes in gene expression. This method is valuable because it allows for rapid gene expression analysis, providing essential insights into the functioning of the gene component or coding region that influences expression patterns.

Transient modification methods implicate DNA incorporation utilizing protoplast electroporation, virus-induced gene silencing, agroinfiltration and particle bombardment. A specific procedure has been fruitfully established utilizing particle bombardment to incorporate large quantities of promoter components in lima bean (Gunadi *et al.*, 2019) (Table 8.3). Transient modification has consequently been utilized to study the huge amount of DNA constructs as the primary selection step before choosing genes for stable incorporation and crop retrieval. Terpenoid has a vital role in crop protection against

Table 8.3. DNA markers and reported significant results of the evaluation of diversity based on molecules in lima bean (Author's own table.)

Marker	Study results	Reference
RAPD (random amplified polymorphic DNA)	The genetic distance of germplasm resources was evaluated in 65 accessions that included four large-seeded and seven small-seeded cultivars and 54 germplasm accessions from the Caribbean and North, Central and South America. Genetic diversity estimated by RAPD markers was revealed. Obtained results showed that *P. lunatus* were found in a large group that belongs to the Mesoamerican heritage, followed by the Andean and Fordhook.	Nienhuis *et al.* (1995)
SCAR (sequence characterized amplified region)	Fourteen primers were used, and a SCAR resulted in 90 bands, of which 83 were polymorphic. The dendrogram shows a clear separation into three groups, corresponding to each of the gene pools and an intermediate group of *Phaseolus*.	Freyre *et al.* (1996)
RFLP (restriction fragment length polymorphism)	Mesoamerican and Andean gene pools and intermediate types were analysed for variability at 17 isoenzyme loci. Some accessions were also examined for RFLP at the rDNA level. These data were used to construct two dendrograms that show a clear separation into two distinct groups corresponding to each of the gene pools and an intermediate one probably representing a transition group.	Lioi *et al.* (1998)
AFLP (amplified fragment length polymorphism)	Cyanogenesis (the release of hydrogen cyanide (HCN)) was quantified as a direct defence, and volatile organic compound (VOC) emission as an indirect defence, against herbivores. To elucidate whether compensations occur at the genetic or phenotypic level, cultivated and wild-type accessions of lima bean and different stages of leaf development were investigated. Genetic relationships between accessions were studied using AFLP analysis.	Ballhorn *et al.* (2008)
ESTs (expressed sequence tags)	A study on the defence of plants against herbivores by emitting specific mixtures of volatiles is described. In particular, we characterized a plastid terpene synthase gene, PITPS2, from lima bean, which produces precursors of a volatile compound (E, E)-4,8,12-trimethyltrideca-1,3,7,11-tetraene (TMTT). Transgenic plants expressing PITPS2 or its *Medicago truncatula* homologue TPS3 (MtTPS3) were produced and used for bioassays with herbivorous and predatory mites. Results indicate that TMTT tampering is an ideal platform for pest control that attracts generalist and specialist predators in different ways.	Brillada *et al.* (2013)

Continued

Table 8.3. Continued.

Marker	Study results	Reference
SNP (single nucleotide polymorphism)	Studies using SNP markers, clustering and applied Bayesian approaches confirmed the existence of three gene pools in wild lima bean: the Mesoamerican (MI), the two Mesoamerican (MII) and the Andean (AI) gene pools, with mainly non-overlapping values. The provenance according to geographic ranges was verified, and a population of *P. lunatus* that belongs to another Andean gene pool (AII) was found in central Colombia.	Chacón-Sánchez and Martínez-Castillo (2017)
Quantitative trait loci (QTL) mapping	Successful identification of 27 QTL intervals governing essential agronomic traits in lima bean. These traits encompass germination percentage, shoot height, seed mass, yield, blossoming period, inflorescence site, growth arrangement and hydrogen cyanide potential. The QTL mapping aids in understanding the genetic basis of these traits.	Dohle (2017)
MABC (molecular marker-assisted backcrossing)	Implementation of an MABC strategy utilizing genome-wide SSR markers for selecting and transferring desired traits. This approach offers precision and speed in developing improved varieties of lima bean.	Pratap *et al.* (2017)
Transient genetic modification	Utilization of transient modification techniques, including protoplast electroporation, virus-induced gene silencing (VIGS), agroinfiltration and particle bombardment, to study gene function and expression. A successful example includes particle bombardment for introducing promoter constructs in lima bean cotyledons.	Gunadi *et al.* (2019)
SSR (simple sequence repeat)	Six SSR markers (BM211, BM160, AGCS91, AG1, BM164 and BM141) were used to analyse the diversity and population structure of 28 lima bean landraces. Three highly differentiated subpopulations were obtained in the analysed collection.	Fagbédji *et al.* (2022)
ISSR (inter-simple sequence repeats)	The ISSR marker was used, revealing high levels of genetic variation, even with a limited set of primers. The present study also shows that ISSR markers can be a valuable tool for assessing genetic diversity among French bean genotypes.	Panda and Paul (2023)
SRAP (sequence-related amplified polymorphism)	A set of genetic diversity and germplasm ratios of *P. lunatus* were determined using morphological markers and SRAP. Twenty-six quantitative and 18 quality traits of 22 lima bean germplasms were evaluated based on morphological and molecular markers SRAP. Twenty-two germplasms were obtained that were rich in genetic diversity. Clustering based on morphological and SRAP molecular markers essentially agreed with classification by pod size.	Yuanzhen *et al.* (2023)

herbivores by appealing to predatory wasps. Seven terpene synthase genes from *P. lunatus* were used to create transgenic *Oryza* to combat the rice pest *Chilo suppressalis*. The transgenic rice plants and resulting terpenoid volatiles showed the potential to improve plant protection indirectly by recruiting natural enemies of the pest (Li *et al.*, 2018). Between 2008 and 2013, non-natural hybridization was utilized for crop improvement. Techniques for applying artificial crosses were conducted at the Federal University of Piauí in 2014. Bi-parental crosses using germplasm from Brazil, the USA, Mexico and Argentina were achieved at Davis in the USA in 2015. Such innovative breeding tools were assessed, along with populations that had evolving determinate and indeterminate growth tendencies (Martínez-Castillo *et al.*, 2022). The programme was prolonged to consider facets linked to growth habits and response to anthracnose and grain shapes with progressive populations through the revised bulk method in 2019. The resultant lines are still under evaluation (Martínez-Castillo *et al.*, 2022). Table 8.4 summarizes the genetics and improvement research in *P. lunatus*.

8.7   Loss of Genetic Variability from Traditional Regions and Limitations in Germplasm Use

To obtain the plants they needed, farmers often employed their abilities and resources. The method was instigated with the domestication of wild taxa and persisted with the vigilant choice of crops with varying and habitually indeterminate growth requirements and preferences. The result is a diverse and continually evolving collection of indigenous species accessions, shaped by interactions with wild taxa, adaptations to diverse agricultural environments, and responses to economic and cultural influences that influence farmers' priorities. The richness, as well as the extent of diversity of these local accessions, is now threatened by the variable nature of farming production. A crucial factor is the widespread selection of improved cultivars that are the outcomes of formal plant breeding. These varieties usually supply yield upsurges and other benefits that lead to their use in monocultures. Technological modifications, such as

the usage of amendments, irrigation and drainage, restrict the use of local accessions adapted to marginal growth requirements.

Furthermore, agricultural production is becoming increasingly market-oriented, leading farmers to select species based on traits other than those that align with historical significance to local customs and culture, leaving them with limited options. However, while the adoption and utilization of improved varieties and other agricultural techniques offer important advantages to farmers, the abandonment of local accessions and species genetic diversity remains a significant concern. Indeed, progress in crop improvement has been made possible by an extensive assortment of genetic materials supplied through local accessions. A triumph of modern crop breeding currently threatens the genetic diversity on which future improvements depend, as farmers find it less gratifying to keep the varied assortment of local accessions. In addition, the extensive utilization of improved varieties increases queries about the reliability of agricultural production and the menace of disease or pest attacks (Tripp and van der Heide, 1996).

Improved varieties are more constant than local accessions, which might raise their vulnerability to pests. Several improved varieties, however, have been approved because they are more tenacious than those they substitute. The primary threat to restoring stability caused by the adoption of enhanced varieties lies in the gradual uniformity and continuous cropping that their use generates. Large-scale monoculture of a lone accession is eventually a concern, regardless of the origin of the accession. In addition, the protection and usage of the genetic variability of plant species are of specific meaning to the more unimportant and different agrarian environments where current crop improvement has been less successful. Public research and extension institutions typically underserve farmers in these regions. These regions are regular centres of diversity for several plant varieties. None the less, snowballing poverty forces several of these farmers to become more reliant on off-farm income, which limits their ability to exploit and preserve the assortment of local accessions they have been accustomed to using. Lastly, the conservation of an extensive and evolving assortment of local accessions is endangered by the development of intellectual property protection

Table 8.4. Lima bean breeding research. (Author's own table.)

Accessions	Breeding aims	Techniques	Organization	Sources
Cultivated landrace in Piauí state	Growth form, erect stance, unvarying maturation, short cycle, anthracnose tolerance, commercial grain pattern	Artificial hybridization, artificial crosses	Federal University of Piauí	Martínez-Castillo *et al.* (2022)
Germplasm Brazil, USA, Mexico, Argentina	Determinate and indeterminate growth habits, etc.	Biparental crosses	Universities of California (UC) and Piauí	Martínez-Castillo *et al.* (2022)
Germplasm Brazil, USA, Mexico, Argentina	Growth habit, reaction toward anthracnose and seed patterns	Modified bulk method	Universities of California and Piauí	Martínez-Castillo *et al.* (2022)
America	Pod dehiscence, as well as seed development	Biparental crosses	Universities or research groups in Colombia, the USA, Mexico	García *et al.* (2021)
35 germplasm accessions	Downy mildew resistance (DMR) (Race E, race F)	Screening of F_2 individuals and RIL populace tolerant	University of Delaware (UD) at Eastern US	Santamaria *et al.* (2018)
313 foreign landraces, UC Haskel, UC92	Root-knot nematode, lygus resistance	Backcrossing, SSR expansion, SNP chart, GBS	University of California (UC) Davis	Dohle (2017)
256 landraces	DMR	Charting race F tolerance locus in F_2 individuals. BSA utilizing GBS on bi-parental F_2 individuals	University of Delaware at Eastern US	Mhora *et al.* (2016)
Fordhook cultivars	DMR	Crossing and plants selection (F_2, F_4, F_6) and screening of Fordhook varieties	UD	Ernest and Kee (2008), Ernest *et al.* (2011)
Baby lima cultivars	Downy mildew resistance	Homozygous formation, screening of baby lima varieties	UD	Ernest and Kee (2008)
Hybrid Pole limas	Propagation ability	Hybridization	UD	Ernest and Kee (2008)
F_1, F_2, $F_{2:3}$, $F_{2:7-9}$ from L-136 × Henderson Bush	Root-knot nematode resistance	Crossing and plant selection	UC Davis	Robert *et al.* (2008)
Green baby lima cultivars	Heat tolerance, enhanced growth habit, downy mildew resistance	Crossing as well as screening of plants	UD	Ernest *et al.* (2006)

BSA, bulk segregant analysis; DMR, downy mildew resistance; GBS, genotyping-by-sequencing; RIL, recombinant inbred line; SNP, single nucleotide polymorphism; SSR, simple-sequence repeats or microsatellites.

for crop cultivars, urged by the World Trade Organization (WTO). An escalating enforcement of crop breeders' privileges is significant for crop genetic variability. The novel cultivar must be clearly described, and is required to be distinctive, constant and steady to be given formal protection. This discourages the use of innately varied local accessions or combinations of cultivars. A supplementary discussion concerns the aptitude of farmers to save seed from an accession, to trade it with their neighbours and to accommodate their growing patterns. When the cultivars grown are supported with rigorous legal backing, seed enterprises may contest these methods. Even scenarios in which the land accessions of farmers could be lawfully defended and then denied to the farmers who developed them are conceivable. As a result, the development of plant variety protection highlights the necessity to find methods of maintaining plant genetic diversity (Tripp and van der Heide, 1996).

8.8   Cultivation Practices

8.8.1   Seed selection

The growers generally utilize certified seeds of *P. lunatus* to guarantee good seed features, decreasing the probability of infection establishment as well as consequently attaining better harvests (Long *et al.*, 2014). Seeds are usually accessible in research organizations, seed enterprises and retail outlets (Temegne *et al.*, 2020).

8.8.2   Tillage and sowing

The ploughing and sowing are automated in developed states. Ploughing prepares a good seedbed for planting (Temegne *et al.*, 2020). Primary ploughing comprises carving, disking, soil razing and creating raised beds. The soil is well-defined in two ways: to expose the land configuration and to fragment some caliche. First, the stubble disc is used, followed by the finish disc to create the ideal seedbed. Finally, the land is levelled into two sections using the triplane. Land should be well-drained, aerated and receive full sunlight. *P. lunatus* cannot tolerate extreme wetlands and might wilt. Sowing equipment implicates

flex-planters. Nevertheless, in several developing states, ploughing and sowing are not automated, and sowing is realized by hand, utilizing a seed-drill planter (Temegne, 2011). To protect them from decaying, treated seeds are dusted with a fungicide. According to Gawande *et al.* (2020), a 1.0 m × 0.75 m spacing gave maximum yield in *P. lunatus* plants. In southern Delaware, Sirait *et al.* (1994) found that 10 cm in a row and 38 cm between row spaces (258,000 individuals per hectare) provide a maximum pecuniary *P. lunatus* harvest.

8.8.3   Seed germination

Germination and sprouting of *P. lunatus* occurs 4–10 days after planting (Baudoin, 2006; Ibeawuchi, 2007). The seeds emerged slowly, otherwise scarcely, upon sowing in freezing conditions. *P. lunatus* is typically sowed a week later in the spring than snap bean. The seeds must be sowed as quickly as feasible because earlier sowing commonly gives considerably more yield than later sowings. Cotyledons emerge over the soil with a juvenile seedling, underscoring the importance of sowing seeds in meticulously prepared land to ensure proper growth. Seeds of the bush type are generally sowed in rows 2– 3 feet (~60–100 cm) apart and 1–2 inches (2.5–5 cm) deep. Pole landraces need more spacing; the sowing interval is defined by the technique used to support them. For 1 acre, approximately 1 bushel (~30 l) of small to medium-seeded landraces is required, whereas 1 pint (~ 0.5 l) is sufficient for 100 feet (30 m) of rows (Funchess, 1936).

8.8.4   Vegetative growth

Vegetative growth accelerates after 30 days, and blossoms begin to appear between 35 and 70 days after sowing, as noted by Baudoin (2006) and Ibeawuchi (2007). Unlike snap bean, the crop is resistant to cold damage and thrives in high-humidity environments during flowering and early pod development. Cloudy nights are especially beneficial for pod growth. *P. lunatus* has deep roots, allowing it to withstand dry spells as long as the soil remains moist enough for the deeper roots to access water. It is advisable to minimize or entirely cease tilling from the flourishing period until the main crop is harvested. Plant growth slows during the main development phase of the crop, and some leaves may naturally fall off (Funchess, 1936).

8.8.5 Fertilization

The soil for cultivating *P. lunatus* should have adequate levels of nitrogen (N), phosphorus (P) ranging from 0.0011 to 0.002%, potassium (K) ranging from 0.008 to 0.012%, sulfur (S) and zinc (Zn). Although *P. lunatus* can form symbiotic relationships with nitrogen-fixing bacteria, sometimes this symbiosis is not enough to provide the required nitrogen quantity for achieving a high yield. Hence, additional nitrogen input is often necessary during the early growth phase to ensure that the crop receives the required quantity for rapid development and optimal yield, as emphasized by Long *et al.* (2014). Supplementing nutrients, including 25 mg/dm^3 of N, 157 mg/dm^3 of P_2O_5, 90 mg/dm^3 of K_2O, 30 mg/dm^3 of S, 1.27 cmole/dm^3 of Ca, and 0.5 cmole/dm^3 of Mg, leads to increased yield and improved nutritional status of *P. lunatus*, as observed by Santiago *et al.* (2019). In moderately fertile soils, applying 1 lb (~500 g) of 8-24-24 manure or its equivalent per 20 feet (~6 m) of a row is recommended. Excessive nitrogen can cause the plant to become excessively viny and result in dropped blooms instead of pod formation. *P. lunatus* produces its highest yield on moderately heavy, rich land, even though it matures early on light, sandy, well-enriched land. *P. lunatus* responds well to a generous use of fertilizer. In low fertility areas in Auburn, yields have been significantly improved by applying 1000–1500 pounds (450–680 kg) per acre of manure with a composition of 4-8-4. Exceptionally high yields have been achieved using 6 tons of fertilizer. To ensure optimal results, it is advisable to apply the manure in the furrows 10–14 days prior to sowing, because direct contact with the seeds should be avoided to prevent poor germination. Additionally, a portion of nitrogen can be withheld and applied as a side dressing once *P. lunatus* has started growing, as suggested by Funchess (1936). *P. lunatus* forms symbiotic relationships with bacteria such as *Bradyrhizobium elkani*, *Rhizobium etli*, *Rhizobium phaseoli* (Araujo *et al.*,

2015) and *Mesorhizobium* (Santos *et al.*, 2011). The response of *P. lunatus* to *Rhizobium* inoculation varies depending on the selected strain. *P. lunatus* nodulated by *Bradyrhizobium* exhibits superior nodule and growth parameters, chlorophyll levels and nitrogen fixation values compared to those nodulated by *Rhizobium* strains (Costa Neto *et al.*, 2017). For organic manure, it is recommended to incorporate 3–5 tons per acre of animal or chicken fertilizer several weeks before planting, as suggested by Long *et al.* (2014).

8.8.6 Seed maturation

Perennial types of *P. lunatus* take 6–8 months to mature and can be gathered around 11 months after planting, whereas bushy types develop speedily and provide seeds in around 2–3 months (Ecocrop, 2011). It requires dry seasons for seed maturation (Temegne *et al.*, 2020). Bush types provide enormous amounts of beans sufficient for use as green beans around 2 months 20 days after sowing. The duration of the green bean harvest is close to 1 month and 20 days. Pole varieties of *P. lunatus* develop vines around 10–14 days after planting and continue to grow for a longer period compared to bush types. Beans intended for consumption as green beans, for canning, or as dehydrated green beans should be harvested when they have reached their peak size but before they turn white.

Crops require time to mature and yield higher returns if beans are harvested before reaching full maturity. For dehydrated beans, harvesting should occur at intervals of 14–21 days throughout the growth cycle. This is especially crucial for cultivars with prominent seeds, as the seedpods become exposed and the beans can burst if not promptly harvested upon maturity. Mechanical bean cutters are used for harvesting in extensive plantings, whether for green or dehydrated beans. Green beans retain their freshness if left in the seedpods after harvesting. Beans inside the pods are susceptible to heat, however, and can quickly deteriorate unless stored in a cool, well-ventilated area. Allowing the seedpods to wilt slightly after harvesting can aid in their preservation. If green beans are stored for extended periods, they should be unrolled briefly after harvesting to allow the bean surface to dry, preventing damage from mould growth, as highlighted by Funchess (1936).

8.8.7 Harvesting

The harvesting process for green and mature seedpods of upright types usually involves manual picking. In warm tropical regions, crops are sometimes collected and dried in the field, with seedpods detached, and stems and leaves used as livestock feed. Erect varieties, which mature uniformly and have seedpods standing above the ground, can be mechanically harvested, as described by Baudoin (1991, 2006). Harvesting should take place in suitable weather conditions to prevent seedpod damage from dry winds, fungus or rain-induced staining. The process involves several steps, including cutting, swathing and separating, each requiring specific equipment. Cutting occurs when about two-thirds of the seedpods have completed their cycle and turned yellow. The crop is cut using cutters attached to a tractor, positioned two to three inches below ground level. Swathing is done either at night or early in the morning, depending on the weather and crop condition, to avoid bursting, shrinking and sunburn. A thrasher with two or three cylinders separates the seedpods from the stems with minimal damage. The beans are bulk transported and then loaded onto pick-up trucks for transfer to storage. After cleaning processes, using an air-screen sieve, gravity table, aspirator and destoner to remove gravel, soil particles and plant fragments, the moisture content of the beans is assessed before packing, following the guidelines provided by Long *et al.* (2014). In some hot regions, seeds are occasionally stored in containers or baskets and protected with a layer of shingles or ashes (PROTA, 2014). *P. lunatus* may produce up to 2–8 tons of fresh seeds; however, harvests fluctuate depending on the kind of varieties and developing environments. In tropical regions, uphill forms developed in pure stands may produce up to 3–4 tons per hectare of dehydrated seeds, and the bushy forms produce 2–2.5 tons per hectare in investigated environments (Baudoin, 2006). Significantly greater harvests in great warmth and high moisture environments may be obtained from some African landraces (Ibeawuchi, 2007). Seed harvest is around 0.2–0.6 kg per hectare in companion planting (Ecocrop, 2011), whereas, in Madagascar, fresh matter yield was 15 tons per hectare (Baudoin, 2006).

8.8.8 Cropping systems

Phaseolus lunatus can be grown separately or in conjunction with other plant species. *P. lunatus* can be developed in rotation with *Solanum lycopersicon*, *Gossypium hirsutum*, *Carthamus tinctorius*, *Zea mays*, *Triticum* sp., *Medicago sativa*, *Oryza sativa*, and further vegetable and seed plants such as Cucurbitaceae and *Helianthus* spp., as well as fruit-trees species. They are well-suited for companion planting in young orchards until tree growth affects dehydrated bean gathering engines. Numerous disease-causing pathogens are passed on through the land or result from plant residues. A three-year plant rotation helps reduce the transmission of pathogens that attack *P. lunatus*. A good rotation requires that dry beans will not be grown in consecutive years. An interval of 2–3 years between bean plants is strongly advised. In small crop rotations, dry beans yield better results than Poaceae crops, as noted by Long *et al.* in 2014.

8.8.9 Weed management

When the land has been adequately prepared before sowing, weeds can be controlled easily through shallow tilling. As the first crop of *P. lunatus* matures, the plants experience a slowdown in growth and often shed some leaves. This slowdown provides an opportunity for undesired plants and grasses to establish themselves in the spaces between the rows and within the drill marks, coinciding with the beginning of the second growth and flowering phase of *P. lunatus*. During this period, it is essential to eliminate the unwanted plants and grasses in the spaces between the rows using shallow tilling, and those within the drill marks should be removed, as advised by Funchess in 1936.

8.8.10 Pest and disease management

The most effective method to control various types of beetles, including flea, bean leaf and

Mexican bean beetles, involves maintaining clean borders and removing debris where the beetles hibernate. This can be achieved by eliminating climbing plants after harvesting *P. lunatus* and systematically covering both sides of the leaves, especially the underside, with a sprayer or dust. Sulfur can help regulate spiders, hoppers and some infections. Severe damage caused by nematodes can be avoided by rotating *P. lunatus* with corn or using nematode-resistant cultivars of *Vigna unguiculata*. It's also beneficial to practice regular rotation and use uncontaminated seeds. If necessary, crops can be treated with sprinkling sulfur (Funchess, 1936). In *P. lunatus*, various insecticides such as organophosphates, pyrethroids, or carbamates are commonly used. Natural regulation involving predators like *Pediobius foveolatus* can also be employed for insect control. The use of early-maturing trap plants can be effective for managing beetles in late-planted *P. lunatus* crops.

Effective control of downy mildew can be achieved with fungicide preparations containing mefanoxam/Cu and azoxystrobin. White mould poses a significant threat to young *P. lunatus* plants. Fungicides such as thiophanate methyl, Endura and the biological fungicide *Coniothyrium minitans* are currently used to manage this mould (Everts and Zhou, 2006). Fungus attacks fading blooms and injured tissues, thriving in prolonged moist conditions. Proper irrigation management and rotation practices are essential for control (Blessing *et al.*, 2003). Chemicals used to combat downy mildew include Ridomil Gold/Coppers, fixed copper and Phostrol (Mulrooney *et al.*, 2006); however, copper alone is a preventive measure. Cultural practices such as rotation, burying old crop residues and planting resistant *P. lunatus* varieties are non-chemical methods used to manage the infection. Planting rows aligned with prevailing winds can reduce water accumulation and decrease infection rates (Blessing *et al.*, 2003). Partial regulation of *P. lunatus* seedpod blight can be achieved through irrigation management and avoiding nighttime watering. Growers should avoid fields where Cucurbitaceae and peppers were cultivated in the previous season. Fixed copper sprayers used for downy mildew control can be helpful in fields with a history of *Phytophthora* blight. Acrobat is effective against downy mildew and can aid in resistance management when used with Ridomil, potentially reducing seedpod blight (Blessing *et al.*, 2003).

To control anthracnose, rotating crops and avoiding *P. lunatus* cultivation for at least 24 months can be beneficial. TopsinM can be used for regulation if necessary. Varietal differences in sensitivity to anthracnose have been observed. Root rots, including damping-off and seedling injury, are caused by pathogens like *R. solani*, *Pythium* spp. and *Fusarium* spp. Ridomil Gold PC11G is effective against *Pythium* but ineffective in controlling *Rhizoctonia* at specified doses (Blessing *et al.*, 2003). Plant management practices include rotation with non-leguminous plants and planting in well-drained or very dry soils, as well as tilling below plant debris. *Rhizoctonia* seedpod rot leads to seedpod burning and staining. Proper plant architecture and irrigation management can help control this problem (Blessing *et al.*, 2003).

Maize earworm can cause extensive damage to *P. lunatus* by injuring pin and flat seedpods, leading to total bean loss. Damage to mature seedpods results in harvest loss and decreased seed value due to cracked seeds. Regular inspections are recommended for effective plant management. Pyrethroids and carbamates are used to manage earworms (Blessing *et al.*, 2003). Lygus sprays can eliminate worms in *P. lunatus* farms (Long *et al.*, 2014). Stink bugs and lygus bug species can be managed using insecticides such as Dimethoate, Lannate, Capture and Mustang (Blessing *et al.*, 2003).

Spider mites, arthropods that attack *P. lunatus*, feed on sap and thrive in hot, arid conditions. Control measures include Dimethoate and bifenthrin. França *et al.* (2018) observed that *P. lunatus* cultivars should possess distinct resistance levels to mites. Treatment against potato leafhoppers, which attack leaf tips, should be applied during vegetative, blossoming and pod expansion stages based on the level of infestation. The initial symptoms include the rising and falling of leaves during sprouting.

Implementing plant management practices before planting is crucial to minimize financial losses caused by seed corn maggot. A combination of cultural strategies can be employed, including tilling cover crops at least one month before sowing, burying shelter plants entirely to reduce fly attraction, and avoiding heavy fertilizer application near sowing. Insecticides such as Diazinon, Chlorpyrifos, Disyston and Thimet can be used to control seed corn maggots (Blessing *et al.*, 2003). Unwanted plants often host pests and infections that attack *P. lunatus*, and many herbicides are utilized for weed control (Vangessel *et al.*, 2000; Temegne *et al.*, 2020).

8.9 Future Directions

Phaseolus lunatus offers promising prospects in tropical environments because of its high yield potential, extensive root system, and resilience to water scarcity and disease. There has been significant evolution through genetic enhancement. However, much work is yet to be done in many tropical regions, particularly in developing more reliable and fruitful varieties for the humid, subhumid and semi-arid tropics. Most studies on the genetic variability of lima bean showed a loss of variabilities. The evidence indicating a decline in the genetic diversity of plant varieties is a global issue and poses a significant threat to future food security. The substantial genetic variation observed within accessions underscores the need for strategic approaches that enable the preservation of the broadest possible range of accessions. This is essential to ensure the conservation of the genetic diversity of *P. lunatus* within its primary regions worldwide. While the findings from various studies are crucial for the preservation of lima bean in areas of production, further research is necessary to address the challenges faced by *P. lunatus* in traditional agricultural systems globally. It is imperative to develop comprehensive programmes for both *ex-situ* and *in-situ* conservation to prevent the loss of these varieties. These efforts are vital to safeguard against future reductions in genetic and biological diversity within existing cultivated populations. Conventional breeding methods have been well-proven in the development of *P. lunatus* cultivars. This selection is, however, still largely a function of subjective evaluation and empirical selection. Molecular-marker-assisted breeding, unlike traditional breeding methods that are labour-intensive, time-consuming and challenging,

offers significant opportunities, breakthroughs and perspectives for conventional improvement. These innovative techniques have been employed for enhancing *P. lunatus* and should be encouraged for further research endeavours in the future. Given its extensive assortment of ecological acclimations as well as widespread phenotypic convergence, *P. lunatus* is certainly a useful example for evolutionary researchers fascinated with studying the adaptations and limitations that shape plant phenotypic variability. Upcoming research should focus on:

- evaluating the presence or absence of adaptive diversity in both wild and cultivated lima bean and examining its correlation with environmental factors, which could be crucial in the context of future climate change;
- generating *in-situ* collections of *P. lunatus* in the different production regions;
- exploring genomic evidence to identify parallel selection patterns in genomic regions throughout the domestication processes in Mesoamerica and the Andes;
- assessing the evolution of similar domestication traits of local Mesoamerican and Andean accessions;
- understanding the relationship between the local accessions of the different pools using population demographic and genetic approaches;
- conducting ethnobiological research on the domesticated gene pool in the traditional agricultural zones of the different regions;
- assessing the adverse effect of breeding programmes on the genetic variability determining local *P. lunatus* accessions grown by subsistence farmers;
- specifically orienting improvement programmes towards production difficulties linked to climate change;
- determining the responses of rhizosphere microorganisms to wild and cultivated varieties of lima bean;
- studying the influence of the reproduction of *P. lunatus* on the microbial flora in the nodules and rhizosphere; and
- determining the ability of different strains of *Rhizobium* to increase the growth and harvest of *P. lunatus*.

References

Adebo, J.A. (2023) A review on the potential food application of Lima beans (*Phaseolus lunatus* L.), an underutilized crop. *Applied Sciences* 13, 1996. DOI: 10.3390/app13031996

Alcázar-Valle, M., García-Morales, S., Mojica, L., Morales-Hernández, N., Sánchez-Osorio, E. *et al.* (2021) Nutritional, antinutritional compounds and nutraceutical significance of native bean species (*Phaseolus* spp.) of Mexican cultivars. *Agriculture* 11, 1031. DOI: 10.3390/agriculture11111031

Andueza-Noh, R.H., Serrano-Serrano, M.L., Chacón Sánchez, M.I., Sanchéz del Pino, I., Camacho-Pérez, L. *et al.* (2013) Multiple domestications of the Mesoamerican gene pool of Lima bean (*Phaseolus lunatus* L.): evidence from chloroplast DNA sequences. *Genetic Resources and Crop Evolution* 60, 1069–1086. DOI: 10.1007/s10722-012-9904-9

Aragão, F.J.L., Brondani, R.P.V. and Burle, M.L. (2011) *Phaseolus*. In: Kole, C. (ed) *Wild Crop Relatives: Genomic and Breeding Resources: Legume Crops and Forages*. Springer, Berlin, Heidelberg, pp. 223–236. DOI: 10.1007/978-3-642-14387-8_11

Araujo, A.S.F., Lopes, A., Gomes, R.L.F., Junior, J.E.A.B., Antunes, J.E.L., Pereira de Lyra, Md.C.C. and Figueiredo, M.V.B. (2015) Diversity of native rhizobia-nodulating *Phaseolus lunatus* in Brazil. *Legume Research* 38(5), 653–657. DOI: 10.18805/lr.v38i5.5946

Ballhorn, D.J., Kautz, S., Lion, U. and Heil, M. (2008) Trade-offs between direct and indirect defences of Lima bean (*Phaseolus lunatus*). *Journal of Ecology* 96(5), 971–980. DOI: 10.1111/j.1365-2745.2008.01404.x

Baudet, J.C. (1977) Taxonomic status of the cultivated types of Lima bean (*Phaseolus lunatus* L.). *Tropical Grain Legume Bulletin* 7, 29–30.

Baudoin, J.P. (1991) La culture et l'amélioration de la légumineuse alimentaire *Phaseolus lunatus* L. en zones tropicales. CTA (Centre Technique de Coopération Agricole et Rurale, Ede, Pays-Bas) et FSAGx (Faculté des Sciences Agronomiques de Gembloux, Belgique), Gembloux, Belgique, pp. 209.

Baudoin, J.P. (1993) Lima bean: *Phaseolus lunatus* L. In: Kalloo, G. and Bergh, B.O. (eds) *Genetic Improvement of Vegetable Crops*. Pergamon, Amsterdam, pp. 391–403. DOI: 10.1016/B978-0-08-040826-2.50031-X

Baudoin, J.P. (2002) Amélioration des plantes protéagineuses. Les légumineuses alimentaires (*Phaseolus, Vigna, Cajanus*, etc.). In: Demol, J. (Coordinator) *Amélioration des Plantes. Application aux Principales Espèces Cultivées en Régions Tropicales*. Les Presses Agronomiques de Gembloux, Gembloux, Belgique, pp. 351–392.

Baudoin, J.P. (2006) *Phaseolus lunatus* L. In: Brink, M. and Belay, G. (eds), *Record from Protabase*. PROTA (Plant Resources of Tropical Africa/Ressources végétales de l'Afrique tropicale), Wageningen, Netherlands.

Bitocchi, E., Rau, D., Bellucci, E., Rodriguez, M., Murgia, M.L. *et al.* (2017) Beans (*Phaseolus* spp.) as a model for understanding crop evolution. *Frontiers in Plant Science* 8, 722. DOI: 10.3389/fpls.2017.00722

Blessing, B., Carey, D.W., Carlisle, K., DuBois, H., Hill, R., Joseph, H. and Ritter, B. (2003) *Lima beans in Delaware, New Jersey and Maryland Eastern shore pest management strategic plan*. Workgroup IPM Center, Delaware/Maryland/New Jersey, p. 36. Available at: https://ipmdata.ipmcenters.org/documents/pmsps/MidAtlLimaBean.pdf (accessed 16 May 2024).

Brillada, C., Nishihara, M., Shimoda, T., Garms, S., Boland, W., Maffei, M.E. and Arimura, G. (2013) Metabolic engineering of the C16 homoterpene TMTT in *Lotus japonicus* through overexpression of (E,E)-geranyllinalool synthase attracts generalist and specialist predators in different manners. *New Phytologist* 200(4), 1200–1211. DOI: 10.1111/nph.12442

Calles, T. (2016) The international year of pulses: what are they and why are they important. *Agriculture for Development* 26, 40–42.

Camacho-Pérez, L., Martínez-Castillo, J., Mijangos-Cortés, J.O., Ferrer-Ortega, M.M., Baudoin, J.P. and Andueza-Noh, R.H. (2018) Genetic structure of Lima bean (*Phaseolus lunatus* L.) landraces grown in the Mayan area. *Genetic Resources and Crop Evolution* 65, 229–241. DOI: 10.1007/s10722-017-0525-1

Casquero, P.A., Lema, M., Santalla, M. and De Ron, A.M. (2006) Performance of common bean (*Phaseolus vulgaris* L.) landraces from Spain in the Atlantic and Mediterranean Environments. *Genetic Resources and Crop Evolution* 53, 1021–1032. DOI: 10.1007/s10722-004-7794-1

Chacón Sánchez, M.I. (2009) Darwin y la domesticación de plantas en las américas: el caso del maíz y el fríjol. *Acta Biológica Colombiana* 14, 351–364.

Chacón-Sánchez, M.I. and Martínez-Castillo, J. (2017) Testing domestication scenarios of Lima bean (*Phaseolus lunatus* L.) in Mesoamerica: insights from genome-wide genetic markers. *Frontiers in Plant Science* 8, 1551. DOI: 10.3389/fpls.2017.01551

Chacón-Sánchez, M.I., Martínez-Castillo, J., Duitama, J. and Debouck, D.G. (2021) Gene flow in *Phaseolus* beans and its role as a plausible driver of ecological fitness and expansion of cultigens. *Frontiers in Ecology and Evolution* 9, 618709. DOI: 10.3389/fevo.2021.618709

CIAT (2022) Gene banks. Available at: https://alliancebioversityciat.org/services/genebanks or http://ciat.cgiar.org (accessed 27 October 2023).

CNPAF-EMBRAPA (2017) Catalog of fava (*Phaseolus lunatus* L.) preserved at Embrapa. Available at: https://www.embrapa.br/en/busca-de-publicacoes/-/publicacao/1074983/catalogo-de-fava-phaseolus-lunatus-l-conservada-na-embrapa (accessed 28 October 2023).

Costa Neto, da V.P., Mendes, J.B.S., de Araújo, A.S.F., de Alcântara Neto, F., Bonifacio, A. and Rodrigues, A.C. (2017) Symbiotic performance, nitrogen flux and growth of Lima bean (*Phaseolus lunatus* L.) varieties inoculated with different indigenous strains of rhizobia. *Symbiosis* 1–9. DOI: 10.1007/s13199-017-0475-6

Cueva, V.D.C. (2018) Guía para el cultivo urbano del pallar Moche, *Phaseolus lunatus*. *Pueblo Continente* 28, 393–417.

Cuny, M.A.C., Shlichta, G.J. and Benrey, B. (2017) The large seed size of domesticated Lima beans mitigates intraspecific competition among seed beetle larvae. *Frontiers in Ecology and Evolution* 5. DOI: 10.3389/fevo.2017.00145

da Silva, J.L., Mendes, L.W., Rocha, S.M.B., Antunes, J.E.L., Oliveira, L.M.D.S. *et al*. (2022) Domestication of Lima Bean (*Phaseolus lunatus*) changes the microbial communities in the rhizosphere. *Microbial Ecology* 1–11. DOI: 10.1007/s00248-022-02028-2

da Silva, V.B., Gomes, R.L.F., Lopes, Â.C.D.A., Dias, C.T.D.S. and Silva, R.N.O. (2015) Genetic diversity and promising crosses indication in Lima bean (*Phaseolus lunatus*) accessions. *Semina Ciências Agrárias* 36, 683. DOI: 10.5433/1679-0359.2015v36n2p683

Dadther-Huaman, H., Zamata-Guzman, R. and Casa-Coila, V.H. (2022) Caracterización morfológica y evaluación agronómica de accesiones de frijol Lima (*Phaseolus lunatus*) de la Colección Nacional de Germoplasma del INIA, Ica, Perú. *Bioagro* 35 (1), 59–68. DOI: 10.51372/bioagro351

David, L. (1984) *Appropriate Technology: Traditional Field Crops*. Information, Collection and Exchange Manual, Peace Corps, p. 234.

Debouck, D.G. and Morales de Borja, A.J. (2012) A report on Bioprospecting for Phaseolus species in El Salvador, Central America. Cooperative work between the International Center for Tropical Agriculture (CIAT), and the Centro Nacional de Tecnología Agropecuaria y Forestal (CENTA), with the support of the Global Crop Diversity Trust. CIAT Reports. Available at: https://hdl.handle.net/10568/81412 (accessed 14 May 2024).

Delgado-Paredes, G.E., Vásquez-Díaz, C., Esquerre-Ibañez, B., Bazán-Sernaqué, P., Rojas-Idrogo, C. *et al*. (2021) In vitro tissue culture in plants propagation and germplasm conservation of economically important species in Peru. *Scientia Agropecuaria* 12, 337–349. DOI: 10.17268/sci.agropecu.2021.037

Delgado-Salinas, A., Turley, T., Richman, A. and Lavin, M. (1999) Phylogenetic analysis of the cultivated and wild species of *Phaseolus* (Fabaceae). *Systematic Botany* 24, 438–460.

Delgado-Salinas, A., Bibler, R. and Lavin, M. (2006) Phylogeny of the genus *Phaseolus* (Leguminosae): A recent diversification in an ancient landscape. *Systematic Botany* 31(4), 779–791.

Dohle, S. (2017) Development of resources for Lima bean (*Phaseolus lunatus*) breeding and genetics research. PhD Thesis, University of California, California, USA.

Ecocrop (2011) Ecocrop database. FAO, Rome. Available at: https://gaez.fao.org/pages/ecocrop (accessed 14 May 2024).

Ernest, E.G. and Kee, E. (2008) Lima bean breeding and genetics research at the University of Delaware. *Annual Report of the Bean Improvement Cooperative* 51, 54–55.

Ernest, E.G., Kee, W.E., Santamaria, L. and Evans, T.A. (2006) Inheritance of resistance to Lima bean downy mildew (*Phytophthora phaseoli*) and preliminary Lima improvement efforts. *Annual Report of the Bean Improvement Cooperative* 49, 37–38.

Ernest, E.G., Evans, T.A. and Gregory, M.F. (2011) An investigation of the sources of resistance to Lima bean downy mildew (*Phytophthora phaseoli*) races E and F and identification of fordhook lima bean cultivar with downy mildew resistance. *Annual Report of the Bean Improvement Cooperative* 54, 98–99.

Espinoza, L., Tarazona Delgado, R. and Pablo, C. (2022) Leguminosas proteaginosas y oleaginosas. In: *Las Leguminosas y su Microbioma en la Agricultura Sostenible*. Universidad Nacional de Barranca, Peru, pp. 43–72.

Everts, K.L. and Zhou, X.-G. (2006) Managing white mold of Lima beans with reduced risk fungicides and biofungicides. *Reports of Bean Improvement Cooperative and National Dry Bean Council Research Conference* 49, 27–28. Available at: https://handle.nal.usda.gov/10113/IND43805311 (accessed 14 May 2024).

Ezeagu, I.E. and Ibegbu, M.D. (2010) Biochemical composition and nutritional potential of ukpa: a variety of tropical Lima beans (*Phaseolus lunatus*) from Nigeria - a short report. *Polish Journal of Food and Nutrition Sciences* 60.

Fagbédji, R.F., Djedatin, L.G., Nanoukon, C., Kinhoegbe, G., Havivi, A., Yédomonhan, H. and Agbangla, C. (2022) Insight into genetic diversity of cultivated Lima bean (*Phaseolus lunatus* L.) in Benin. *American Journal of Molecular Biology* 13(1). DOI: 10.4236/ajmb.2023.131003

Federal Research Center (2023) Collection of plant genetic resources VIR plant genetic resources gene bank. Available at: http://www.vir.nw.ru/unu-kollektsiya-vir/ (accessed 28 October 2023).

Ferreira-Miani, P. (2004) Exchange of genetic resources. In: IUCN (ed) *Accessing Biodiversity and Sharing the Benefits: Lessons from Implementing the Convention on Biological Diversity*. IUCN, (No. 54), p. 79.

Ferrer-Vilca, T.H. and Valverde-Rodríguez, A. (2020) Rendimiento del frejol (*Phaseolus vulgaris* L.) variedad canario con tres fuentes de abonos orgánicos en el distrito de Cholón, Huánuco-Perú. *Revista de Investigacion Agraria y Ambiental* 2, 33–44. DOI: 10.47840/ReInA.2.3.901

França, S.M., Silva, P.R.R., Gomes-Neto, A.V., Gomes, R.L.F., da Silva Melo, J.W. and Breda, M.O. (2018) Resistance of Lima bean (*Phaseolus lunatus* L.) to the red spider mite *Tetranychus neocaledonicus* (Acari: Tetranychidae). *Frontiers in Plant Science* 9, 1466. DOI: 10.3389/fpls.2018.01466

Freyre, R., Ríos, R., Guzmán, L., Debouck, D.G. and Gepts, P. (1996) Ecogeographic distribution of *Phaseolus* spp. (Fabaceae) in Bolivia. *Economic Botany* 50(2), 195–215.

Funchess, M.J. (1936) Lima Beans. *Agricultural Experiment Station of the Alabama Polytechnic Institute* 14, 1–4.

García, T., Duitama, J., Zullo, S., Gil, J., Ariani, A. *et al.* (2021) Comprehensive genomic resources related to domestication and crop improvement traits in Lima bean. *Nature Communications* 12. DOI: 10.1038/s41467-021-20921-1

Gawande, S.S., Ningot, E.P., Rathod, D. and Parkhe, S. (2020) Effect of different spacings on growth and yield of Lima bean (*Phaseolus lunatus* L.). *International Journal of Chemical Studies* 8(6), 202–205. DOI: 10.22271/chemi.2020.v8.i6c.10770

Gepts, P. (2000) A Phylogenetic and genomic analysis of crop germplasm: a necessary condition for its rational conservation and use. In: Gustafson, J.P. (Ed.), *Genomes, Stadler Genetics Symposia Series*. Springer US, Boston, Massachusetts, pp. 163–181. DOI: 10.1007/978-1-4615-4235-3_13

Gepts, P. (2006) Plant genetic resources conservation and utilization: the accomplishments and future of a societal insurance policy. *Crop Science* 46, 2278–2292. DOI: 10.2135/cropsci2006.03.0169gas

Greene, S.L., Khoury, C.K. and Williams, K.A. (2018) Wild plant genetic resources in North America: an overview. North American crop wild relatives. *Conservation Strategies* 1, 3–31.

Gross, B.L. and Olsen, K.M. (2010). Genetic perspectives on crop domestication. *Trends in Plant Science* 15(9), 529–537.

Gunadi, A., Dean, E.A. and Finer, J.J. (2019) Transient transformation using particle bombardment for gene expression analysis. In: Kumar, S., Barone, P. and Smith, M. (eds) *Transgenic Plants: Methods and Protocols, Methods in Molecular Biology*. Springer/Business Media, LLC, pp. 67–79. DOI: 10.1007/978-1-4939-8778-8_5

Gutiérrez Salgado, A., Gepts, P. and Debouck, D.G. (1995) Evidence for two gene pools of the Lima bean, *Phaseolus lunatus* L., in the Americas. *Genetic Resources and Crop Evolution* 42, 15–28. DOI: 10.1007/BF02310680

Hacquenghem, A.-M. (1984) El hombre y el pallar en la iconografía moche. *Anthropologica* 2, 403–411.

Heredia-Pech, M., Chávez-Pesqueira, M., Ortiz-García, M.M., Andueza-Noh, R.H., Chacón-Sánchez, M.I. and Martínez-Castillo, J. (2022) Consequences of introgression and gene flow on the genetic structure and diversity of Lima bean (*Phaseolus lunatus* L.) in its Mesoamerican diversity area. *PeerJ* 10, e13690. DOI : 10.7717/peerj.13690

Ibeawuchi, I. (2007) Landrace legumes: synopsis of the culture, importance, potentials and roles in agricultural production systems. *Journal of Biological Sciences* 7(3), 464–474. DOI: 10.3923/jbs.2007.464.474

ICGR-CAAS (2023) National Gene bank. Available at: https://www.cgris.net/icgr/icgrgb.html (accessed 28 October 2023).

IITA-CGIAR (2023) Our genetic resources. Available at: https://www.cgiar.org/research/center/iita/ (accessed 28 October 2023).

INIA (2023a) Germplasm Bank. Available at: https://www.inia.gob.pe/banco-de-germoplasma/ (accessed 28 October 2023).

INIA (2023b) Germplasm Bank Network. Available at: https://www.recursosgeneticos.com/#solicitudes-de-germoplasma (accessed 27 October 2023).

INIFAP (2018) Germplasm Bank. Available at: https://www.gob.mx/inifap (accessed 28 October 2023).

IPK (2023) Germplasm Bank. Available at: https://www.ipk-gatersleben.de/en/ (accessed 28 October 2023).

Jackson, M.A. (2021) The symbolic value of food in Moche iconography. In: Staller, J.E. (ed.) *Andean Foodways: Pre-Columbian, Colonial, and Contemporary Food and Culture. The Latin American Studies Book Series*. Springer, Cham, Switzerland, pp. 257–279. DOI: 10.1007/978-3-030-51629-1_10

Jarvis, D.I., Zoes, V., Nares, D. and Hodgkin, T. (2004) *On-farm Management of Crop Genetic Diversity and the Convention on Biological Diversity Programme of Work on Agricultural Biodiversity*. International Plant Genetic Resources Institute, Rome, Italy.

Kaplan, L. (1965) Archeology and domestication in American *Phaseolus* (beans). *Economic Botany* 19, 358–368. DOI: 10.1007/BF02904806

Kaplan, L. and Kaplan, L.N. (1988) *Phaseolus* in archaeology. In: Gepts, P. (ed.) *Genetic Resources of Phaseolus Beans: Their Maintenance, Domestication, Evolution and Utilization, Current Plant Science and Biotechnology in Agriculture*. Springer Dordrecht, Netherlands, pp. 125–142. DOI: 10.1007/978-94-009-2786-5_7

Kaplan, L. and Lynch, T.F. (1999) *Phaseolus* (Fabaceae) in archaeology: AMS radiocarbon dates and their significance for pre-Colombian agriculture. *Economic Botany* 53, 261–272.

Kelly, J.D. (2010) The story of bean breeding. White paper prepared for Bean CAP & PBG Works on the topic of dry bean production and breeding research in the U.S. Michigan State University. U.S. Michigan State University, Michigan.

Li, F., Li, W., Lin, Y-J., Pickett, J.A., Birkett, M.A. *et al.* (2018) Expression of lima bean terpene synthases in rice enhances recruitment of a beneficial enemy of a major rice pest. *Plant, Cell and Environment* 41 (1), 111–120.

Lioi, L. (1994) Morphotype relationships in Lima bean (*Phaseolus lunatus* L.) deduced from variation of the evolutionary marker phaseolin. *Genetic Resources and Crop Evolution* 41, 81–85.

Lioi, L. and Piergiovanni, A.R. (2013) 2 - European common bean. In: Singh, M., Upadhyaya, H.D. and Bisht, I.S. (eds) *Genetic and Genomic Resources of Grain Legume Improvement*. Elsevier, Oxford, pp. 11–40. DOI: 10.1016/B978-0-12-397935-3.00002-5

Lioi, L., Lotti, C. and Galasso, I. (1998) Isozyme diversity, RFLP of the rDNA and phylogenetic affinities among cultivated Lima beans, *Phaseolus lunatus* (Fabaceae). *Plant Systematics and Evolution* 213(3), 153–164. DOI: 10.1007/BF00985196

Long, R., Temple, S., Meyer, R., Schwankl, L., Godfrey, L. *et al.* (2014) *Lima production in California*. ARN publication, California, USA.

López-Alcocer, J. de J., Lépiz-Ildefonso, R., González-Eguiarte, D.R., Rodríguez-Macías, R. and López-Alcocer, E. (2016) Morphological variability of wild *Phaseolus lunatus* L. from the western region of México. *Revista Fitotecnia Mexicana* 39, 49–58.

Lustosa-Silva, J.D., Ferreira-Gomes, R.L., Martinez-Castillo, J., Carvalho, L.C.B., de Oliveira, L.F. *et al.* (2022) Genetic diversity and erosion in Lima bean (*Phaseolus lunatus* L.) in Northeast Brazil. *Genetic Resources and Crop Evolution* 69(8), 2819–2832. DOI: 10.1007/s10722-022-01402-w

Martínez-Castillo, J., Camacho-Pérez, L., Villanueva-Viramontes, S., Andueza-Noh, R.H. and Chacón-Sánchez, M.I. (2014) Genetic structure within the Mesoamerican gene pool of wild *Phaseolus lunatus* (Fabaceae) from Mexico as revealed by microsatellite markers: implications for conservation and the domestication of the species. *American Journal of Botany* 101(5), 851–864. DOI: 10.3732/ajb.1300412

Martínez-Castillo, J., Araujo, A.S.F., Chacón-Sánchez, M.I., Santos, L.G., Lopes, A.C.A. *et al.* (2022) Lima Bean International Network: From the origin to the plant breeding. *Research Square* 1–10. DOI: 10.21203/rs.3.rs-2310209/v1

Martínez-Nieto, M.I., Estrelles, E., Prieto-Mossi, J., Roselló, J. and Soriano, P. (2020) Resilience capacity assessment of the traditional Lima Bean (*Phaseolus lunatus* L.) landraces facing climate change. *Agronomy* 10, 758. DOI: 10.3390/agronomy10060758

Martínez-Nieto, M.I., González-Orenga, S., Soriano, P., Prieto-Mossi, J., Larrea, E. *et al.* (2022) Are traditional lima bean (*Phaseolus lunatus* L.) landraces valuable to cope with climate change? Effects of drought on growth and biochemical stress markers. *Agronomy* 12, 1715. DOI: 10.3390/agronomy12071715

Matilla, A., Gallardo, M. and Puga-Hermida, M.I. (2005) Structural, physiological and molecular aspects of heterogeneity in seeds: a review. *Seed Science Research* 15, 63–76. DOI: 10.1079/SSR2005203

Matsubara, M. and Zúñiga-Dávila, D. (2015) Phenotypic and molecular differences among rhizobia that nodulate *Phaseolus lunatus* in the Supe valley in Peru. *Annals of Microbiology* 65, 1803–1808. DOI: 10.1007/s13213-015-1054-9

Melgar, L.M.E. (2021) Caracterización morfológica de 30 genotipos de *Phaseolus lunatus* L. de la costa del Perú. *Ciencia Latina Revista Científica Multidisciplinar* 5, 10039–10053. DOI: 10.37811/cl_rcm.v5i5.1051

Mhora, T.T., Ernest, E.G., Wisser, R.J., Evans, T.A., Patzoldt, M.E. *et al.* (2016) Genotyping-by-sequencing to predict resistance to Lima bean downy mildew in a diversity panel. *Phytopathology* 106 (10), 1152–1158. DOI: 10.1094/PHYTO-02-16-0087-FI

Montero-Rojas, M., Ortiz, M., Beaver, J.S. and Siritunga, D. (2013) Genetic, morphological and cyanogen content evaluation of a new collection of Caribbean Lima bean (*Phaseolus lunatus* L.) landraces. *Genetic Resources and Crop Evolution* 60, 2241–2252. DOI: 10.1007/s10722-013-9989-9

Mulrooney, R.P., Davey, J.F., Riverhead, N.Y. and Evans, T.A. (2006) Chemical control strategies for downy mildew (*Phytophthora phaseoli*) of baby Lima bean. *Annual Report of the Bean Improvement Cooperative* 49, 29–30.

Myers, J.R., Formiga, A.K. and Janick, J. (2022) Iconography of beans and related legumes following the Columbian exchange. *Frontiers in Plant Science*, 13. DOI: 10.3389/fpls.2022.851029

Nasir, L.N.L., Feyissa, T. and Asfaw, Z. (2021) Genetic diversity analysis of Lima bean (*Phaseolus lunatus* L.) landrace from Ethiopia as revealed by ISSR marker. *SINET: Ethiopian Journal of Science* 44(1), 81–90. DOI: 10.4314/sinet.v44i1.8.

Nienhuis, J., Tivang, J., Skroch, P. and dos Santos, J.B. (1995) Genetic relationships among cultivars and landraces of Lima bean (*Phaseolus lunatus* L.) as measured by RAPD markers. *Journal of the American Society for Horticultural Science* 120, 300–306. DOI: 10.21273/JASHS.120.2.300

Oliveira-Silva, R.N., Lobo-Burle, M., Gomes-Pádua, J., Almeida-Lopes, A.C. de, Ferreira-Gomes, R.L. *et al.* (2017) Phenotypic diversity in Lima bean landraces cultivated in Brazil, using the Ward-MLM strategy. *Chilean Journal of Agricultural Research* 77, 35–40. DOI : 10.4067/S0718-58392017000100004

Ormeño-Orrillo, E., Vinuesa, P., Zúñiga-Dávila, D. and Martínez-Romero, E. (2006) Molecular diversity of native *Bradyrhizobia* isolated from Lima bean (*Phaseolus lunatus* L.) in Peru. *Systematic and Applied Microbiology* 29, 253–262. DOI: 10.1016/j.syapm.2005.09.002

Panda, A. and Paul, A. (2023) Genetic divergence in French bean accessions. *Legume Research* 1, 9. DOI: 10.18805/LR-5018.

Piergiovanni, A.R. and Lioi, L. (2010) Italian common bean landraces: History, genetic diversity and seed quality. *Diversity* 2, 837–862. DOI: 10.3390/d2060837

Pires, C.deJ., Costa, M.F., Zucchi, M.I., Ferreira-Gomes, R.L., Pinheiro, J.B. *et al.* (2021) The Brazilian Lima bean landraces: conservation and breeding (preprint). *In Review*. DOI: 10.21203/rs.3.rs-242568/v1

Pires, C. deJ., Costa, M.F., Zucchi, M.I., Ferreira-Gomes, R.L., Pinheiro, J.B. *et al.* (2022) Genetic diversity in accessions of Lima bean (*Phaseolus lunatus* L.) determined from agro-morphological descriptors and SSR markers for use in breeding programs in Brazil. *Genetic Resources and Crop Evolution* 1–14. DOI: 10.1007/s10722-021-01272-8

Plucknett, D.L. and Smith, N.J. (2014) *Gene Banks and the World's Food* (Vol. 457). Princeton University Press, New Jersey.

Plucknett, D.L., Smith, N.J.H., Williams, J. and Anishetty, N.M. (1983) Crop germplasm conservation and developing countries. *Science* 220(4593), 163–169.

Porch, T.G., Beaver, J.S., Debouck, D.G., Jackson, S.A., Kelly, J.D. and Dempewolf, H. (2013) Use of wild relatives and closely related species to adapt common bean to climate change. *Agronomy* 3, 433–461. DOI: 10.3390/agronomy3020433

Pratap, A., Chaturvedi, S.K., Tomar, R., Rajan, N., Malviya, N. *et al.* (2017) Marker-assisted introgression of resistance to *Fusarium* wilt race 2 in Pusa 256, an elite cultivar of *desi* chickpea. *Molecular Genetics and Genomics* 292, 1237–1245. DOI: 10.1007/s00438-017-1343-z.292

PROTA (2014) Plant Resources of Tropical Africa. Grubben G.J.H. and Denton O.A. (eds). PROTA Foundation, Wageningen, Netherlands.

Renzi, J.P., Coyne, C.J., Berger, J., von Wettberg, E., Nelson, M. *et al.* (2022) How could the use of crop wild relatives in breeding increase the adaptation of crops to marginal environments? *Frontiers in Plant Science* 13, 886162. DOI: 10.3389/fpls.2022.886162

Reynoso Camacho, R., Ríos Ugalde, M. del C., Torres Pacheco, I., Acosta Gallegos, J.A., Palomino Salinas, A.C. *et al.* (2007) El consumo de frijol común (*Phaseolus vulgaris* L.) y su efecto sobre el cáncer de colon en ratas Sprague-Dawley. *Agricultura Técnica en México* 33, 43–52.

Roberts, P.A., Matthewsa, W.C., Ehlersb, J.D. and Helms, D. (2008) Genetic determinants of differential resistance to root-knot nematode reproduction and galling in Lima bean. *Crop Breeding and Genetics* 48 (2), 553–561. DOI: 10.2135/cropsci2007.07.0384

Saini, P., Saini, P., Kaur, J.J., Francies, R.M., Gani, M. *et al.* (2020) Molecular approaches for harvesting natural diversity for crop improvement. In: Salgotra, R.K., Zargar, S.M. (eds) *Rediscovery of Genetic and Genomic Resources for Future Food Security*. Springer, Singapore, pp. 67–169. DOI: 10.1007/978-981-15-0156-2_3

Salgado, A., Gepts, P. and Debouck, D. (1995) Evidence for two gene pools of the Lima bean, *Phaseolus lunatus* L., in the Americas. *Genetic Resources and Crop Evolution* 42, 15–28. DOI: 10.1007/BF02310680

Santamaria, L., Ernest, E.G., Gregory, N.F. and Evans, T.A. (2018) Inheritance of resistance in Lima bean to *Phytophthora phaseoli*, the causal agent of downy mildew of Lima bean. *Horticultural Sciences* 53 (6), 777–781. DOI: 10.21273/HORTSCI12748-18

Santiago, F.L.A., Santiago, F.E.M., Aguiar, A.L. Jr, Nóbrega, J.C.A., Dias, B.O. and Nóbrega, R.S.A. (2019) Macronutrient fertilization on the yield components and nutrition of Lima bean. *The Journal of Agricultural Science* 11(5), 395–407. DOI: 10.5539/jas.v11n5p395

Santos, J.O., Antunes, J.E.L., Araújo, A.S.F., Lyra, M.C.C.P., Gomes, R.L.F., Lopes, A.C.A. and Figueiredo, M.V.B. (2011) Genetic diversity among native isolates of rhizobia from *Phaseolus lunatus*. *Annals of Microbiology* 61, 437–444. DOI: 10.1007/s13213-010-0156-7

Serrano-Serrano, M.L., Hernández-Torres, J., Castillo-Villamizar, G., Debouck, D.G. and Sánchez, M.I.C. (2010) Gene pools in wild Lima bean (*Phaseolus lunatus* L.) from the Americas: evidences for an Andean origin and past migrations. *Molecular Phylogenetics and Evolution* 54, 76–87. DOI: 10.1016/j.ympev.2009.08.028

Sirait, Y., Pill, W.G. and Kee W.E. Jr (1994) Lima bean (*Phaseolus lunatus* L.) response to irrigation regime and plant population density. *HortScience* 29(2), 71–73.

Smartt, J. (1985) Evolution of Grain Legumes. IV. Pulses in the Genus *Phaseolus*. *Experimental Agriculture* 21, 193–207. DOI: 10.1017/S0014479700012552

Smith, C.E. (1968) The new world centers of origin of cultivated plants and the archaeological evidence. *Economic Botany* 22(3), 253–266.

Smith, W.K., Vogelmann, T.C., DeLucia, E.H., Bell, D.T. and Shepherd, K.A. (1997) Leaf form and photosynthesis. *BioScience* 47(11), 785–793. DOI: 10.2307/1313100

Smýkal, P., Coyne, C.J., Ambrose, M.J., Maxted, N., Schaefer, H. *et al.* (2015) Legume crops phylogeny and genetic diversity for science and breeding. *Critical Reviews in Plant Sciences* 34(1–3), 43–104. DOI: 10.1080/07352689.2014.897904

Smýkal, P., Nelson, M.N., Berger, J.D. and Von Wettberg, E.J.B. (2018) The Impact of genetic changes during crop domestication. *Agronomy* 8, 119. DOI: 10.3390/agronomy8070119

Sood, S. and Gupta, N. (2017) Lima bean. In: Rana, M.K. (ed.) *Vegetable Crops Science*. CRC Press, Boca Raton, Florida, pp. 701–714. DOI: 10.1201/9781315116204

Temegne, N.C. (2011) Analysis of early indicators involved in water stress tolerance of four leguminous species. Master thesis, University of Yaoundé I, Faculty of Science Yaounde, Cameroon.

Temegne, N.C., Tsoata, E., Ngome, A.F.E., Tonfack, L.B., Agendia, A.P. and Youmbi, E. (2020) Chapter 7 - Lima bean. In: Pratap, A., Gupta, S. (eds) *The Beans and the Peas*. Elsevier, Netherlands, pp. 133–152. DOI: 10.1016/B978-0-12-821450-3.00001-9

Tian, S., Lu, P., Zhang, Z., Wu, J.Q., Zhang, H. and Shen, H. (2021) Chloroplast genome sequence of Chongming Lima bean (*Phaseolus lunatus* L.) and comparative analyses with other legume chloroplast genomes. *BMC Genomics* 22(1), 194. DOI: 10.1186/s12864-021-07467-8

Tripp, R. and van der Heide, W. (1996) The erosion of crop genetic diversity: challenges, strategies and uncertainties. *Natural Resources Perspectives* 7, 1–10.

Tsoata, E., Temegne, C.N. and Youmbi, E. (2016) Analysis of early physiological criteria to screen four Fabaceae plants for their tolerance to water stress. *International Journal of Recent Scientific Research* 7(11), 14334–14338.

Tsoata, E., Temegne, C.N. and Youmbi, E. (2017a) Analysis of early biochemical criterion to screen four Fabaceae plants for their tolerance to drought stress. *International Journal of Current Research* 9(1), 44568–44575.

Tsoata, E., Temegne, C.N. and Youmbi, E. (2017b) Analysis of early growth criterion to screen four Fabaceae plants for their tolerance to drought stress. *Research Journal of Life Sciences, Bioinformatics, Pharmaceutical and Chemical Sciences* 2(5), 94–109.

USDA-ARS (2023) National Germplasm Resources Laboratory. Available at: https://www.ars.usda.gov/northeast-area/beltsville-md-barc/beltsville-agricultural-research-center/national-germplasm-resources-laboratory/ (accessed 28 October 2023).

Vangessel, M.J., Monks, D.W. and Johnson, Q.R. (2000) Herbicides for potential use in lima bean (*Phaseolus lunatus*) production. *Weed Technology* 14 (2), 279–286. DOI: 10.1614/0890-037X(2000)014[0279:HF-PUIL]2.0.CO;2

Wilker, J., Humphries, S., Rosas-Sotomayor, J.C., Gómez Cerna, M., Torkamaneh, D. *et al.* (2020) Genetic diversity, nitrogen fixation, and water use efficiency in a panel of Honduran common bean (*Phaseolus vulgaris* L.) landraces and modern genotypes. *Plants* 9, 1238. DOI: 10.3390/plants9091238

Yuanzhen, G.U.O., Qiang, H., Xinru, Y.E., Zhi, C., Zhi H. *et al.* (2023) Morphological and SRAP markers-based genetic diversity determination on *Phaseolus lunatus* L. germplasms. *Fujian Journal of Agricultural Sciences* 38(2), 166–173. DOI: 10.19303/j.issn.1008-0384.2023.02.006

9 Runner Bean (*Phaseolus coccineus* L.)

T. Basavaraja[1]*, Anupam Tripathi[2], S. Gurumurthy[3], C. Mahadevaiah[2,4], Kanishka Chandora[6], M. Devindrappa[5] and Rahul Chandora[6]

[1]*ICAR-Indian Institute of Oilseed Research, Hyderabad, Telangana, India;* [2]*Acharya Narendra Deva University of Agriculture and Technology, Kumarganj, Ayodhya, Uttar Pradesh, India;* [3]*ICAR-National Institute of Abiotic Stress Management, Baramati, Maharashtra, India;* [4]*ICAR-Indian Institute of Horticultural Research Hesaraghatta Lake Post, Bangalore, India;* [5]*ICAR-Central Institute for Cotton Research, Nagpur, Maharashtra, India;* [6]*ICAR-National Bureau of Plant Genetic Resources-Regional Station, Shimla, Himachal Pradesh, India*

Abstract

Runner bean, a legume species originating from Mesoamerica, has a close genetic relationship with the common bean. Although classified as perennial, it is typically cultivated as an annual crop for its edible immature pods and dry seeds. Domestication of *Phaseolus* and its related species has been observed several times, indicating their usefulness and suitability for domestication. The genus *Phaseolus* contains various groups or natural gene pools. While genetic research on the common bean is extensive, runner bean has received comparatively less attention. Approximately 4685 runner bean accessions have been reported worldwide, encompassing a wide range of phenotypic variation and distinct genetic traits. This extensive genetic diversity is critical for crop improvement, providing a reservoir of disease resistance and cold-tolerance features. Nevertheless, the limited characterization of the germplasm hinders its potential as a donor species for inter-specific hybridization, thereby restricting its application in *Phaseolus* breeding efforts, such as for the improvement of common bean. Achieving the development of cultivars that are adaptable to a wide range of conditions and have a fixed growth pattern, which would make mechanical harvesting easier, is a significant challenge and opportunity for the future. The diversity and population differentiation levels of runner bean suggest that it has the potential to serve as an excellent source of variability for breeding several plant traits.

Keywords: Runner bean, genetic resource, gene pool, characterization, adaptive traits

9.1 Introduction

Runner bean (*Phaseolus coccineus* L.; 2n = 2x = 22) is an allogamous legume species and is the third most economically important species in the *Phaseolus* genus (Debouck, 2014). It is only exceeded in the genus by the common bean (*Phaseolus vulgaris* L.) and the lima bean (*Phaseolus lunatus* L.) in terms of economic importance (Arriagada *et al.*, 2021). Runner bean is grown in several regions worldwide, including South America, the UK, southern Europe and Asia, particularly China and Sri Lanka. It serves a dual purpose as both a grain and a vegetable crop. In Eastern Africa, Latin America and various other regions across the globe, the cultivation of runner

*Corresponding author: basu86.gpb@gmail.com

bean primarily revolves around the production of grain and the development of vegetable green beans (Kimani *et al.*, 2008). These beans are commonly consumed in their fresh state as a vegetable, and when fully matured, dried grain serves as an essential source of pulses. Kenya and Zimbabwe have emerged as prominent regional stakeholders in cultivating and exporting vegetable runner beans to the European market. These nations have established themselves as leaders in Africa's production of this specific leguminous crop (Kimani *et al.*, 2019). Further, runner bean is recognized for its high nutritional content (Tables 9.1 and 9.2), which includes carbohydrates, ash, lipids, crude fibre, protein and minerals (high iron and zinc). Runner bean is also high in phenolic compounds. Furthermore, these beans contain substantial amounts of protease inhibitors, saponins and lectins, all of which have been classified as anti-nutrients (Olaleke *et al.*, 2006; Celmeli *et al.*, 2018; Lopez-Martinez *et al.*, 2020). Alvarez *et al.* (1998) discovered that the average protein content in common bean (20.48%) aligned with values previously documented for this species and surpassed the average observed in runner bean landraces (16.33%).

Potassium emerged as the predominant mineral in the seed of *P. coccineus*, consistent with its status as the most prevalent mineral in agricultural commodities. Magnesium and calcium followed suit as the subsequent most abundant minerals, whereas copper had the lowest abundance (Table 9.2). None the less, recent runner bean research has primarily focused on agronomic approaches and crop improvement, largely neglecting its nutritional value. As a result, studies are urgently required to assess the bioactive component content and biological activity of this underutilized bean.

Runner bean has a considerable degree of genetic diversity, which is of great importance in the context of breeding efforts, especially because it serves as a valuable source of traits associated with resistance to diseases (fungi, bacteria and viruses) and tolerance to cold climates. The International Center for Tropical Agriculture (CIAT) has the most extensive *Phaseolus* collection globally, comprising around 40,000 accessions.

Table 9.1. Nutritional composition of *Phaseolus coccineus* grain (g/100 g).

Variety	Ash	Protein	Lipids	Carbohydrate	Crude fibre
White	4.12	18.2	1.63	67.1	6.53
Purple (dark)	4.11	18.9	1.62	67.7	6.74
Black	4.04	18.5	1.61	67.7	6.72
Purple (light)	3.77	21.9	3.78	64.30	6.22
Black	3.50	23.8	3.50	64.41	4.79
Dark brown	4.11	22.9	3.42	65.29	4.21
Light brown	4.08	21.7	3.25	65.76	5.18
Hepho	3.9	24.6	0.90	65.8	4.63

Source: Lopez-Martinez *et al.* (2020). (Table used with permission.)

Table 9.2. Mineral composition of *Phaseolus coccineus* grain (mg/100 g).

Mineral	Variety				
	Black	Purple	Brown	White	Hepho
Calcium	120.54	122.96	124.11	123.06	145.21
Zinc	2.55	2.63	2.59	2.45	1.70
Manganese	1.54	1.45	1.43	1.51	NR
Copper	0.56	0.55	0.58	0.53	NR
Iron	5.61	6.11	5.93	5.68	8.20
Sodium	12.09	10.54	12.54	11.33	NR
Potassium	1443.32	1311.81	1238.47	1325.64	NR
Magnesium	156.34	146.74	146.96	152.24	NR

Source: Lopez-Martinez *et al.* (2020). (Table used with permission.)

This is followed by the USDA-ARS gene bank, which holds nearly 18,000 samples, and the Leibniz-Institut für Pflanzengenetik und Kulturpflanzenforschung (IPK), with over 10,000 samples. Genesys (2023) lists *Phaseolus* accessions in the following distribution: 86% *P. vulgaris*, 4% *P. coccineus*, 6% *P. lunatus*, 1% *P. acutifolius* and 0.5% *P. dumosus*. Traditional cultivars or landraces account for around 65% of the accessions, with improved cultivars accounting for 15% and wild relatives accounting for 3% of the accessions (Debouck, 2014; Schwember *et al.*, 2017). However, inadequate characterization of the germplasm of *P. coccineus* hampers its potential as a species suitable for inter-specific hybridization. Therefore, national and international gene banks and germplasm research centres serve as valuable resources for breeders exploring tolerance traits to address stresses associated with climate change (Bomers *et al.*, 2022). Modern genomic and phenomic techniques are necessary for characterizing plant genetic resources (PGRs) and effectively harnessing their vast diversity. As a first step, the genetic enhancement of common bean through interspecific hybridization requires identifying whole germplasm banks of the donor species to find the best populations (Schwember *et al.*, 2017).

Despite the significant advancements in molecular marker technology and genome sequencing in various legumes, runner bean has not received adequate attention (Guerra-García *et al.*, 2022). Prior research has been mainly focused on elucidating the genetic diversity and evolutionary lineage of this particular species. The investigations have primarily utilized molecular markers such as simple-sequence repeat (SSR) and inter-simple sequence repeat (ISSR) markers (Alvarez *et al.*, 1998). During the past decade, significant progress has been made in developing common bean genome technology, leading to novel perspectives in resource creation and acquiring crucial scientific knowledge. This advancement also facilitates the feasible incorporation of biotechnological techniques into the practices of runner breeding programmes (de Souza *et al.*, 2023). A recent study retrieved genotyping by sequencing data from domesticated cultivars of two closely associated species, *P. vulgaris* and *P. dumosus*, and 242 individuals of *P. coccineus*, resulting in 42,548 single nucleotide polymorphisms (SNPs). A total of 24 SNPs

associated with the process of domestication were identified. Among them, 13 SNPs were found to be linked to the diversification of cultivars, whereas eight SNPs were observed to be subject to natural selection. This study enhances our comprehension of the domestication history of *P. coccineus*, revealing light on the notable differences in genetic evidence for domestication within closely related species (Guerra-García *et al.*, 2022). Furthermore, the primary objective of whole genome sequencing (WGS) initiatives focused on the common bean is to uncover new perspectives on its evolutionary history and genetic diversity. These endeavours contribute substantially to developing valuable genomic resources, thereby accelerating the improvement of *Phaseolus* species and boosting overall efficiency in its production.

9.2 Origin, Distribution and Botanical Description

9.2.1 Origin

The runner bean is positioned within the group of five domesticated *Phaseolus* species. It originates from the Americas and has been cultivated by indigenous cultures for an extensive period, maybe thousands of years before European colonization. This plant possesses various beneficial properties, encompassing medical and other applications. The utilization of beans, specifically *Phaseolus coccineus*, by humans in Mesoamerica can be traced back to a period ranging from 5500 to 7000 years ago. It accounts for a close relationship with the common bean (*Phaseolus vulgaris*) and year bean (*Phaseolus dumosus*), both of which also underwent domestication in Mesoamerica (Kaplan, 1965; Debouck and Smartt, 1995). Moreover, Mesoamerica serves as the primary centre of origin for the domestication of *P. coccineus*. In this region, wild forms coexist with cultivated forms in cool and humid uplands at elevations ranging from 1500 to 3000 m above sea level, spanning Chihuahua to Panama (Salinas, 1988). According to Santalla *et al.* (2004), the scarlet runner bean is ranked as the third most important bean species in terms of its economic significance, followed by *P. vulgaris* and *P. lunatus*. The climbing perennial crop is commonly cultivated as an annual to produce green pods or harvest dried seeds. Furthermore, it is cultivated for aesthetic reasons as a climber.

9.2.2 Distribution

Runner bean, a perennial climbing plant, grows in mid to high elevations (1000–3000 m above sea level) in northern Mexico, specifically Chihuahua and Panama (Salinas, 1988). According to Giurcă (2009), the economic significance of runner bean is region specific, with cultivation observed mainly in Central America, South America, Africa and Europe. According to Mullins *et al.* (1999), its significance in the USA is quite low. According to Westphal (1974), the introduction of *P. coccineus* as a crop in Europe occurred during the discovery of the Americas. It was perhaps enabled by the introduction of *P. vulgaris* by the Spaniards. It is predominantly grown in the UK, the Netherlands, Italy and Spain in Europe. The significance of runner bean in the UK arises from its higher adaptability to cooler climates compared to the common bean (Rodriguez *et al.*, 2013). The process of domestication of this species continues to be observed in contemporary times, occurring within the regions of the Americas and Europe, where it was brought by the Spanish (Salinas, 1988; Piñero, 2017).

9.2.3 Botanical description

Runner bean belongs to the Leguminosae family. It is indigenous to Mesoamerica and has been widely cultivated in Europe (Giupponi *et al.*, 2018). It is a long-day climbing perennial leguminous species commonly grown on a small scale as an annual crop to produce edible immature pods or dried seeds. Further, this fast-growing climbing herb can reach heights up to 3 m. The leaves are trifoliate, with oval folioles tapering to a pointy tip and a rounded base. The inflorescence is a multi-flowered raceme that extends beyond the leaves and measures 25–35 cm in length. It is distinguished by many-flowered racemes that are longer than its leaves. The flowers have a red corolla with white

wings, a keel and, on rare occasions, an all-white corolla (Santalla *et al.*, 2004; Palmero *et al.*, 2008; Giupponi *et al.*, 2018). Furthermore, the flowers of the scarlet runner bean, typically exhibiting a red hue, find application as a culinary element for consumption or adornment in salads, soups, appetisers, desserts and beverages. Taxonomists have delineated three botanical varieties of runner bean based on the coloration of their flowers: *P. coccineus* var. *albiflorus*, characterized by white flowers; *P. coccineus* var. *coccineus*, distinguished by red flowers; and *P. coccineus* var. *bicolour*, featuring both white and red flowers (Fig. 9.1) (Arriagada *et al.*, 2021). Wild and domesticated *P. coccineus* exhibit characteristics of short-day plants. However, specific genotypes display neutrality or only mild sensitivity to day length, showcasing variability within the species. The seeds of most runner bean cultivars are noticeably larger than those of sieva bean and common bean. The fruit is textured and about 15 cm long, with 4–5 seeds per pod and seeds measuring around 2.5 cm in length. The seed coat can be pink, purple, beige or white, and usually has brown or black streaks. Typically, these seeds exhibit a purple or multicoloured purple form, while certain cultivars possess white seeds. The blooms of purple-seeded variations exhibit a striking red colour, whereas the flowers of white-seeded sorts display an immaculate white colour (Table 9.3) (Sinkovič *et al.*, 2019a).

9.3 Crop Gene Pool, Evolutionary Relationships and Systematics

9.3.1 Crop gene pool

The *Phaseolus* genus is native to the Americas, with wild species growing as far north as Connecticut in the USA. In fact, 70–80 wild *Phaseolus* species were reported (Gepts and Debouck, 1991; Rendón-Anaya *et al.*, 2017). *Phaseolus* is a unique example of multiple parallel and independent domestications. The five species that were affected by domestication are *P. acutifolius* A. Gray (tepary bean), *P. coccineus* L. (scarlet runner bean), *P. dumosus* Macfady (year bean), *P. lunatus* L. (lima bean) and *P. vulgaris* L. (common bean) (Debouck and Smart,

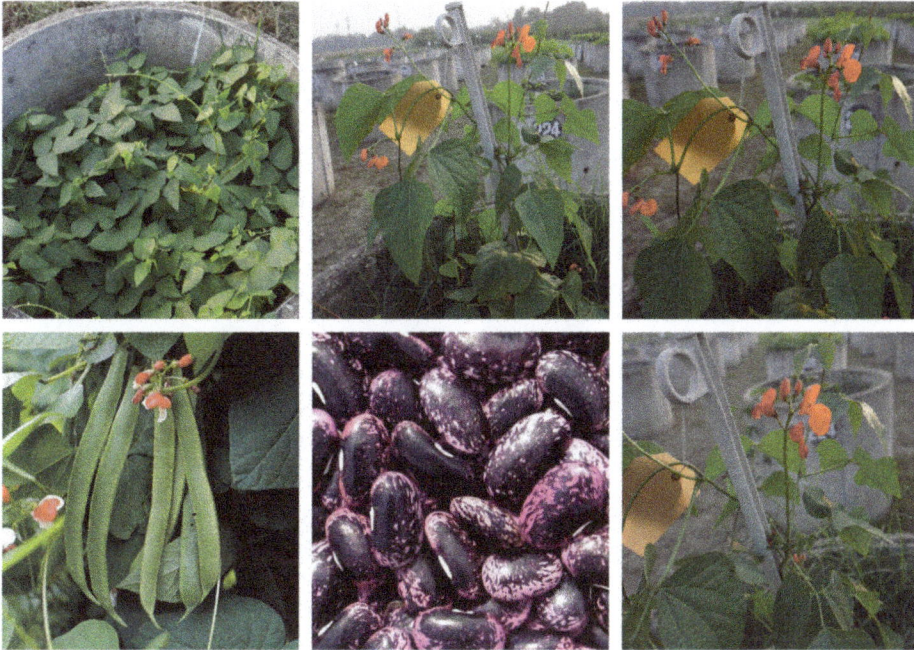

Fig. 9.1. Runner bean plant morphology. (Author's own figure.)

Table 9.3. Phenotypic variation for seed attributes in 953 accessions of common bean and 47 accessions of runner bean.

Variables	Cultivated common bean	Wild runner bean
Quantitative seed characteristic		
L (mm)	12.84	24.38
T (mm)	7.36	9.71
W (mm)	8.73	14.66
L/W	1.48	1.69
W/T	1.19	1.51
100-seed weight (g)	55.15	20.08
Qualitative seed characteristic		
Seed colour	Dark colour (83.70*)	Colour mixture (100.00*)
Number of seed colours	One (100.00*)	Two (100.00*)
Primary/main seed colour	Brown (51.98*)	Beige (66.67*)
Predominant secondary seed colour	None (99.56*)	Brown (66.67*)
Distribution of secondary seed colour	Without secondary colour 100.00*	On the entire seed (100.00*)
Seed veining	Weak (75.33*)	Weak (83.33*)
Seed shape	Oval/circular to elliptic (55.07*)	Cuboid/elliptic (83.33*)
Seed colour and coat pattern	One-colour: brown (48.90*)	Bi-colour: pinto T type (50.00*)

L, seed length; T, seed thickness; W, seed width; L/W, length/width ratio; W/T, width/thickness ratio; * % of main characteristic. Source: Sinkovič et al. (2019a). (Table used with permission.)

1995; Delgadosalinas et al., 2006; Parker and Gepts, 2021). The common bean evolved two distinct gene pools through evolution and domestication: the Mesoamerican gene pool and the Andean gene pool. Throughout its worldwide domestication, various modifications in its physical traits took place, including increased seed and leaf size, changes in growth patterns and sensitivity to daylight duration, and differences in seed colour and seed coat patterns (Chacón *et al.*, 2005; De Ron *et al.*, 2016). Two independent domestications occurred for both lima bean and common bean, in Mesoamerica and the Andes of South America. In contrast, a single domestication occurred for the tepary, the scarlet runner and the year-bean, all of them in Mesoamerica. *P. lunatus* and *P. vulgaris*, which have the most widespread distribution within the genus, have undergone dual domestication (Gepts and Debouck, 1991; Butare *et al.*, 2012). In contrast, *P. acutifolius*, *P. coccineus* and *P. dumosus* each had a single domestication. Further, each domesticated species has unique features that distinguish them from the others. Additionally, the two *Phaseolus* gene pools have been divided into seven races. The Mesoamerican gene pool consists of four races: Durango (D), Jalisco (J), Central America (M) and Guatemala (G), while the Andean gene pool comprises three races: Chile (C), New Granda (N) and Peru (P) (Blair *et al.*, 2009, 2012; Parker and Gepts, 2021). Additionally, *Phaseolus* members can be categorized into different gene pools according to their hybridization capabilities (Fig. 9.2). All members of *P. vulgaris* are in the primary gene pool, whereas *P. dumosus* and *P. coccineus* are in the secondary gene pool, and *P. acutifolius* is in the tertiary gene pool, as hybridization

necessitates sophisticated methods such as embryo rescue. Species that are more distantly related and cannot hybridize are placed in a separate quaternary gene pool, which includes *P. lunatus* (Singh *et al.*, 1998).

9.3.2   Evolutionary relationships

Phaseolus underwent its initial separation some 4.6 million years ago (Evans, 1976). Among the ~70 species that belong to the *Phaseolus* genus, most are geographically distributed in Mesoamerica (Rendón-Anaya *et al.*, 2017). Phylogenetic analysis has shown that there are two main sister clades of *Phaseolus* species. One clade, A, includes the *Pauciflorus*, and the other, B, includes the Filiformis, Vulgaris, Lunatus, Leptostachyus and Polystachios groups. Some species, such as *P. glabellus*, *P. macrolepis*, *P. microcarpus* and *P. oaxacanus*, have not been well resolved (Esquivel *et al.*, 1990; Dohle *et al.*, 2019). Out of the eight subgroups within the genus, domestication was observed in only two: the *Lunatus* group, which had two separate instances of domestication, and the *Vulgaris* group. Only seven species from the *Vulgaris* group were responsible for five domestications. The species *P. lunatus* and *P. vulgaris*, which have the broadest geographical

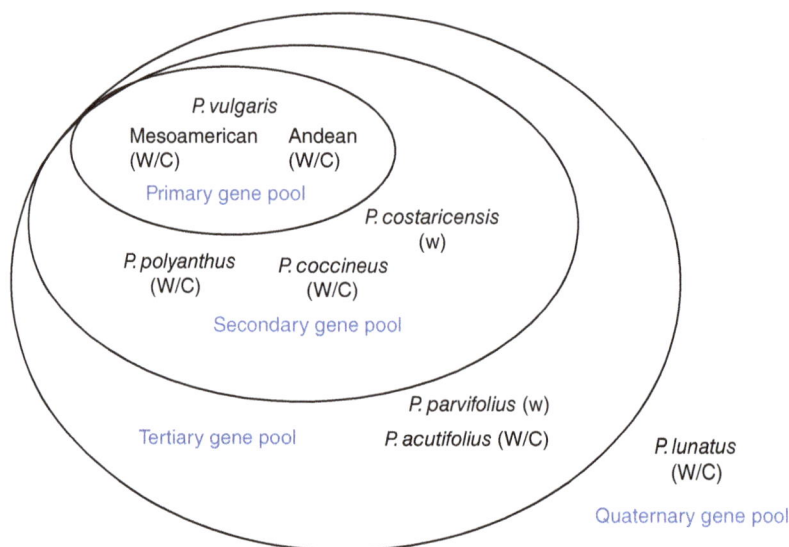

Fig. 9.2. *Phaselous* gene pool classification. (Author's own figure.)

distributions within the genus, underwent domestication on two separate occasions. *Phaseolus acutifolius*, *P. coccineus* and *P. dumosus* underwent domestication on separate occasions (Table 9.4). Every domesticated species has distinct traits that set it apart from other domesticated species (Dohle *et al.*, 2019; Chacon-Sanchez *et al.*, 2021; Parker and Gepts, 2021). Further, *P. acutifolius* A. Gray, *P. coccineus* L., *P. dumosus* M., and *P. vulgaris* L., belonging to the *Vulgaris* group, along with *P. lunatus* from the *Lunatus* group, form the five cultivated species comprising clade B (Chacon-Sanchez *et al.*, 2021). Genetic studies have shown that lima bean and common beans are diploid species with 22 chromosomes (2n = 2x = 22). These crops have a high degree of homozygosity throughout their genomes, mainly due to frequent self-pollination (Arriagada *et al.*, 2021; Chacon-Sanchez *et al.*, 2021). The widespread distribution of wild *Phaseolus* species throughout the Americas, coupled with numerous instances of domestication and the global dissemination of cultivated varieties, positions the *Phaseolus* genus as an excellent choice for investigating the influence and extent of gene flow in primary and secondary centres of diversity, as well as exploring its evolutionary implications.

9.3.3 Systematics

The *Phaseolus* genus belongs to the Fabaceae family, with around 80 cultivated and wild species. However, *P. vulgaris* is the most extensively cultivated species (Gepts and Debouck, 1991; Chacón *et al.*, 2005). The wild runner bean (*P. coccineus*) is found in a range extending from Chihuahua in Northern Mexico to Guatemala and Honduras. It is suited for the uppermost mean altitudes where *Phaseolus* domesticates thrive, ranging from 1100 to 2900 m (Bitocchi *et al.*, 2012; Beebe *et al.*, 2013). *P. dumosus* and *P. coccineus* are part of the secondary gene pool of the common bean, and several disease resistance traits have been transferred into the common bean through introgression (Parker and Gepts, 2021). The species is distinguished by its hypogeal germination, characterized by the cotyledons remaining underground during germination. Flowering commences during a span of 4–6 weeks following sowing, depending on the specific variety. Harvesting dry beans occurs when the pods have reached a yellow colour, and the seeds have fully grown.

9.4 Germplasm Collection, Characterization, Evaluation and Conservation

9.4.1 Germplasm collection and conservation

Runner bean holds the position of being the third most economically important bean species on a global scale, following the common bean and lima bean. Approximately 6658 reported accessions of runner bean are currently stored in *ex-situ* collections worldwide (Table 9.5) (Bomers *et al.*, 2022; Genesys, 2023). The world's largest and most diverse collection of beans is stored at the Germplasm Bank and Genetic Resources Program at the International Center for Tropical Agriculture (CIAT) in Cali, Colombia. There are more than 36,000 *Phaseolus* spp. accessions in the collection, representing 44 taxa collected from 110 different countries. In addition, there are around 958 runner bean accessions preserved in the repository (Schwember *et al.*, 2017). A collection of 2674 accessions is maintained by the European Cooperative Programme for Plant Genetic Resources, which is controlled by the Austrian Agency for Health and Food Safety in Linz, Austria (Debouck *et al.*, 2014). Further, the Agricultural Institute of Slovenia (AISLJ) maintains a collection that involves 995 accessions of runner bean (Sinkovič *et al.*, 2019a). The national germplasm bank in Chapingo, Mexico, controlled by INIFAP (Instituto Nacional de Investigaciones Forestales, Agrícolas y Pecuarias), preserves 798 accessions of *P. coccineus* in Mexico (Kimani *et al.*, 2019). Other organizations assisting in maintaining *P. coccineus* include the German Leibniz Institute of Plant Genetics and Crop Plant Research (IPK), which has 439 accessions, the USDA, which has 374 accessions, and a number of germplasm banks in Europe and South America (Normah *et al.*, 2012). Most of these materials remain primarily uncharacterized, with some exceptions, particularly involving European materials (Santalla *et al.*, 2004; Rodiño

Table 9.4. Phylogenetic relationships of the five domesticated *Phaseolus* species and salient their features.

Species	No. of domestications	Presumed domestication locations	Reproductive systems	Life history	Wild elevation range (masl)	Adaptation	Growth habit	Genome size (mbp)
Common bean (*P. vulgaris*)	2	Mexican Pacific region and Southern Andes	Predominantly autogamous	Annual (medium)	600–3000	Moist	Bush determinate to climbing	587
Lima bean (*P. lunatus*)	2	Western Mexico, Ecuador and N Peru	Mixed auto- and allogamous	Annual (long)	0–2100	Hot, dry to humid	Bush determinate to climbing	685
Runner bean (*P. coccineus*)	1	Central Mexico	Predominantly allogamous	Perennial	1100–2900	Cool and moist	Climbing (some bush)	660
Tepary bean (*P. acutifolius*)	1	NW Mexico	Auto- to cleistogamous	Annual (short)	0–2300	Hot and dry	Bush prostrate	734
Year bean (*P. dumosus*)	1	Western Guatemalan highlands	Leaning allogamous	Pluri-annual	1400–2000	Intermediate between runner and common bean	Climbing	709

Source: Parker and Gepts (2021). (Table used with permission.)

Table 9.5. Runner bean genetic resources availability globally.

Sr. No.	Country	No. of accessions	Institute / programme
1.	Colombia	949	Centro Internacional de Agricultura Tropical (CIAT), Columbia
2.	USA	623	Western Regional Plant Introduction Station, USDA-ARS, Washington State University
		140	Seed Savers Exchange, Non-profit, membership organization with the mission of conserving agricultural biodiversity
3.	Bulgaria	35	Institute for Plant Genetic Resources 'K. Malkov' Bulgaria
4.	Bosnia and Herzegovina	3	Institute of Genetic Resources, University of Banja Luka, Bosnia and Herzegovina
5.	United Kingdom	2	Millennium Seed Bank – Royal Botanic Gardens Kew
6.	Austria	2674	The European Cooperative Programme for Plant Genetic Resources
7.	Slovenia	995	Agricultural Institute of Slovenia
8.	Mexico	798	National Germplasm Bank of INIFAP (Instituto Nacional de Investigaciones Forestales, Agrícolas y Pecuarias)
9.	Germany	439	Leibniz Institute of Plant Genetics and Crop Plant Research (IPK)
	Total	6658	

Source: Schwember *et al.* (2017) and https://www.genesys-pgr.org/a/overview/v2p5OPwlQl9 (2023). (Table used with permission.)

et al., 2006; Rodriguez *et al.*, 2013). Most of these runner bean genetic resources remain predominantly uncharacterized, with limited exceptions, notably in the case of European materials, where a more detailed understanding of their genetic traits and characteristics has been undertaken. The comprehensive characterization of these genetic resources is crucial for unlocking valuable information about their genetic diversity, adaptability and potential contributions to breeding programmes. Enhanced characterization efforts could shed light on traits related to resistance to diseases, tolerance to environmental stresses and other agronomically important features. This in-depth analysis is essential for harnessing the full potential of genetic resources in advancing agricultural practices, ensuring food security, and promoting sustainable crop production.

9.4.2   Characterization and evaluation

Understanding the genetic structure of wild runner bean populations is essential for elucidating the evolutionary history, adaptive traits and ecological interactions of the species (Osorno *et al.*, 2007). It provides a foundation for conservation efforts, aiding in identifying unique

genetic resources (Table 9.6). Further, assessment of the genetic diversity within cultivated accessions enables breeders to make informed decisions, promoting the development of improved varieties with desirable traits such as disease resistance, increased yield, and adaptability to diverse environmental conditions (Beaver *et al.*, 2005; Schwember *et al.*, 2017). Therefore, assessing the extent of genetic variability and relationships between various gene pools are essential stages in any genetic enhancement initiative. Several investigations have established that *P. coccineus* differs significantly in morpho-agronomic features and has low to moderate genetic diversity in both wild and cultivated populations. Furthermore, it appears to be little affected by the domestication process (Nalupya *et al.*, 2021). Identifying introgression from wild to domesticated populations represents a significant breakthrough in genetic research, holding promising implications for the revitalization of genetic diversity. This phenomenon, wherein genes from wild counterparts infiltrate domesticated populations, introduces a dynamic element to the genetic landscape, potentially bolstering the overall resilience and adaptability of cultivated species (Guerra-García *et al.*, 2022). A study by Santalla *et al.* (2004) focused on analysing the diversity of 31 runner bean accessions

Table 9.6. Traits identified from runner bean germplasm for bean improvement. (Author's own table.)

Trait	Genetic stock /donor	Source
Aluminium toxicity	G35346 (*P. coccineus*, from Oaxaca)	Butare *et al.* (2012)
Low phosphorus	G19227A; Chaucha Chuga G19833	Beebe *et al.* (2006)
Low temperature seedling growth	*P. coccineus* G35171 from Rwanda	Rodiño *et al.* (2006)
Angular leaf spot resistance	G10613 from Guatemala	Pastor-Corrales *et al.* (1998)
Angular leaf spot resistance	Interspecific hybrids with *P. coccineus*; G4691	Pastor-Corrales *et al.* (1998); Mahuku *et al.* (2003)
White mould resistance	*P. coccineus* PI 175829 from Turkey	Abawi *et al.* (1978)
White mould resistance	Interspecific hybrids with *P. coccineus* G35172	Singh *et al.* (2009)
Ophiomyia beanfly resistance	*P. coccineus* G35023 and G35075 and interspecific hybrids	Kornegay and Cardona (1991)
High seed weight (per 100 g)	PHA-1031 and PHA-1028	Santalla *et al.* (2004)
High grain yield	PHA-0311and PHA-0352	Santalla *et al.* (2004)

from the Iberian Peninsula (Spain and Portugal). This analysis was based on evaluating morphological characteristics, agronomical attributes and seed quality. Significant variations were noted among landraces in the majority of agronomical and seed quality traits examined, except for seeds per pod, water absorption, seed-coat softness and floury texture. Similarly, Negri and Tosti (2002) used three amplified fragment length polymorphism (AFLP) primer combinations to assess genetic variation among 36 collected germplasm materials, a wild accession of *P. vulgaris* and *P. coccineus* commercial varieties of both species. They revealed quite a high percentage of polymorphism (90.2% of polymorphic bands on average). A broad genetic variation was observed among collected materials, and each accession showed a unique polymorphism pattern. Santalla *et al.* (2004) noted that various selection forces may have influenced the genetic diversity of runner bean in different locations of the Iberian Peninsula. Furthermore, they have found an extra-large runner bean germplasm with a high grain yield. This germplasm is a valuable reservoir of genetic variation and holds great potential for generating enhanced cultivars that can be selected for commercialization. Sinkovič *et al.* (2019b) reported a total of 142 accessions of runner bean from south-eastern Europe were characterized using 28 quantitative and qualitative morpho-agronomic descriptors for *Phaseolus* spp.. The results revealed that the highest variability, at 76.39%, was between the different countries, representing different geographic origins, while

the variability within the countries was 23.61%. Cluster analysis based on these collected morpho-agronomic data also classified the accessions into three groups according to genetic origins. Arriagada *et al.* (2021) characterized ten Chilean lines of runner bean from different origins (Central and Southern Chile) using selected morphological and agronomical traits. The results revealed that the Central lines produced more pods per plant than the Southern lines even though the size and weight of their seeds were smaller, and this population had a low level of genetic diversity (He = 0.251). Thus, genomic technologies can be utilized to identify patterns of variety and differentiation across genomes. Regions or variations that depart from neutral preferences may be created by selective pressures and identified as possible candidates. Applying this approach to crop species and their wild relatives allows distinguishing loci modified during domestication, while comparisons across landraces and improved cultivars determine the effect of later selection (Gepts, 2014; Guerra-García *et al.*, 2017).

9.5   Use of Germplasm in Crop Improvement by Conventional Approaches

Runner bean offers a source of diversity for various agronomic and disease resistance features that can be used to improve common bean (Hardwick, 1972). During the initial phase

(1993), the main objectives in bean breeding focused on enhancing yield, boosting protein content, improving flavour, developing an optimal plant ideotype, and enhancing resistance against diseases. Exploring distant hybridization involving *P. coccineus* led to the identification of varying degrees of reproductive isolation (Kalloo, 1993). Farmers have conserved their traditional or heirloom cultivars for generations by participating in seed exchanges within local markets and across their communities. This tradition has persisted for generations (Beebe *et al.*, 2006; Gepts *et al.*, 2014). Consequently, the edible components of the plant (pod and grain) exhibit distinct variations in size, shape, tenderness and cooking quality. Runner bean has undergone secondary diversification, which has led to the development of new recombinant types resulting from the breeding between Andean and Mesoamerican genetic pools (Delgado-Salinas *et al.*, 2006; Guerra-García *et al.*, 2022). This provides a good opportunity for making genetic advances in this species. Now, the breeders have the chance to perform crosses not just within the Mesoamerican and Andean gene pools but also between other races. It is important to emphasize that specific genetic difficulties, such as blocked cotyledon lethal (BCL), crinkle leaf dwarf (CLD) and dwarf lethal (DL), have been considered hindrances to the effective breeding of Mesoamerican and Andean germplasm (Kornegay and Cardona, 1991). Although facing these difficulties, the continuous investigation of these genetic boundaries presents significant prospects for improving the genetic variety and broad adaptability of common bean.

Runner bean provides a range of genetic variations that can be utilized to enhance multiple traits in the common bean. This species has a few interesting agronomic features, including lodging resistance due to thick stem bases, cold tolerance, the presence of tuberous roots, which allowed for a perennial cycle as well as drought resistance, and a potentially large number of pods per inflorescence (Table 9.7) (Schwartz *et al.*, 2006; Schwember *et al.*, 2017). These traits reflect not only adaptability of runner bean to different ecological niches but also present opportunities for enhancing the resilience and productivity of the common bean through selective breeding programmes. The economic significance and agronomic potential of runner bean are hindered by the absence of molecular characterization in germplasm repositories, limiting its use as a donor species for interspecific hybridization in *Phaseolus* breeding initiatives (Arriagada *et al.*, 2021).

The magnitude of divergence between the Andean and Mesoamerican gene pool has implications for bean breeding that have not yet been fully explored. Despite their partial reproductive isolation, crosses between the Middle American and Andean gene pools are easily accomplished, although differences in flowering time can make crossing difficult. Nevertheless, limited research has concentrated on enhancing *P. coccineus*, resulting in a lack of global breeding initiatives dedicated to this species. Transferring quantitative features between the Mesoamerican and Andean gene pools seems more challenging (Palmero *et al.*, 2011; González *et al.*, 2014). Therefore, understanding the genetic linkages within and across races and gene pools and their performance will give bean breeders a foundation for creating crosses utilizing parents with opposing and complementary traits. This will help to expand the genetic diversity within different commercial classes. In this direction, González *et al.* (2009) achieved a breakthrough in successful interracial and inter-pool crosses for common bean development, representing a pivotal step forward in the ongoing quest for improved crop varieties. Their findings not only contribute to the scientific understanding of bean genetics but also offer tangible possibilities for advancing sustainable and resilient agriculture in Europe. The breeding of heat-tolerant runner bean varieties involves the identification of heat-tolerant accessions and the development of molecular markers to assess heat tolerance. This is crucial for ensuring the continued cultivation of runner bean in the face of changing climatic conditions. Therefore, developing resilience to the impacts of climate change through selective breeding is a multifaceted process that necessitates the incorporation of genetic variability from either wild or domesticated species. Suárez *et al.* (2023) produced 112 breeding materials by crossing *P. vulgaris*, *P. coccineus*, *P. dumosus* and *P. acutifolius* species to achieve certain features. The breeding lines consisted of 46 *P. vulgaris* lines (6 Andean and 40 Mesoamerican), 5 lines from the cross of *P. vulgaris* × *P. acutifolius*, 1 line from the cross of *P. vulgaris* × *P. acutifolius* × *P. parvifolius*,

Table 9.7. Breeding approaches and traits improvement in runner bean. (Author's own table.)

Traits	Type of biological materials	Breeding methods	Donor identified	Reference
Agronomic traits				
Early flowering	Advanced breeding lines (*F6.7*)	Pedigree	KAB-RB13-321-190/1, KAB-RB13-301-174, SUB-OL-RB13-312-252 (42–49 days)	Kimani *et al.* (2019)
High pod / plant	Advanced breeding lines (*F6.7*)	Pedigree	KAB-RB13-108-125 and KAB-RB13-301-171/2	Kimani *et al.* (2019)
High grain yield		Pedigree	KAB-RB13-155-122 (5117 kg/ha), KAB-OL-RB13-426-228 (4686 kg/ha), and KAB-RB13-120-123/1 (4611 kg/ha)	Kimani *et al.* (2019)
Biotic stress				
Bean golden yellow mosaic virus (BGYMV)	Germplasm	Germplasm evaluation	G35172	Osorno *et al.* (2007)
Bean golden yellow mosaic virus (BGYMV)	Germplasm	Germplasm evaluation	G35127	Beaver *et al.* (2005)
Anthracnose	Differential standard cultivars	Germplasm evaluation	G10843, G3514, G4327, G1320, G19021, G19021, G19021, G68, G21134	Nalupya *et al.* (2021)
White mould resistance	Breeding lines	Interspecific crosses	PI 433246 and PI 439534	Schwartz *et al.* (2006)
Abiotic stress				
Low-temperature tolerance	Commercial cultivars	Pedigree	Lady Bi-color, PHA-0013, PHA-0133, PHA-0311, PHA-0664, and PHA-1025	Rodiño *et al.* (2006)
Heat tolerance	Local landrace	Genetic and phenotypic characterization	BVAL-610181, BVAL-610348 and BVAL-610637	Bomers *et al.* (2022)
Nutritional quality				
High phenolic content	Pre-breeding lines	Pre-breeding	RRA 79 (19.65 mg g^{-1}, red), RRA 81 (16.03 mg g^{-1}, red), SER 213 (19.03 mg g^{-1}, red), SER 212 (17.58 mg g^{-1}, red), SEF 71 (13.87 mg g^{-1}, red), and ALB 60 (12.49 mg g^{-1}, red)	Suárez *et al.* (2023)
High carotenoid content	Pre-breeding lines	Pre-breeding	SMR 139 (11.74 mg g–1), RRA 81 (10.28 mg g–1), SMR 133 (6.20 mg g–1), SER 271 (5.45 mg g–1), and RRA 69 (4.77 mg g–1)	Suárez *et al.* (2023)

19 from the cross of *P. vulgaris* × *P. coccineus*, 13 from the cross of *P. vulgaris* × *P. acutifolius* × *P. coccineus*, 5 from the cross of *P. vulgaris* × *P. acutifolius* × *P. coccineus* × *P. dumosus*, and 1 line from the cross *P. vulgaris* × *P. dumosus*. There are 22 accessions of *P. acutifolius*, including 16 cultivated accessions of *P. acutifolius* var. *acutifolius*, four wild accessions of *P. acutifolius* var. *acutifolius* and two wild accessions of *P. acutifolius* var. *tenuifolius*. These lines are breeding resources with strong agronomic potential, characterized by their tolerance to both biotic and abiotic stress conditions, as well as enhanced grain and nutritional quality. Likewise, Mulanya *et al.* (2019) investigated the inheritance of short-day photoperiod in runner bean by generating seven single crossings between female parents, White Emergo (long-day imported variety) and seven short-day local landraces (Kin 2, Kin 3, Nyeri, Dwarf 1, Dwarf 2 and Dwarf 3). Therefore, promoting the development of short-day cultivars clearly benefits local seed production. These cultivars decrease reliance on external seed supplies by synchronizing with the inherent day lengths in specific locations. Consequently, this strengthens the local agricultural economies, fosters self-reliance and improves the ability of small-scale farmers to withstand challenges. Further, interspecific crossing with *P. coccineus* has been demonstrated to be beneficial for producing new common bean varieties (Table 9.7). This was evidenced by the successful breeding work aimed at imparting disease resistance. Researchers have achieved successful introgression from runner bean to common bean, resulting in the incorporation of moderate levels of resistance against *Xanthomonas* (Miklas *et al.*, 1994), *Fusarium* root rot (Wallace and Wilkinson, 1965) and white mould in dry bean (Miklas *et al.*, 1998; Schwartz *et al.*, 2006). This accomplishment represents a significant breakthrough in the ongoing efforts to enhance the defence mechanisms of common bean against various plant diseases.

9.6 Integration of Genomic and Genetic Resources for Crop Improvement

The genus *Phaseolus* has experienced substantial advancements in the expansion of genomics tools. The sophisticated molecular tools and techniques have significantly broadened our understanding of *Phaseolus* species' genetic make-up, diversity and functional aspects, especially of runner bean (*P. coccineus*), encompassing several economically important plants, including common bean (Schwember *et al.*, 2017). These comprise superior reference genomes, a comprehensive collection of genetic markers, assays and databases, gene expression data derived from RNA sequencing, a diverse range of genotyped diversity panels and biparental populations, and a well-established track record of genetic research.

Previous studies have investigated the domestication history of the scarlet runner bean using less detailed molecular markers, which resulted in the proposal of numerous domestication events. More precisely, study focuses on the chloroplast and nuclear SSRs of runner bean accessions, which include European domesticated populations, Mesoamerican landraces, and wild samples from Mexico, Guatemala and Honduras. Only a limited number of studies have thoroughly investigated the genetic diversity of runner bean using molecular markers (Alvarez *et al.*, 1998). In the research findings by Nowosielski *et al.* (2002), polymorphic fragments were derived from six distinct primers through the random amplified polymorphic DNA (RAPD) method and 15 distinct primers through the AFLP method. Distinct groups were observed for *P. vulgaris* and *P. coccineus* accessions. Both RAPD and AFLP analyses facilitated the unique differentiation of all accessions in the study. Rodriguez *et al.* (2013) used six chloroplast microsatellites (cpSSRs) to analyse the cytoplasmic diversity in 331 European domesticated scarlet runner bean accessions. Their research uncovered that both demographic and selective forces shape the genetic make-up of the European runner bean population.

In a study conducted by Mercati *et al.* (2015), a total of 61 different genotypes was analysed to determine the genetic structure and distinctiveness of a runner bean landrace called Fagiolone. These samples included 20 populations from Central Italy and 41 accessions from Italy and Mesoamerica. The analysis was performed using 14 nuclear SSRs. The advancement and utilization of genomic tools enable the characterization of diversity and provide insights into differentiation patterns across entire genomes. Researchers can comprehensively explore and analyse genetic material variations using advanced techniques such as next-generation sequencing, comparative genomics and high-throughput genotyping. Members of *Phaseolus* spp. have a genomic size of approximately 500–600 Mb (2n = 2x = 22), which is relatively compact and consistent (De Ron *et al.*, 2019). There is no inherent variation in ploidy (Lobaton *et al.*, 2018). The first high-quality reference genome was sequenced with an inbred landrace line of common bean (G19833) obtained from the Andean pool (Race Peru) using a whole-genome shotgun sequencing method (Schmutz *et al.*, 2014). Furthermore, there have been significant advancements in the development of reference genomes, especially for the Middle American common bean breeding line BAT93 (Vlasova *et al.*, 2016). This useful information will be used as a future reference for sequencing new legume species, such as runner and lima bean, to conduct comparative genomic studies between these species.

The combination of next-generation sequencing technology, genetic mapping and synteny approaches proves to be a robust methodology for identifying functional SNPs and indels linked to agriculturally significant traits (Xanthopoulou *et al.*, 2019). A database has been established, presenting a comprehensive collection of markers, including legacy and genomics-based ones

such as SSRs, SNPs and indels, particularly in *Phaseolus*, focusing on the common bean (Table 9.8). Guerra-García *et al.* (2017) obtained genotyping by sequencing data consisting of 42,548 SNPs from a total of 242 individuals. These individuals included the species *P. coccineus* and two closely related domesticated forms: *P. vulgaris* (20 individuals) and *P. dumosus* (35 individuals). Eight genetic groups were identified, with half representing wild populations and the remaining clusters representing domesticated plants. Likewise, Xanthopoulou *et al.* (2019) conducted RNA-Seq analysis using *de novo* transcriptome assembly in two landraces of *P. coccineus*, specifically 'Gigantes' and 'Elephantes', in order to detect and describe the genomic variation present in these two closely related cultivars of runner bean. These findings greatly increase the polymorphic markers in the *P. coccineus* genus, allowing for accurate identification of runner bean cultivars, creating detailed genetic maps and facilitating genome-wide association studies. They ultimately contribute to the genetic pool by enhancing the closely related and inter-crossable *P. vulgaris*. The Guerra-García *et al.* (2022) study assessed the demographic history of traditional varieties from various regions of Mexico. It investigated the presence of introgression across sympatric wild and cultivated populations using genotyping by sequencing data. This data consisted of 79,286 single-nucleotide

Table 9.8. Markers availability and utilization for runner bean improvement. (Author's own table.)

Markers	Salient findings	Material	Reference
12 ISSR	Germplasm characterization	Ten germplasm used for molecular characterization	Arriagada *et al.* (2021)
3 AFLP	Germplasm characterization	31 germplasm used to assess genetic variation	Negri and Tosti (2002)
1190 SNPs	Genome-wide association analysis for heat tolerance	113 runner bean genotypes	Bomers *et al.* (2022)
13 SNPs	Genetic signatures of domestication (flower and pods)	242 genotypes of *P. coccineus* for domestication events of *P. coccineus*	Guerra-García *et al.* (2022)
19,281 SNPs	Characterized the underlying genomic variation of two very closely related runner bean cultivars, and new SNP markers identified	Two landraces of *P. coccineus*, 'Gigantes' and 'Elephantes'	Xanthopoulou *et al.* (2019)
8278 SSR	New SSR markers identified	Two landraces of *P. coccineus*, 'Gigantes' and 'Elephantes'	Xanthopoulou *et al.* (2019)

variations obtained from 237 cultivated and wild plant samples. Despite a significant initial reduction in genetic diversity, traditional types exhibit substantial variation owing to subsequent population expansion. These newly developed tools will be valuable in aiding breeding operations for *Phaseolus* species, including *P. acutifolius* and *P. coccineus*. Highly saturated mapping is a crucial technique for identifying genes and polymorphisms to gain insight into the genetic foundation of key agronomic and nutritional qualities. It also helps expedite the progress of breeding programmes through marker-assisted selection (MAS).

9.7 Erosion of Genetic Diversity from Traditional Areas and Limitations in Germplasm Use

Genetic erosion refers to the reduction in crop diversity resulting from the industrialization of agriculture (van de Wouw et al., 2010; Wallace et al., 2018). The modernization process is closely linked to the genetic erosion of cultivated diversity, which refers to the gradual reduction in genetic variability within cultivated crops over time (Alvarez et al., 1998; Akimoto et al., 1999). The degradation of variety levels in the crop has been clearly observed throughout its history. With the development of contemporary agricultural practices and technologies, agriculture has undergone significant changes, including a noticeable shift towards monoculture and the emphasis on high-yielding, homogeneous varieties (Khoury et al., 2022).

The *Phaseolus* spp. exhibit a variety of breeding systems and life history features, ranging from annual autogamous to perennial allogamous. This allows for the assessment of whether these characteristics have immediate implications for the impact of domestication on the physical and genetic structure of the cultivated plants (Wallace et al., 2018). The distribution of genetic diversity within a species is inconsistent across different climatic and environmental situations. Hence, it relies not solely on demographic processes such as genetic drift but also on adaptation, which describes the capacity to manage specific environmental situations (Bitocchi et al., 2017). Bean markets and customers in various countries and locations have distinct preferences for grain size, shape, colour and cooking duration. Out of the numerous bean varieties cultivated globally, there are 62 recognized market classifications for dry beans (De Ron et al., 2016). Therefore, traditional bean types are a rich source of suitable genetic material for common bean, as shown by the wide range of observable physical characteristics in existing genetic collections. Replacing traditional varieties with superior cultivars in developed nations has caused genetic erosion, negatively affecting the species. The first step to understanding the current diversity is carefully analysing different physical and genetic traits in collections and gene banks (Blair et al., 2009). Executing breeding programmes to develop new and improved plant varieties with desired characteristics such as quality, productivity, and resilience to environmental and biological challenges is important. The Mesoamerican beans that reached the Iberian Peninsula probably had restricted genetic diversity owing to small population size, and the establishment of new populations was based on a few individuals chosen by farmers, potentially leading to increased genetic drift.

9.8 Cultivation Practices

Runner bean possesses a taproot that has undergone tuberization and is abundant in starch. Additionally, it has vegetative buds located in the neck region. The stem, measuring 2–7 m in height, is tall, robust, and somewhat contorted, making it well-suited for support with poles. The fruit is elongated, with a linear-oblong to oblong shape, and can reach a length of up to 40 cm. The pods are flattened. The seeds are large, averaging around 20–25 mm in length, 13–14 mm in width and 8 mm in thickness. *P. coccineus* seeds are utilized throughout Central America, and unripe and ripe seeds are consumed. In other countries like Ethiopia, predominantly ripe seeds are eaten and also preserved in parts of Europe (Palmero et al., 2011). The primary method of preparation is mainly through the process of boiling. In temperate locations, the underdeveloped seed pods are typically consumed as a vegetable after slicing and cooking (Kimani et al., 2008; Lobaton et al., 2018).

To obtain high-quality pods, runner bean is grown on trellises, poles, fence lines, plants (e.g. maize) or other support structures. Previous research determined that optimal economic seeding rate of runner bean was 153 kg/ha. It was also suggested that the beans be spaced 20 cm apart within rows and 60 cm apart between rows during sowing, resulting in a density of 83,300 plants/ha. It is worth noting, however, that a greater density of 100,000 plants/ha in Mexico is advised to get maximum yields. In a study conducted by Stefan *et al.* (2013), it was demonstrated that the inoculation of runner bean seeds with two specific strains of rhizobacteria (S4 and S7) led to notable enhancements in photosynthetic activities, water-use efficiency and chlorophyll content. As a result, when both strains were applied, there was a significant increase of 28% in grain yield compared to seeds that were not inoculated. Further, intercropping offers effective resource exploitation, decreases environmental risk and production costs, and ensures more financial stability, making it ideal for labour-intensive, smallholder farmers. Contreras-Govea *et al.* (2009) demonstrated that including three distinct varieties of climbing beans, including *P. coccineus*, in maize resulted in an elevation of crude protein levels. However, this addition also had an impact on the concentration of fibre and the fermentation profile. In another study in India, the practice of intercropping maize with runner bean at a row ratio of 2:2 resulted in the highest yield of maize-equivalent (63.69 q/ha), production efficiency (217.5%), land equivalent ratio (LER) of 1.35, as well as favourable economic return and energy-use efficiency indicators. This intercropping system proved to be the most productive, economical and energetically viable system. The second most productive system was maize + cowpea at a row ratio of 1:1 (Padhi *et al.*, 2001). Runner bean is typically harvested for its green pods when they reach maximum length but before rapid seed development begins, with a common picking interval of 4–5 days. In contrast, for dry seed production, plants are pulled or cut when most pods are dry. Afterwards, they are left to dry for a few days before the seeds are threshed. Due to asynchronous flowering and ripening, pod hand-picking may occur in multiple rounds.

9.9 Conclusions and Future Prospects

The in-depth knowledge of the genetic make-up of agronomic and domestication-related traits would be beneficial for enhancing the quality of *Phaseolus* crops. Much of the variability available within the genus is not well suited for commercial cultivation. Often, this also applies to lesser-known species, including those that have yet to be tamed. Conventional population genetic tests are frequently unsuitable for analysing germplasm that is more distantly related, because they can exhibit substantial variations in SNPs, repetitive sequences, insertions and deletions (indels), and chromosomal rearrangements. The agronomic advancements of runner bean have primarily targeted small-scale cultivators because a significant amount of its genetic material exhibits uncertain growth. This emphasis has, however, led to a significant amount of work and resources needed to cultivate the crop with the aid of support structures. Further investigation is required to ascertain the target demographics and the feasibility of employing support systems.

Additional research is required to investigate other agronomic measures, such as fertilization, irrigation and crop protection because the current suggestions are primarily based on common bean and may not be directly applicable to runner bean. In the past ten years, numerous genetic resources have been created. Presently, molecular marker technologies are shifting from predominantly sequential methods that involve separating DNA fragments based on their size (such as SSR and AFLP) to advanced techniques that utilize hybridization and can analyse hundreds to tens of thousands of genetic variations simultaneously, especially in genes. Furthermore, the availability of fully sequenced and annotated genomes for model legumes and certain grain legume crops and ongoing efforts to sequence the genomes of other legume crops presents a significant opportunity for comparative genomics. This field of study has the potential to facilitate specific gene/allele identification and enable more comprehensive diversity analyses of legume germplasm collections. Runner bean has been subject to limited study from a genetic standpoint. An incentive for conducting agronomic and genetic studies is to capitalize on the lucrative market for runner bean as an export product to European markets. Research on the potential occurrence of heterosis in runner bean supports the cultivation of hybrids, providing advantages to growers regarding early maturity, uniformity, disease resistance and increased crop yields.

References

Abawi, G.S., Provvidenti, R., Crosier, D.C. and Hunter, J.E. (1978) Inheritance of resistance to white mold disease in *Phaseolus coccineus*. *Journal of Heredity* 69, 200–202.

Akimoto, M., Shimamoto, Y. and Morishima, H. (1999) The extinction of genetic resources of Asian wild rice, *Oryza rufipogon* Griff: a case study in Thailand. *Genetic Resources and Crop Evolution* 46, 419–425.

Alvarez, M., Sáenz de Miera, L. and Pérez de la Vega, M. (1998) Genetic variation in common and runner bean of the Northern Meseta in Spain. *Genetic Resources and Crop Evolution* 45, 243–251. DOI: 10.1023/A:1008604127933

Arriagada, O., Schwember, A.R., Greve, M.J., Urban, M.O., Cabeza, R.A. and Carrasco, B. (2021) Morphological and molecular characterisation of selected Chilean runner bean (*Phaseolus coccineus* L.) genotypes shows moderate agronomic and genetic variability. *Plants* 10, 1688.

Beaver, J.S., Munoz Perea, C.G., Osorno, J.M., Ferwerda, F.H. and Miklas, P.N. (2005) Registration of bean golden yellow mosaic virus resistant dry bean germplasm lines PR9771-3-2, PR0247-49, and PR0157-4-1. *Crop Science* 45, 2126.

Beebe, S., Rao, I., Mukankusi, C. and Buruchara, R. (2013) Improving resource use efficiency and reducing risk of common bean production in Africa, Latin America, and the Caribbean. In: Hershey, C.H. (ed.) *Eco-efficiency: From Vision to Reality*. Cali: Centro Internacional de Agricultura Tropical (CIAT), pp. 117–134.

Beebe, S.E., Rojas-Pierce, M., Yan, X., Blair, M.W., Pedraza, F. *et al*. (2006) Quantitative trait loci for root architecture traits correlated with phosphorus acquisition in common bean. *Crop Science* 46, 413–423.

Bitocchi, E., Nanni, L., Bellucci, E., Rossi, M., Giardini, A. *et al*. (2012) Mesoamerican origin of the common bean (*Phaseolus vulgaris* L.) is revealed by sequence data. *Proceedings of the National Academy of Sciences USA* 109, 788–796.

Bitocchi, E., Rau, D., Bellucci, E., Rodriguez, M., Murgia, M.L. *et al*. (2017) Beans (*Phaseolus* ssp.) as a model for understanding crop evolution. *Frontiers in Plant Science* 8, 722.

Blair, M., Pantoja, W. and Muñoz, L.C. (2012) First use of microsatellite markers in a large collection of cultivated and wild accessions of tepary bean (*Phaseolus acutifolius* A. Gray). *Theoretical and Applied Genetics* 125, 1137–1147. DOI: 10.1007/s00122-012-1900-0

Blair, M.W., Díaz, L., Buendía, H. and Duque, M. (2009) Genetic diversity, seed size associations and population structure of a core collection of common beans (*Phaseolus vulgaris* L.). *Theoretical and Applied Genetics* 119, 955–972. DOI: 10.1007/s00122-009-1064-8

Bomers, S., Sehr, E.M., Adam, E., Von Gehren, P., Prat, N. and Ribarits, A. (2022) Towards heat tolerant runner bean (*Phaseolus coccineus* L.) by utilizing plant genetic resources. *Agronomy* 12, 612. DOI: 10.3390/agronomy12030612

Butare, L., Rao, I., Lepoivre, P., Cajiao, C., Polania, J. *et al*. (2012) Phenotypic evaluation of interspecific recombinant inbred lines (RILs) of *Phaseolus* species for aluminium resistance and shoot and root growth response to aluminium–toxic acid soil. *Euphytica* 186, 715–730. DOI: 10.1007/s10681-011-0564-1

Celmeli, T., Sari, H., Canci, H., Sari, D., Adak, A., Eker, T. and Toker, C. (2018) The nutritional content of common bean (*Phaseolus vulgaris* L.) landraces in comparison to modern varieties. *Agronomy* 8, 166.

Chacón, S.M.I., Pickersgill, B. and Debouck, D.G. (2005) Domestication patterns in common bean (*Phaseolus vulgaris* L.) and the origin of the Mesoamerican and Andean cultivated races. *Theoretical and Applied Genetics* 110, 432–444. DOI: 10.1007/s00122-004-1842-2

Chacon-Sanchez, M.I., Martínez-Castillo, J., Duitama, J. and Debouck, D.G. (2021) Gene flow in Phaseolus beans and its role as a plausible driver of ecological fitness and expansion of cultigens. *Frontiers in Ecology and Evolution* 9, 618709.

Contreras-Govea, F.E., Muck, R.E., Armstrong, K.L. and Albrecht, K.A. (2009) Nutritive value of corn silage in mixture with climbing beans. *Animal Feed Science and Technology* 150, 1–8.

De Ron, A.M., González, A.M., Rodiño, A.P., Santalla, M., Godoy, L. and Papa, R. (2016) History of the common bean crop: its evolution beyond its areas of origin and domestication. *Arbor* 192, a317. DOI: 10.3989/arbor.2016.779n3007

De Ron, A.M., Kalavacharia, V., Alvarez-Garcia, S., Casquero, P.A., Carro-Huelga, G. *et al*. (2019) Common bean genetics, breeding, and genomics for adaptation to changing to new agri-environmental conditions. In: Kole, C. (ed.) *Genomic Designing of Climate-Smart Pulse Crops*. Springer, Cham, Switzerland, pp. 1–106.

De Souza, I.P., de Azevedo, B.R., Coelho, A.S.G., Oliveiro de Souza, T.L.P., Ribeiro Valdisser, P.A.M. *et al*. (2023) Whole-genome resequencing of common bean elite breeding lines. *Scientific Reports* 13, 12721. DOI: 10.1038/s41598-023-39399-6

Debouck, D.G. (2014) *Conservation of Phaseolus Beans Genetic Resources: A strategy*. Global Crop Diversity Trust, Rome, Italy.

Debouck, D.G. and Smartt, J. (1995) Beans, *Phaseolus* spp. (Leguminosae-Papilionoideae). In: Smartt, J. and Simmonds, N.W. (eds) *Evolution of Crop Plants*. Longman Scientific and Technical, London, UK, pp. 287–294.

Delgado-Salinas, A., Bibler, R. and Lavin, M. (2006) Phylogeny of the genus *Phaseolus* (Leguminosae): a recent diversification in an ancient landscape. *Systematic Botany* 31, 779–791. DOI: 10.1600/036364406779 695960

Dohle, S., Berny Mier, Y., Teran, J.C., Egan, A., Kisha, T. and Khoury, C.K. (2019) Wild beans (*Phaseolus L.*) of North America. In: Greene, S., Williams, K., Khoury, C., Kantar, M. and Marek, L. (eds) *North American Crop Wild Relatives*. Springer, Cham, Switzerland, pp. 99–127.

Esquivel, M., Castiñeiras, L. and Hammer, K. (1990) Origin, classification, variation and distribution of lima bean (*Phaseolus lunatus* L.) in the light of Cuban material. *Euphytica* 49, 89–97. DOI: 10.1007/bf00027258

Evans, A.M. (1976). Beans - *Phaseolus* spp. (Leguminosae - Papilionatae). In: Simmonds, N.W. (ed.) *Evolution of Crop Plants*. Longman, London, pp. 168–172.

Freytag, G.F. and Debouck, D.G. (2002) *Taxonomy, Distribution, and Ecology of the Genus Phaseolus (Leguminosae – Papilionoideae) in North America, Mexico and Central America*. Botanical Research Institute of Texas, Fort Worth, Texas.

Genesys (2023) Genesys - an online platform with the information about Plant Genetic Resources for Food and Agriculture (PGRFA) conserved in genebanks worldwide. Available at: https://www.genesys-pgr.org/a/overview/v2p5OPwlQl9 (accessed 17 May 2024).

Gepts, P. (2014) The contribution of genetic and genomic approaches to plant domestication studies. *Current Opinion in Plant Biology*. 18, 51–59. DOI: 10.1016/j.pbi.2014.02.001

Gepts, P. and Debouck, D.G. (1991) Origin, domestication, and evolution of the common bean, *Phaseolus vulgaris*. In: Voysest, O. and Van Schoonhoven, A. (eds) *Common Beans: Research for Crop Improvement*. CABI, Wallingford, Oxfordshire, pp. 7–53.

Giupponi, L., Tamburini, A. and Giorgi, A. (2018) Prospects for broader cultivation and commercialization of Copafam, a local variety of *Phaseolus coccineus* L., in the Brescia Pre-Alps. *Mountain Research and Development* 38, 24–34.

Giurcă, D.M. (2009) Morphological and phenological differences between the two species of the Phaseolus genus (*Phaseolus vulgaris* and *Phaseolus coccineus*). *Cercetari Agronomice Moldova* 42, 39–45.

González, A.M., Rodiño, A.P., Santalla, M. and De Ron, A.M. (2009) Genetics of intra-gene pool and inter-gene pool hybridization for seed traits in common bean (*Phaseolus vulgaris* L.) germplasm from Europe. *Field Crops Research* 112, 66–76.

González, A.M., De Ron, A.M., Lores, M. and Santalla, M. (2014) Effect of the inbreeding depression in progeny fitness of runner bean (*Phaseolus coccineus* L.) and its implications for breeding. *Euphytica* 200, 413–428.

Guerra-García, A., Suárez-Atilano, M., Mastretta-Yanes, A., Delgado-Salinas, A. and Piñero, D. (2017) Domestication genomics of the open-pollinated scarlet runner bean (*Phaseolus coccineus* L.). *Frontiers in Plant Science* 8, 1891.

Guerra-García, A., Rojas-Barrera, I.C., Ross-Ibarra, J., Papa, R. and Piñero, D. (2022) The genomic signature of wild-to-crop introgression during the domestication of scarlet runner bean *(Phaseolus coccineus* L.). *Evolution Letters* 6, 295–307.

Hardwick, R.C. (1972) The emergence and early growth of French and runner beans (*Phaseolus vulgaris* L. and *Phaseolus coccineus* L.) sown on different dates. *Journal of Horticultural Science and Biotechnology* 47, 395–410.

Kalloo, G. (1993) Runner bean *Phaseolus* (*Coccineus* L.). In: Kalloo, G. and Bergh, B.O. (ed.) *Genetic Improvement of Vegetable Crops*. Pergamon Press Limited, USA, pp. 405–407.

Kaplan, L. (1965) Archaeology and domestication in American Phaseolus (Beans). *Economic Botany* 19, 358–368. DOI: 10.1007/bf02904806

Khoury, C.K., Brush, S., Costich, D.E., Curry, H.A., De Haan, S. *et al.* (2022) Crop genetic erosion: understanding and responding to loss of crop diversity. *New Phytologist* 233, 84–118.

Kimani, P.M., Beebe, M., Musoni, A., Ngulu, F., Van Rheenen, H. *et al.* (2008) Progress in the development of snap and runner beans for smallholder production in East and Central Africa. *Improving Beans for the Developing World. Annual Report SBA-1*. CIAT, Cali, Colombia, pp. 208–212.

Kimani, P.M., Njau, S., Mulanya, M. and Narla, R.D. (2019) Breeding runner bean for short-day adaptation, grain yield, and disease resistance in eastern Africa. *Food and Energy Security* 8:e171. DOI: 10.1002/fes3.171

Kornegay, J. and Cardona, C. (1991) Breeding for insect resistance in beans. In: van Schoonhoven, A. and Voysest, O. (eds) *Common Beans: Research for Crop Improvement*. CABI, Wallingford, UK, pp. 619–648.

Lobaton, J., Miller, T., Gil, J., Ariza, D., de la Hoz, J. *et al.* (2018) Re-sequencing of common bean identifies regions of inter-gene pool introgression and provides comprehensive resources for molecular breeding. *Plant Genome* 11, 170068. DOI: 10.3835/plantgenome2017.08.0068

Lopez-Martinez, L.X. (2020) Bioactive compounds of runner bean (*Phaseolus coccineus* L.). In: Murthy, H.N. and Paek, K.Y. (ed.) *Bioactive Compounds in Underutilized Vegetables and Legumes*. Reference Series in Phytochemistry. Springer, Cham, Switzerland, pp. 1–17.

Mahuku, G.S., Jara, C., Cajiao, C. and Beebe, S. (2003) Sources of resistance to angular leaf spot (*Phaeoisariopsis griseola*) in common bean core collection, wild *Phaseolus vulgaris*, and secondary gene pool. *Euphytica* 130, 303–313.

Mercati, F., Catarcione, G., Paolacci, A.R., Abenavoli, M.R., Sunseri, F. *et al.* (2015) Genetic diversity and population structure of an Italian landrace of runner bean *(Phaseolus coccineus* L.): inferences for its safeguard and on-farm conservation. *Genetica* 143, 473–485. DOI: 10.1007/s10709-015-9846-1

Miklas, P.N., Beaver, J.S., Grafton, K.F. and Freytag, G.F. (1994) Registration of TARS VCI-4B multiple disease resistant dry bean germplasm. *Crop Science* 34, 1415.

Miklas, P.N., Grafton, K.F., Kelly, J.D., Steadman, J.R. and Silbernagel, M.J. (1998) Registration of four white mold resistant dry bean germplasm lines: I9365-3, I9365-5, I9365-31, and 92BG-7. *Crop Science* 38, 1728.

Mulanya, M.M., Kimani, P.M., Narla, R.D. and Ojwang, P.O. (2019) Genetic inheritance of photoperiod sensitivity in runner bean (*Phaseolus coccineus* L.). *Australian Journal of Crop Science* 13, 1511–1519.

Mullins, C.A., Allen Straw, R., Stavely, J.R. and Wyatt, J.E. (1999) Evaluation of half runner bean breeding lines. *Annual Report Bean Improvement Cooperative* 42, 113–114.

Nalupya, Z., Hamabwe, S., Mukuma, C., Lungu, D., Gepts, P. and Kamfwa, K. (2021) Characterization of *Colletotrichum lindemuthianum* races in Zambia and evaluation of the CIAT Phaseolus core collection for resistance to anthracnose. *Plant Disease* 105, 1–31. DOI: 10.1094/pdis-02-21-0363

Negri, V. and Tosti, N. (2002) *Phaseolus* genetic diversity maintained on-farm in central Italy. *Genetic Resources and Crop Evolution* 49, 511–520.

Normah, M.N., Chin, H.F. and Reed, B.M. (2012) *Conservation of Tropical Plant Species*. Springer-Verlag, New York.

Nowosielski, J., Podyma, W. and Nowosielska, D. (2002) Molecular research on the genetic diversity of Polish varieties and landraces of *Phaseolus coccineus* L. and *Phaseolus vulgaris* L. using the RAPD and AFLP methods. *Cellular & Molecular Biology Letters* 7, 753–762.

Olaleke, A.M., Olorunfemi, O. and Akintayo, T.E. (2006) Compositional evaluation of cowpea (*Vigna unguiculata*) and scarlet runner bean (*Phaseolus coccineus*) varieties grown in Nigeria. *Journal of Food Agriculture and Environment* 4, 39–43.

Osorno, J.M., Muñoz, C.G., Beaver, J.S., Ferwerda, F.H., Bassett, M.J. *et al.* (2007) Two genes from *Phaseolus coccineus* confer resistance to *Bean Golden Yellow Mosaic Virus* in common bean. *Journal of the American Society for Horticultural Science* 132, 530–533. DOI: 10.21273/JASHS.132.4.530

Padhi, A.K. (2001) Effect of vegetable intercropping on productivity, economics and energetics of maize (*Zea mays*). *Indian Journal of Agronomy* 46, 204–210.

Palmero, D., Iglesias, C., de Cara, M., Tello, J.C. and Camacho, F. (2011) Diversity and health traits of local landraces of runner bean (*Phaseolus coccineus* L.) from Spain. *Journal of Food, Agriculture & Environment* 9, 290–295.

Parker, T.A. and Gepts, P. (2021) Population genomics of *Phaseolus* spp.: A domestication hotspot. *Population Genomics* [Preprint]. DOI:10.1007/13836_2021_89

Pastor-Corrales, M.A., Jara, C. and Singh, S.P. (1998) Pathogenic variation in, sources of, and breeding for resistance to *Phaeoisariopsis griseola* causing angular leaf spot in common bean. *Euphytica* 103, 161–171.

Piñero, D. (2017) Domestication genomics of the open-pollinated scarlet runner bean (*Phaseolus coccineus* L.). *Frontiers in Plant Science* 8, 299557. DOI: 10.3389/fpls.2017.01891

Rendón-Anaya, M., Montero-Vargas, J.M., Saburido-Álvarez, S., Vlasova, A., Capella-Gutierrez, S. *et al.* (2017) Genomic history of the origin and domestication of common bean unveils its closest sister species. *Genome Biology* 18, 60. DOI: 10.1186/s13059-017-1190-6

Rodiño, A.P., Lema, M., Pérez-Barbeito, M., Santalla, M. and De Ron, A.M. (2006) Assessment of runner bean (*Phaseolus coccineus* L.) germplasm for tolerance to low temperature during early seedling growth. *Euphytica* 155, 63–70. DOI: 10.1007/s10681-006-9301-6

Rodriguez, M., Rau, D., Angioi, S.A., Bellucci, E., Bitocchi, E. *et al.* (2013) European *Phaseolus coccineus* L. landraces: population structure and adaptation, as revealed by cpSSRs and phenotypic analyses. *PLOS ONE* 8, e57337. DOI: 10.1371/journal.pone.0057337

Salinas, A.D. (1988) Variation, taxonomy, domestication, and germplasm potentialities in *Phaseolus coccineus*. In: Gepts, P. (ed.) *Genetic Resources of Phaseolus Beans. Current Plant Science and Biotechnology in Agriculture*, Springer, Dordrecht, Netherlands, pp. 441–463.

Santalla, M., Monteagudo, A.B., González, A.M. and De Ron, A.M. (2004) Agronomical and quality traits of runner bean germplasm and implications for breeding. *Euphytica* 135, 205–215.

Schmutz, J., McClean, P.E., Mamidi, S., Wu, G.A., Cannon, S.B. *et al.* (2014) A reference genome for common bean and genome-wide analysis of dual domestications. *Nature Genetics* 46, 707–713.

Schwartz, H.F., Otto, K., Terán, H., Lema, M. and Singh, S.P. (2006) Inheritance of white mold resistance in *Phaseolus vulgaris* × *P. coccineus* crosses. *Plant Disease* 90, 1167–1170. DOI:10.1094/pd-90-1167

Schwember, A.R., Carrasco, B. and Gepts, P. (2017) Unravelling agronomic and genetic aspects of runner bean (*Phaseolus coccineus* L.). *Field Crops Research* 206, 86–94. DOI: 10.1016/j.fcr.2017.02.020

Singh, S.P., Debouck, D.G. and Roca, W.M. (1998) Interspecific hybridization between *Phaseolus vulgaris* L. and *P. parvifolius* Freytag. *Annual Report Bean Improvement Cooperative* 41, 7–8.

Singh, S.P., Terán, H., Schwartz, H.F., Otto, K. and Lema, M. (2009) Introgressing white mould resistance from *Phaseolus* species of the secondary gene pool into the common bean. *Crop Science* 49, 1629–1637.

Sinkovič, L., Pipan, B., Sinkovič, E. and Meglič, V. (2019a) Morphological seed characterization of common (*Phaseolus vulgaris* L.) and runner (*Phaseolus coccineus* L.) bean germplasm: a Slovenian gene bank example. *BioMed Research International* 6376948. DOI: 10.1155/2019/6376948

Sinkovič, L., Pipan, B., Vasić, M., Antić, M., Todorović, V. *et al.* (2019b) Morpho-agronomic characterisation of runner bean (*Phaseolus coccineus* L.) from south-eastern Europe. *Sustainability* 11(21), 6165.

Stefan, M., Munteanu, N., Stoleru, V., Mihasan, M. and Hritcu, L. (2013) Seed inoculation with plant growth-promoting rhizobacteria enhances photosynthesis and yield of runner bean (*Phaseolus coccineus* L.). *Scientia Horticulturae* 151, 22–29.

Suárez, J.C., Polanía-Hincapié, P.A., Saldarriaga, S., Ramón-Triana, V.Y., Urban, M.O. *et al.* (2023) Bioactive compounds and antioxidant activity in seeds of bred lines of common bean developed from interspecific crosses. *Foods* 12, 2849.

Van de Wouw, M., Kik, C., van Hintum, T., van Treuren, R. and Visser, B. (2010) Genetic erosion in crops: concept, research results, and challenges. *Plant Genetic Resources* 8, 1–15. DOI: 10.1017/S1479262109990062

Vlasova, A., Capella-Gutiérrez, S., Rendón-Anaya, M., Hernández-Oñate, M., Minoche, A.E., Erb, I. *et al.* (2016) Genome and transcriptome analysis of the Mesoamerican common bean and the role of gene duplications in establishing tissue and temporal specialisation of genes. *Genome Biology* 17, 1–18. DOI: 10.1186/s13059-016-0883-6

Wallace, D.H. and Wilkinson, R.E. (1965) Breeding for Fusarium root rot resistance in beans. *Phytopathology* 55, 1227–1231.

Wallace, L., Arkwazee, H., Vining, K. and Myers, J.R. (2018) Genetic diversity within snap beans and their relation to dry beans. *Genes* 9, 587.

Westphal, E. (1974) Phaseolus. In: Westphal, E. (ed.) *Pulses in Ethiopia, their Taxonomy and Agricultural Significance*. Centre for Agricultural Publishing and Documentation (PUDOC), Wageningen, Netherlands, pp. 129–176.

Xanthopoulou, A., Kissoudis, C., Mellidou, I., Manioudaki, M., Bosmali, I. *et al.* (2019) Expanding *Phaseolus coccineus* genomic resources: de novo transcriptome assembly and analysis of landraces 'Gigantes' and 'Elephantes' reveals rich functional variation. *Biochemical Genetics* 57, 747–766. DOI: 10.1007/s10528-019-09920-6

10 Hyacinth Bean (*Lablab purpureus* L.)

Lakshmi Gangavati[1]*, Vinita Ramtekey[2,3]* and Nandan L. Patil[1]

[1]*Department of Genetics and Plant Breeding, University of Agricultural Sciences, Dharwad, Karnataka, India;* [2]*ICAR-Indian Institute of Seed Science, Mau, Uttar Pradesh, India;* [3]*Plant Breeding Department, University of Bonn, Bonn, Germany*

Abstract

Hyacinth bean (*Lablab purpureus* L. Sweet) is a multipurpose legume crop grown as a pulse, vegetable, fodder, green manure and intercrop. Genus *Lablab*, being monospecific, is comparatively diverse. The wide genetic diversity has contributed to its adaptability and benefitted the crop in terms of sustaining itself over various environments. Conventional breeding approaches, namely pure line selection, pedigree breeding and single seed descent breeding methods, are major breeding approaches for varietal development in *Lablab*. The integration of genomics tools has helped in studying the genetics of *Lablab*, domestication events, marker-trait associations, and the identification of candidate genes for numerous important traits in *Lablab*. The sustainable utilization of these genomic tools will enable researchers to understand comprehensively genetic makeup, gene expression patterns, and protein profiles of *Lablab* beans. Future *Lablab* improvement programmes should focus on developing multipurpose elite cultivars exhibiting resistance to numerous biotic and abiotic stresses. Meanwhile, care should be taken to preserve the germplasm, which is a reservoir of critical novel genes for future *Lablab* improvement.

Keywords: *Lablab*, multipurpose legume, pure line, resilient, gene pools, core collection, omics

10.1 Introduction

Hyacinth bean (*Lablab purpureus* L. Sweet) is an underutilized legume crop widely distributed in India, Africa, South-east Asia, Sudan and other tropical regions of the world (Maass *et al.*, 2010). It is called by various names, such as lablab bean, bonavist bean, Egyptian kidney bean, field bean, dolichos bean and lubia bean in different areas of the world (Vaijayanthi *et al.*, 2019a). The monotypic genus *Lablab* belongs to the tribe Phaseoleae, sub-tribe Phaseolinae, and the sub-family Faboideae of the family Fabaceae. The genus *Lablab* is monospecific, with three subspecies. The crop originated in Africa and has been disseminated to India and other Asian regions of the world (Maass, 2016). It is grown as a pulse crop in India and South-east Asia. It is widely cultivated in Africa by smallholder farmers and is well-identified as a farmer crop in Sudan (George, 2011). Lablab bean is a perennial herb cultivated annually or biennially for pulse, vegetables, fodder, green manure and ornamental purposes (Raghu *et al.*, 2018; Naeem *et al.*, 2020). The legume is a reliable source of protein (20–28% protein), carbohydrates, fibres, fats and minerals. It is used as an intercrop with maize in Southern Africa and sorghum and ragi in India under

*Corresponding author: lakshmirg8@gmail.com; s79vramt@uni-bonn.de

© CAB International 2024. *Potential Pulses: Genetic and Genomic Resources*
(eds R. Chandora, T. Basavaraja and A. Pratap)
DOI: 10.1079/9781800624658.0010

adverse climatic conditions (Rapholo *et al.*, 2020). The has numerous medicinal properties. The flavonoids and alkaloids in the bean are used for drug preparation, and the tyrosinase enzyme in hyacinth bean is a remedy for hypertension in humans (Naeem *et al.*, 2020). Compared to other legumes, hyacinth bean has wider adaptability and greater potential to grow under adverse climatic situations, including drought and salinity stresses. The crop can also be used as a source of organic matter and to fix soil nitrogen, thereby improving the yields of subsequent crops in an economical and environmentally friendly way. Despite its numerous benefits, the crop is underutilized owing to a lack of understanding of its nutritional value, limited market, unavailability of improved varieties, imperfect management practices, and replacement by other potential crops (Letting *et al.*, 2021). Research should focus on expanding the knowledge of this neglected legume crop and exploring genetic diversity to develop potential multipurpose *Lablab* varieties. The dissemination of knowledge about these underutilized crops is crucial to prevent the erosion of possible gene diversity from the traditional cultivated areas. If that is the case, *Lablab* will emerge as a resilient, multipurpose legume in the future.

10.2   Origin, Distribution and Botanical Description

Limited understanding regarding the crop diversity of hyacinth bean led to speculation that the crop is either Indian (Nene, 2006; Kukade and Tidke, 2014) or African in origin (Maass, 2016). Some reports suggested that the crop had been introduced to West Asia and Africa from India (Ayyangar and Nambiar, 1935). The floristic crop surveys (Verdcourt, 1971; Fuller, 2003) and genetic studies reported that wide variations of wild and cultivated forms were found in East Africa (Fuller, 2003; Robotham and Chapman, 2017), proving Africa to be the centre of origin. Robotham and Chapman (2017) confirmed the origin of *Lablab* as East Africa with the help of microsatellite markers designed using the transcriptome sequence of *Lablab* (Chapman, 2015). They found much relatedness between the wild

subspecies, subsp. *uncinatus* and East African lines. The crop was then disseminated to India and later to other parts of Asia. Studies have reported India as the secondary centre of diversity based on the intermediate status of the Indian accessions between African wild forms and cultivars (Maass *et al.*, 2005; Maass, 2016). The crop is now well established throughout the tropics and is grown mainly in warmer parts of the world. It is grown for food purposes in India, South-east Asia, Africa, including Sudan and Egypt, and Japan. The bean is popularized as 'Fuji mame' in China after its introduction from Japan in 1654.

Hyacinth bean, with perennial growth habit, is cultivated as an annual or biennial as a pulse crop, fresh vegetable or fodder crop. The herb grows up to 6 m with a usual growth habit of a twining type; however, bushy, semi-erect or branching types are also observed (Fig. 10.1). Trifoliate leaves with pointed edges exhibit sparse pubescence. The inflorescence is an axillary raceme with many flowers. The compressed and glabrescent peduncle is 2–24 cm long and 1–5 flowers are clustered together to form tubercles. The flower stalk (pedicle) is short and sparsely pubescent. Some cultivars bear white flowers, and some might bear red, pink or purple flowers (Fig. 10.2). Stamens are joined to form a diadelphous condition (9+1). A sessile and sparsely pubescent ovary is 10 mm long, the style is around 8 mm long, abruptly upturned, and laterally compressed. The stigma is capitate and glandular. The legume pods vary in size, shape and colour, depending on the cultivar (Fig. 10.3). They may be inflated or flat, smooth or pubescent, curved or straight in shape, and pale green to bright purple. Each legume pod carries 3–6 ovoid or flattened seeds, the colour of which could be white, red, brown or black, depending on the variety. Mottled seeds are a characteristic feature of this wild species (CABI, 2022). Flower opening in hyacinth bean is from 9 am to 5 pm and anther dehiscence occurs between 5 am and 2 pm. Pollen is viable on the day of anthesis and remains viable for 48 hours provided the temperature is 28.5°C and 85–90% relative humidity (Rai *et al.*, 2019). It is strictly a self-pollinated crop (Kukade and Tidke, 2014). Emasculation is performed between 4.30 pm and 5.30 pm, and pollination occurs on the following day between 7 am and 9 am (Rai *et al.*, 2019).

Determinate   Semi-determinate   Indeterminate

Fig. 10.1. Variability for growth habit in hyacinth bean. (Author's own figure.)

Fig. 10.2. Variability for flower colour in hyacinth bean. (Author's own figure.)

10.3 Crop Gene Pool, Evolutionary Relationships and Systematics

Hyacinth bean, with a chromosome number of 2n = 22, is in the genus *Lablab* in the family Leguminosae, which includes beans, chickpeas, soybeans, groundnuts and garden peas, among many other legumes. The genus *Lablab* is closely related to *Dolichos* and *Macrotyloma* and was previously in the same genus, i.e. *Dolichos*. Later, hyacinth bean (*Dolichos lablab*) and horse gram (*Dolichos uniflorus*) were allotted to different genera. Hyacinth bean was assigned to a monotypic genus *Lablab* (*L. purpureus* or *Lablab niger*) (Verdcourt, 1971), while horse gram was to *Macrotyloma* (as *M. uniflorum*). Genus *Lablab* is monospecific, which includes *L. purpureus*. *L. purpureus* comprises three subspecies: (i) subsp.

Fig. 10.3. Variability for pod shape, size and colour in hyacinth bean. (Author's own figure.)

uncinatus, a wild ancestral form distributed mainly in East Africa; (ii) subsp. *purpureus*, a cultivated type distributed in Asian countries; and (iii) subsp. *bengalensis*, a developed form distributed primarily in Africa and South-east Asia (Verdcourt, 1971). Important characteristic features of the three subspecies are briefly described in Table 10.1.

Lablab originated in East Africa and was later domesticated in various parts of the world. During domestication, the wild ancestral form underwent numerous changes in seed-attributing traits, plant growth habits and other morphological traits (Pengelly and Maass, 2001; Islam, 2008; Maass, 2016). Large-seeded variants were preferred for grain and longer pods for vegetable purposes. Wild forms were indeterminate, and the selection was made for determinate/bushy types.

A comparison between wild and domesticated hyacinth bean for important morphological traits is illustrated in Table 10.2, which demonstrates the apparent domestication syndrome for the lablab bean. Studies have reported the existence of two domestication events for hyacinth bean (Maass, 2016; Robotham and Chapman, 2017). The study by Robotham and Chapman (2017) clearly distinguished the wild accessions into two categories, namely 2-seeded and 4-seeded types. Most of the present-day cultivars had a close resemblance to the 4-seeded wild types. The cultigens subsp. *perpureus* and subsp. *bengalensis* distributed in India, Africa and other Asian countries have been domesticated from the 4-seeded wild types (Maass, 2016); however, a few 2-seeded cultivated forms from Ethiopia revealed genetic similarity to the 2-seeded wild accessions (subsp. *uncinatus*) (Maass, 2016). These two major wild phenotypic

forms, 2-seeded and 4-seeded, were confirmed by a polymorphism at a chloroplast locus and grouped as haplotype A and haplotype B, respectively (Robotham and Chapman, 2017). Hence, these studies clearly explained the two domestication events for hyacinth bean and suggested Ethiopia as one of the important centres of domestication (Maass, 2016).

Little published information is available regarding the crop gene pools of hyacinth bean. However, some studies reported the existence of components/sub-gene pools of the primary gene pool (GP 1) as GP 1A (domesticated) and GP 1B (wild) for the lablab bean (Hepper, 1963; Smartt, 1984). The gene pool includes the cultigens, which intermate easily. Two hyacinth bean cultivars, Typicus and Lignosus, are easily crossable (Shivashankar *et al.*,1971) and can be assigned to the primary gene pool, considering them as two varieties of *Lablab*. The former is mainly cultivated for vegetable purposes due to its soft edible pods, and the latter is a field bean for pulses.

The two subspecies, *purpureus* and *bengalensis*, are taxonomically similar and freely interbreed despite their morphological differences (Verdcourt, 1971). Hybridization is possible among these subspecies to get a viable hybrid progeny. Hence, these subspecies can be assigned to the crop's secondary gene pool.

Previous studies have reported the development of the F_1 hybrid through the hybridization of *Lablab niger* with *Macrotyloma uniflorum* (Rajendra, 1978). A detailed description of the hybrid would enable the inclusion of the species as a wild relative, either under a secondary or tertiary gene pool or reconsider assigning these two species to the same genus (Smartt, 1984).

Table 10.1. Comparison between subspecies of hyacinth bean. (Author's own table.)

Sr. No.	Feature	subsp. *uncinatus*	subsp. *purpureus*	subsp. *bengalensis*
1	Distribution	East Africa	Asia	India and South-east Asia
2	Form	Wild	Cultivated	Cultivated
3	Pod size	Small (40 mm x 15 mm)	Large (100 mm x 40 mm)	Medium (140 mm x 10–25 mm)
4	Pod shape	Scimitar-shaped	Scimitar-shaped	Linear-oblong to oblong, falcate
5	Number of seeds per pod	2	4	Up to 6
6	Kernel shape	Flattened	Kidney shape	Kidney shape
7	Diversity	Ancestral form distinct from both the cultivated forms	Both the subspecies are genetically very similar	
8	Purpose	Ornamental and vegetable	Pulse and green manure	Pulse (Asia) and forage (Africa)

Table 10.2. Comparison between wild and domesticated *Lablab* for important morphological traits. (Author's own table.)

Sr. No.	Character	Wild	Domesticated	Reference
1	Seed shape	Flattened	Elongated, rounded, kidney shaped	Maass (2006)
2	Seed colour	Uniformly greyish-brown and mottled	Cream, tan, greyish-brown and black	
3	Seed length (mm)	5.7–8.4	10.3–11.4	
4	Seed width (mm)	4.0–5.7	6.3–7.1	
5	Seed mass (g/1000 seeds)	56–130	228–494	
6	Length of pod (cm)	2 to 6	4 to 17	Pengelly and
7	Pod colour	Purple, green	Light green, green and purple	Maass
8	Number of seeds per pod	2 to 3	1 to 7	(2001);
9	Days to 50% flowering	Intermediate to very late	Very early to mid-late	Islam
10	Growth habit	Indeterminate	Indeterminate/twining type/ bushy/erect	(2008); Maass
11	Flower colour	Purple, white	Purple, white	(2016)

Crossing between cultivated *Lablab* and wild 4-seeded types was reported (Venkatesha, 2013; Maass, 2016), while 2-seeded wild types were probably not tested (Maass, 2016). Hence, the 4-seeded wild types can be included under the tertiary gene pool.

10.4 Germplasm Collection, Characterization, Evaluation and Conservation

Germplasm is the base material requisite to plan any crop improvement programme. It is the source of various novel genes and a prerequisite to broadening the genetic base of a crop species.

Lablab purpureus L. is comparatively diverse, and around 3000 germplasm accessions have been collected worldwide (Maass *et al.*, 2010). Most of the diverse wild accessions were collected in southern and eastern Africa. As a secondary centre of diversity, India has contributed to some of the semi-domesticated or cultivated forms of the bean. All these accessions are conserved globally at different seed gene banks and as working germplasm collections at various institutions for breeding purposes. They include the International Livestock Research Institute (ILRI, Ethiopia), Bangladesh Agricultural Research Institute (BARI), International Institute of Tropical Agriculture (IITA, Nigeria), National Bureau of Plant Genetic Resources (NBPGR-New Delhi), University of Agricultural Sciences, Bengaluru

(India) and many others. The germplasm conservation status at different international organizations is summarized in Table 10.3.

Genesys PGR (Plant Genetic Resources) is an online global portal (https://www.genesys-pgr.org) that provides access to information on world crop diversity conserved in various parts of the world, providing information on passport data of the accession, core collections, evaluation, characterization data sets and a list of crop descriptors. The portal holds around 1350 accessions of hyacinth bean collected worldwide (Mihailovic *et al.*, 2016).

Germplasm characterization includes the description of the germplasm accessions for highly heritable traits or the recording of the observations for different crop descriptors. The characterization will assist the breeders in varietal identification and the maintenance of the genetic purity of the varieties. Such identified morphological markers are useful for indirectly selecting agronomically important traits during crop improvement programmes. These descriptors can also be used for performing DUS testing (distinctness, uniformity and stability), a compulsory requirement for protecting

varieties under the International Union for the Protection of New Varieties of Plants. Germplasm of hyacinth bean is being characterized for different descriptors (a list of crop descriptors is available at Biodiversity International (BI) (2008) and Byregowda *et al.*, 2015). The important descriptors include twining or bushy growth habits, determinate or indeterminate growth, pod size, shape and colour, flower colour, seed colour, shape, etc. The descriptors, especially pod morphology, are used as important markers by breeders and farmers to select promising genotypes. Farmers in Nepal's low land 'terai' zone differentiate among season-adapted varieties using descriptors such as pod shape, size and colour as morphological markers (Sunwar *et al.*, 2006). Once the germplasm accessions are characterized, the subsequent data could also be used to develop a core collection reflecting most of the variation in the entire crop germplasm, facilitating the handling of germplasm and effective selection during crop improvement programmes.

The evaluation of germplasm is different from characterization. The evaluation includes the assessment of the agronomic potentiality of

Table 10.3. Hyacinth bean germplasm conservation status at different international organizations. (Author's own table.)

Sr. No.	Institute	Number of accessions	Source
1	Agricultural Research Service, US Department of Agriculture, North America	94	USDA (2023)
2	International Institute of Tropical Agriculture, Nigeria	70	IITA (2023)
3	National Bureau of Plant Genetic Resources, India	1723	NBPGR (2023)
4	South Australian Research and Development Institute (SARDI)	261	SARDI (2023)
5	International Livestock Research Institute, Ethiopia	340	ILRI (2023); Genesys (2023)
6	University of Agricultural Sciences, Bengaluru, India	650	Vaijayanthi *et al.* (2019a)
7	Asian Vegetable Research and Development Centre, Taiwan	557	AVRDC (2023)
8	Genetic Resources Research Institute, Kenya Agricultural and Livestock Research	366	Genesys (2023)
9	International Center for Tropical Agriculture, Colombia	150	
10	Council for Scientific and Industrial Research (CSIR)- Plant Genetic Resources Research Institute (PGRRI), Ghana	57	
11	Agricultural Plant Genetic Resources Conservation and Research Centre, Sudan	34	
12	Tropical Agricultural Research and Higher Education Center, Central America	32	

the accession for metric or quantitative traits and reaction to numerous biotic and abiotic stresses (Kumar and Kaur, 2010). The evaluation of hyacinth bean accessions for grain yield should focus on major yield-attributing traits, such as number of pods/plant, grains per pod, length of pod, average pod weight, and 1000-grain weight, among other traits (Letting et al., 2021; Singh et al., 2021). Stable grain yield of Lablab across the environments should be evaluated for earliness and a neutral photo-periodical reaction coupled with uniform and synchronous pod and grain maturity (Mihailovic et al., 2016). The percentage of crude protein, digestibility, and calcium to phosphorus ratio of the forage dry matter (Fuquay et al., 2011) are considered for forage Lablab, along with traits such as the interval between sowing and maturity, and number of side branches and photosynthetically active leaves (Mihailovic et al., 2016). A few recent research findings on germplasm characterization and evaluation are illustrated in Table 10.4, providing insights into the different germplasm accessions, studied descriptors, evaluation methods and significant findings.

10.4.1   Core collection

As discussed earlier, the core collection should include representative accessions from the available germplasm and reflect the genetic and ecological diversity of a crop species with minimum repetitions. While developing a core collection, it is rather important to select the plant types of high potentiality rather than only focusing on the morphological diversity (Pengelly and Maass, 2001). Many studies have reported the development of core collections in hyacinth bean using germplasm accessions collected from Africa, where both wild and semi-domesticated types are available and few collections from the centre of secondary diversity (India). Pengelly and Maass (2001) developed a core collection (n = 47) out of 249 germplasm accessions (n = 249) collected from southern Asia. They considered parameters such as mean, range, and standard deviation for quantifying various quantitative traits and assessed morphological diversity using crop descriptors for Lablab.

Similarly, Vaijayanthi et al. (2015) developed a core collection set of 64 accessions representing more than 90% of the variability in the whole collection (648 accessions) following the standard clustering procedure and heuristic approach. Further, they evaluated the core collection for productivity traits across environments and reported FPB 35, GL 250 and Kadalavare as widely adapted lines with higher fresh pod yield (Vaijyayanthi et al., 2016). Integrating germplasm characterization with molecular markers would effectively understand the genetic diversity and facilitate the development of core collections. The attempt was made by Maass et al. (2005), who developed core subsets based on the structure of diversity by amplified fragment length polymorphism (AFLP). More studies are needed to enlarge the core collections by including additional germplasm accessions.

10.5   Use of Germplasm in *Lablab* Crop Improvement by Conventional Approaches

Germplasm collection, evaluation and utilization of Lablab are being conducted in many countries, mainly in India, Indonesia, Australia and Bangladesh, to develop promising multipurpose Lablab varieties. India is the most prolific in breeding activities for hyacinth bean. The primary goals of Lablab breeding include breeding for higher yield, a multipurpose bean (pulse, vegetable and fodder), short-duration types, photo-insensitive varieties, and disease- and pest-tolerant genotypes. So far, there are three crucial Lablab cultivars: Rongai, Highworth and Endurance. The decision to conduct different breeding programmes for the genetic enhancement of Lablab is crucial for developing potential varieties. Breeding methods applied to a self-pollinated crop can also be used here. Individual plant selection (IPS), pure line, pedigree and single seed descent methods are appropriate breeding methods for hyacinth bean improvement (Kalloo and Pandey, 1993) because it is a self-pollinated pulse crop.

Single/individual plant selection (IPS) favours the isolation of a desirable genotype from a genetically diverse population. Crop improvement in hyacinth bean should be focused on the

Table 10.4. A summary of studies on germplasm collection, characterization and evaluation in *L. perpureus*. (Author's own table.)

Sr. No.	Germplasm collection	Characterization	Evaluation method/design	Results	References
1	Seventeen hyacinth bean genotypes from the Patuakhali region	Days to maturity, plant height, primary branches/plant, pod weight, number of pods/plants, weight of 100 seeds, pod yield/plant, total pod yield	Augmented design	High variability was found for pod size and colour. Lines, namely LPPK015, LPPK011, LPPK001, LPPK005 and LPPK016, were high pod yielders. LPPK001 and LPPK014 lines with creamy white seeds (for pulses).	Ratna *et al.* (2021)
2	Germplasm accessions from Bangladesh	Qualitative traits: leaf shape and colour, vein colour, seed shape and colour, cotyledon colour, stem colour, growth habit, pod traits, e.g. pod colour, pod beak, pod constriction and pod curvature, standard, wing and keel petal colour. Quantitative traits: no. of spikes/plant, spike length, number of flower/spikes, average pod weight, length, and width and seed length	Principal component analysis	Variability of 66.38% for qualitative traits was contributed by five components and 74.49% by six components to the total variation. Germplasm lines BD-10804, BD-10807, BD-11091, BD-10808, BD-10815, BD-11089, and the variety Goal Goda lablab were promising for higher pod weight, pod length, width and thickness.	Khatun *et al.* (2022)
3	Three-pole type dolichos bean germplasm	Days to first flowering and 50% flowering, plant height, number of branches, number of pods/cluster, number of seeds/pod, pod traits, e.g. pod length, pod girth, pod weight, pod yield/plant and pod yield/hectare	Randomized complete block design	Identified promising line PSRJ-13021-2 for pod length, pod yield/plant, and pod yield/hectare.	Mamatha *et al.* (2022)
4	277 accessions of lablab bean from Tanzania	Thirty-eight morpho-agronomic traits (24 qualitative and 14 quantitative)	Clustering based on agglomerative hierarchy and principal component analysis	Fourteen morpho-agronomic traits clearly distinguished the accessions and revealed a wide diversity. Five principal components cumulatively accounted for 61.89% of the total variability, and four clusters were found based on clustering.	Letting *et al.* (2021)

5	20 diverse genotypes of lablab bean from different regions of India	12 quantitative traits related to yield and yield attributes	Single linkage cluster analysis and Principal component analysis	Five principal components explained 82.53% of the total variability. Superior genotypes for no. of pods/cluster (CISH-DC-13), pod length (CISH-DC-6), pod weight (CISH-DC-7). No. of seeds/pod varied (CISH-DC-16), and pod yield/plant (CISH-DC-7) were identified. Single linkage cluster analysis grouped all genotypes into three clusters.	Singh et al. (2021)
6	16 germplasm of VRS, Hyderabad and a few local collections (Adilabad & Warangal districts, and IARI, Pusa, New Delhi)	Leaf, floral, pod and seed-related traits	Randomized block design	The highest pod yield was recorded in RND-1, and ADP-10 was good for no. of seeds per pod.	Preetham et al. (2020)
7	Reference varieties (HA 4 and HA 3), one advanced breeding line (HA 11–3), and Kadalavare and GL 66- candidate varieties	25 morphological descriptors	Randomized block design	Clear distinctness was found between candidate varieties and reference varieties for at least one trait under DUS testing, and the uniformity of reference and candidate varieties was confirmed as the off types were absent.	Shrikrishna and Ramesh (2020)

evaluation of germplasm accessions worldwide, mainly focusing on the African and Asian regions for high-yielding traits coupled with tolerance to major diseases and pests. In this way, the IPS is made for white or slightly cream-coloured seeds over dark-seeded types for pulses/seeds (Smartt, 1984; Pengelly and Maass, 2001; Duke, 2012). Similarly, direct selection can be made for accessions with long tender pods for vegetable *Lablab*, late flowering/maturity, high drought tolerance, and higher biomass yield for forage purposes.

Pure line selection is another important conventional approach for varietal development in *L. purpureus*. Efforts are being made at UAS, Bengaluru (India), Bangladesh, and the Indian Grass Land and Fodder Research Institute (IGFRI, India) to develop determinate and photoperiod-insensitive pure line varieties of dolichos bean for dual purposes. The first released *Lablab* cultivar Rongai (late-flowering, summer-growing perennial genotype for forage) followed by Highworth (an early-flowering and high seed yield variety) are the pure line selections from landrace material CPI 17883 and CPI 30212, respectively (Wilson and Murtagh, 1962; Liu, 1998). The available germplasm accessions need to be screened for important novel traits for developing superior pure line varieties following a systematic breeding procedure.

Hybridization, followed by selection, favours the identification of desirable genotypes in segregating generations following pedigree or single seed descent breeding methods. These methods of varietal development are appropriate for genetic enhancement in *L. purpureus*. However, the selection of parents is essential during the initial hybridization. It facilitates the breeder to include one of the parents as a donor parent carrying any novel trait for transferring into other genetic backgrounds. Later, desirable stabilized lines will be selected at advanced generations. Many photo-insensitive and pole/bushy high-yielding varieties have been developed following selection in advanced generations. The cultivar Endurance, an advanced selection from a cross of Rongai with a perennial wild accession, CPI 24973, is grown for forage during summer in crop–pasture farming systems in Queensland. Arka Jay and Arka Vijay are photo-insensitive bushy types, which are the selections from advanced generations developed at the Indian Institute of Horticultural Research (IIHR), Bengaluru, India. Susmita *et al.* (2020) developed segregating populations from crosses, namely IIHR 178 × Arka Swagath and IC 556824-IPS-2 × Arka Swagath, separately and isolated the superior transgressive segregants from both the crosses and later crossed the segregants with Arka Amogh to impart superior pod traits. Following the pedigree selection method, they identified high-yielding, pole-type and photo-insensitive promising lines at the F_7 generation.

Germplasm is the source of important genes, and returning to the germplasm to enhance the genetic potentiality of the crop is a nonstop activity. Germplasm screening for abiotic and biotic stress tolerance is the best method for identifying resistance sources. Major diseases of *Lablab* include anthracnose and dolichos yellow mosaic (DoYM). Plants infested with yellow mosaics produce bright yellow patches and cause a significant reduction in pod yield. Efforts were made by Narayan and Poonan (1987) to identify the resistance source in germplasm accessions, and they recognized HD-104, HD-91, HD-66, HD-98, HD-10 and HD-93 as tolerant under both field and greenhouse for DoYM. Similarly, out of 300 hyacinth bean genotypes, VRSEM 860, VRSEM 894 and VRSEM 887 were devoid of DoYM virus infection under laboratory and field conditions (Singh *et al.*, 2012). The identified resistant lines can be used as potential sources for developing yellow-mosaic-resistant *Lablab* varieties.

Pod borer (*Adisura atkinsoni*) is the major and most severe pest of lablab bean. During storage, the major infestation is caused by bruchid beetles, *Callosobruchus* spp. The losses caused by both these pests can reach up to 100%. Breeding for resistance to these major pests is very necessary. Screening germplasm accessions to search for resistance is appropriate, and many studies have reported sources of resistance. Paikra *et al.* (2019) screened 118 cultivars and identified 14 cultivars, namely IS-1, IS-26, IS-2, IS-51, IS-32, IS-52, IS-88, IS-81, IS-103, IS-107, IS-74, IS-115, IS-113 and IS-84 as resistant to pod borer. Germplasm accessions GL 77, GL 233, and GL 63 were identified as resistant to pod borer with 13.4%, 14.69% and 18.34% seed damage, respectively (Rajendra, 2015).

Hyacinth bean is reportedly well adapted to dry environments and other abiotic stresses (Rangaiah and Dsouza, 2016; Naeem *et al.*, 2020). Few studies have helped in understanding the traits of elevated levels of hydrogen peroxide, phenols, proline and ascorbic acid, leading to drought tolerance (Rangaiah and Dsouza, 2016) in *Lablab*. Research on breeding for abiotic stress resistance is limited and needs to be expanded further. In that case, hyacinth bean will emerge as a resilient legume crop for adverse climatic situations.

The grain- and forage-purpose varieties of hyacinth bean developed in various parts of the world following other breeding methods are summarized in Table 10.5.

Table 10.5. A summary of varieties of hyacinth bean developed in various parts of the world. (Author's own table.)

Sr. No.	Cultivar	Pedigree	Salient features	Developing centre
Grain purpose *Lablab* varieties				
1	Hebbal Avare-1	Local landrace with photoperiod sensitivity × Red typicus with photoperiod insensitivity	Photo-insensitive bushy type with small and soft pods.	University of Agricultural Sciences (UAS), Bengaluru, India
2	HA3	Cross between HA 1 and US 67–31 (an introduction from USDA	Bush and photo-insensitive variety, white flowers, green pods, 2–3 seeds/pod, and brown, round, and small seeds.	
3	HA4	HA 3 × CO 8 (photoperiod insensitive)	Photo-insensitive bushy variety with soft pods and can be harvested in 5 pickings.	
4	Arka Jay	F_7 selection from Hebbal Avare 3 (bushy type) x IIHR 99 (pole-recurrent parent)	Photo-insensitive bush variety: pods are long, light green and slightly curved.	Indian Institute of Horticultural Research, Bengaluru
5	Arka Vijay	Selection at F_7 from Hebbal Avare 3 (bushy) x Pusa Early Prolific (pole)	Photo-insensitive bush variety: Pods are short, dark green and have a characteristic aroma.	
6	IPSA Seam-1 and IPSA Seam-2	One of the parents is JER	Photo-insensitive year-round cultivars (grown during hot and humid seasons), early-flowering types.	BSMRAU, Gazipur, Bangladesh
7	Koala	Selection from introduced accession of France to Australia as Q 6680	Koala is an early maturing, short-season *Lablab* suitable for food and fodder production. Seeds are white-to-cream coloured with good export quality.	Australia
Forage purpose *Lablab* varieties				
8	Rongai	Pure line selection in CPI 17883, which was introduced from Kenya	It is a late-flowering type, with white flowers and light brown seeds, grows upright, grown for forage.	Australia
9	Highworth	Pure line selection in CPI 30212 introduced from south India to Australia	Purple flowers and black seeds are known for more than 1.5 tons of seed per hectare in commercial cultivation and 5–11 tons (dry weight) of forage. The protein content in forage is 22%.	Australia

Continued

Table 10.5. Continued.

Sr. No.	Cultivar	Pedigree	Salient features	Developing centre
10	Endurance	Selection from the cross Rongai × African wild perennial germplasm accession CPI 24973	White flowers and black seeds, forage *Lablab*.	Australia
11	Rio Verde	–	Flowers are white lavender, seeds are black to brown, grown for forage.	Texas A&M AgriLife Research and Extension Center at Overton, Texas
12	IGFRI-1-S-2218 and IGFRI-1- S- 2214	–	Grown for forage purposes.	Indian Grassland and Fodder Research Institute, India

10.6 Integration of Genomics and Genetic Resources for Crop Improvement

The varieties developed so far are mostly genetically uniform and lack genetic variability and diversity, making them more prone to emerging biotic and abiotic stress. These modern varieties have reached a plateau where further improvement is questionable. Unlike other legumes, *Lablab* is an underutilized and unexploited crop. In such conditions, integrating omics tools with genetic resources can considerably enhance the efficiency and efficacy of efforts made for lablab bean improvement. The advent of omics tools includes various high-throughput technologies such as molecular markers, next-generation sequencing, transcriptomics and proteomics, which are used to study different biological components as well as to explore untapped genetic variation hidden in these crops to develop stress-resilient varieties (Vaijayanthi *et al.*, 2019a). Integration of these tools with genetic resources enables researchers to understand comprehensively the genetic make-up, gene expression patterns, and protein profiles of lablab beans. This information can then be used to identify and specify key genes and molecular markers associated with desirable traits and develop improved varieties through targeted breeding or genetic engineering technologies. The status of lablab bean as an orphan crop, however, means only genomic approaches have contributed much to its improvement. A few recent findings are summarized in Table 10.6, which reveals the usage and availability of different molecular markers in hyacinth bean.

10.6.1 Advancements in the genomics of hyacinth bean

The most recent advancement includes next-generation sequencing and assembly of the entire lablab bean genome. This information blueprints the genetic material of lablab bean, including its genes, regulatory elements, and repetitive sequences. Comparative genomics can pinpoint the genes associated with important traits and evolutionary relationships with related species. In addition, recent developments in *Lablab* genomics indicate a smaller and less complex genome size of 367 Mb, making it suitable to characterize functionally the essential genes that could lead to adaptation to biotic and abiotic stress. Missanga *et al.* (2021) reported utilizing genomic information to develop drought-tolerant genotypes in *Lablab*. Likewise, Sserumaga *et al.* (2021) used DArT-seq-based single-nucleotide polymorphisms

Table 10.6. A summary of the availability of molecular markers and significant findings from such molecular markers-based studies in hyacinth bean. (Author's own table.)

The marker used/ reported	Genetic material	Significant outcome	References
RAPD markers	Indian lablab bean accessions	Reported significant polymorphism (74%) among the accessions and could effectively be used for varietal selection.	Singh *et al.* (2020)
SSR markers	91 accessions (10 exotic + 22 from Ethiopia + 59 accessions from Konso zone)	Identified Konso and West Wellega zones as hot spots for *Lablab* genetic diversity.	Tamiru Workneh (2020)
DArTseq-based SNPs and 15,719 Silico-Dart markers	65 lablab bean accessions	Developed high-density markers representing good genome recovery and used effectively for estimating relatedness among the accessions.	Sserumaga *et al.* (2021)
Cross-transferable EST-SSR markers from cowpea	Exotic and indigenous accessions	Reported eight allele-specific SNP and InDel markers for future genomics-assisted breeding in *Lablab*.	Venkatesha and Prakash (2022)
SSR markers	474 cultivated and 19 wild *Lablab* accessions	Two independent domestication paths for *Lablab* were reported, and gene diversity was higher in wild accessions than in cultivated accessions.	Kongjaimun *et al.* (2022)
DArTseq genotyping-by-sequencing	203 *Lablab* accessions from ILRI	Identified eight SNPs and ten SilicoDArT markers across eight chromosomes and found significant association with leaf traits and economic traits, namely pod width, pod ratio, and 1000-seed weight traits.	Njaci *et al.* (2023)
SSR markers	F_2 mapping population	Identified novel alleles at two SSR markers linked to QTLs controlling flowering time in lablab bean.	Gonal *et al.* (2023)
SSR markers	96 diverse accessions	Reported many genomic SSRs and their effectiveness in studying novel alleles, gene diversity and population dynamics of dolichos bean.	Spoorthi *et al.* (2023)

RAPD, random amplified polymorphic DNA; SSR, simple sequence repeat.

(SNPs) and SilicoDart markers to study genetic and structural variation to implement in the forage breeding programme. This was helpful in the development of high-density markers that cover a good proportion of the genome to associate marker-trait studies. Additionally, the trypsin inhibitor gene, a crucial anti-nutritional component, was organized using the high-quality chromosome-scale assembly of the *Lablab* genome. This case study was useful in analysing genomic loci linked to significant agronomic traits and domestication events, genetics and phenotypic diversity in a whole germplasm collection (Njaci *et al.*, 2023).

Teshome *et al.* (2022) performed whole genome sequencing and developed a genomics database for *Lablab* to unlock the genetic potential to advance forage breeding strategies such as marker-assisted selection and genomic selection. *Lablab* is a multipurpose crop often used as an alternative to vegetables. Fresh pods blended with fragrance are always of interest to farmers and consumers. Therefore, the breeding programme could expand if molecular markers linked to essential genes for fragrance were uncovered.

An orphan crop like lablab bean has benefitted greatly from a comparative genomic approach. Savitha and Ravikumar (2009) performed RAPD analysis to identify molecular diversity with the help of markers from lima bean in pendal and non-pendal types of lablab bean. Comparative genomic analysis was used by Wang *et al.* (2018) to identify drought-inducible regulatory factors such as GRAS, a WRKY transcription factor, using other legumes such as common bean, soybean and lucerne. The authors also predicted a protein–protein interaction (PPI) network for drought-induced root-related unigenes and emphasized that MAPK4 (mitogen-activated protein kinase) and an EF-hand calcium-binding protein (CML24) module might be the major signalling route to this PPI network. Likewise, Venkatesha and Patil (2022) developed novel expressed sequence tag (EST)-derived SNPs and InDel markers through amplicon sequencing of cowpea EST–SSR primers, namely CP2, CP8, CP29 and CP435, with a high transferability in lablab bean. Based on the sequencing of 18 genic fragments, 59 SNPs and 16 InDels were identified and used to develop eight allele-specific primers. These novel allele-specific markers were utilized for trait mapping in the parental genotype and F_2 segregating population. Such comparative studies are of utmost importance for initial breeding improvement.

Other than such genomic tools, the application of molecular markers has greatly improved the understanding of population structure and genetic variation available in *Lablab* gene pools. Maass *et al.* (2005) used AFLP to understand better the existence of genetic diversity in both wild and domesticated *Lablab*, as well as domestication and distribution patterns. Likewise, Esther *et al.* (2012) identified low levels of molecular and genetic diversity present in Kenyan *Lablab* collected from farmers' fields with the help of AFLP markers. Microsatellite genotyping and cpDNA sequencing were performed to explore the origin and genetic variation in the germplasm collection (Robotham and Chapman, 2017). Vaijayanthi *et al.* (2019b) performed genome-wide association analysis using SSR markers to link multiple agronomically important traits such as flowering, fresh pod yield, etc. Gene-specific SSR markers have been used to identify the terminal flowering locus (TFL) associated with photoperiod-responsive flowering in the F_2 segregating population. Homology analysis suggested that TFL markers shared a similarity of 89%, 78% and 89% in common bean, soybean and cowpea loci, respectively. Later, it was discovered that PvTFL1y was responsible for splice site mutation for lablab bean growth habit differences (Kaldate *et al.*, 2021),

thanks to the candidate gene approach, sequencing, multiple sequence alignment and protein structure analysis. Similarly, Krishna *et al.* (2022) performed phylogenetic analysis to understand the evolution of the PHY A (phytochrome A) gene responsible for photoperiod flowering in *Lablab* via primers designed from the *GmPHYA3* locus of soybean. Henceforth, by integrating these genomics tools with genetic resources, it would be possible to generate vast amounts of genetic data that could uncover valuable insights into the genetics of lablab bean. This information could be used in breeding programmes to accelerate the creation of superior genotypes with enhanced agronomic traits, nutritional value, and resistance to biotic and abiotic stresses.

10.6.2 Transcriptomics of hyacinth bean

RNA sequencing technologies enable identifying and quantifying gene expression levels in different tissues at various developmental stages and under different environmental conditions. By using comparative transcriptomics between different genotypes or related legumes, researchers could investigate the genetic basis of adaptive and desired traits (Chapman, 2015). Chapman (2015) sequenced, assembled and annotated the transcriptomes of lablab bean and three other legumes and reported the highest percentage of hits for lablab bean and the putative orthologue *Phaseolus vulgaris.* The study also identified many potential microsatellites and conserved orthologous sets (COS) markers. Only a few of them, however, were tested for amplification in cross families. The information generated from such studies is useful in studying genetic diversity and evolutionary relationships, identifying QTLs controlling critical economic traits, and providing a good platform for comparative genomics. Thilagavathy and Devaraj (2016) conducted a miRNA expression study by real-time PCR and reported miRNA 156 as an appropriate reference gene under abiotic stresses such as drought and salinity. In a transcriptome study, Rai *et al.* (2017) found that the γECS gene was downregulated at the mature stage, which was closely correlated with the decline in enzymatic antioxidants towards the end of maturation and the rise in oxidative stress brought on by the inflation of reactive oxygen species.

Numerous studies are required to understand the connection between the genome and cellular function. Transcriptome investigations can offer a glimpse of the genes and transcripts actively expressed under varied circumstances.

10.7 Erosion of Genetic Diversity from Traditional Areas and Limitations in Germplasm Use

Modern agriculture focuses on growing more uniform cultivars to feed more people on less available land, leading to genetic erosion of diverse traditional landraces. Hyacinth bean is a crop in which self-pollination is the natural reproduction mechanism, and varietal development is the preferred breeding method to develop high-yielding varieties. Alleles may be quickly fixed into homozygous genotypes because of this activity. However, the wild relatives or landraces exposed to open pollination are usually heterogeneous and carry alleles adapted to adverse conditions even when complete self-pollination occurs (Upadhyaya *et al.*, 2011). However, such precious entities are neglected in various parts of the world.

For example, in countries such as Africa, the research activities focus on improving *Lablab* as an animal feed (Ewansiha *et al.*, 2007; Maass *et al.*, 2010;) and green manure for soil conservation (Whitbread and Pengelly, 2004). This would result in the disappearance of potential genetic resources from the gene pool in Africa, where the vast diversity of *Lablab* is scattered (Minde *et al.*, 2021). The decreased cultivation area, lack of knowledge regarding the crop and replacement by other economically important crops are also significant factors contributing to the erosion of genetic diversity from the traditional areas. Wild forms (2-seeded and 4-seeded types) have already been identified, as well as feral escapes of subsp. *purpureus* from India, and cultivated and wild forms of subsp. *uncinatus* need our serious attention. Conservation of these wild forms is paramount to preserve the novel alleles for future *Lablab* crop improvement programmes.

There is, however, little research on the germplasm in the adjacent parts of Asia and other African countries, and its characterization and evaluation, which, coupled with the unavailability of the passport data of the accessions, are limiting the usage of available germplasm. Meanwhile, duplicate accessions are imposing difficulties during germplasm management in the gene banks, which need further study through molecular techniques to identify novel genes or exclude them, making it more convenient for gene banks to manage the germplasm.

10.8 Cultivation Practices

Lablab can be produced as a stand-alone crop or intercropped with cereals such as maize, sorghum and ragi (finger millet) (Naeem *et al.*, 2020). It is propagated through seeds. *Lablab* grows well on a well-prepared seedbed or even broadcast on roughly managed land. Seeds are sown (up to 4–5 cm deep) on beds (60–70 cm apart) following 30–45 cm spacing between the plants, and a seed rate of 25–30 kg is required per hectare. Stakes should be provided to support the twining types. During soil preparation, farmyard manure of 12.5 t/ha is advised. Later, NPK is supplied as a basal dose before sowing at the rate of 40 g/m^2. Applying molybdenized superphosphate at 250–500 kg/ha and some potash is recommended in poor sandy soils (George, 2011; CABI, 2022). Mild irrigation immediately after sowing and on the third day is recommended. During flowering and pod formation stages, irrigation is compulsory. Weed management during the initial stages of growth is necessary. Pre-emergence herbicides such as chlorthal dephenamid, chloramben and trifluralin have been suggested (CABI, 2022). In India, pre-emergence application of pendimethalin (2 litres/ha) is recommended three days after sowing and should be followed by one hand weeding at 40–45 days after sowing. For vegetable purposes, the green pods of reasonable size are picked by hand once a week. For grain, pods are harvested once they are dry. After harvest, the green pods and fresh beans can be stored at 0–2°C with a relative humidity of 85–90% for 21 and 7 days, respectively. Maintaining the moisture content below 10% is effective for storage. The dried grain can be stored in metallic containers (South-east Asia) or enclosed with a protective layer of sand (India).

10.9 Future Directions

To fulfil the global demands for food, along with the cultivation of regular crops, preference should be given to underexploited multipurpose crops such as lablab bean. As a legume crop, *Lablab* has numerous health benefits and establishes itself well under adverse climates. Future crop improvement programmes should focus on exploring, characterizing, and evaluating wider germplasm accessions and utilizing them to develop potential *Lablab* varieties. Attention should be given to exploring the centres of secondary diversity and domestication, exchanging exotic collections or collecting local landraces. The mutual sharing of information and research collaboration between countries should be useful for the improvement of *Lablab*. The integration of conventional approaches with genomic tools allows the genetic study of *Lablab*, identifying novel alleles governing numerous desirable traits and transferring them into other potential cultivars with the help of genomic tools. Apart from scientific improvement, focus should be directed towards increasing awareness on the importance of the crop to the community. Serious attempts should be made to conserve germplasm accessions, which are reservoirs of important traits to be exploited for the *Lablab* crop improvement in the future. Different strategies for the overall improvement of *Lablab* are summarized in Fig. 10.4.

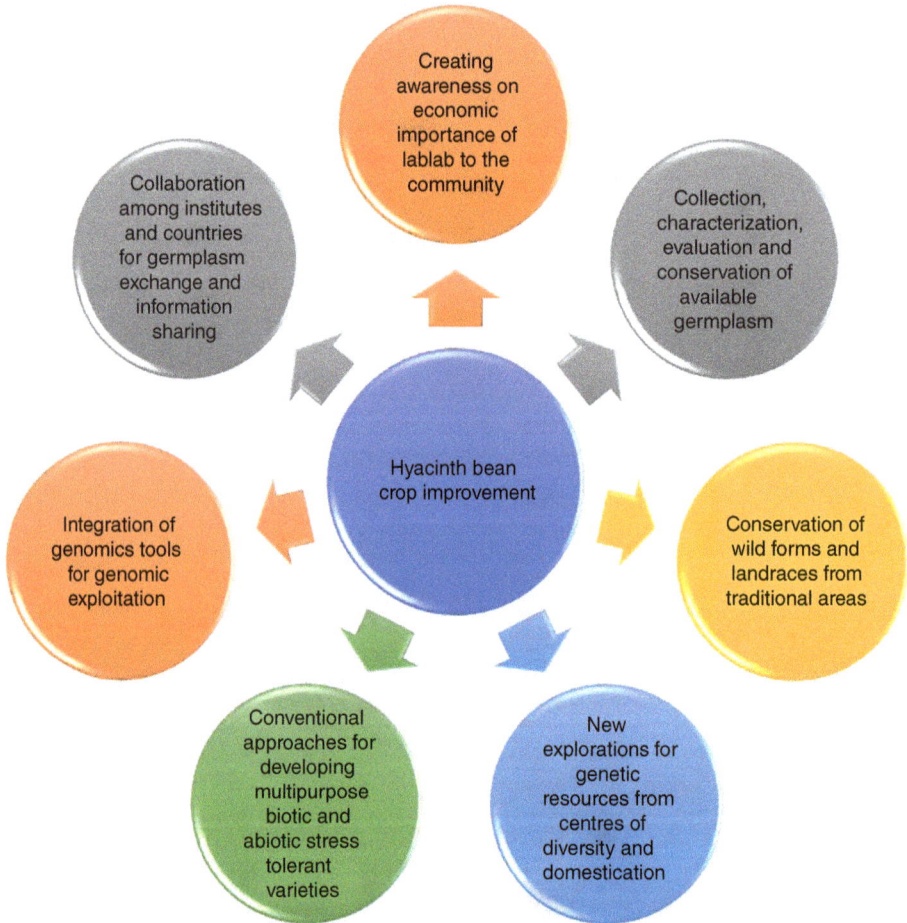

Fig. 10.4. Strategies for improvement of *Lablab*. (Author's own figure.)

References

AVRDC (2023) WorldVeg Genebank. Available at: https://genebank.worldveg.org/#/ (accessed 25 August 2023).

Ayyangar, G.R. and Nambiar, K.K.K. (1935) Studies in *Dolichos lablab* (Roxb.) and (L.)-The Indian field and garden bean. 1. In: *Proceedings of the Indian Academy of Sciences-Section B, India* 4, 411–433. DOI: 10.1007/BF03051442

BI (Bioversity International) (2008) Bioversity directory of germplasm collections. Available at: https://alliancebioversityciat.org/ (accessed 31 July 2023).

Byregowda, M., Girish, G., Ramesh, S., Mahadevu, P. and Keerthi, C.M. (2015) Descriptors of Dolichos bean (*Lablab purpureus* L.). *Journal of Food Legumes* 28(3), 203–214.

CABI International (2022) *Lablab purpureus* (hyacinth bean). Available at: https://www.cabidigitallibrary.org/ (accessed 31 July 2023).

Chapman, M.A. (2015) Transcriptome sequencing and marker development for four underutilised legumes. *Applications in Plant Sciences* 3(2), 1400111.

Duke, J. (2012) *Handbook of Legumes of World Economic Importance*. Springer Science & Business Media, USA.

Esther, N.K., Francis, N.W. and Miriam, G.K. (2012) Molecular diversity of Kenyan lablab bean (*Lablab purpureus* (L.) Sweet) accessions using amplified fragment length polymorphism markers. *American Journal of Plant Sciences* 3, 313–321. DOI: 10.4236/ajps.2012.33037

Ewansiha, S.U., Chiezey, U.F., Tarawali, S.A. and Iwuafor, E.N.O. (2007) Potential of *Lablab purpureus* accessions for crop-livestock production in the West African savanna. *The Journal of Agricultural Science* 145(3), 229–238. DOI: 10.1017/S0021859606006599

Fuller, D.Q. (2003) African crops in prehistoric South Asia: a critical review. *Food, Fuel, and Fields: Progress in African Archaeobotany*. Heinrich Barth Institut, Cologne, Germany, pp. 239–271.

Fuquay, J.W., McSweeney, P.L. and Fox, P.F. (2011) *Encyclopedia of Dairy Sciences*. Academic Press, Cambridge, Massachusetts.

Genesys (2023) Available at: https://www.genesys-pgr.org/ (accessed 25 August 2023).

George, A.A.T. (2011) *Tropical Vegetable Production*. CABI, Wallingford, UK.

Gonal, B., Ramesh, S., Ranjitha, G.V., Kalpana, M.P., Siddu, B.C. *et al.* (2023) Selective genotyping for the discovery of QTL controlling flowering time in dolichos bean (*Lablab purpureus* L.). *Crop Breeding and Applied Biotechnology* 23, e44182327. DOI: 10.1007/s10722-022-01535-y

Hepper, F.N. (1963) The bambara groundnut (*Voandzeia subterranea*) and Kersting's groundnut (*Kerstingiella geocarpa*) wild in West Africa. *Kew Bull* 16(3), 395–407. DOI: 10.2307/4114681

IITA (2023) Genetic Resources Centre. Available at: http://www.iita.org/research/genetic-resources/ (accessed on 20 August 2023).

ILRI (2023) Genebank. Available at: https://genebank.ilri.org/gringlobal/search (accessed 25 August 2023).

Islam, M.T. (2008) Morpho-agronomic diversity of hyacinth bean (*Lablab purpureus* (L.) Sweet) accessions from Bangladesh. *Bulletin de Ressources Phytogénétiques* 156, 73–78.

Kaldate, S., Patel, A., Modha, K., Parekh, V., Kale, B. *et al.* (2021) Allelic characterisation and protein structure analysis reveals the involvement of splice site mutation for growth habit differences in *Lablab purpureus* (L.) Sweet. *Journal of Genetic Engineering and Biotechnology* 19(1), 1–12. DOI: 10.1186/s43141-021-00136-z

Kalloo, G. and Pandey, S.C. (1993) Hyacinth bean: *Lablab purpureus* (L) Sweet. In: *Genetic Improvement of Vegetable Crops*. Pergamon, Oxford, UK, pp. 387–389.

Khatun, R., Uddin, M.I., Uddin, M.M., Howlader, M.T.H. and Haque, M.S. (2022) Analysis of qualitative and quantitative morphological traits related to yield in country bean (*Lablab purpureus* L. sweet) genotypes. *Heliyon* 8(12), e11631. DOI: 10.1016/j.heliyon.2022.e11631

Kongjaimun, A., Takahashi, Y., Yoshioka, Y., Tomooka, N., Mongkol, R. *et al.* (2022) Molecular analysis of genetic diversity and structure of the Lablab (*Lablab purpureus* (L.) Sweet) gene pool reveals two independent routes of domestication. *Plants* 12(1), 57. DOI: 10.3390/plants12010057

Krishna, S., Modha, K., Parekh, V., Patel, R. and Chauhan, D. (2022) Phylogenetic analysis of phytochrome A gene from *Lablab purpureus* (L.) Sweet. *Journal of Genetic Engineering and Biotechnology* 20(1), 1–11. DOI: 10.1186/s43141-021-00295-z

Kukade, S.A. and Tidke, J.A. (2014) Reproductive biology of *Dolichos lablab* L. (*Fabaceae*). *Indian Journal of Plant Sciences* 3, 22–25.

Kumar, A. and Kaur, V. (2010) *Characterisation and Evaluation of PGR: Principles and Techniques. Division of Germplasm Evaluation.* ICAR-NBPGR, New Delhi, India, pp. 139–145.

Letting, F.K., Venkataramana, P.B. and Ndakidemi, P.A. (2021) Breeding potential of lablab [*Lablab purpureus* (L.) Sweet]: a review on characterization and bruchid studies towards improved production and utilisation in Africa. *Genetic Resources and Crop Evolution* 68, 3081–3101. DOI: 10.1007/s10722-021-01271-9

Liu, C.J. (1998) Lablab cv. Endurance. *Plant Variety Journal* 11, 26–27.

Maass, B.L. (2006) Changes in seed morphology, dormancy and germination from wild to cultivated hyacinth bean germplasm (*Lablab purpureus*: Papilionoideae). *Genetic Resources and Crop Evolution* 53, 1127–1135. DOI:10.1007/s10722-005-2782-7

Maass, B.L. (2016) Origin, domestication and global dispersal of *Lablab purpureus* (L.) Sweet (Fabaceae): Current understanding. *Seed* 10, 14. DOI: 10.1007/s10722-005-2782-7

Maass, B.L., Jamnadass, R.H., Hanson, J. and Pengelly, B.C. (2005) Determining sources of diversity in cultivated and wild *Lablab purpureus* related to the provenance of germplasm by using amplified fragment length polymorphism. *Genetic Resources and Crop Evolution* 52, 683–695. DOI: 10.1007/s10722-003-6019-3.

Maass, B.L., Knox, M.R., Venkatesha, S.C., Angessa, T.T., Ramme, S. *et al.* (2010) *Lablab purpureus*-A crop lost for Africa? *Tropical Plant Biology* 3, 123-135.

Mamatha, A., Lavanya, A.V.N., Nayak, M.H. and Anitha, D. (2022) Studies on evaluation of dolichos bean–pole type germplasm (*Dolichos lablab* L.) for growth and yield characteristics under Central Telangana zone. *The Pharma Innovation Journal* 11(7), 142–145.

Mihailovic, V., Mikic, A., Ceran, M., Cupina, B., Dordevic, V. *et al.* (2016) Some aspects of biodiversity, applied genetics and agronomy in hyacinth bean (*Lablab purpureus*) research. *Legume Perspectives* 13, 9–15.

Minde, J.J., Venkataramana, P.B. and Matemu, A.O. (2021) *Dolichos lablab*-an underutilized crop with future potentials for food and nutrition security: a review. *Critical Reviews in Food Science and Nutrition* 61(13), 2249–2261. DOI: 10.1080/10408398.2020.1775173

Missanga, J.S., Venkataramana, P.B. and Ndakidemi, P.A. (2021) Recent developments in *Lablab purpureus* genomics: a focus on drought stress tolerance and use of genomic resources to develop stress-resilient varieties. *Legume Science* 3(3), e99. DOI: 10.1002/leg3.99

Naeem, M., Shabbir, A., Ansari, A.A., Aftab, T., Khan, M.M.A. *et al.* (2020) Hyacinth bean (*Lablab purpureus* L.)–An underutilised crop with future potential. *Scientia Horticulturae* 272, 109551. DOI: 10.1016/j.scienta.2020.109551

Narayan, R. and Poonan, D. (1987) Bean mosaic and bean yellow mosaic diseases of hyacinth bean (*Lablab purpureus* (L.) Sweet) and screening of available genotypes to find sources of resistance. *Indian Journal of Plant Pathology* 5, 63–68.

NBPGR (2023) Seed gene bank. Available at: http://genebank.nbpgr.ernet.in/SeedBank/Default.aspx (accessed 26 August 2023).

Nene, Y.L. (2006) Indian pulses through the millennia. *Asian Agri-History* 10(3), 179–202.

Njaci, I., Waweru, B., Kamal, N., Muktar, M.S., Fisher, D. *et al.* (2023) Chromosome-level genome assembly and population genomic resource to accelerate orphan crop lablab breeding. *Nature Communications* 14(1), 1915. DOI: 10.1038/s41467-023-37489-7

Paikra, M., Kolhekar, S., Gupta, K., Chandrakar, G.K. and Nagvanshi, P.K. (2019) Screening of field bean genotypes for tolerant to aphid (*Aphis craccivora*). *Journal of Pharmacognosy and Phytochemistry* 8, 320–322.

Pengelly, B.C. and Maass, B.L. (2001) *Lablab purpureus* (L.) Sweet–diversity, potential use and determination of a core collection of this multi-purpose tropical legume. *Genetic Resources and Crop Evolution* 48, 261–272. DOI: 10.1023/A:1011286111384

Preetham, R., Suchitra, V. and Ashwini, D. (2020) Evaluation of Dolichos genotypes (*Dolichos lablab* L.) to qualitative and quantitative traits in Northern Telangana Zone. *Journal of Pharmacognosy and Phytochemistry* 9(5), 2717–2720.

Raghu, B.R., Samuel, D.K., Mohan, N. and Aghora, T.S. (2018) Dolichos bean: An underutilized and unexplored crop with immense potential. *International Journal of Recent Advances in Multidisciplinary Research* 5(12), 4338–4341.

Rai, K.K., Rai, N. and Rai, S.P. (2017) Downregulation of γECS gene affects antioxidant activity and free radical scavenging system during pod development and maturation in *Lablab purpureus* L. *Biocatalysis and Agricultural Biotechnology* 11, 192–200. DOI: 10.1016/j.bcab.2017.07.005

Rai, K.K., Rai, N. and Pandey-Rai, S. (2019) Enhancing abiotic stress tolerance in hyacinth bean (*Lablab purpureus* L.). Nova Science Publishers, Inc. Hauppauge, USA.

Rajendra, B.R. (1978) Genetic studies of micro-morphological traits in some interspecific and intergeneric crosses: i. Triticeae and ii. Leguminosae. PhD Thesis. Kansas State University, Manhattan, Kansas.

Rajendra, S.P. (2015) Identification of sources and mechanisms of resistance to pod borers in dolichos bean *(Lablab purpureus* L. Sweet. PhD Thesis. University of Agricultural Sciences, Bengaluru, India.

Rangaiah, D.V. and Dsouza, M. (2016) Hyacinth bean *(Lablab purpureus)*: An adept adaptor to adverse environments. *Legume Perspectives* 13, 20.

Rapholo, E., Odhiambo, J.J., Nelson, W.C., Rotter, R.P., Ayisi, K. *et al.* (2020) Maize–lablab intercropping is promising in supporting the sustainable intensification of smallholder cropping systems under high climate risk in southern Africa. *Experimental Agriculture* 56(1), 104–117. DOI: 10.1017/S0014479719000206

Ratna, M., Kayum, M.A. and Ali, M.Z. (2021) Evaluation of country bean germplasm collected from Patuakhali region. *Malaysian Journal of Halal Research* 4(2), 52–54. DOI: 10.2478/mjhr-2021-0010

Robotham, O. and Chapman, M. (2017) Population genetic analysis of hyacinth bean *(Lablab purpureus* (L.) Sweet, Leguminosae) indicates an East African origin and variation in drought tolerance. *Genetic Resources and Crop Evolution* 64, 139–148. DOI: 10.1007/s10722-015-0339-y

SARDI (2023) Australian Pasters Genebank. Available at: https://apg.pir.sa.gov.au/gringlobal/search.aspx (accessed 25 August 2023).

Savitha, B.N. and Ravikumar, R.L. (2009) Comparative analysis of phenotypic and molecular diversity in selected pendal and non-pendal genotypes of field bean [*Lablab purpurem* (L.) Sweet]. *Indian Journal of Genetics and Plant Breeding* 69(03), 232–236.

Shivashankar, G., Srirangasayi, I., Kempanna, C. and Viswanatha, S.R. (1971) Day-neutral varieties of *Dolichos lablab* L. *Mysore Journal of Agricultural Sciences* 5, 216–218.

Shrikrishna, P.D. and Ramesh, S. (2020) Visually assayable morphological descriptors-based establishment of distinctiveness [D], uniformity [U] and stability [S] of dolichos bean *(Lablab purpureus* L. Sweet var. Lignosus) genotypes. *Plant Genetic Resources* 18(2), 105–108. DOI: 10.1017/S147926212000009X

Singh, P.K., Kumar, A., Rai, N. and Singh, D.V. (2012) Identification of host plant resistant to Dolichos yellow mosiac virus (DYMV) in Dolichos bean *(Lablab purpureus)*. *Journal of Plant Pathology and Microbiology* 3, 130. DOI: 10.4172/2157-7471.1000130

Singh, S., Rajan, S., Kumar, D. and Soni, V.K. (2021) Genetic diversity assessment in Dolichos bean *(Lablab purpureus* L.) based on principal component analysis and single linkage cluster analysis. *Legume Research* 1–7. DOI: 10.18805/ LR-4561

Singh, V., Kudesia, R. and Bhadauria, S. (2020) Assessment of genetic diversity in some Indian *Lablab purpureus*, L. Bean genotypes based on RAPD marker. *Proceedings of the National Academy of Sciences, India Section B: Biological Sciences, India* 90, 855–861.

Smartt, J. (1984) Gene pools in grain legumes. *Economic Botany* 38(1), 24–35. DOI: 10.1007/BF02904413

Spoorthi, V., Ramesh, S., Sunitha, N.C., Anilkumar, C. and Vedashree, M.S. (2023) Revisiting population genetics with new genomic resources in dolichos bean *(Lablab purpureus* L. Sweet): an orphan crop. *Genetic Resources and Crop Evolution* 71, 1–11. DOI: 10.1007/s10722-023-01655-z

Sserumaga, J.P., Kayondo, S.I., Kigozi, A., Kiggundu, M., Namazzi, C. *et al.* (2021) Genome-wide diversity and structure variation among lablab [*Lablab purpureus* (L.) Sweet] accessions and their implication in a forage breeding program. *Genetic Resources and Crop Evolution* 68, 2997–3010. DOI: 10.1007/s10722-021-01171-y

Sunwar, S., Thornstrom, C.G., Subedi, A. and Bystrom, M. (2006) Home gardens in western Nepal: opportunities and challenges for on-farm management of agrobiodiversity. *Biodiversity & Conservation* 15(13), 4211–4238. DOI: 10.1007/s10531-005-3576-0

Susmita, C., Mohan, N. and Aghora, T.S. (2020) Breeding for evolution of photo-insensitive pole type vegetable dolichos *(Lablab purpureus* L.) varieties to suit year round cultivation. *Electronic Journal of Plant Breeding* 11(02), 633–637. DOI: 10.37992/2020.1102.103

Tamiru Workneh, S. (2020) Ethnobotanical knowledge of Lablab *(Lablab purpureus* L.) Sweet Fabaceae) in Konso zone and genetic diversity of collections from Ethiopia using SSR markers. PhD Thesis. Addis Ababa University, Ethiopia.

Teshome, A., Hailu, A., Habte, E., Muktar, M.S., Mekasha, A. *et al.* (2022) Unlocking the genetic potential of an indigenous forage, lablab *(Lablab purpureus* L.) collection by the whole genome sequencing approach. Poster. International Livestock Research Institute, Nairobi, Kenya.

Thilagavathy, A. and Devaraj, V.R. (2016) Evaluation of appropriate reference gene for normalization of microRNA expression by real-time PCR in *Lablab purpureus* under abiotic stress conditions. *Biologia* 71(6), 660–668. DOI: 10.1515/biolog-2016-0091

Upadhyaya, H.D., Dwivedi, S.L., Ambrose, M., Ellis, N., Berger, J. *et al.* (2011) Legume genetic resources: management, diversity assessment, and utilization in crop improvement. *Euphytica* 180, 27–47. DOI: 10.1007/s10681-011-0449-3

USDA (2023) U. S. National germplasm system. Available at: https://www.grin-global.org (accessed 16 August 2023)

Vaijayanthi, P.V., Ramesh, S., Gowda, M.B., Rao, A.M., Gowda, J. *et al.* (2015) Development and validation of a core set of dolichos bean germplasm. *International Journal of Vegetable Science* 21(5), 419–428. DOI: https://doi.org/10.1080/19315260.2014.883459

Vaijayanthi, P.V., Ramesh, S., Gowda, M.B., Rao, A.M., Ramappa, H.K. *et al.* (2016) Identification of selected germplasm accessions for specific/wide adaptation coupled with high pod productivity in dolichos bean (*Lablab purpureus* L. Sweet). *Mysore Journal of Agricultural Sciences* 50(2), 376–380.

Vaijayanthi, P.V., Chandrakant and Ramesh, S. (2019a) Hyacinth bean (*Lablab purpureus* L. Sweet): Genetics, breeding and genomics. *Advances in Plant Breeding Strategies: Legumes: Volume* 7, Springer, US, pp. 287–318. DOI: 10.1007/978-3-030-23400-3_8

Vaijayanthi, P.V., Ramesh, S., Gowda, M.B., Rao, A.M. and Keerthi, C.M. (2019b) Genome-wide marker-trait association analysis in a core set of Dolichos bean germplasm. *Plant Genetic Resources* 17(1), 1–11. DOI: 10.1017/S1479262118000163

Venkatesha, S.C. and Patil, P.G. (2022) Development of novel SNP/InDel markers through amplicon sequencing in dolichos bean (*Lablab purpureus* L.). *Journal of Theoretical Biological Forum*. DOI: 10.5281/zenodo.7629782

Venkatesha, S.C., Ganapathy, K.N., Byre Gowda, M., Ramanjini Gowda, P.H., Mahadevu, P. *et al.* (2013) Variability and genetic structure among lablab bean collections of India and their relationship with exotic accessions. *Vegetos* 26, 121–130. DOI: 10.5958/j.2229-4473.26.2s.130

Verdcourt, B. (1971) Studies in the *Leguminosae-Papilionoideae* for the 'Flora of Tropical East Africa. *Kew Bulletin* 65–169. DOI: 10.2307/4102845

Wang, B., Zhao, M., Yao, L., Babu, V., Wu, T. *et al.* (2018) Identification of drought-inducible regulatory factors in *Lablab purpureus* by a comparative genomic approach. *Crop and Pasture Science* 69(6), 632–641. DOI: 10.1071/CP17236

Whitbread, A. and Pengelly, B.C. (2004) Tropical legumes for sustainable farming systems in southern Africa and Australia. In: *ACIAR Proceedings,* Centre for International Agricultural Research, Canberra, Australia, pp.180.

Wilson, G.P. and Murtagh, G.J. (1962) Lablab-New forage crop for the north coast. *New South Wales Agricultural Gazette* 73, 460–462.

11 Horse Gram (*Macrotyloma uniflorum* (Lam.) Verdc.)

W.S. Philanim[1]*, Amit Kumar[1], Ingudam Bhupenchandra[2], Umakanta Ngangkham[3], Letngam Touthang[1], Binay Kumar Singh[1], Sandeep Jaiswal[1], Rumki Ch. Sangma[1], N. Raju Singh[1] and Kanishka Chandora[4]

[1]*ICAR-RC-NEH Region, Umiam, Meghalaya, India;* [2]*ICAR-KVK, Tamenglong, ICAR-RC-NEH Region, Manipur Centre, India;* [3]*ICAR-RC-NEH Region, Manipur Centre, India;* [4]*ICAR-NBPGR, RS, Shimla, India*

Abstract

Horse gram is a relatively unexplored legume with a plethora of nutritional benefits, a high protein content, inherent resilience to biotic and abiotic stress, and adaptability to climate change. This legume, with its myriad advantages, has yet to gain importance nationally and internationally in terms of its genetic improvement and popularization. In this chapter, we attempt to appeal to the altruistic traits of this underutilized legume by underlining its significance in terms of nutraceutical properties, sustenance as a feed and fodder, high protein content and health benefits. The chapter includes deliberation on the origin and distribution of horse gram, gene pool and genetic stocks available in the wild species and relatives, germplasm collection, characterization and conservation, utilization of the germplasm through conventional plant breeding approaches and advanced molecular techniques such as mapping and genome editing, its cultivation practices and gene erosion due to interference such as environmental depletion and varietal replacement. The chapter ends with a focus on the need for collaborative efforts between different stakeholders and policy makers to allow this potential legume to flourish and stand on the global stage to support nutritional security.

Keywords: Horse gram, crop gene pool, genetic resources, genetic diversity

11.1 Introduction

Horse gram is an underutilized legume belonging to the family Fabaceae and grouped under the species *Macrotyloma uniflorum* (Herath *et al.*, 2020). Horse gram is an important pulse, being fifth among grown legumes in India (Fuller and Murphy, 2018). It is a resourceful pulse used as animal feed, fodder, food and cover crops. It offers myriad nutraceutical benefits encompassing good protein sources (17.9–25.3%), carbohydrates (51.9–60.9%), essential amino acids, and small lipid proportion (0.58–2.06%) (Herath *et al.*, 2020). Horse gram has a rich composition of micronutrients, including iron, molybdenum and phosphorus, amino acids (Gautam *et al.*, 2020a), essential vitamins (carotene, riboflavin, thiamine) (Sodani *et al.*, 2006), dietary fibre and antioxidants. It is a multipurpose crop consumed as sprouts, green leafy vegetables, animal feed and fodder (Bhartiya *et al.*, 2015). It contains a high amount of minerals

*Corresponding author: philanim09@gmail.com

© CAB International 2024. *Potential Pulses: Genetic and Genomic Resources*
(eds R. Chandora, T. Basavaraja and A. Pratap)
DOI: 10.1079/9781800624658.0011

(3.2%) including phosphorus (311 mg/100 g), calcium (287 mg/100 g), iron (8.4 mg/100 g), crude fibre (5.3%) and vitamins such as vitamin C (1.0 mg/100 g), thiamine (0.42 mg), riboflavin (0.2 mg) and niacin (1.5 mg) (Gopalan *et al.*, 2006). Horse gram has a higher composition of manganese, iron, molybdenum, calcium and crude fibre than pigeon pea and chickpea (Longvah *et al.*, 2017; Sadawarte *et al.*, 2018). The nutrient-rich compounds can aid in many complex ailments, such as coronary heart disease, kidney stones, urinary diseases, piles, obesity and diabetes (Yadava and Vyas, 1994; Bazzano *et al.*, 2001; Gautam *et al.*, 2020b). It is a healthy diet option for the diabetic population (Bazzano *et al.*, 2001). It is considered palatable for milk animals owing to its highly edible protein and few digestive issues (Mahesh *et al.*, 2021). Horse gram is a source of the lectin haemagglutinin, which can have health benefits at low concentrations (Chahota *et al.*, 2013). The presence of antinutrient phytochemicals such as trypsin inhibitors, tannins and phytate, however, limits its consumption and diminishes its food value (Sreerama *et al.*, 2012). Horse gram, also known as kulthi, is deemed as a 'poor man's pulse'. The crop can grow under adverse climatic conditions and harbours good adaptability in stressful environments. It can survive in marginal land areas with minimal inputs and is drought resistant with an innate nitrogen-fixing capacity that adds nutrients to the soil (Yadava and Vyas, 1994; Yadav *et al.*, 2004), rendering it a climate-resilient crop. Despite its importance since time immemorial to the diet of a significant proportion of India, bias against horse gram use still exists, mainly because it is viewed as food for those of humbler status, predominantly in southern parts of India. As Smartt (1985) stated, 'there has been minimal incentive to study domestication and evolution of horse gram'. Scientific research on the crop is still at a nascent stage, and much research attention needs to be diverted towards this legume crop (Fuller and Murphy, 2018).

11.2   Origin, Distribution and Botanical Description

The origination of present-day horse gram is considered to be in India (Vavilov, 1951; Zohary, 1970; Purseglove, 1974; Smartt, 1985),

specifically south-western India (Arora and Chandel, 1972). Other researchers suggested, however, that horse gram originated from both Africa and India (Mehra and Magoon, 1974), South Asia (Hooker, 1879) and Africa (Verdcourt, 1971). Wild species of *M. uniflorum* are considered to exist both in Africa and India (Verdcourt, 1971). The secondary centre of origin is considered to be the north-western Himalayan region (Verdcourt, 1971). Fuller and Harvey (2006) postulate horse gram to be of south Indian derivation, assessing herbarium specimens and evidence; however, archaeobotanical evidence suggests north-west India is the likely origin of horse gram. The old World tropical region is considered to have the maximum genetic diversity, especially in the Himalayas and the southern part of India (Zeven and de Wet, 1982). Horse gram derived its common English name from its significant use as silage for horses and cattle for eras (Watt, 1889–1893). Its importance during the Neolithic era is proven by its ubiquitous archaeological evidence. The proof advocates that it was domesticated and cultivated in north-west India and south Deccan (Kingwell-Banham and Fuller, 2014). Horse gram now covers subtropical and temperate areas in India, the Philippines, Sri Lanka, Bhutan, Australia, China and Pakistan (Krishna, 2010).

The primitive archaeological evidence of horse gram derives from three regions of India: the Saurashtra peninsula, which lies towards the western part of Gujarat, the north-west of Haryana state, and specifically the south Deccan Plateau. In the north-west regions, the primary occurrence of horse gram dates back to the Early Harappan period (3000–2600 BC), including Banawali, Balu and Masudpur VII (Bates, 2015). Findings corroborate the Southern Neolithic (covering parts of Tamil Nadu and the present states of Karnataka and Andhra Pradesh) as the earliest domesticated sites of horse gram (Fuller, 2011; Fuller *et al.*, 2014). Horse gram finds a place in regions of the Indian peninsula, including Tamil Nadu, Andhra Pradesh and Karnataka (Nezamuddin, 1970; Sundararaj and Thulasidas, 1993). Of the acreage under horse gram, 90% is concentrated in Tamil Nadu and Andhra Pradesh. Observations on seed morphology indicate smaller seed sizes in length and width in the primitive populations, and the seed size increased around the centre of the second millennium BC, suggesting that horse gram domestication was initiated about 4000 years ago. A conspicuous size increase was evident between 3500 years ago (1500 BC) and 3000 years ago (1000 BC). Wild horse gram (*Macrotyloma uniflorum* var. *stenocarpum*) is native and concentrated in the Acacia grove that ranges from Rajasthan (Aravalli hills), through Gujarat and the Southern Peninsula (the savannahs) (Asouti and Fuller, 2008; Fuller and Murphy, 2014), further extending towards borders of deciduous dry woodlands, to the central and eastern hills of India. Horse gram is now cultivated in various South-east Asian countries, such as India, Sri Lanka, Bangladesh, Bhutan and Myanmar. This crop is also grown for fodder and manure in various tropical countries such as Australia and Africa. The horse gram wild relatives are grown mainly in Australia, India, Papua New Guinea and Africa. However, no report is found on horse gram cultivation as a pulse crop in eastern, southern and central Africa (Blumenthal *et al.*, 1989).

Botanically, horse gram is an annual herb that grows up to a stature of 30–40 cm (Nezamuddin, 1970; Smartt, 1985; Sundararaj and Thulasidas, 1993; Neelam *et al.*, 2014). Horse gram is considered native to India's arid climatic portions (Asouti and Fuller, 2008). In southern India, it grows in regions where there is a smaller amount of precipitation per annum ranging from 40 cm to 90 cm during its growing season from August to October, mostly on poor or lateritic soils, usually with no irrigation (Nezamuddin, 1970; Kingwell-Banham and Fuller, 2014). Horse gram is an annual herb with slender and cylindrical tomentose stems. It usually requires the support of a structure because of its weak stem. As a pure crop, it forms a compressed mat of 30–60 cm in height; in mixed cropping with cereals, it climbs to a height of 60–110 cm. The leaves are trifoliate (Fig. 11.1) with long persistent stipules of 7–10 mm and petiole of 3–7 cm. Flowers have two- to four-flowered axillary racemes and are short, sessile or subsessile, 10–12 mm long, and greenish yellow in colour on the

Fig. 11.1. Close-up view of horse gram plants depicting the trifoliate leaf shape. (Author's own figure.)

standard petal. The calyx is densely covered with short, matted woolly hairs with 2–3 mm long tubes, and the lobes are 3–8 mm long and are lanceolate setaceous. The flowers of horse gram are papilionaceous in nature, as common with all legumes. The standard petal is oblong, slightly serrated at the apex, length 9–10.5 mm, width 7–8 mm, with two linear appendages of about 5 mm in length and wings of 8–9.5 mm in length comparable to the keel petal. Dense white hairs surround a solid ovary, vitiate style, and a circle of short, dense hairs surrounds the stigma. Pods are stipitate, tomentose, slightly curved, with a 4.5–6 cm length and a width of about 6 mm. There are six or seven seeds per pod, with a length of 6–8 mm, width of 4–5 mm, pale fawn, sometimes with small scattered black spots or faint mottles with hilum placed centrally (Purseglove, 1974) (Fig. 11.2).

11.3   Crop Gene Pool, Evolutionary Relationships and Systematics

Horse gram was initially incorporated in the genus *Dolichos* by Linnaeus, but later, Verdcourt (1980) reassigned horse gram to the genus *Macrotyloma*. The variation in standard petal, style and pollen characteristics distinguished the

Fig. 11.2. Variability in horse gram seed colour. (Author's own figure.)

genus *Macrotyloma* from *Dolichos* (Verdcourt, 1970). Horse gram belongs to the kingdom Plantae, subkingdom Tracheobionta, division Magnoliophyta, class Magnoliopsida and family Fabaceae. The genus *Macrotyloma* (Wight & Arn.) Verdc. have diploid chromosome numbers $2n = 2x = 20$ and $2n = 2x = 22$ (Lackey, 1981). The distribution of the genus *Macrotyloma* indicates the localized spread of species across continents. A total of 25 species of the genus *Macrotyloma* exist. Most of the *Macrotyloma* species are concentrated in Africa, including *M. ciliatum* (Willd.) Verdc. and *M. axillare* (E.Mey.) Verdc. (Dikshit *et al.*, 2014). In Africa, the species *M. stenophyllum*, *M. africanum*, *M. stipulosum*, *M. dewildemanianum*, *M. oliganthum*, *M. ellipticum*, *M. rupestre*, *M. geocarpa*, *M. fimbriatum*, *M. brevicaule*, *M. decipiens*, *M. bieense*, *M. densiforum*, *M. hockii*, *M. schweinfurthii*, *M. coddii*, *M. daltonii*, *M. maranguense*, *M. kasaiense*, *M. prostratum* and *M. tenuiforum* were distributed, whereas *M. axillare* appears both in Africa and Australia and *M. ciliatum* appears in Africa and Asia, and *M. uniforum* in Australia, Asia and Africa (Dikshit *et al.*, 2014).

Superior gene sources from the gene pool of horse gram are available for different traits that can be transferred into the elite lines through a prebreeding programme. *M. axillare*, a wild species of horse gram, harbours a gene for yellow mosaic virus (YMV) disease resistance (Kumar, 2007), and *M. sar-gharwalensis* (an accession IC 212722 is registered with NBPGR) that was collected and identified from the hills of Uttarakhand contains higher protein levels (38.37±1.03%) than cultivated horse gram (Yadav *et al.*, 2004). Similarly, three *Macrotyloma* species, namely *M. axillare*, *M. africanum* and *M. daltonii* from Australia, have potential for silage purposes (Kumar, 2005). Different gene pools embody expedient genes that are otherwise rendered unavailable in the existing superior cultivars. Traits include seed yield, pod number, powdery mildew and YMV resistance, profuse flowering and cold and drought tolerance (*M. axillere*) (Staples, 1966), higher biomass (*M. axillare* and *M. africanum* (Brenan ex R.Wilczek) Verdc.) (Chahota *et al.*, 2013), and increased oil and protein content (*M. sar-garhwalensis* R.D.Gaur & L.R.Dangwal) (Yadav *et al.*, 2004). The transfer of genes from *M. sar-garhwalensis* helps to improve the lipid (10.85%) and seed protein (38.37%) portion in the horse gram (Yadav *et al.*, 2004).

11.4 Germplasm Collection, Characterization, Evaluation and Conservation

Global endeavours to collect, conserve and utilize horse gram germplasm are still lagging behind, despite horse gram being an important pulse. Only a few organizations are collecting germplasm: the Australian Tropical Crops and Forages Genetic Resources Centre, Biloela, Queensland, which hosts 38 accessions of horse gram; the Germplasm Resources Information Network (GRIN) of the US Department of Agriculture (USDA) which has conserved 35 accessions; and the Kenya Agricultural Research Institute (KARI), Kikuyu, Kenya, which has 21 accessions (Brink, 2006; Chahota *et al.*, 2013). The National Bureau of Plant Genetic Resources (NBPGR) in New Delhi, however, is leading the way in systematically collecting this important legume (Table 11.2). Germplasm collection and conservation can be traced back to the 1970s when the efforts were initiated with the inception of the PL480 scheme (a collaborative scheme between the USDA project on food security and the Indian Council of Agricultural Research (ICAR) in Haiti, using Public Law

480). A total of 1627 accessions of horse gram germplasm are conserved in the national gene bank repository as a result of the vast exploration and collection programmes. Of the total conserved germplasm, 1161 accessions were characterized from 1999 to 2004 (Chahota *et al.*, 2013).

Phenotypic characterization of the collected horse gram germplasm (Fig. 11.3) from different growing regions and pockets of the country has indicated the presence of huge diversity among horse gram accessions (Gupta *et al.*, 2010; Sahoo *et al.*, 2014; Bhartiya *et al.*, 2017). Characterization of the germplasm revealed promising accessions for different traits, namely IC 16976, PLKU 20, BC 18037, PLKU 49A and IC 18679 for early maturity and high yield; IC 120825, NIC 11387, 120837, IC 94591, IC 120838, NIC 11095 and 120844 for long pod; and IC 94592, IC 120837 and IC 110286 for high pod number (Patel *et al.*, 1995). ICAR-NBPGR has published a catalogue for 1426 accessions with the provision of descriptors for 11 economically important traits. Characterization for various qualitative and quantitative traits of about 506 accessions and 920 accessions were conducted at

Table 11.1. Transfer of usable genes from wild species to cultivated horse gram. (Author's own table.)

Sr. No	Wild species	Trait transferred	Reference
1	*Macrotyloma axillare*	High number of pods per plant, increased number of seeds per plant, cold and drought tolerance	Dikshit *et al.* (2014)
2	*Macrotyloma axillare*	Forage	Dikshit *et al.* (2014)
3	*Macrotyloma africanum*	Forage	Dikshit *et al.* (2014)
4	*Macrotyloma sargarhwalensis*	High protein content	Chahota *et al.* (2020a)

Table 11.2. Germplasm conservation at the global gene banks.

Sr No.	Country	Name of the organization	Accessions
1	India	National Bureau of Plant Genetic Resources, New Delhi	1627
2	USA	Germplasm Resources Information Network of the US Department of Agriculture	35
3	Australia	Tropical Crops and Forages Genetic Resources Centre, Biloela, Queensland	38
4	Kenya	National Gene Bank of Kenya, Crop Plant Genetic Resources Centre, KARI, Kikuyu	21

Source: Chahota *et al.* (2013). (Table used with permission.)

Fig. 11.3. Field view: characterization of horse gram germplasm. (Author's own figure.)

New Delhi and NBPGR regional station Akola (Patel *et al.*, 1995). Horse gram germplasm was evaluated for various agro-morphological traits in different institutions and research stations, namely The Vivekanand Parvartiya Krishi Anusandhan Sansthan (VPKAS Almora) for 10 lines (Mahajan *et al.*, 2007), CSK Himachal Pradesh Agricultural University, Palampur for 63 horse gram accessions that were procured from NBPGR, Phagli, Shimla (Chahota *et al.*, 2005), and 22 accessions for various agronomically beneficial traits that deciphered five high-yielding horse gram lines for various nutritional and antinutritional factors (Subba Rao and Sampath, 1979; Sudha *et al.*, 1995). Morphological diversity was explored in different studies, including in 100 horse gram lines (Prakash *et al.*, 2010) at the NBPGR regional centre, Thrissur, in 500 germplasm accessions and landraces maintained in AICRP on arid legumes at University of Agricultural Sciences, Bangalore (Viswanathan *et al.*, 2016), on 56 genotypes for ten quantitative traits (Singh *et al.*, 2019), and in 48 horse gram genotypes (Sahoo *et al.*, 2014). Horse gram genotypes were assessed for different traits (Durga *et al.*, 2015; Ingle *et al.*, 2020; Sirisha *et al.*, 2022). In addition, much work has been conducted in different areas of horse gram research by several researchers (Savithriamma *et al.*, 1990; Dobhal and Rana, 1994a, b; Joshi *et al.*, 1994; Patil *et al.*, 1994; Sharma, 1995; Lad *et al.*, 1999; Nagaraja *et al.*, 1999; Tripathi, 1999; Jayan and Maya, 2001; Rana, 2010).

11.5 Use of Germplasm in Crop Improvement by Conventional Approaches

Research in horse gram through selection and breeding interventions is meagre. Despite horse gram having the potential for improving agriculture and providing economic products (Bhag and Joshi, 1991), germplasm evaluation and documentation for most horse gram germplasm is not up to date. Thus, the utilization of the germplasm by breeders concerned with improving this crop is hindered. The crop improvement programme in horse gram based on conventional breeding approaches mainly utilizes selection approaches based on morphology to identify targeted and altruistic attributes either from naturally present germplasm or from segregating populations or advanced generations after hybridization. This may be attributed to insufficient genetic and genomic information in horse gram (Chahota *et al.*, 2020b). There are approximately 1800 accessions of horse gram germplasm in India, but only 912 have been evaluated and documented. The genetic development of horse gram for different traits has been limited to just a handful of institutions in India, and that number is even more scarce globally (Table 11.3).

Table 11.3. Horse gram varieties developed and released for cultivation in India. (Author's own table.)

Sr No.	Varieties	Year	Trait improved	Breeding method employed	Organization/institute	Source
1	CO 1	1953	For rainfed conditions in South India	Pureline selection from Mudukulathur	Tamil Nadu Agricultural University, Coimbatore	https://agritech.tnau.ac.in/agriculture/CropProduction/Pulses/pulses_horsegram.html
2	Paiyur 1	1988	For rainfed conditions in South India	Pureline selection from Mettur local	Tamil Nadu Agricultural University, Coimbatore	https://agritech.tnau.ac.in/agriculture/CropProduction/Pulses/pulses_horsegram.html
3	Paiyur 2	1998	For rainfed conditions in South India	Gamma irradiation of CO 1	Tamil Nadu Agricultural University, Coimbatore	https://agritech.tnau.ac.in/agriculture/CropProduction/Pulses/pulses_horsegram.html
4	Maru Kulthi	1989	Semi-spreading type with light brown seeds	Selection	Central Arid Zone Research Institute	Kumar (2007)
5	Hebbel Hurali-1	1976	Photosensitive	Selection from breeding line PLKU32	Central Arid Zone Research Institute	Kumar (2007)
6	Hebbel Hurali-1	1976	Photosensitive	Selection from breeding line EC7460	Central Arid Zone Research Institute	Kumar (2007)
7	Birsa Kulthi-1	1985	Moderately resistant to *Macrophomina* leaf blight	Selection from local germplasm	Central Arid Zone Research Institute	Kumar (2007)
8	VLG-1	1983	Semi-spreading type with tolerance to blight	Selection from local material	Central Arid Zone Research Institute	Kumar (2007)
9	KBH-1 (BGM-1)	1990	Tolerance to Yellow Mosaic Virus	Selection from Bailhongal local	Central Arid Zone Research Institute	Kumar (2007)
10	PHG-9	1997	For southern India	Selection from Palampur local	Central Arid Zone Research Institute	Kumar (2007)
11	AK-21	1999	Early high-yielding variety	–	Central Arid Zone Research Institute	Kumar (2007)
12	AK-42	2004	Early maturing variety	Improved over PHG-9	Central Arid Zone Research Institute	Kumar (2007)
13	CRHG-18R	2009	Released for south India with 40% yield superiority over local variety	Mutation breeding	Central Research Institute for Dryland Agriculture	Maheshwari et al. (2019)
14	CRHG-4	2010	Released for south India with 40% yield superiority over local variety	Mutation breeding	Central Research Institute for Dryland Agriculture	Maheshwari et al. (2019)
15	CRHG-19	2014	Released for south India with 65% yield superiority over local variety	Mutation breeding	Central Research Institute for Dryland Agriculture	Maheshwari et al. (2019)
16	CRHG-22 (CRIDA VARDHAN)	2016	Released for south India with 37–60% yield superiority over local variety	Mutation breeding	Central Research Institute for Dryland Agriculture	Maheshwari et al. (2019)

The commercial acceptance of horse gram in India is limited because the crop has many unwelcome traits such as photosensitivity, delayed maturity, an indeterminate growth habit, a twining nature and asynchronous maturity (Chahota *et al.*, 2020a). An exploration of the genetics of growth habit in horse gram identified two genes conferring the trait, and they had an inhibitory epistasis mode of gene action with indeterminacy dormant over determinancy (Chandana *et al.*, 2021). Hybridization studies conducted to study the genetics of day-neutral and photosensitive varieties deciphered two genes responsible for photoperiod response. Inheritance of seed colour was also determined, wherein the black seed colour was dominant over brown seed colours. These two controlling seed colours were found to have a polymeric mode of gene action (Sreenivasan, 2003). In India, most horse gram varieties are derived through selection following specific and effective screening from local germplasm and landraces. Different horse gram varieties were developed for various states including: VZM 1 and PDM 1 from Andhra Pradesh; BR 5 and BR 10 from Bihar; 35-5-122, 35-5-123 and Co-1 from Tamil Nadu; Birsa Kulthi and K82 from Jharkhand; KS 2, Maru Kulthi, AK 21 and AK 42 from Rajasthan; HPK-2 and HPK-4 from Himachal Pradesh; S8, S27, S39 and S1264 from Orissa; PHG 9, KBH 1 and Hebbal Hurali 2 from Karnataka; and VLG 1 from Uttarakhand. Single plant selection approach varieties included No 35-5-122, 123, Co-1, Hebbal Hurali 1 and 2 (Kumar, 2005, 2013; Chahota *et al.*, 2020a).

11.5.1 Wide hybridization

The absence of pre-breeding activities and lack of variability in cultivated germplasm at phenotypic and genetic levels pose major limitations in the development of horse gram. Therefore, wide hybridization is an effective strategy for increasing variability in conditions where variability is low to broaden the species' genetic base. The narrow genetic base in horse gram can be broadened through distant hybridization (Aditya *et al.*, 2019). Increasing variability helps in the effective selection and introduction of new traits in breeding programmes. The genus *Macrotyloma* consists of more than 25 species; however, the

genetic utility of these species has not been explored. A phylogenetic study to understand the ancestry of horse gram by using descriptors revealed species *M. axillare* as the closer relative having similar and desirable traits to the present-day cultivated horse gram, such as tolerance to biotic and abiotic stresses, greater counts of pods/plant, and higher seed yield/plant (Morris, 2008). The wild species *M. sargarhwalensis* is a source of high protein of about 39.5% that can be bred to introgress the trait in elite genotypes (Yadav *et al.*, 2004). A systematic hybridization programme has been initiated at CSK, Himachal Pradesh Agricultural University, Palampur, India, to evaluate wild species of *Macrotyloma* and introgress desirable attributes from *M. axillare* and *M. sargarhwalensis* to an elite background. *Macrotyloma axillare* harbours valuable traits such as high seed yield per plant, a high number of pods per plant, cold tolerance and drought tolerance (Staples, 1966, 1982). Cultivated horse gram species *M. uniflorum* is susceptible to diseases such as *Pellicularia* root rot, *Fusarium* wilt, *Aschochyta* blight, *Cercospora* leaf spot and anthracnose, particularly in high rainfall areas. Wild species *M. axillare* is a source of resistant genes reported against many diseases, and therefore efforts have been made to transfer those genes from the former to *M. uniflorum*. However, hybridization between *M. uniflorum* and *M. axillare* resulted in a more extended vegetative phase in F1 plants, prolonging the breeding process. Cytogenetic differences were observed among different species. The karyotype study revealed variations in chromosome sizes and secondary constrictions among the species. A larger chromosome was observed in *M. sargarhwalensis* followed by *M. axillare* and *M. uniflorum*. This could be the primary reason that led to meiosis laggards, resulting in barren pollen grains in the cross between *M. axillare* and *M. uniflorum* (Chahota *et al.*, 2020a).

11.5.2 Mutation

Mutagens create variation by altering the nucleotide sequence of a gene, creating diversity within a population by introducing new alleles. With the intention of creating variability for the effective selection of traits, researchers utilize both physical and chemical mutagens. Mutagens such as ethylmethane sulfonate (EMS; 0.2–0.6%) and gamma radiation (100–600 Gy) were combined to define their effect on seed germination. Seed germination was observed to be inversely proportional to the dosage of mutagen. Viable and usable macro mutations were scored for primary branches, plant height, pod length, pods per plant, seeds per pod, yield per plant and 1000-grain weight in the M2 population (Bolbhat *et al.*, 2009). Gamma radiation and EMS and their combinations resulted in desirable variations for a higher number of pods, early maturity, grains and grain weight, profuse branching, increased leaf area, and tall and dwarf habits (Bolbhat *et al.*, 2012). Similarly, Ramakanth *et al.* (1979) used five doses of gamma rays to induce mutation. Gupta *et al.* (1994) induced variability through mutagenesis for the number of pods per plant, number of seeds per pod, pod length, biological yield per plant, seed yield per plant and 100-seed weight. Two varieties of horse gram, namely PAIYUR 2 and CRIDA 1-18R that are photosensitive and indeterminate, were utilized for mutation screening in horse gram. Gamma rays and EMS used individually and in combination were statistically analysed to study the extent of variability and frame selection strategy for trait improvement (Priyanka *et al.*, 2019). A photosensitive variety, Dapali-1, was developed through irradiation with 20kR of three photo-insensitive mutants (Jamadagani and Birari, 1996). Chahota (2009) used gamma radiation to treat HPKC-2, a promising line that induced variation for horse gram agronomic traits. Mutagens such as gamma radiation, sodium azide (SA) and EMS were employed to generate genetic changes in various cultivars of horse gram. Variations after the treatment were observed in M2 and M3 generations, such as chlorophyll mutations, yield characteristics, seed mutations, plant growth habit, pod mutations (tall, small pod and early) and maturity mutations (early and late) (Datir, 2016). SINA (K-42) and KS-2, two horse gram varieties, were treated with three concentrations of the mutagens (EMS, SA and N-nitroso N-ethyl urea. Different parameters were considered to study the effectiveness of the mutagens at different concentrations. The mutagenic effectiveness, mutagenic efficiency and mutagenic rate were analysed in M2 generation and SA was found to be the most effective mutagen with the highest mutagenic rate (Shirsat

et al., 2010). A search to find alternative and more precise forms of mutagens to induce mutation in horse gram led to a pioneer report on nanoparticles capable of inducing mutation in the phenotype and, therefore, could be an alternative source for conventional mutagens. The most significant advantage of Cu-nanoparticles are the cost-effectiveness and easy synthesis in the laboratory (Halder *et al.*, 2015). In addition, a study to understand the variability produced by high doses of gamma rays on different traits, such as morphological, flowering and palynological traits in horse gram (*M. uniflorum*) was conducted (Priyanka *et al.*, 2021). Significant variability was induced through mutagenesis for yield-attributing traits such as the number of clusters, plant height, number of pods per plant and single plant yield (Sudhagar *et al.*, 2023). Two horse gram varieties, PAIYUR 2 and CRIDA1-18R, were treated with gamma rays, gamma rays and electron beam, and G and EMS to determine the strength of mutagenic activity in breeding programme on traits such as root length, seed germination, shoot length, plant height, pollen fertility, plant survival and seed fertility (Priyanka *et al.*, 2020).

11.6 Integration of Genomic and Genetic Resources for Crop Improvement

The inception of molecular markers in the early 1980s paved the way for the dissection of plants at the molecular level and the succinct understanding of the genetics of crops. The study of genetic diversity in horse gram was initiated using random amplified polymorphic DNA (RAPD) and inter-simple sequence repeat (ISSR) markers to understand the genetic architecture and diversity of *M. uniflorum* and its interrelationships and genetic phylogeny with two other species, namely *M. axillare* and *M. sar-gharwalensis* (Table 11.4). A total of 25 polymorphic primers were amplified with band size from 300 bp to 3 kb. OPR-22 recorded the maximum polymorphic information content) (PIC) value of 0.499. A structure-based understanding of the species elucidated two discrete gene pools based on their geographic origin. A dendrogram study assembled the genotypes into two clusters, revealing that *M. axillare* is

Table 11.4. Genetic enhancement of horse gram through genomic resource utilization. (Author's own table.)

S. No.	Type of marker	Type of material	Genetic enhancement	References
1	RAPD, ISSR	Germplasm	Genetic diversity and population structure	Sharma *et al.* (2015a)
2	SSR, ILP	Variety and germplasm	Development of markers	Sharma *et al.* (2015b)
2	SSR	Germplasm	Development and characterization of SSR markers	Chahota *et al.* (2017)
3	SSR, COS, EST-SSR	RILs	Linkage map construction and QTL study for drought and yield	Chahota *et al.* (2020a)
4	COS, RAPD, EST-SSR, SSR, dsSSR	RILs	QTL study for phenological and morphological traits	Katoch *et al.* (2022)
5	SSR, EST-SSR, RAPD, COS	RILs	QTL mapping for drought tolerance	Katoch and Chahota (2021)
6	SSR	RILs	Mapping for dwarfing and determinate growth habit	Modi *et al.* (2023)
7	SNP	Variety	Genome sequencing to decipher functional variation	Mahesh *et al.* (2021)
8	SNP	Germplasm	Genome-wide association mapping for drought tolerance	Choudhary *et al.* (2022)
9	SSR	Genotypes	Development of transcriptome-wide SSR markers	Kumar *et al.* (2020a)
10	SNP	Germplasm	Genome-wide association mapping for days to flowering, days to maturity plant height and number of branches per plant	Sharma and Chahota (2023)

COS, conserved orthologue set; EST, expressed sequence tag; ILP, intron length polymorphism; ISSR, inter-simple sequence repeat; QTL, quantitative trait loci; RAPD, random amplified polymorphic DNA, RIL, recombinant inbred line; SNP, single nucleotide polymorphism; SSR, single sequence repeat.

more closely related to *M. uniflorum* than *M. sar-gharwalensis* (Sharma *et al.*, 2015a). Novel SSR and intron length polymorphism (ILP) markers were developed and characterized by the genome sequence database in horse gram. Nineteen of these primer pairs delivered maximum (100%) cross-transferability. In comparison, minimum (8.3%) transferability was obtained in 27 primers of 12 closely related legume species *M. gharwalensis*, *Macrotyloma axillare*, *Vigna mungo*, *Vigna umbellata*, *Glycine max*, *V. radiata*, *Trifolium pratense*, *Cicer arietinum*, *Pisum sativum*, *Phaseolus vulgaris*, *Vigna unguiculata* and *Lens culinaris* (Sharma *et al.*, 2015b). Efforts were made to design novel SSR markers (2458) from next generation sequencing (NGS) data and characterize 117 microsatellites in 48 diverse lines of horse gram. The cross-species transferability of these markers was studied in nine related legume species. Fifteen showed amplification in *V. unguiculata*, 16 markers amplified in *P. vulgaris* and *M. sar-gharwalesis*, 19

markers in *V. umbellata*, 17 markers each in *C. arietinum*, *L. culinaris* and *V. mungo*, 25 markers in *T. pratense* and 43 markers in *P. sativum*. Maximum cross transferability was observed in *P. sativum* with 91.49% transferability, followed by *T. pratense* with 53.19%, *V. umbellata* with 40.43%, and *V. mungo*, *C. arietinum*, *L. culinaris* and *M. sar-gharwalensis* with 36.17% each. The lowest transferability was observed in *P. vulgaris*, 34.04%, and *V. unguiculata*, 31.91% (Kaldate *et al.*, 2017). In flow with the development of microsatellites in horse gram, Chahota *et al.* (2017) designed 5755 primers through NGS, of which 1425, 1310, 856, 1276 and 888 were of di-, tri-, tetra-, penta- and hexanucleotide repeats respectively. A study of genetic diversity and population architecture with 30 polymorphic SSR primers for 24 morphological traits in 360 horse gram genotypes revealed the demarcation of the accessions into two subpopulations, namely southern India and Himalayan origin, and the presence of huge diversity among the genotypes studied. In addition, 7352 SSRs markers were developed from the transcriptome data, of which 150 SSRs were synthesized and validated alongside a diversity study in a panel of 58 horse gram genotypes (Kumar *et al.*, 2020). Similarly, the assessment of genetic diversity in horse gram was also investigated by other researchers (Singhal *et al.*, 2010; Varma *et al.*, 2013; Viswanatha *et al.*, 2016).

Important agronomic traits are mostly quantitative in nature and, therefore, complex. These polygenic traits are conferred by quantitative trait loci (QTLs), which control the expression of phenotypic variation. Mapping these QTLs will help define the chromosomal locations that substantially affect the expression of a phenotype. This ultimately aids in the identification of responsible genes and deciphering genetic mechanisms of variation (Zeng, 1993). The first successful attempt at determining QTL traits related to drought tolerance and yield used 211 molecular markers (157 SSR, 39 RAPD, 8 ISSR and 7 conserved orthologue set (COS) markers) in 190 recombinant inbred lines, which were a derivation of a cross between two parents, HPKM249 and HPK4. The markers were distributed across 13 chromosomes that stretched for 1423.4 cM, with a mean marker interval of 9.6 cM. Five QTLs spanning over five linkage groups for four drought-conferring traits (root length and days to temporary wilting) and yield (days to maturity and numbers of seeds per plant) were identified with a LOD threshold of 4.0 (Chahota *et al.*, 2020a). Further, QTLs associated with morphological traits were mapped in recombinant inbred lines (RILs) using expressed sequence tag (EST) SRRs, SSRs, RAPD and COS markers; 295 polymorphic markers were mapped, straddling a total distance spanning 1,541.7 cM on ten linkage groups with LOD 3.5 and an average distance between markers of 5.20 cM. Composite interval mapping revealed four significant QTLs (LOD ≥2.5) for phenological traits (days to 50% flowering, reproductive period and days to maturity) and seven significant QTLs (LOD ≥2.5) for morphological traits (plant height, secondary branches and primary branches) that explained phenotypic variation from 6.36% to 47.53% (Katoch *et al.*, 2022). Mapping efforts have been made to elucidate genetic entities for stress resistance as horse gram has a tolerance for salt, drought, and

heavy metals. Genome-wide association studies (GWAS) identified seven QTLs conferring relative water content in horse gram and eight and four QTLs for root length and volume, respectively. Further, genes present in these markers involved in myriad biochemical pathways associated with plant abiotic stresses were identified using the database of Kazusa DNA Research Institute Japan (Choudhary *et al.*, 2022). A CpG methylation study in two genotypes of *M. uniflorum* (Lam.) Verdc., HPK4 (drought-tolerant) and HPKC2 (drought-sensitive) through methylation-sensitive amplified polymorphism revealed higher methylation in the sensitive genotype HPKC2 (10.1%) than in the tolerant genotype HPK4 (8.6%). Moreover, sequence homology was found with the POZ/BTB protein, DRE binding factor (cbf1) and the Ty1-copia retrotransposon (Bharadwaj *et al.*, 2013). The major transcription factors associated with drought tolerance in horse gram were *C2H2*, *bHLH*, *WRKY*, *ERF*, *NAC*, *bZIP* and *MYB* (Mahesh *et al.*, 2021).

The pioneer global protein–protein interactome (PPI) map was established to give an insight into the drought tolerance mechanism of horse gram. The characterization of PPI through gene ontology analysis elucidated that cellular processes such as kinase and transferase activities involved in signal transduction, localization of molecules, nucleocytoplasmic transport, and protein ubiquitination were found to be most responsive to drought stress. Information compiled can be retrieved from the database (*Mau*PIR) for accurate biology information on horse gram genes, proteins and involved processes (Bhardwaj *et al.*, 2016). Two horse gram genotypes, HPKM317 and HPK4, were used to assess nine candidate reference genes for six abiotic stresses, namely cold, drought, methyl viologen-induced oxidative stress, abscissic acid, heat and salinity identified from horse gram RNA-seq data. The outputs were assessed using delta-delta Ct, Bestkeeper, Normfinder and geNorm methods, as well as RankAggreg and RefFinder software, to assign a ranking. The study identified TCTP as the most stable gene in tolerant genotype HPK4, while in sensitive genotype HPKM317, profilin was found to be the most stable (Sinha *et al.*, 2021). The first draft genome sequence of horse gram was developed by Shirasawa *et al.* in 2021 for HPK-4 through the Illumina platform. The draft genome size is 259.2 Mb, covering 89% of the assembled sequences. Comparative analysis was conducted with *Vigna angularis*, *Phaseolus vulgaris*, *Lotus japonicus* and *Arabidopsis thaliana* using 36,105 genes predicted with 14,736 horse gram specific genes. Macrosynteny analysis revealed a closer affinity of the genome structure of horse gram to *V. angularis* than that of *P. vulgaris*. Employing dd-RAD-Seq analysis to understand the diversity in the 91 horse gram accessions indicated narrow genetic diversity in the crop. Further, genome sequencing of 279.12Mb, which covered 83.53% of the total horse gram genome with 24,521 annotated genes, was developed using a combination of short- and long-read sequencing technologies. The close relationship between horse gram and common bean was deciphered through maximum likelihood phylogenetic and synonymous substitution analyses that indicate diversion between the two crops roughly 10.17 million years ago. The double-digested restriction-associated DNA sequencing of 40 germplasms identified 3942 high-quality SNPs in the horse gram genome. A genome-wide association mapping study for powdery mildew resistance identified ten significant associations similar to the *MLO* and *RPW8.2* genes (Mahesh *et al.*, 2021).

11.7 Erosion of Genetic Diversity from Traditional Areas and Limitations in Germplasm Use

Genetic erosion implies the depletion of genetic diversity within a species. This exhaustion could cause the loss of individual genes, or a combination of genes referred to as gene complexes manifested in traditional and locally adapted landraces. The pursuit to augment agricultural production has now expanded from major crops to minor crops. In the Indian farming system, genetic erosion became conspicuous in the early 1960s with the advent of a shift in cropping pattern and the introduction of new crops. Significant factors leading to this depletion are over-exploitation of species, variety replacement, environmental degradation, land cleaning, policy, overgrazing and changing agricultural systems. These have led to the disappearance of both gene-rich landraces and wild relatives.

Horse gram is an overlooked crop grown by marginal farmers and people with low incomes in tribal villages and drought susceptible regions of India (Jansen, 1989). Internationally, there are no pooled endeavours for varietal development in this crop, unlike other significant crops, except for in a few institutes in India. The main reason for the loss of genetic variability in horse gram is the replacement of horse gram cultivated areas with commercial crops. The changing economic and socio-cultural dimensions of the farming community led to a significant decrease in India's total cultivated area under horse gram (Bhartiya *et al.*, 2017). The scope of the plant genetic resource (PGR) includes ancient forms of cultivated plant species, breeding lines, landraces, genetic stocks, obsolete cultivars, modern cultivars, related wild species and weedy types (IPGRI, 1993). Genetic variation, once contemplated boundless, is ebbing away as recent varieties replace landraces and traditional cultivars over sizeable areas, and the habitat of the wild relatives of present-day species is demolished. With the current demand to develop climate-resilient varieties, new genes for biotic and abiotic stress tolerance need to be retained and utilized. Genetic diversity present within and between plant species propels breeders to choose carefully genotypes that can be incorporated in a hybridization programme to develop desired varieties (Salgotra and Chauhan, 2023). As the gene stocks become depleted, the possibility of harnessing functional genes from the traditional genetic background and their utilization in crop improvement programmes will narrow. This will lead to current elite varieties becoming susceptible to various biotic and abiotic stress that will invariably reduce their performance in terms of production and productivity as a consequence.

11.8 Cultivation Practices

Horse gram grows at an optimum temperature of 20–30°C but can adapt to even a temperature of 40°C (Krishna, 2010; Mehra and Upadhyaya, 2013). It thrives well at an altitude of 1800 m above mean sea level with annual rainfall of 300–600 mm (Haq, 2011). Horse gram is a resilient pulse that thrives on soils with low soil organic matter and low nitrogen at a pH range of

5–7.5. It grows well in alfisols, inceptisols and coastal sandy soils in East Africa, and in oxisols in Central Africa, whereas in Asia, it flourishes well in alfisols, laterites, inceptisols and sandy soils mostly as a cover crop (Krishna, 2013). Horse gram has a crop duration of 120–180 days (Cook et al., 2005), a relatively short duration crop and therefore congenial for cropping sequences between two main crops (Krishna, 2010). It is grown during the kharif season, especially in the dry regions of India (Bhardwaj et al., 2013; Bhardwaj and Yadav, 2015). Horse gram is susceptible to frost and thus sown during the summer season and harvested during dry and cool weather.

11.9 Conclusion

The genetic improvement of horse gram is still at the budding stage compared to major pulses because of the scant variability for different agro-morphological traits, diminishing genetic resources and lack of genetic information on the crop. Efforts to increase the diversity of the crop through wide hybridization, mutation and tissue culture techniques can be explored to make the selection of useful, targeted traits in crop improvement of horse gram more rewarding. Furthermore, the systematic evaluation and characterization of the wild species, except *M. axillare* and *M. sar-gharwalensis*, which have been the subject of some studies, must be researched in depth for their valuable traits. These traits are imperative gene stocks for breeding superior horse gram genotypes. Horse gram is nutrient dense and inherently has the capacity to combat and thrive in input-scarce, drought and heat-stress environments. Thus, it can potentially evolve as a significant climate-resilient crop. The focus should be on curbing the antinutritional factors and ensuring quality value additions to release useful and relevant products to the market. Functional markers developed for various traits in horse gram can be a vital tool to map genes conferring targeted traits. The implementation of advanced molecular tools, such as genome-wide association studies for deciphering marker-trait associations, speed breeding, genomic selection, pangenomes, whole-genome resequencing, CRISPR/Cas9-based genome editing techniques and high-throughput phenotyping could help achieve fast genetic gain in horse gram (Kamenya et al., 2021). Intentional and community endeavours to popularize and increase public awareness on the importance of horse gram, complementary political support and framed policies bolstering cultivation, value addition and marketing, thus covering holistic aspects of this orphan legume, could augment its cultivation to aid in addressing impending issues of nutrition and food security (Aditya et al., 2019). Elaborate germplasm exploration, germplasm conservation in gene banks for both long-term and immediate use, characterization and documentation, strengthening of the seed system, marketing and dissemination among the farmer community by firming extension activities in developing and underdeveloped countries is needed to accelerate horse gram genetic improvement (Jha et al., 2022). Considering the potential importance of horse gram, it becomes imperative to develop the genetic architecture of the crop to advance its myriad benefits through conventional breeding methods, modern molecular approaches and -omics approaches such as genome selection and gene editing tools for the sustainable improvement of this underutilized legume.

References

Aditya, J.P., Bhartiya, A., Chahota, R.K., Joshi, D., Chandra, N., Kant, L. and Pattanayak, A. (2019) Ancient orphan legume horse gram: a potential food and forage crop of future. *Planta* 250, 891–909.

Arora, R.K. and Chandel, K.P.S. (1972) Botanical source areas of wild herbage legumes in India. *Tropical Grasslands* 6, 213–221.

Asouti, E. and Fuller, D.Q. (2008) *Trees and Woodlands of South India. Walnut Creek, California: Archaeological Perspectives*. Left Coast Press, San Francisco, California.

Bates, J. (2015) Social organisation and change in bronze age South Asia: a multi-proxy approach to urbanisation deurbanisation and village life through phytolith and macrobotanical analysis. PhD Dissertation, Cambridge University, Cambridge, UK.

Bazzano, L.A., He, J., Ogden, L.G., Loria, C., Vupputuri, S. *et al.* (2001) Legume consumption and risk of coronary heart disease in US men and women: NHANES I Epidemiologic Follow-up Study. *Archives of Internal Medicine* 161, 2573–2578.

Bhag, M. and Joshi, V. (1991) Underutilized plant resources. In: Paroda, R.S. and Arora, R.S. (eds) *Plant Genetic Resources, Conservation and Management.* Malhotra Publication House, New Delhi, India, pp. 211–239.

Bhardwaj, J. and Yadav, S.K. (2015) Drought stress tolerant horse gram for sustainable agriculture. *Sustainable Agriculture Reviews* 15, 293–328.

Bhardwaj, J., Mahajan, M. and Yadav, S.K. (2013) Comparative analysis of DNA methylation polymorphism in drought-sensitive (HPKC2) and tolerant (HPK4) genotypes of horse gram (*Macrotyloma uniflorum*). *Biochemical Genetics* 51, 493–502.

Bhardwaj, J., Gangwar, I., Panzade, G., Shankar, R. and Yadav, S.K. (2016) Global de novo protein–protein interactome elucidates interactions of drought-responsive proteins in horse gram (*Macrotyloma uniflorum*). *Journal of Proteome Research* 15, 1794–1809.

Bhartiya, A., Aditya, J.P. and Kant, L. (2015) Nutritional and remedial potential of an underutilised food legume horse gram (*Macrotyloma uniforum*): a review. *Journal of Animal and Plant Sciences* 25, 908–920.

Bhartiya A., Aditya, J.P., Pal, R.S. and Bajeli, J. (2017) Agromorphological, nutritional and antioxidant properties in horse gram [*Macrotyloma uniforum* (Lam.) Verdc.] germplasm collection from diverse altitudinal range of Northwestern Himalayan hills of India. *Vegetos* 30, 1.

Blumenthal, M.J., O'Rourke, P.K., Hilder, T.B. and Williams, R.J. (1989) Classification of the Australian collection of the legume *Macrotyloma*. *Australian Journal of Agricultural Research* 40, 591–604.

Bolbhat, S.N. and Dhumal, K.N. (2009) Induced macromutations in horsegram [*Macrotyloma uniflorum* (Lam.) Verdc]. *Legume Research - An International Journal* 32, 278–281.

Bolbhat, S.N., Gawade, B.B., Bhoge, V.D., Wadavkar, D.S., Shendage, V.S. and Dhumal, K.N. (2012) Effect of mutagens on frequency and spectrum of viable mutations in horsegram (*Macrotyloma uniflorum* (Lam.) Verdc). *IOSR Journal of Agriculture and Veterinary Science* 1, 45–55.

Brink, M. (2006) *Macrotyloma uniflorum* (Lam.) Verdc. In: Brink, M. and Belay, G. (eds) PROTA1: *Cereals and Pulses.* Wageningen, the Netherlands.

Chahota, R.K. (2009) Induction of early flowering and other important traits in horse gram (*Macrotyloma uniflorum*) using gamma irradiation. Final project report submitted to Bhabha Atomic Research Centre, Trombay, India.

Chahota, R.K., Sharma, T.R., Dhiman, K.C. and Kishore, N. (2005) Characterisation and evaluation of horse gram (*Macrotyloma uniflorum* Roxb.) germplasm from Himachal Pradesh. *Indian Journal of Plant Genetic Resources* 18, 221–223.

Chahota, R.K., Sharma, T.R., Sharma, S.K., Kumar, N. and Rana, J.C. (2013) Horsegram. In: Singh, M., Upadhyaya, H.D. and Bisht, I.S. (eds) *Genetic and Genomic Resources of Grain Legume Improvement.* Elsevier, London, pp. 293–305.

Chahota, R.K., Shikha, D., Rana, M., Sharma, V., Nag, A. *et al.* (2017) Development and characterization of SSR markers to study genetic diversity and horse gram germplasm population structure (*Macrotyloma uniflorum*). *Plant Molecular Biology Reporter* 35, 550–561.

Chahota, R.K., Sharma, V., Rana, M., Sharma, R., Choudhary, S. *et al.* (2020a) Construction of a framework linkage map and genetic dissection of drought-and yield-related QTLs in horse gram (*Macrotyloma uniflorum*). *Euphytica* 216, 1–11.

Chahota, R.K., Thakur, N. and Sharma, R. (2020b) Efficient improvement in an orphan legume: Horsegram, *Macrotyloma uniflorum* (Lam.) Verdi, using conventional and molecular approaches. In: Gosal, S.S. and Wani, S.H. (eds) *Accelerated Plant Breeding, Volume 3: Food Legumes.* Springer, Cham, Switzerland, pp. 369–388.

Chandana, B.R., Ramesh, S., Basanagouda, G., Kirankumar, R. and Ashwini, K.V.R. (2021) Genetics of growth habit in horse gram [*Macrotyloma uniflorum* (Lam.) Verdc.]. *Genetic Resources and Crop Evolution* 68, 2743–2748.

Choudhary, S., Isobe, S. and Chahota, R.K. (2022) Elucidation of drought tolerance potential of horse gram (*Macrotyloma uniflorum* Var.) germplasm using genome-wide association studies. *Gene* 819, 146241.

Cook, B.G., Pengelly, B.C., Brown, S.D., Donnelly, J.L., Eagles, D.A. *et al.* (2005) Tropical Forages: an interactive selection tool. CSIRO, DPI&F (QLD), CIAT and ILRI, Brisbane, Australia.

Datir, S.S. (2016) Genetic improvement in horse gram [(*Macrotyloma uniflorum* (lam) Verdec.](syn. *Dolichos biflorus* L.) through induced mutations. *Journal of Food Legumes* 29, 174–179.

Dikshit, N., Katna, G., Mohanty, C.S., Das, A.B. and Sivaraj, N. (2014) *Horse Gram. Broadening the Genetic Base of Grain Legumes*. Springer, New Delhi, pp. 209–215.

Dobhal, V.K. and Rana, J.C. (1994a) Multivariate analysis in horse gram. *Legume Research* 17, 157–161.

Dobhal, V.K. and Rana, J.C. (1994b) Genetic analysis for yield and its components in horsegram (*Macrotyloma uniflorum*). *Legume Research* 17, 179–182.

Durga, K.K., Ankaiah, R. and Ganesh, M. (2015) Characterisation of horse gram cultivars using plant morphological characters. *Indian Journal of Agricultural Research* 49, 215–221.

Fuller, D.Q. (2011) Finding plant domestication in the Indian subcontinent. *Current Anthropology* 52, 347–362. DOI: 10.1086/658900

Fuller, D.Q. and Harvey, E.L. (2006) The archaeobotany of Indian pulses: identification, processing and evidence for cultivation. *Environ Architecture* 11, 219–246.

Fuller, D.Q. and Murphy, C. (2014) Overlooked but not forgotten: India as a centre for agricultural domestication. *General Anthropology* 21, 4–8.

Fuller, D.Q. and Murphy, C. (2018) The origin and early dispersal of horse gram (*Macrotyloma uniflorum*), a major crop of ancient India. *Genetic Resources and Crop Evolution* 65, 285–305.

Gautam, M., Katoch, S. and Chahota, R.K. (2020a) Comprehensive nutritional profiling and activity directed identification of lead antioxidant, antilithiatic agent from *Macrotyloma uniflorum* (Lam.) Verdc. *Food Research International* 137, 109600.

Gautam, M., Datt, N. and Chahota, R.K. (2020b) Assessment of calcium oxalate crystal inhibition potential, antioxidant activity and amino acid profiling in horse gram (*Macrotyloma uniflorum*): high altitude farmer's varieties. *3 Biotech* 10, 402.

Gopalan, C., Ramasastry, B.V. and Balasubramanian, S.C. (2006) *Nutritive Value of Indian Foods* (revised and updated). National Institute of Nutrition, Hyderabad, India.

Gupta, A., Bhartiya, A., Singh, G., Mahajan, V. and Bhatt, J.C. (2010) Altitudinal diversity in horse gram [*Macrotyloma uniforum* (Lam.) Verdc.] land races collected from hill region. *Plant Genetic Resources* 8, 214–216.

Gupta, V.P., Sharma, S.K. and Rathore, P.K. (1994) Induced mutagenic response of gamma irradiation on horsegram (*Macrotyloma uniflorum*). *Indian Journal of Agricultural Sciences* 64, 160–164.

Halder, S., Mandal, A., Das, D., Gupta, S., Chattopadhyay, A.P. and Datta, A.K. (2015) Copper nanoparticle induced macromutation in *Macrotyloma uniflorum* (Lam.) Verdc. (Leguminosae): a pioneer report. *Genetic Resources and Crop Evolution* 62, 165–175.

Haq, N. (2011) Underutilised food legumes: potential for multipurpose uses. In: Pratap, A. and Kumar, J. (eds) *Biology and Breeding of Food Legumes*. CABI International, Wallingford, UK, pp. 335–336.

Herath, H.T., Samaranayake, M.D.W., Liyanage, S. and Abeysekera, W.K.S.M. (2020) Horse gram: an incredible food grain as a potential source of functional and nutritional food ingredients. *International Journal of Food Sciences and Nutrition* 5, 93–101.

Hooker, J. (1879) *Flora of British India*, vol 2. L. Reeve & Co., London.

Ingle, K.P., Al-Khayri, J.M., Chakraborty, P., Narkhede, G.W. and Suprasanna, P. (2020) Bioactive compounds of Horse gram (*Macrotyloma uniflorum* Lam. [Verdc.]). In: Murthy, H.N. and Paek, K.Y. (eds) *Bioactive Compounds in Underutilized Vegetables and Legumes. Reference Series in Phytochemistry*. Springer, Cham, Switzerland, pp.1–39.

IPGRI (1993) *Diversity for Development*. International Plant Genetic Resources Institute, Rome, Italy.

Jamadagani, B.M. and Birari, S.P. (1996) Modification in photoperiodic response of flowering by gamma-rays induced mutation in horse gram. *Journal of Maharashtra Agricultural University* 21, 384–386.

Jansen, P.C.M. (1989) *Macrotyloma uniflorum* (Lam.) Verdc. In: Westphal, E., Jansen, P.C.M. and den Outer, R.W. (eds) *Plant Resources of South-East Asia, Pulses*. Wageningen, the Netherlands, pp. 53–54.

Jayan, P.K. and Maya, C.N. (2001) Studies on germplasm collections of horse gram. *Indian Journal of Plant Genetic Resources* 14, 443–447.

Jha, U.C., Nayyar, H., Parida, S.K. and Siddique, K.H. (2022) Horse gram, an underutilized climate-resilient legume: breeding and genomic approach for improving future genetic gain. In: Jha, U.C., Nayyar, H., Agrawal, S.K. and Siddique, K.H.M. (eds) *Developing Climate Resilient Grain and Forage Legumes*. Springer Nature, Singapore, pp. 67–178.

Joshi, S.S., Chikkadevaiah and Shashidhas, H.E. (1994) Classification of horse gram germplasm. *Crop Research* 7, 375–391.

Kaldate, R., Rana, M., Sharma, V., Hirakawa, H., Kumar, R. *et al*. (2017) Development of genome-wide SSR markers in horse gram and their use for genetic diversity and cross-transferability analysis. *Molecular Breeding* 37, 1–10.

Kamenya, S.N., Mikwa, E.O., Song, B. and Odeny, D.A. (2021) Genetics and breeding for climate change in orphan crops. *Theoretical and Applied Genetics* 23, 1–29.

Katoch, M. and Chahota, R.K. (2021) Mapping of Quantitative Trait Locus (QTLs) that contribute to drought tolerance in a recombinant inbred line population of horse gram (*Macrotyloma uniflorum*). *bioRxiv* 2021-03.

Katoch, M., Mane, R.S. and Chahota, R.K. (2022) Identification of QTLs linked to phenological and morphological traits in RILs population of horsegram (*Macrotyloma uniflorum*). *Frontiers in Genetics* 12, 2710.

Kingwell-Banham, E. and Fuller, D.Q. (2014) Horse gram: origins and development. In: Smith, C. (ed.) *Encyclopedia of Global Archaeology*. Springer, New York, pp. 3490–3493.

Krishna, K.R. (2010) *Agroecosystems of South India: Nutrient Dynamics, Ecology and Productivity*. Brown-Walker Press, Boca Raton, Florida.

Krishna, K.R. (2013) *Agroecosystems: Soils, Climate, Crops, Nutrient Dynamics and Productivity*. Apple Academic Press, New York, pp. 170–174.

Kumar, D. (2005) Status and direction of arid legumes research in India. *Indian Journal of Agricultural Sciences* 75, 375–391.

Kumar, D. (2007) *Production Technology for Horse Gram in India*. Central Arid Zone Research Institute, Jodhpur, India.

Kumar, R., Kaundal, S.P., Sharma, V., Sharma, A., Singh, G. *et al.* (2020) Development of transcriptome wide SSR markers for genetic diversity and structure analysis in *Macrotyloma uniflorum* (Lam.) Verdc. *Physiology and Molecular Biology of Plants* 26, 2255–2266.

Lackey, J.A. (1981) Phaseoleae In: Polhill, R.M. and Raven, P.H. (eds) *Advances in Legume Systematics*. Royal Botanic Gardens, Kew, UK, pp. 301–327.

Lad, B.D., Chavan, S.A. and Dumbre, A.D. (1999) Genetic variability and correlation studies in horse gram. *Journal of Maharashtra Agricultural University* 23, 268–271.

Longvah, T., Ananthan, R., Bhaskarachary, K. and Venkaiah, K. (2017) *Indian Food Composition Tables*. National Institute of Nutrition, Hyderabad, India.

Mahajan, V., Shukla, S.K., Gupta, N.S., Majumdera, D., Tiwari, V. and Piiasad, S.V.S. (2007) Identifying phenotypically stable genotypes and developing strategy far yield improvement in horse gram (*Macrotyloma uniflorum*) for mid-altitudes of Northwestern Himalayas. *Indian Journal of Agricultural Sciences* 3, 19–22.

Mahesh, H.B., Prasannakumar, M.K., Manasa, K.G., Perumal, S., Khedikar, Y. *et al.* (2021) Genome, transcriptome, and germplasm sequencing uncovers functional variation in the warm-season grain legume horsegram *Macrotyloma uniflorum* (Lam.) Verdc. *Frontiers in Plant Science* 12, 758119.

Maheswari, M., Sarkar, B., Salini, K., Nirmala, G., Anshida Beevi *et al.* (2019) Horsegram varieties developed for increasing farm income in the dryland. TOT/ FFP/ CRIDA/ 2019/ 2. Central Research Institute for Dryland Agriculture, Hyderabad, India.

Mehra, A. and Upadhyaya, M. (2013) *Macrotyloma uniforum* Lam. A traditional crop of Kumaun Himalaya and ethnobotanical perspectives. *International Journal of Agriculture and Food Science* 3, 148–150.

Mehra, K.L. and Magoon, M.L. (1974) Gene centres of tropical and sub-tropical pastures legumes and their significance in plant introduction. Proceedings of the XII International Grassland Congress, Moscow, pp. 251–256.

Modi, M.R., Katoch, M., Thakur, N., Gautam, M., Choudhary, S. and Chahota, R.K. (2023) Mapping of genomic regions associated with dwarfing and the determinate growth habit in horsegram (*Macrotyloma uniflorum*). *Czech Journal of Genetics and Plant Breeding* 59, 196–204.

Morris, J.B. (2008) *Macrotyloma axillare* and *M. uniflorum*: descriptor analysis, anthocyanin indexes, and potential uses. *Genetic Resources Crop Evolution* 55, 5–8.

Nagaraja, N., Nehru, S.D. and Manjunath, A. (1999) Plant types for high yield in horse gram as evidenced by path coefficients and selection indices. *Karnataka Journal of Agricultural Sciences* 12, 32–37.

Neelam, S., Kumar, V., Natarajan, S., Venkateshwaran, K. and Pandravada, S.R. (2014) Evaluation and diversity observed in horsegram (*Macrotyloma uniflorum* (Lam.) Verdc.), India. *Plant* 4, 17–22.

Nezamuddin, S. (1970) *Pulse Crops of India*. Indian Council of Agricultural Research, New Delhi.

Patel, D.P., Dabas, B.S., Sapra, R.S. and Mandal, S. (1995) Evaluation of horsegram (*Macrotyloma uniflorum*) (Lam.) germplasm. National Bureau of Plant Genetic Resources Publication, New Delhi, India.

Patil, J.V., Deshmukh, R.B. and Singh, S. (1994) Selection of genotypes in horse gram based on genetic diversity. *Journal of Maharashtra Agricultural University* 19, 412–414.

Prakash, B.G., Channayya Hiremath, P., Devarnavdgi, S.B. and Salimath, P.M. (2010) Assessment of genetic diversity among germplasm lines of horse gram (*Macrotyloma uniflorum*) at Bijapur. *Electronic Journal of Plant Breeding* 1, 414–419.

Priyanka, S., Sudhagar, R., Vanniarajan, C., Ganesamurthy, K. and Souframanien, J. (2019) Combined mutagenic ability of gamma-ray and EMS in horse gram (*Macrotyloma uniflorum* (Lam) Verdc.). *Electronic Journal of Plant Breeding* 10, 1086–1094.

Priyanka, S., Sudhagar, R., Vanniarajan, C., Ganesamurthy, K., Souframanien, J. and Jeyakumar, P. (2020) Comparative studies on mutagenic effectiveness and efficiency in horsegram [*Macrotyloma uniflorum* (Lam) Verdc.]. *Legume Research* 1, 8.

Priyanka, S., Sudhagar, R., Vanniarajan, C., Ganesamurthy, K. and Souframanien, J. (2021) Gamma rays induced morphological, flowering and palynological modifications in horse gram (*Macrotyloma uniflorum*). *Journal of Environmental Biology* 42, 1363–1369.

Purseglove, J.W. (1974) Dolichos uniflorus. *Tropical Crops: Dicotyledons.* Longman, London and New York.

Ramakanth, R.S., Setharama, A. and Patil, N.M. (1979) Induced mutations in *Dolichos lablab*. *Mutation Breeding Newsletter* 14, 1–2.

Rana, J.C. (2010) Evaluation of horsegram germplasm. Annual Progress Report of the NBPGR, Shimla, India.

Sadawarte, S.K., Pawar, V.S., Sawate, A.R., Thorat, P.P., Shere, P.D. and Surendar, J. (2018) Effect of germination on vitamin and mineral content of horse gram and green gram malt. *International Journal of Communication Systems* 6, 1761–1764.

Sahoo, J.L., Das, T.R., Baisakh, B., Nayak, B.K. and Panigrahi, K.K. (2014) Assessment of genetic diversity in horse gram [*Macrotyloma uniflorum* (Lam.) Verdec.]. *E-planet* 12, 31–35.

Salgotra, R.K. and Chauhan, B.S. (2023) Genetic diversity, conservation, and utilization of plant genetic resources. *Genes* 14, 174.

Savithriamma, D.L., Shambulingappa, K.G. and Rao, M.G. (1990) Evaluation of horse gram germplasm. *Current Research* 19, 150–153.

Sharma, R.K. (1995) Nature of variation and association among grain yield and yield components in horse gram. *Crop Improvement* 22, 73–76.

Sharma, A. and Chahota, R.K. (2023) Genome-wide association studies for identification of novel QTLs related to four agronomic traits in horsegram (*Macrotyloma uniflorum*). *Research Square* rs-3366692. DOI: 10.21203/rs.3.rs-3366692/v1

Sharma, V., Sharma, T.R., Rana, J.C. and Chahota, R.K. (2015a) Analysis of genetic diversity and population structure in horse gram (*Macrotyloma uniflorum*) using RAPD and ISSR markers. *Agricultural Research* 4, 221–230.

Sharma, V., Rana, M., Katoch, M., Sharma, P.K., Ghani, M. *et al.* (2015b) Development of SSR and ILP markers in horsegram (*Macrotyloma uniflorum*), their characterisation, cross-transferability and relevance for mapping. *Molecular Breeding* 35, 1–22.

Shirasawa, K., Chahota, R., Hirakawa, H., Nagano, S., Nagasaki, H., Sharma, T. and Isobe, S. (2021) A chromosome-scale draft genome sequence of horse gram (*Macrotyloma uniflorum*). *BioRxiv* 2021-01.

Shirsat, R.K., Mohrir, M.N., Kare, M.A. and Bhuktar, A.S. (2010) Induced mutations in horse gram: mutagenic efficiency and effectiveness. *Recent Research in Science and Technology* 2, 20–23.

Singh, R., Salam, J.L., Mandavi, N.C., Saxena, R.R. and Sao, A. (2019) Genetic diversity estimation in horse gram [*Macrotyloma uniflorum* (L.) Verdcout] genotypes collected from Bastar plateau. *International Journal of Current Microbiology and Applied Sciences* 8, 613–620.

Singhal, H.C., Tomar, S.S., Baraiya, B.R., Sikarwar, R.S. and Tomar, I.S. (2010) Study of genetic divergence in horse gram (*Macrotyloma uniflorum* L.). *Legume Research* 33, 119–123.

Sinha, R., Bala, M., Prabha, P., Ranjan, A., Chahota, R.K., Sharma, T.R. and Singh, A.K. (2021) Identification and validation of reference genes for qRT-PCR-based studies in horse gram (*Macrotyloma uniflorum*). *Physiology and Molecular Biology of Plants* 27, 2859–2873.

Sirisha, A.B.M. (2022) Exploration, collection and characterisation of horse gram [*Macrotyloma uniflorum* (Lam.) Verd.] germplasm. *Legume Research* 45, 1235–1240.

Smartt, J. (1985) Evolution of grain legumes. II. Old and new world pulses of lesser economic importance. *Experimental Agriculture* 21, 1–18.

Sodani, S., Paliwal, R. and Jain L. (2006) Phenotypic stability for seed yield in rainfed horsegram [*Macrotyloma uniflorum* (Lam.) Verdc.] *Journal of Arid Legumes* 4, 340.

Sreenivasan, E. (2003) Hybridization studies in horse gram. In: Henry, A., Kumar, D. and Singh, N.B. (eds) *Advances in Arid Legumes Research.* Scientific Publishers Jodhpur, India, pp. 115–121.

Sreerama, Y.N., Sashikala, V.B., Vishwas, P.M. and Singh, V. (2012) Nutrients and antinutrients in cowpea and horse gram flours compared to chickpea flour: evaluation of their functionality. *Food Chemistry* 131, 462–468.

Staples, I.B. (1966) A promising new legume. *Queensland Agricultural Journal* 92, 388–392.

Staples, I.B. (1982) *Intra and interspecific crosses in Macrotyloma axillare. Proceedings of the Second Australian Agronomy Conference*, Wagga, pp. 340.

Subba Rao, A. and Sampath, S.R. (1979) Chemical composition and nutritive value of horsegram (*Dolichos biflorus*). *Mysore Journal of Agricultural Sciences* 13, 198–205.

Sudha, N., Mushtari Begum, J., Shambulingappa, K.G. and Babu, C.K. (1995) Nutrients and some anti-nutrients in horse gram *(Macrotyloma uniflorum* (Lam.) Verdc.*). Food and Nutrition Bulletin* 16, 81–83.

Sudhagar, R., Pushpayazhini, V., Vanniarajan, C., Hepziba, S.J., Renuka, R. and Souframanien, J. (2023) Characterisation of horse gram mutants for yield, nutrient, and anti-nutrient factors. *Journal of Environmental Biology* 44(1), 99–107.

Sundararaj, D.D. and Thulasidas, G. (1993) *Botany of Field Crops* (2nd edn). MacMillan India Limited, New Delhi.

Tripathi, A.K. (1999) Variability in horse gram. *Annals of Agricultural Research* 20, 382–383.

Varma, V.S., Durga, K.K. and Ankaiah, R. (2013) Assessment of genetic diversity in horse gram (*Dolichos uniflorus*). *Electronic Journal of Plant Breeding* 4, 1175–1179.

Vavilov, N.I. (1951) Phytogeographic basis of plant breeding. *Chronica Botanica* 13, 13–54.

Verdcourt, B. (1970) Studies in the Leguminosae-Papilionoideae from the 'flora of tropical East Africa' III. *Kew Bulletin* 24, 379–447.

Verdcourt, B. (1971) Phaseoleae. In: Gillet, J.B., Polhill, R.M. and Verdcourt, B. (ed.) *Flora of East Tropical Africa. Leguminosae sub family Papilionoideae.* Crown Agents, pp. 581–594.

Verdcourt, B. (1980) The classification of *Dolichos* L. emend. Verde, *Lablab* Adans., *Phaseolus* L., *Vigna* Savi and their allies. In: Summerfield, R.J. and Bunting, A.H. (eds) *Advances in Legume Science.* Kew Royal Botanic Gardens, Kew, UK, pp. 45–48.

Viswanatha, K.P., Yogeesh, L.N. and Amitha, K. (2016) Morphological diversity study in horsegram (*Macrotyloma uniflorum* (Lam.) Verdc) based on Principal Component Analysis (PCA). *Electronic Journal of Plant Breeding* 7, 767–770.

Watt, G. (1889–1893) *A Dictionary of the Economic Products of India*, vols I–VI. Cosmo Publications, Delhi, India.

Yadav, S., Negi, K.S. and Mandal, S. (2004) Protein and oil-rich wild horse gram. *Genetic Resources and Crop Evolution* 51, 629–633.

Yadava, N.D. and Vyas, N.L. (1994) *Arid Legumes.* Agro Publishers, India, pp. 64–78.

Zeng, Z.B. (1993) Theoretical basis of separation of multiple linked gene effects on mapping quatitative trait loci. *Proceedings of the National Academy of Sciences, USA* 90, 10972–10976.

Zeven, A.C. and de Wet, J.M.J. (1982) *Dictionary of Cultivated Plants and their Regions of Diversity.* Centre for Agricultural Publishing and Documentation, Wageningen, the Netherlands.

Zohary, D. (1970) Centers of diversity and centers of origin. In: Frankel O.H. and Bennett, E. (eds) *Genetic Resources in Plants – their Exploration and Conservation.* Blackwell Science Publishing, Oxford, pp. 33–42.

12 Bambara Groundnut (*Vigna subterranea* (L.) Verdc.)

K.M. Boraiah[1]*, Siddhanath Shendekar[2], Chetan Shinde[2], P.S. Basavaraj[1], Aliza Pradhan[1], C.B. Harisha[1], Hanamant M. Halli[1], K.K. Pal[1] and K. Sammi Reddy[1]

[1]*ICAR-National Institute of Abiotic Stress Management Malegaon, Baramati, Maharashtra, India;* [2]*Mahatma Phule Krishi Vidyapeeth, Rahuri, Ahmednagar, Maharashtra, India*

Abstract

Crop diversification is an essential strategy to bolster the income of small-scale farmers while simultaneously addressing food and nutritional security. The Bambara groundnut, an indigenous African legume, holds great promise as a versatile crop suitable for cultivation in various agroecological regions characterized by tropical climates and marginal soils. Its multifaceted applications in food, nutrition and medicine, and its resilience to drought and high temperatures make it an attractive option for cultivation as an alternative crop, particularly in arid and semi-arid areas. High levels of antinutritional factors such as high phytate, tannin and phenol content pose a problem, for increased use of Bambara groundnut. Despite its potential in terms of climate resilience and nutritional value, this crop has been neglected by the scientific community. It is imperative to bring crops like Bambara groundnut to the forefront, positioning them alongside significant crops such as groundnut, cowpea, soybean, maize, wheat and rice for their nutritional significance and adaptability to various stress conditions. In this book chapter, we have provided comprehensive information on Bambara groundnut, encompassing its origin, distribution, botanical characteristics, the current status of germplasm collection, characterization, utilization, conservation and nutritional values. Additionally, we have highlighted the pivotal stages in crop improvement, including collecting information on landraces and molecular markers. The chapter concludes by offering insights into essential cultivation practices required to unlock the crop's genetic potential. Scientific knowledge of underutilized crops like Bambara groundnut plays a crucial role in accelerating ongoing crop improvement efforts, enabling the exploitation of genetic and genomic resources to develop high-yielding varieties of the crop, which can thrive well in climate extremes such as drought and high temperature.

Keywords: Bambara groundnut, genetic and genomic resources, collections, gene pools, omics

12.1 Introduction

To achieve zero hunger in all its forms and eliminate malnutrition through food and nutritional security, many countries globally target the most significant Sustainable Development Goals (SDGs).

Achieving these goals demands adequate, healthy and diverse food to improve diets and nourish individuals. Although many countries are currently self-sufficient in food production to feed their population, they cannot provide adequate, diverse and healthy food at a reasonable

*Corresponding author: bors_km@yahoo.co.in

© CAB International 2024. *Potential Pulses: Genetic and Genomic Resources*
(eds R. Chandora, T. Basavaraja and A. Pratap)
DOI: 10.1079/9781800624658.0012

cost. In recent decades, there has been a notable rise in the occurrence of hunger and undernourishment, alongside the emergence of issues related to childhood overweight and adult obesity. This is substantiated by a significant and steep increase in the prevalence of undernourishment (PoU) on a global scale, even though there was a more gradual increase between 2020 and 2021, primarily owing to the impact of the Covid-19 pandemic (FAO, 2022). The severity of this issue is most pronounced in Africa, where PoU rates are the highest compared to all other regions. Additionally, Asian countries face a substantial challenge, because they are home to 55% of the world's population, in suffering from hunger and Asia holds the second-highest position in terms of PoU. Despite the global increase in dietary energy supply through increased yield and production, it does not translate to the nutritional quality of our food, nor does it afford food to vulnerable populations.

To achieve the SDG of eradicating hunger by 2030 and eliminating all forms of malnutrition, the primary objective is to enhance the bioavailability and accessibility of a wide range of essential nutrients. Despite the abundant agrobiodiversity on our planet, human reliance has gradually shifted towards a few crops such as rice, wheat and maize to meet their nutritional needs. These crops are not only vulnerable to climate change but also may threaten food security. With the predicted increase in frequency and intensity of climate extremes owing to climate change, there will be a possible reduction in the cultivation of traditional staple food crops, which will probably make the food chain more vulnerable than ever. With the concerns for all these issues, crop diversification will be the ultimate approach to cope with the problems of food and nutritional security under changing climate scenarios. Adopting and cultivating climate-resilient and nutrient-rich crops like Bambara groundnut provides one of the solutions to the issues mentioned earlier. Because of its distinctive characteristics, including its ability to tolerate drought and high temperature, adaptability to marginal and degraded soils, and being suitable for cultivation under sub-optimal inputs, Bambara groundnut can be employed as a component of climate-resilient agriculture (Mayes *et al.*, 2019). Owing to its ability to fix biological nitrogen and its short plant architecture, it is suitable for intercropping systems with cereals. Given its versatility, it is crucial to understand its origin, distribution, adaption and acclimatization, botanical types, gene pools, evolutionary linkages, cultivation methods, and information on the characterization, evolution and conservation of germplasm. Hence, in this chapter, we attempted to understand this potential and underutilized crop in general, and from the genetic and breeding point of view.

12.1.1   Importance and scope

Bambara groundnut is a remarkably adaptable crop, thriving in a wide range of environmental conditions and exhibiting resilience to various abiotic stresses. This adaptability can be attributed to its possession of all three essential mechanisms for drought tolerance: avoidance, escape and tolerance. Moreover, this resilient crop has effectively acclimated to diverse environments, including those characterized by cold nights and scorching daytime temperatures, even in arid regions. Thus, it is grown extensively and considered a vital and indigenous crop in dryland areas of Africa, but it is also now grown in more humid areas of Indonesia. The significance of Bambara groundnut lies in its potential for promotion in drought-prone regions and its ability to produce substantial yield even in regions where climate change predictions indicate an increased frequency and intensity of droughts. Mabhaudhi *et al.* (2018) forecasted that, in response to projected climate change in South Africa, Bambara groundnut could see a substantial increase of approximately 37.5% in yield and a 33% improvement in water productivity. This crop resilience may contribute to expanding suitable areas for Bambara groundnut production in South Africa in changed climate scenarios (Mayes *et al.*, 2019).

Bambara groundnut is the third most important legume crop in dryland areas of Africa but it is neglected and underutilized by scientific communities. It plays an important role in the daily diets of African people, besides the utility of its aboveground biomass as fodder. Although it is often termed as 'women's crop' or 'poor man's' crop, it has recently been declared as 'crops for the new millennium' (Ahmad *et al.*, 2013; Khan *et al.*, 2021) because it is often termed as a 'complete balanced diet', being a nutrient-rich legume containing carbohydrates (64.4%), protein (23.6%), fat (6.5%) and fibre (5.5%), as well as rich in micronutrients such as K (11.44–19.35 mg/100 g), Fe (4.9–48 mg/100 g), Na (2.9–12.0 mg/100 g) and Ca (95.8–99 mg/100 g), on average (Tan *et al.*, 2020; Paliwal *et al.*, 2021). However, the composition of seeds varies from location to location and among varieties, but the range is 51–70% carbohydrate, 18–24% protein with high lysine and methionine, 4–12% crude oil, 3–5% ash and 3–12% fibre. It is used as a source of protein for those people who cannot afford animal protein because of high cost. It is not eaten as a fresh pod. The young fresh seeds are fried or boiled with salt and eaten as snacks. Sometimes, dry seeds can be roasted and eaten. The dried seeds are ground into flour and can be used as a supplement with millet to prepare products such as 'akara', 'moi-moi' and 'okpa'. It is used in the preparation of the local food drink 'Kunu' and other dishes such as 'tuwo' (Linnemann, 1988), and in making bread in Zambia. Brough *et al.* (1993) reported on the potential of Bambara groundnut in preparing a vegetable milk with a preferred flavour to milk made from cowpea, pigeon pea and soybean. It has a myriad of potential applications in the pharmaceutical and food industries. The by-products of Bambara groundnut possess valuable attributes such as emulsion stabilizers, thickening agents and compatible fortifiers for delivering beneficial active compounds. These qualities are particularly advantageous for creating appealing food products. Additionally, the nanocomposite derived from Bambara groundnut exhibits novel physicochemical, antioxidant and functional properties, positioning it as a promising crop for the nutri-pharmaceutical industries. For instance, combining Bambara groundnut soluble dietary fibre with groundnut starch effectively addresses the undesirable traits of both biopolymers in their natural state while preserving their favourable characteristics. Biopolymers such as the Bambara groundnut starch–soluble dietary fibre nanocomposite can potentially bring about transformative changes in the food industry (Maphosa *et al.*, 2022a, b).

Despite its potential for climate resilience and nutritional significance, several threats affect the cultivation of Bambara groundnut. The grain yield of the crop stagnated around 1.00 t ha⁻¹ despite a potential yield of over 3 t ha⁻¹ (Smýkal *et al.*, 2013). Unforeseen abiotic stresses arising from extreme moisture and temperature conditions, coupled with a range of biotic stresses, primarily weeds (Musango *et al.*, 2022), pests such as ladybird beetles, grasshoppers, aphids and leaf borers, as well as attacks by rats during the pod development stage, along with the presence of diseases including the bean common mosaic virus, soybean common mosaic virus, peanut rosette virus and leaf spot, collectively exert adverse effects on productivity. Bambara groundnut is highlighted as one of the most neglected crops, with its rich and diverse plant genetic resources remaining underutilized in breeding programmes. In the primary cultivation regions of African countries, the crop is typically grown using unimproved varieties or landraces developed through mass selection techniques from their wild relatives. Hence, there is a scope to unlock the potential yield of up to 3 t/ha by properly utilizing available plant genetic resources and genetic enhancement by integrating conventional and modern breeding tools. Besides improving production by unlocking the potential yield, there is an urgent need to address other challenges such as a hard-to-cook problem, antinutritional factors, value addition and marketing constraints, and so on. With this background, the chapter will cover different aspects of the crop's genetic improvement, guiding breeders to develop Bambara groundnut varieties with more climate-resilient, pest- and disease-resistant high yields and better nutritional quality.

12.2 Origin, Distribution and Botanical Description

Bambara groundnut (*Vigna subterranea* (L.) Verdc.) is a drought-tolerant African legume widely grown in semi-arid and tropical regions (Ogwu *et al.*, 2018). It is a self-pollinated crop with a chromosome number of 2n = 22 (genome size of approximately 1.02 Gb). The suffix 'groundnut' is added to its name because it sets pods like groundnut (underground), although some genotypes partially set pods on- or just sub-surface. Hence, it is commonly referred to as 'Bambara groundnut'. It is known by different names in various regions of Africa, such as Bambara bean, Earth pea, Madagascar groundnut, Congo goober, Jugoboon

(South Africa), Bogor Nut (Indonesia), Okpa (Nigeria), Kwaruru (Niger), Ntoyo cibemba (Zambia) and Jugu beans in Kenya (Ogwu *et al.*, 2018; Soumare *et al.*, 2022).

The exact origin of Bambara groundnut has yet to be discovered; however, being an indigenous crop to Africa, the centre of origin of Bambara groundnut is believed to be 'Bambara', near Timbuktu in Central Mali, West Africa (Holm and Marloth, 1940). In contrast, Goli (1997) proposed that wild crop relatives of Bambara groundnut are present in north-eastern Nigeria and northern Cameroon regions, and that these areas are considered the primary centre of genetic diversity. Similarly, based on crop diversity, Sri Lanka, Malaysia, the Philippines and India are considered secondary centres of diversity (Mohammed, 2023). Olukolu *et al.* (2012) used diverse arrays technology (DArT) marker analysis along with phenotypic descriptors to determine the origin and concluded that Nigeria and Cameroon were likely to be the centres of origin for Bambara groundnut. This crop is widely distributed and grown across several agroecological regions, mainly concentrated in semi-arid parts of Africa and rarely spread out of Africa, particularly in Asia (especially Indonesia, Malaysia, Philippines, India and Thailand) and South America (particularly Brazil). Though Bambara groundnut (the third most important crop in Africa after groundnut and cowpea) is widely distributed throughout Africa, it spreads further to Madagascar, South Africa and low-lying Ethiopia. Bambara groundnut-producing countries in West Africa include Benin, Burkina Faso, Côte d'Ivoire, Gambia, Ghana, Guinea, Mali, Niger, Nigeria, Senegal and Togo (Goli, 1997). In Southern Africa, important producing countries include Botswana, Madagascar, Malawi, Zambia, South Africa, Swaziland, Tanzania and Zimbabwe, and East and Central Africa include Burundi, Cameroon, Central African Republic, Congo, Ethiopia and Sudan (Goli, 1997). In the 18th century, it was introduced in Indonesia through trade ties and then spread across Asian countries such as Malaysia from Ethiopia, where it became one of the important crops (Khan *et al.*, 2021).

The optimal conditions for the growth of Bambara groundnut are: 30–35°C daytime temperature for germination, average day temperatures ranging from 20 to 28°C under full sun, average annual rainfall of 600–750 mm (optimum yields are obtained when rainfall is higher), good P and K content in soil and light sandy loams with a pH of 5.0–6.5. Bambara groundnut is adapted to hot, dry, marginal soils from sea level to an altitude of 2000 m. It can grow well in more humid conditions (annual rainfall > 2000 mm) and in every type of soil provided it is well drained and not too calcareous. The inherent and adaptive drought tolerance mechanisms and traits (Table 12.1) make this crop grow well in harsh conditions, better than most drought-tolerant crops such as sorghum, maize and peanuts (groundnuts). Owing to its short duration and dwarf plant architecture, it can be intercropped with other root, tuber and cereal crops.

Table 12.1. The mechanisms and traits associated with drought tolerance in Bambara groundnut. (Author's own table.)

Traits/mechanisms	References
Maintenance of leaf turgor pressure by osmotic adjustment, stomatal regulation and reduced leaf area	Mayes *et al.* (2019)
Canopy size and architecture	
Biomass accumulation and its partitioning	
Phenological plasticity	
Traits associated with gas exchange	
Regulation of photosynthesis	
Regulation of transpiration (leaf orientation/ paraheliotropism and epicuticular wax content)	
Dark seeds (tannins) with better emergence rates	
Genotypes and their place of origin	Shareef *et al.* (2014)
High water use efficiency, SPAD chlorophyll meter reading, specific leaf area, specific leaf N and gas exchange traits	Singh and Basu (2005)

Bambara groundnut is a leguminous crop with two major botanical types, i.e. cultivated type *Vigna subterranea* var. *subterranea* and wild type *Vigna subterranea* var. *spontanea*. It is an annual herbaceous pulse, having an intermediate-type creeping stem with a plant height of 25–40 cm. It possesses a well-developed tap root system and dense lateral roots containing nodules for biological nitrogen fixation (Fig. 12.1).

The lateral stems grow from the root. Trifoliate leaves with a length of about 5 cm having stipules about 3 mm long are affixed to the stem by the petiole. The petioles have a base that is either green or purple and are around 15 cm long, rigid and grooved. At each node, leaf and flower buds appear in alternating order. The pinnately trifoliate, glabrous leaves have an upright petiole that is enlarged at the base. The terminal leaflet has two stipels, but the two lateral leaflets each have just one. The rachis is joined to the oval leaflets by conspicuous pulvini. The terminal leaflet has a median length of 6 cm and a median breadth of 3 cm, making it more significant than the lateral leaflets (Fig. 12.2). Like groundnut, pod development is an exciting feature of this crop. It bears cleistogamous flowers on its stems. The flowers have a tube calyx about 1 mm long, five lobes about 1 mm long, and a whitish-yellow corolla 4–7 mm long. The calyx-covered ovary folds back at the base of the flower and points towards the glandular apex during fertilization. The peduncle lengthens to bring the fertilized ovary to or below the soil surface, while the fertilized ovary remains unchanged for several days. The beginning of pod production is the time when the peduncle reaches its maximum length. The ovary at the apical end, which contains the fertilized ovules, grows into a pod (Basu *et al.*, 2007a). The pods are round and wrinkled, and each contains 1–3 seeds that are round, smooth and very hard when dried. Most varieties have only one seed (Fig. 12.2), but pods with three seeds have been found in Congo ecotypes. The ovate-shaped seeds may be cream-coloured, brown, red, mottled or black-eyed, and about 1 cm long, depending on genotype or landraces (Fig. 12.3). Husk or shell constitutes about 20% of the seed. The growth cycle is about 120–150 days to pod maturity. First flowers appear during 40–60 days. The pod matures about 30 days after fertilization, and the seed develops in the next ten days. Harvest index is about 45%. The mechanism determining that photoperiodism influences pod filling is unknown and needs to be understood.

Fig. 12.1. Nodulation in Bambara groundnut (cv. SB42). (Author's own figure.)

12.3   Crop Gene Pool, Evolutionary Relationships and Systematics

Various genetic features and variations distinguish the Bambara groundnut gene pool from others. There is a need to study gene pools because they provide solutions to key challenges such as disease resistance, insect resistance, drought, salinity and many more. Guillenem, Perrottet and Richard reported a wild plant of Bambara groundnut near Senegal in 1958 (Hepper, 1963). Similarly, according to Linnemann (1987), several wild plants occur in Jos Plateau, Yola (Northern Nigeria), Garua (Cameroon), and the Central African Republic. According to Doku and Karikari (1971), Bambara groundnut was found in its wild form in the

Fig. 12.2. Morphology of Bambara groundnut. (A) Plant at pod development stage. (B) Trifoliate leaves with terminal leaf with elongated petiole. (C) Bambara groundnut pod. (D) Bambara groundnut seed. (Author's own figure.)

North Yola province of Nigeria. Cultivated Bambara groundnut may be evolved from these wild landraces which later adapted to semi-arid regions of Africa and became popular with African farmers (Pasquet *et al.*, 1999; Basu *et al.*, 2007a). Begemann (1988) confirmed that the origin of Bambara groundnut lies in the northeastern region of Nigeria and northern Cameroon, a finding consistent with its presence in the North Yola province of Nigeria.

Furthermore, Pasquet (2004) identified wild Bambara groundnut varieties with extreme spreading growth habits, which probably played a role in domestication. Notably, the wild line VSSP11 exhibits a spreading growth trait as a dominant feature, unlike the domesticated bunchy types, as Basu *et al.* (2007a) demonstrated. Gbaguidi *et al.* (2018) showcased significant morphological diversity in Bambara groundnut accessions collected from Benin, underscoring the rich genetic variety within the West African region. However, Aliyu *et al.* (2016) suggested the possibility of a secondary centre of domestication

or diversity for Bambara groundnut in the Southern/Eastern African region based on a thorough analysis of genetic population structure. Their findings classified West African and Southern/Eastern African accessions into distinct clusters, raising the notion of multiple centres of diversity for this crop. Furthermore, Bonny *et al.* (2019) established a clear link between the agro-morphological diversity of Bambara groundnut and its geographical origins, emphasizing the importance of considering regional genetic variations. These collective insights shed light on the intricate history of the genetic diversity and domestication of Bambara groundnut, providing valuable knowledge for improving this vital crop.

Bambara groundnut is a member of the genus *Vigna*, which encompasses other legume crops such as cowpea and mung bean. It is a member of the kingdom Plantae, phylum Angiosperms, class Eudicots, order Fabales, family Fabaceae, subfamily Faboideae, tribe Phaseoleae, the genus *Vigna* and the species *subterranea*.

Fig. 12.3. Genetic variation in seed coat colour in Bambara groundnut. (Figure used with permission.)

The Bambara groundnut belongs to the subgenus *Ceratotropis* of the genus *Vigna*. It has a close link with the common bean and a partial relationship with mung bean (*Vigna radiata*) and adzuki bean (*Vigna angularis*) (Ho *et al.*, 2017), whereas, in terms of mean gene diversity, it is more diverse than mung bean but less than adzuki bean. The taxonomy of *Vigna* is complex, and it has changed over time. The classification of the important domesticated crop species of genus *Vigna* and corresponding primary, secondary and tertiary gene pools were compiled by Nair *et al.* (2023). The information on knowledge of the extent of genetic diversity and evolutionary relationships helps in tracking and collecting traits specific to diverse germplasm.

12.4   Germplasm Collection, Characterization, Evaluation and Conservation

Germplasms are the basic genetic resources that play a crucial role in plant breeding for the development of suitable varieties to achieve food and nutritional security. Merely having a vast number of germplasms may not solve the purpose of the breeding programme; rather, its efficient collection, grouping and characterization enhance the efficiency of germplasm utilization. As mentioned in the above section, it is crucial to understand the crop's domestication and evolutionary lineage for effective germplasm collection. It is strongly believed that the domestication of Bambara groundnut took place in two distinct African regions. The first is considered to be in Western Africa and the other in Southern and Eastern Africa, leading to the emergence of significant but less prominent diversity centres, depending on the selection pressure and evolutionary rate (Rungnoi *et al.*, 2012). Analysis of the significant germplasm collections of Bambara groundnut using passport data and simple sequence repeat (SSR) and DArT markers revealed the presence of two distinct subgroups, indicating the diverse source and origin of accessions. The first subgroup comprises accessions from West and Central Africa, while the second comprises accessions from Southern and Eastern Africa, including Madagascar.

12.4.1 Collections

According to recent data, about 6145 accessions were conserved at various international, regional and national gene banks (Table 12.2). The landraces constituted a major share of germplasm in Bambara groundnut and that of the collections maintained through *ex-situ* conservation by regional and international gene banks. The landraces were mainly collected from the farmers. Therefore, on-farm collection and conservation should be considered important conservation approaches in Bambara groundnut while executing germplasm exploration and conservation.

The International Institute of Tropical Agriculture Genetic Resource Conservation (IITA-GRC), Nigeria, conserves the highest number of Bambara accessions (2031) collected from 25 countries representing the centres of diversity and origin. Thus, it possesses a greater genetic diversity than other gene banks of plant genetic resources (PGRs). With 1416 accessions, ORSTOM in France is the second largest gene bank in terms of the collection of Bambara groundnut accessions with most of them predominantly from Cameroon, which was considered one of the centres of diversity. The data presented in Table 12.2 is not, however, a true representation of the size of the PGRs for Bambara conserved by gene banks owing to the under-representation and duplication of accessions in the collections. Moreover, most of the accession origins are unknown. Furthermore, some national gene banks, such as those in the Republic of South Africa, Botswana, Kenya, Ethiopia, South Sudan and Côte d'Ivoire, reported the number of accessions as less than ten (Majola *et al.*, 2021). Hence, it is essential to update the data and relevant information, including informal collection and conservation by farming and tribal communities, for better exploitation of PGRs. The predominance of landraces in the composition of PGRs and their nomenclature (Tables 12.2 and 12.3), cultivation under marginal lands by small-scale farmers and tribal populations for fulfilling their calorie requirements and for traditional medicinal purposes indicate strong links of this crop with cultural heritage in the remote regions of African countries. This implies the significance of informal collection and on-farm conservation of local cultivars or landraces.

Despite claiming several places as probable centres of origin and diversity (see Section 12.2), the work on tracing the wild species and their relatives is very limited. Among Bambara collections, wild counterparts make up a mere 1% of the holdings at IITA (Muhammad *et al.*, 2022). To diminish the existing collection gap, the IITA-GRC has initiated exploratory efforts to acquire more diversity from centres of origin and diversity, particularly in Cameroon regions (Snook *et al.*, 2011).

The genetic improvement of any crop depends on the richness of genetic diversity, not merely the number, which is assessed through variability at the phenotype and genotype

Table 12.2. Germplasm of Bambara groundnut conserved at various international, regional and national gene banks.

Institute/gene bank	Number of accessions	% of total	WT	LR	BL	AC	OT
IITA-GRC, Nigeria	2031	33	<1	100	UN	UN	UN
ORSTOM, France	1416	23	UN	100	UN	UN	UN
DAR, Botswana	338	6	UN	2	UN	UN	98
PGRRI, Ghana	296	5	UN	UN	UN	UN	100
NPGRC, Tanzania	283	5	<1	81	UN	UN	18
NPGRC, Zambia	232	4	UN	100	UN	UN	UN
Unspecified	1549	25	1	59	9	1	29
Total	6145	100	<1	79	2	<1	18

IITA-GRC, International Institute of Tropical Agriculture Genetic Resource Conservation; ORSTOM, Office de la Recherche Scientifique et Technique; DAR, Department of Agricultural Research; PGRRI, Plant Genetic Resources Research Institute; NPGRC, National Plant Genetic Resource Centre; WT, wild types; LR, landraces; BL, breeding lines; AC, advanced cultivars; OT, others; UN, unspecified number of accessions. Source: Muhammad *et al.* (2022). (Author's own table.)

Table 12.3. List of landraces evaluated for different traits and their special features. (Author's own table.)

Name of landraces, cultivars/accessions	Special features/traits	Reference
Ci12	Rich in essential amino acids, n-6 fatty acids and minerals, mainly Fe	Yao et al. (2015)
Sukabumi and Bogor	Earlier maturity with black testa and produce higher number and weight of pods	Manggung et al. (2016)
AS-17 and S19-3	Short life cycle and hence early maturing	Singh and Basu (2005)
Black Eye, Nav Red and Nav 4 u	Sensitive to heat stress	Berchie et al. (2012)
Burkina	Tolerance to both heat and drought	
Tiga Necuru and DipC	Day-neutral for flowering and photoperiod-sensitive for pod development	Linnemann et al. (1993)
Ankpa 4	Highly photoperiod sensitive for both flowering and pod-set	
Burkina, Mottled Cream, Zebra Coloured, Tan 1 and 2	Produce pods under both 12 h and 14 h photoperiod	Berchie et al. (2013)
Mtwara, Lindi and Shinyanga	Partial photoperiod-insensitive	Jorgensen et al. (2009)
Tolerant: VMDV4, VMD4, VMD7, VMD8 and VMD10 Sensitive: VMDV1, VMD1, VMD2, VMD11 and N'jibawolo	Genotypic variation in symbiotic nitrogen fixation (SNF) for high phosphorus use efficiency (PUE)	Andriamananjara et al. (2013)
Acc 82 and Acc 96	Identified as drought-tolerant genotypes based on drought indices and seed germination and seedling vigour	Kunene et al. (2022)
Acc 199, Acc 25 and Acc 82	Bold seeded with variation for seed coat	Kunene et al. (2022)
TVSU-455	Superiority over other accessions in terms of yield-attributing traits	Esan et al. (2023)

(molecular marker) level. Since the major portion of the germplasm of Bambara groundnut is constituted of landraces evolved from its wild relatives from diverse regions, under the influence of environmental factors, natural mutations and selection pressure, the variability may be underestimated because the population is not genetically stable and expression of some traits will be masked owing to their recessive allelic nature. Hence, selfing the landraces for two to three generations before their characterization and utilization is advisable. Accordingly, Khan et al. (2022) employed the same approach before assessing the genetic diversity of Bambara groundnut accessions collected from the genebank of the Institute of Tropical Agriculture and Food Security (ITAFoS) at Universiti Putra Malaysia (UPM).

Further, the variability was attributed to diverse agroecological and climatic conditions, which were likely to exert very different selective pressures on these populations. Numerous evaluation studies have consistently found that accessions from the northern populations exhibit higher genetic diversity. The accessions from northern African regions typically have a bunch-type growth habit and display lower phenological, vegetative and yield traits compared to other populations. It has been observed that local environmental factors, particularly rainfall and temperature, play a significant role in driving the differentiation of populations from diverse regions. These findings have valuable implications, as they can inform the development of strategies for collecting and conserving genetic diversity and guide

the design of breeding and improvement programmes specifically tailored to local Bambara
groundnut varieties. However, validating such
findings with molecular markers will help confirm results and better understand the diversity
of Bambara groundnut.

Diverse and extensive genetic variability
within PGRs forms the essential foundation for
crop breeding. In the case of Bambara groundnut, this genetic variation is particularly evident
among landraces (Aliyu *et al.*, 2016; Muhammad *et al.*, 2022). The data presented in Table
12.3, along with assessment and reviews of
various accessions, confirm the presence of significant genetic diversity within Bambara
groundnut. Given this rich genetic diversity, leveraging these accessions to identify superior
parental candidates, pairing them to broaden
the genetic variation further, and aligning desirable traits would be prudent. This process can
lead to the development of a selection of elite
cultivars, as suggested by Pasipanodya *et al.* in
2022. In essence, harnessing the genetic diversity within Bambara groundnut can lead to improved varieties with enhanced agronomic and
morphological characteristics, which is essential
for advancing this important crop.

12.4.2 Characterization

The traits associated with high yield (pods/plant,
seed weight, etc.) and cream seed colour, drought
tolerance, nutrition and low cooking time are most
significant in the genetic improvement of Bambara groundnut, and thus, pre-breeding including germplasm characterization should target
these traits. A well-characterized germplasm
collection that includes high-quality genotypic
data is a valuable resource for plant breeding
and crop improvement initiatives. This information can be effectively utilized by scientists and
breeders to identify and select suitable parental
genotypes for their breeding programmes (Aliyu
et al., 2014, 2016). It is worth noting that accessions collected from different countries or even different regions within the same country may share
a common ancestry or represent identical genotypes with distinct local names, as highlighted by
Massawe *et al.* (2005). To address this potential
issue and enhance germplasm management efficiency, a genome-wide genotyping-by-sequencing

approach is being employed at the International
Institute of Tropical Agriculture (IITA) to characterize the Bambara groundnut collection
thoroughly. The primary objective of this effort
is to identify and manage duplicate accessions,
ensuring good quality and diverse germplasm
collection and its utility in breeding and research
programmes.

Bonny *et al.* (2019) characterized the four
Bambara groundnut populations comprising
101 accessions of four agroecological zones in
Côte d'Ivoire using 19 agro-morphological traits
(phenological, vegetative and yield traits). There
was a significant and high phenotypic variability
for traits like leaves per plant, internode length,
canopy spread, growth habit, number of pods per
plant and 100-seed weight. Khan *et al.* (2022)
characterized 44 accessions. They observed a
positive, strong, significant correlation among
yield components, namely fresh pod weight,
100-seed weight, dry pod weight, and dry seed
weight with high heritability and genetic gain.
They shortlisted 30 promising high-yielding
lines (more than 1.8 tons/ha), which can be used
as parental lines in future breeding programmes.
The genetic variability exists for pod weight, seed
size and testa colour/patterns (Table 12.3 and
Fig. 12.3), and the relation between dark and
thicker testa and cooking time was reported. In
many studies, however, only a small number of
landraces were utilized, and these landraces are
often sourced from markets where the origin of
seed is unknown. This lack of information about
seed sources raises the possibility that the genetic
variation for a specific trait exists within each
landrace and can also vary between different
landraces. This situation can make it challenging
to accurately assess and understand the genetic
basis of specific traits within these landraces.
Despite these limitations, there have been efforts
to investigate the genetic basis of traits. For example, one study identified quantitative trait loci
(QTL) on three linkage groups in a cross between
Tiga Nicuru and DipC, involving 65 F_6 lines. It is
important to note, however, that none of these
QTLs was found to be associated with the trait of
testa colour. This highlights the complexity of
trait inheritance and the need for further research to uncover the genetic factors responsible
for specific characteristics in Bambara groundnut, especially when considering landraces obtained from uncertain sources.

In view of climate change and the frequent occurrence of biotic and abiotic stresses, it is essential to consider the traits associated with pest and disease resistance, photoperiod, drought tolerance and adaptation to marginalized soil conditions (nutrient use efficient lines) during the genetic improvement of the low-input and dry-land-based Bambara groundnut. Hence, several researchers attempted to understand the crop and its accession response to some traits, particularly drought, photoperiod and nutrient-use efficiency.

The Bambara groundnut is considered one of the most drought-tolerant crops; however, studies reported up to 50% yield loss due to drought (Mayes *et al.*, 2019). It can produce an average yield of 1.7 t ha^{-1} of dried pods with a range of 1.3–2.1 t ha^{-1}, even under severe drought conditions in African countries. In contrast, most crops, including chickpea and groundnut, failed to produce reasonable yields (Mwale *et al.*, 2007). Further, the physiology of the crop supports good recovery qualities when subjected to drought, indicating the drought tolerance capacity of the crop. Thus, the crop is generally cultivated in arid to semi-arid regions of sub-Saharan Africa. This exemplar crop performs well under drought conditions with higher productivity, mainly due to its higher resource capture and conversion efficiency compared to soybean and cowpea under minimal soil moisture. It is reported that under drought, Bambara groundnut had a higher efficiency of converting water into dry matter (1.65 g kg^{-1}) than most of the grain legumes grown in low rainfall areas, such as lentil (1.37 g kg^{-1}) and chickpea (1.11 g kg^{-1}). Germplasm evaluation indicated that most of the landraces were tolerant to drought (Table 12.3). Some landraces could withstand 120 days of drought when irrigation was stopped 30 days after sowing (Berchie *et al.*, 2012).

The crop requires a photoperiod of around 12 hours of for optimum vegetative and reproductive growth with maximum pod-set and seed yield, while longer photoperiods enhance vegetative growth. The process of reproduction, such as flowering initiation, flowering, pod formation and development, may be retarded at daylengths above 14 h. Embryo abortion and the defective pod development during long photoperiods is primarily caused by the deficiency of, or competition for, photo-assimilates and nutrients among vegetative and reproductive sink organs. Despite being classified as a facultative short-day crop for pod set, many landraces have adapted to regions with a broad range of day lengths. Hence, genetic diversity exists for photoperiod sensitivity and other traits among landraces of diverse origin (Table 12.3). Similarly, genotypic variation for high phosphorus use efficiency (PUE) was observed among germplasms, and variability was attributed to nodule characters like ratio of the nodule and shoot biomass, nodule permeability, nodule respiration and gene phytase expression. The contrasting cultivars and landraces regarding PUE for symbiotic nitrogen fixation were identified (Table 12.3). The compilation of traits-specific varieties/landraces aids in the orientation of the research programme (Boraiah *et al.*, 2021).

12.4.3 Importance of traditional varieties or landraces in Bambara groundnut

Traditional varieties or landraces, carefully selected and maintained by farmers, are invaluable genetic resources for crops. They serve as a form of insurance for growers, particularly as some landraces exhibit resilience to extreme weather conditions, resistance to various pests and diseases, and their ability to provide stable yields even in severe droughts or other stresses. In this context, Bambara groundnut stands as a model crop that highlights the vital role of landraces in ensuring food and nutritional security, especially in the world's driest agroecological regions. The majority of the presently cultivated improved varieties of Bambara groundnut are, in fact, landraces that have been chosen for their capacity to offer better yields, improved seed and flour quality, and enhanced drought tolerance. Typically, Bambara groundnut is grown as a minor crop, often intercropped with staple crops, by small-scale farmers primarily for household consumption. Given the crop's limited current market demand, achieving yield stability is considered more significant than maximizing grain yield. This emphasis on yield stability is crucial to meet the calorie and nutritional requirements of smallholder farmers in the arid and semi-arid tropics. Moreover, the genetic diversity preserved within the Bambara groundnut gene pool opens avenues to developing some of these landrace accessions into high-protein and high-oil-yielding cultivars. This underpins the potential of the crop to play a pivotal role in enhancing nutritional security in a region affected by drought and other abiotic stresses. By concentrating breeding and improvement efforts on promising landraces, it is possible not only to boost the profitability of Bambara groundnut but also to encourage its adoption in various regions where it is cultivated. This, in turn, can lead to improved food security, better nutrition and increased economic benefits for the communities relying on this important crop.

12.4.4 Utilization

Gene banks store numerous accessions of crop species to serve a range of stakeholders, which includes breeders who aim to enhance specific traits through resource utilization. The effective use of PGRs relies on the identification of proper accessions, determined through the assessment of quantitative and qualitative agronomic characteristics, and molecular and biochemical profiles. The large size of non-characterized PGRs demands time and cost for phenotyping, which limits the direct utilization of PGRs in breeding programmes. To encourage the utilization of preserved PGRs, it is essential to implement strategies to improve the utilization of these preserved resources (Pasipanodya *et al.*, 2022). The different strategies for the utilization of germplasm are discussed here.

The basic strategy is narrowing down the germplasm collection to the specific subset of desired numbers of accessions within a species (Frankel, 1984). This approach increases the efficiency of the breeding programme of underutilized crops such as Bambara groundnut by enhancing the effectiveness of germplasm resource management and their utilization through pre-breeding. Core sets play a valuable role in simplifying research endeavours in the fields of systematic and evolutionary biology, the recognition of breeding groups, and the selection of germplasm for introduction into novel ecosystems. However, a breeding programme based on market segmentation of the produce and diverse use of Bambara groundnut demands

trait-specific smaller sub-core sets, comprising approximately 1% of the core collection size. Additionally, small sets of samples enable the cost-effective evaluation of genotypes across multiple environments across the globe, which is crucial for selecting donors or parents for hybridization and introduction in the new area. Alternatively, a comprehensive core set can be formed by amalgamating the mini- and core-collections, which were established through distinct methods utilizing phenotypic traits, molecular markers and ecological data. By merging entries from separate core sets, identified based on either phenotypic or molecular diversity assessments, the likelihood of capturing the maximum genetic diversity within a species in a composite core set is significantly increased compared to when individual approaches are employed (Mahmoodi *et al.*, 2021; Pasipanodya *et al.*, 2022). In Bambara, where efforts to enhance traits are in their nascent stages, specialized core sets for specific traits have not yet been sufficiently developed. There are few documented reports available, as depicted in Table 12.4. Given the limited financial investments in orphan crops, it is essential that endeavours are undertaken to create functional collections with broader adaptability. This will ensure that researchers can readily share and utilize the identified germplasm across various agroecologies or those with similar conditions.

The focused identification of germplasm strategy (FIGS) is another recent approach employed to enhance the use of PGRs in trait-targeted pre-breeding programmes. The FIGS approach is the most suitable and appropriate strategy to enhance the utility of the genetically and geographically more diverse uncharacterized germplasm of Bambara groundnut as it mainly emphasizes the idea of connecting trait information with climatic data, assuming that traits are linked to the prevalent agroclimatic conditions in which they exist. Consequently, it aids in tracing the origins of germplasm accessions and aligning their adaptive characteristics with the environmental factors that influence the selection traits and development of different versions of diverse landraces from their wild relatives through the process of evolution, acclimatization and domestication. Through this approach, it could become feasible to identify their genomic signatures through

Table 12.4. Core sets of Bambara groundnut.

Accessions in the base population/collections (No.)	Accessions in core set (No.)	Traits	Markers/breeding method employed
223	24	Unspecified	AFLP
123	24	Unspecified	SSR
UN	12	Plant architecture, drought and photoperiod	SSD

AFLP, amplified fragment length polymorphism; SSD, single seed descent; SSR, simple sequence repeat; UN, unspecified number of accessions. Source: Pasipanodya *et al.* (2022). CC BY 4.0 DEED.

machine learning predictions. Implementing the FIGS approach in crop enhancement has the potential to introduce germplasm resources into new environments to uncover adaptive traits related to both biotic and abiotic stresses, such as pest resistance, disease resilience, drought tolerance and cold resistance. For instance, in the case of lentils, this methodology has been successfully employed to create a subgroup of lentil varieties suitable for regions anticipated to become 'wetter' or 'drier' because of the impact of climate change (Khazaei *et al.*, 2013). In view of the extent and nature of diversity in Bambara groundnut linked with geographical origin, centre of diversity, sites of collection and on-farm based conservation, the FIGS strategy, along with *de novo* domestication, may help in the exploitation and introgression of domesticated genes to develop new crops or varieties resilient to biotic and abiotic stresses.

12.5 Use of Germplasm in Crop Improvement by Conventional Approaches

The ultimate objective of PGR conservation and management is their better utilization for the genetic improvement of the crop. Despite the availability of diverse PGRs in Bambara groundnut (Table 12.2), they are underexploited in the breeding programme. The selective breeding has played a crucial role in the selection of preferred local varieties of Bambara groundnut that thrive in specific local farming conditions and align with the preferences of various socio-linguistic groups. Investigations into national plant genetic collections and gene bank holdings have unveiled significant diversity in seed size and weight (Table 12.3). The breeding programme should focus on the development of modern Bambara groundnut varieties that possess the ideal agronomic characteristics, such as improved pod filling, high yield and photoperiod neutrality with resistance to biotic and abiotic stresses.

The primary and secondary introductions facilitate the sharing of germplasm for genetic enhancement through varietal development by means of hybridization and selective breeding, besides the scope to introduce the selected agronomically superior accessions directly as varieties in the new areas outside African countries. For instance, in India, an attempt was made at NRCG, Junagadh, to introduce the crop by evaluating the selected landraces of Bambara groundnut, namely, S16-5A, S19-3, SB-4-2, Gabc, DODR-TZ, DODC-TZ, DIPC, Uniswa Red, AS-17 and AHM-753, mainly to diversify the crop and/or stabilize production in rainfed production systems either as the sole crop or as an intercrop with groundnut (Singh and Basu, 2005). Although selected landraces, namely DODR-TZ, Uniswa red and AHM 753 were shortlisted for large-scale production in targeted environments in Andhra Pradesh, Gujarat, Karnataka, Maharashtra and Rajasthan under rainfed conditions, production did not materialize for several reasons, including phytosanitary measures, e.g. risks associated with the transfer of new pests and diseases, and also policy and financial aspects. It is crucial, however, to revive the attempts with scientific insights because the phenology and morphology (Fig. 12.4) of Bambara groundnut are mostly like groundnut and can be fitted well into the existing cropping systems that prevail in the dry lands of India. Consequently, introducing both wild relatives and domesticated varieties can facilitate gene exchange and bolster the expansion

Fig. 12.4. Comparative morphology of Bambara groundnut (left) and groundnut (right). Plants at pod development stage indicating smaller, dwarf and lesser biomass in Bambara groundnut. (Author's own figure.)

of the primary gene pool. This makes *ex-situ* PGRs a valuable means to broaden the genetic foundations for both public and private breeding initiatives. Beyond their role in breeding programmes, plant introductions also hold significant importance in biosystematics, aiding in exploring the origins and evolutionary history of plant species (Pasipanodya *et al.*, 2022).

The globalization phenomenon followed by relaxations in the regulations of sharing PGRs and the potential of the crop in terms of its importance in the food and nutraceutical industry, besides being resilient to abiotic stresses, mainly drought and heat, facilitates the spread of Bambara groundnut germplasm across the globe. Subsequently, Bambara groundnut germplasm has been used in pre-breeding, hybridization and evaluation programmes in different countries, including the UK, India and Malaysia.

12.6   Integration of Genomic and Genetic Resources for Crop Improvement

The utilization of molecular markers and developing genetic resources play a crucial role in enhancing the understanding of the genetic control of complex traits associated with biotic

and abiotic stress tolerance. This knowledge, in turn, facilitates the development of stable and climate-smart crop varieties through integrated molecular-conventional breeding approaches (Boraiah *et al.*, 2023). Although the first genome sequence of Bambara groundnut was released recently (Chang *et al.*, 2019), the absence of a complete genome sequence in Bambara groundnut (Ho *et al.*, 2017) led to the successful mapping of two individual intraspecific genetic linkage maps developed from two populations (IITA686 × Ankpa4 and Tiga Nicuru × DipC) onto the physical maps of other well-annotated, closely related legume species. This approach identified conserved synteny between underutilized Bambara groundnut and well-characterized *Vigna* species, helping to identify the corresponding flanking regions of QTL controlling various agronomic traits in Bambara groundnut. Furthermore, an expression marker-based genetic map was created through cross-hybridization of Bambara groundnut leaf RNA onto the Affymetrix Soybean Genome GeneChip, which helped in identifying QTLs associated with agronomic and drought-related traits in Bambara groundnut, as well as genome regions potentially harbouring pathway-activating genes for the stress response (Chai *et al.*, 2017).

Translating genetic resources and information from models and related species has played a significant role in genetic improvement efforts for underutilized crops such as Bambara groundnut (Bonthala *et al.*, 2016; Chai *et al.*, 2017; Ho *et al.*, 2017; Khan *et al.*, 2017). For example, analysing the Bambara groundnut transcriptome using cross-species gene expression data led to the identification of 375 and 659 differentially expressed genes within 6 and 7 gene modules, respectively, encoding various stimuli and metabolic processes under different temperature regimes and stress conditions (Bonthala *et al.*, 2016). These identified gene modules can potentially be valuable in breeding for low-temperature stress-tolerant Bambara groundnut varieties. Despite numerous efforts to leverage genetic resources from major, model and *Vigna* species for the genetic improvement of Bambara groundnut, progress has not been substantial. The development of within-species genetic resources in Bambara groundnut has only gained momentum in the past decade, primarily because of limited resources and insufficient efforts to understand crop breeding behaviour. The African Orphan Crop Consortium (AOCC) has included Bambara groundnut in its list of 101 traditional African food crops for further improvement in terms of nutritional potential. Genetic diversity in several Bambara groundnut landraces has been evaluated using various markers, including random amplified polymorphic DNA (RAPD), amplified fragment length polymorphism (AFLP), SSR and next-generation markers (Table 12.5). Odesola *et al.* (2023) recently identified QTLs associated with drought tolerance.

Molecular markers aid in assessing the diversity at the gene/nucleotide level, relative inbreeding within lines, and the composition of landraces in Bambara groundnut. While some molecular marker studies have concluded that individual Bambara groundnut plants within landraces are highly inbred, others have revealed that farmer landraces and even released varieties consist of a mixture of related inbred lines (Ho *et al.*, 2016). Genotyping using a diverse set of markers, including 7894 markers such as silicoDArT markers and co-dominant SNP markers, has shown that the range of heterozygous alleles varies from 0.8% to 5.0% after a round of single plant descent (SPD) purification from a subset of gene bank accessions (n = 229). This subset serves as a good representation of 'pure' genotypes, indicating the importance of these markers in assisting conventional breeding approaches. Despite the limited application of markers, there is a continued need to leverage genome-wide association studies (GWAS) in Bambara groundnut to unravel predictive and causative variations linked to agronomic traits. Molecular markers, including AFLP and SSR, have been utilized during hybridization programmes to identify true hybrids before advancing populations to recombinant inbred lines. Bambara groundnut SSR markers have also been used to assess the heterogeneity of commercial cultivars, aiding in the selection of candidate individual plants for whole genome sequencing (Ho *et al.*, 2016). Furthermore, they have been employed for line label verification, especially in large-scale multi-locational trials (Molosiwa *et al.*, 2015).

12.7 Genetic Improvement of Bambara Groundnut

Bambara groundnut breeding programmes should focus on improving grain yield and nutritional quality under prevailing stress conditions, especially drought and high temperatures, considering the crop's geographical adoption and farming systems across the globe. The genetic improvement of Bambara groundnut has primarily been attributed to the domestication of wild forms through natural and artificial selection. This was followed by pure line and multi-line selection from the highly diverse landraces (Table 12.6). However, improving traits such as pod yield or harvest index, particularly under diverse environmental conditions, including drought, has been challenging due to significant genotype × environment interactions (GEI), often resulting in low heritability. Besides pod yield, breeding efforts for Bambara groundnut could also encompass traits such as seed nutrient quality, disease resistance, heat tolerance, high foliage volume and palatability of the foliage for use as a fodder crop. Understanding the trade-offs between pod yield and nutrient and fodder quality traits is crucial for developing multipurpose varieties.

Modern breeding tools and their application in underutilized crops, including Bambara

Table 12.5. Reported molecular markers/QTLs in Bambara groundnut. (Author's own table.)

Type and number of markers used	Type and number of germplasms used	Output/application	References
Seven selective primer combinations generated 504 amplification products, ranging from 50 to 400 bp	16 landraces	Polymorphisms obtained ranged from 68.0 to 98.0%, indicating the variability among landraces and 16 landraces grouped into 3 clusters, linked to their place of collection or origin	Massawe *et al.* (2003a)
16 RAPD primers	12 landraces	Confirmed the polymorphism (63.2% to 88.2%) among landraces	Massawe *et al.* (2003b)
10 SSR markers developed from microsatellite-enriched genomic library	–	First microsatellite markers derived from Bambara groundnut	Basu *et al.* (2007b)
554 DArT markers	124 accessions representing 25 African countries maintained at the International Institute of Tropical Agriculture	Identified important descriptors for germplasm conservation, confirmed the crop's geographical centre of origin	Olukolu *et al.* (2012)
22 SSR markers used	240 accessions	The mean gene diversity of Bambara groundnut was lower than adzuki bean but higher than mung bean	Somta *et al.* (2011)
68 and 201 polymorphic SSR and DArT markers	24 single individual seeds from a core collection of landraces	Seed derived from a single plant represents an unselected variety and identified five genotypes with good yield potential and adaptability	Molosiwa *et al.* (2015); Aliyu and Massawe (2013)
12 SSR markers used	123 accessions	Reported the inbreeding among accessions	Molosiwa (2012)
19 Bambara-specific SSR markers were used	78 accessions	Reported higher genetic diversity (92%) within landraces than between (8%) different landraces	Minnaar-Ontong *et al.* (2021)
Three selected SSR markers	Eight landraces vary in seed coat colour	Black Gresik and Black Madura landraces were different from other Indonesian landraces	Muhammad *et al.* (2020)
12 SSR markers used	105 accessions from Kenya	Assessed the genetic relationships among accessions	Odongo *et al.* (2015)
Genome-wide association mapping using 5927 DArTs loci	100 accessions from 12 African countries	Identified 8 QTLs associated with drought tolerance	Odesola *et al.* (2023)

groundnut, have led to the development of segregating populations, genetic maps and the identification of QTLs, aiding in trait dissection. Therefore, reliable phenotyping protocols and tightly linked markers are essential for incorporating targeted traits through marker-assisted molecular breeding programmes. Despite these advancements, limited progress has been made in varietal development, with most efforts focused on the selection, evaluation and seed multiplication of a few promising landraces. Hence, there is a need to tap into the diverse genetic resources of

Table 12.6. Crop improvement phases in Bambara groundnut. (Author's own table.)

Major developments/ achievements in crop improvement	References
Development of pure lines (1990s–2000)	
Single plant selection for pod and seed numbers for the generation of pure lines. Six lines with high yield potential (352 g/plant) bearing up to 540 mature pods were selected from landraces. 100-seed weight serves as a criterion for the selection of best lines	Wigglesworth (1997)
Between 1997 and 2000, research focused on evaluating the variation within and between Bambara groundnut landraces. The objective was to identify superior plants with specific traits. To achieve this, several pure lines were developed using the single-seed descent method	Massawe (2000)
Evaluation of pure lines under farmers' fields through participatory breeding approach for selection of desired lines with the emphasis more on early maturity, high yield, large pods, sweet taste, fast cooking, spreading growth habit and cream-coloured seeds	Massawe et al. (2005)
Development of multiline/ mix varieties (2000s)	
The landraces growing in Southern Africa are composed of 3 to 8 different genotypes. The selection of lines to form a mixture based on the qualities (drought/heat tolerant, pest and disease resistant) of each line suitable for different agroecological zones	Singrün and Schenkel (2004)
Trait specific breeding (2005 onwards)	
Characterization of landraces in terms of flower number, days to maturity, leaf number, pod development, yield, along with consumer traits (easy cooking and quality) and traits associated with drought tolerance, namely deep rooting, reduction in leaf area and osmotic adjustment aids in identifying the ideotype (Table12.3)	Mwale et al. (2004)
Hybridization (2000s)	
Number of crosses are generated by crossing between cultivated and wild forms based on yield potential, molecular and morphological marker variation, maturity, and growth habit. This aids in the development of improved varieties and mapping populations, including recombinant inbred lines (RILs)	Massawe et al. (2004)
Molecular breeding	
Developed a Bambara groundnut genetic linkage map using two F_2 populations based on molecular and morphological markers to identify potential and promising RILs. The RILs are important genetic resource materials that assist in the identification of the molecular markers linked to QTL controlling traits of interest	Basu et al. (2004); Massawe et al. (2004)
Introduction and participatory plant breeding	
In Swaziland, a participatory plant breeding approach for Bambara groundnut conservation involves growers and consumers selecting their preferred traits from a range of landraces. This method helps preserve landrace genes while aligning with the preferences of those involved	Sesay et al. (2004)
Efforts have been made to introduce the crop by evaluation of 10 selected landraces in India for their suitability to grow in dryland areas where groundnut is already growing	Singh and Basu (2005)

Bambara groundnut, particularly the numerous landraces conserved by small-scale farmers and informal seed banks across Africa. Collaborative research efforts involving partners from Africa and Europe have initiated hybridization and molecular breeding programmes in Bambara groundnut, which may offer opportunities to develop new high-yielding varieties. The available genomic information and the wider genetic diversity in landraces provide opportunities to breed improved varieties by integrating conventional and molecular breeding approaches. However, the breeding approach has been limited to cross-breeding or hybridization, resulting in only a few improved varieties being released. The pedigree structure of Bambara groundnut is quite complex, being composed of a mixture of genotypes and landraces with non-hierarchical populations.

A pre-breeding approach, coupled with the characterization of available germplasm, especially landraces at the molecular and phenotypic levels, may provide the foundation for crop improvement by identifying potential desirable parents. Nevertheless, the high heterozygosity of landraces presents a challenge, as they are often mixtures of genotypes, complicating the understanding of the genetics of targeted traits and genotype-by-environment interactions. Pedigree-based pure line selection is the preferred method to advance breeding and achieve homozygosity, often employing a single seed descent approach. Integrating modern molecular tools has enhanced Bambara groundnut improvement by complementing conventional breeding efforts to develop improved varieties. The availability of the completed genome sequence, diverse molecular markers, linkage maps and QTLs/markers has further advanced the genetic improvement of Bambara groundnut. Specific markers, such as bgPabg-596774, associated with multiple traits have been identified in Bambara groundnut, offering valuable insights for targeted trait improvement. The availability of the draft reference genome of Bambara groundnut in the National Center for Biotechnology Information (NCBI) database has accelerated genetic gains by facilitating the identification of genetic regions associated with traits of interest. The draft genome is 535.05 Mb in size, with predictions of 31,707 protein-coding genes (Chang *et al.*, 2019).

12.8 Cultivation Practices of Bambara Groundnut

12.8.1 Soil and climate

Bambara groundnut is tolerant to drought and has the potential to grow in poor and marginal soils. However, a levelled, well-drained site of medium soil fertility is suitable for its cultivation. Sandy loam soils are considered best for ease of pod development and harvesting. The crop is fast growing and requires around an evenly distributed rainfall of 500–1200 mm to complete its life cycle of 100–150 days. Some early cultivars may mature in 50 days. The optimum temperature for seed germination is between 30–35°C. Although the crop is resistant to high temperatures, very high temperatures of >38°C during the growing season negatively impact crop growth and development (FAO, 2007). The crop needs at least four months of a frost-free environment. The optimum soil pH for Bambara groundnut cultivation is 5.0–6.5, and a pH of < 4.3 and > 7.0 is unsuitable for its cultivation (FAO, 2007). Bambara needs a day length <14 h for floral initiation, flowering, pod initiation and development (Linneman, 1993).

12.8.2 Land preparation

Bambara groundnut needs a well-prepared levelled land as the pods are formed underground. A deep ploughing followed by two harrowings and levelling is required to get a good tilth, which helps pod development and penetration.

12.8.3 Seed sowing

Due to the hard seed coat, seed priming by soaking seeds in water overnight, followed by drying before sowing, is an important practice in Bambara groundnut for improved germination, seedling establishment and vigour. Seeds at 25–75 kg ha[-1] should be sown 3–5 cm deep with a spacing of 30–60 cm × 10–20 cm recommended. The lower seed rate is usually when the crop is intercropped with maize, millets, sorghum, etc., while the higher rate is for sole cropping. While growing in a new region, efficient rhizobia should also be applied as seed treatment to ensure biological nitrogen fixation.

12.8.4 Intercultural operations

Two hoeing and weeding sessions are recommended for good crop production of Bambara groundnut, the first, three weeks after sowing and the second at flowering. Manual weeding is generally practiced; however, the crop responds well to herbicide applications. Earthing up is an essential cultural practice in Bambara groundnut, which should be done after flowering for better pod formation and higher yield.

12.8.5   Fertilizer management

As a leguminous crop, Bambara groundnut primarily relies on natural nitrogen fixation for its nitrogen needs. However, an initial application of nitrogen and phosphorus is recommended to promote improved root growth and nodulation. Typically, a fertilizer dosage of 25 kg nitrogen, 50–80 kg phosphorus, and 30–40 kg potash per hectare is necessary for achieving a good yield. Different studies have reported varying fertilizer recommendations for Bambara groundnut. For instance, in south-east Nigeria, using 165 kg/ha of single superphosphate fertilizer resulted in improved growth and yield. In north-east Nigeria, an application rate of 60 kg/ha of a single superphosphate was suggested. In the central region of Cameroon, an increase in growth and yield was observed by using 150 kg/ha of single superphosphate. Additionally, applying organic manure at a rate of 8–10 t/ha has shown positive effects on Bambara groundnut growth and yield. These recommendations indicate that the appropriate fertilizer dosage can vary based on the specific growing conditions and regional factors. Applying efficient strains of rhizobia enhances growth and nodulation in Bambara groundnut (Fig. 12.5). Inoculation of efficient strains of Bambara groundnut rhizobia, namely PAS17-2 and PUR 3, resulted in enhancement in the root and shoot length, plant biomass and pod yield over the uninoculated control with cultivars SB-4-2, S-19-3 and DODR at NRCG, Junagadh, India (K.K. Pal, unpublished data).

12.8.6   Harvesting and post-harvest management

The crop is ready for harvest when the leaves start yellowing and fall down, and harvesting is by pulling or uprooting, followed by removing the pods. After harvesting, the pods are sun-dried to

Fig. 12.5. Effect of different rhizobia strains on growth and nodulation in Bambara groundnut. (Author's own figure.)

a 12–14% moisture level, followed by their shelling either by hand or mechanical shellers and safe storage.

12.9 Conclusion

Bambara groundnut, a leguminous crop, emerges as a prime choice for crop diversification, offering a path towards sustainable food production, zero hunger and alleviating malnutrition, especially in African countries grappling with abiotic stresses and malnutrition. Notably, it exhibits resilience to climate variability. It is nitrogen-fixing, making it suitable for challenging environmental conditions of the tropics and sub-tropics, especially in areas affected by drought, degraded soil conditions and high temperatures. Bambara groundnut is a complete food and repository of vital nutrients. It is rich in carbohydrates, crucial amino acids, proteins, minerals and vitamins. Additionally, this crop possesses unique nutraceutical and nutritional attributes, such as biopolymers and starch-soluble dietary fibre nanocomposites, with promising applications in the global food and pharmaceutical industries. Its extensive genetic diversity, characterized by diverse landraces and wild variants, underscores its potential for enhancing

yield and quality through breeding programmes. Nevertheless, the scientific community has encountered hurdles in selecting and incorporating desired agronomic, nutritional and processing traits to achieve genetic gains. Initiatives have been undertaken to improve the landraces and develop modern, genetically uniform varieties with the aim of augmenting productivity in regions with established agricultural practices and expanding their cultivation to previously unexplored areas worldwide. Leveraging molecular tools, including markers and a few QTLs, is essential to unlock the genetic potential of these diverse landraces, ultimately leading to high-yielding varieties and eliminating antinutritional factors. Furthermore, the versatility of the genetic diversity found in Bambara groundnut in response to different photoperiods enables it to extend its geographical reach beyond its native regions, making it a valuable asset in the context of changing climate scenarios, which can be cultivated as well as alternative crops for future food and nutritional security.

Acknowledgements

The authors thank ICAR-NIASM for the in-house project 'Alternative crops'.

References

Ahmad, N.S., Basu, S.M., Redjeki, E.S., Murchie, E., Massawe, F. *et al.* (2013) Developing genetic mapping and marker-assisted breeding techniques in Bambara groundnut (*Vigna subterranea* L.). *Acta Horticulturae* 979, 437–450.
Aliyu, S. and Massawe, F.J. (2013) Microsatellites based marker molecular analysis of Ghanaian bambara groundnut (*Vigna subterranea* (L.) Verdc.) landraces alongside morphological characterisation. *Genetic Resources and Crop Evolution* 60(3), 777–787. DOI: 10.1007/s10722-012-9874-y
Aliyu, S., Massawe, F. and Mayes, S. (2016) Genetic diversity and population structure of Bambara groundnut (*Vigna subterranea* (L.) Verdc.): Synopsis of the past two decades of analysis and implications for crop improvement programmes. *Genetic Resources and Crop Evolution* 63(5), 925–943.
Andriamananjara, A., Rabeharisoa, L., Abdou, M.M., Malam, M., Masse, D. *et al.* (2013) Genotypic variation in phosphorus use efficiency for symbiotic nitrogen fixation in Voandzou (*Vigna subterranea*). *Optimizing Productivity of Food Crop Genotypes in Low Nutrient Soils*, 1721, IAEA International Atomic Energy Agency, 2013, IAEA-TECDOC series, IAEA.
Basu, S.M., Roberts, J.A., Davey, M.R., Azam-Ali, S.N. and Mithen, F.R. (2004) Towards genetic linkage mapping in bambara groundnut (*Vigna subterranea* (L.) Verdc.). *Proceedings of the International Symposium on Bambara Groundnut*, Botswana College of Agriculture, 8–12 September 2003.
Basu, S., Mayes, S., Davey, M., Roberts, J.A., Azam-Ali, S.N., Mithen, R. and Pasquet, R.S. (2007a) Inheritance of 'domestication' traits in bambara groundnut (*Vigna subterranea* (L.) Verdc.). *Euphytica* 157, 59–68.

Basu, S., Roberts, J.A., Azam-Ali, S.N. and Mayes, S. (2007b) Development of microsatellite markers for bambara groundnut (*Vigna subterranea* (L.) Verdc.)—an underutilised African legume crop species. *Molecular Ecology Notes* 7, 1326–1328. DOI: 10.1111/j.1471-8286.2007.01870.x

Begemann, F. (1988) Ecogeographic differentiation of bambara groundnut (*Vigna subterranea*) in the collection of the International Institute of Tropical Agriculture (IITA). Wissenschaftlicher Fachverlag Dr Fleck, Niederkleen.

Bonny, B.S., Adjoumani, K., Seka, D., Koffi, K.G., Kouonon, L.C., Koffi, K.K. and Zoro Bi, I.A. (2019) Agro morphological divergence among four agroecological populations of Bambara groundnut (*Vigna subterranea* (L.) Verdc.) in Côte d'Ivoire. *Annals of Agricultural Sciences* 64(1), 103–111. DOI: 10.1016/j.aoas.2019.04.001

Berchie, J.N., Opoku, M., Adu-Dapaah, H., Agyemang, A., Sarkodie-Addo, J. *et al.* (2012) Evaluation of five bambara groundnut (*Vigna subterranea* (L.) Verdc.) landraces to heat and drought stress at Tono-Navrongo, Upper East Region of Ghana. *African Agricultural Research* 7(2), 250–256.

Berchie, J.N., Amelie, G., McClymont, S., Raizada, M., Adu-Dapaah, H. and Sarkodie-Addo, J. (2013) Performance of 13 bambara groundnut (*Vigna subterranea* (L.) Verdc.) landraces under 12 h and 14 h photoperiod. *Journal of Agronomy* 12(1), 20–28.

Bonthala, V.S., Mayes, K., Moreton, J., Blythe, M., Wright, V. *et al.* (2016). Identification of gene modules associated with low temperatures response in bambara groundnut by network-based analysis. *PLoS ONE* 11(2), e0148771. DOI: 10.1371/journal. pone.0148771

Boraiah, K.M., Basavaraj, P.S., Harisha, C.B., Kochewad, S.A., Khapte, P.S. *et al.* (2021) Abiotic stress tolerant crop varieties, livestock breeds and fish species. In: *Technical Bulletin No. 32*. ICAR-National Institute of Abiotic Stress Management, Baramati, Pune, Maharashtra, India.

Boraiah, K.M., Basavaraj, P.S., Kakade, V.D., Harisha, C.B., Khapte, P. *et al.* (2023) Abiotic stress-tolerant crop varieties in India: status and a way forward. *Updates in Plant Breeding*. IntechOpen. DOI:10.5772/intechopen.1001916

Brough, S.H., Azam-Ali, S.N. and Taylor, A.J. (1993) The potential of bambara groundnut (*Vigna subterranea*) in vegetable milk production and basic protein functionality systems. *Food Chemistry* 47, 277–283.

Chai, H.H., Ho, W.K., Graham, N., May, S., Massawe, F. and Mayes, S. (2017) A cross-species gene expression marker-based genetic map and QTL analysis in Bambara groundnut. *Genes* 8, 84. DOI: 10.3390/genes8020084

Chang, Y., Liu, H., Liu, M., Liao, X., Sahu, S.K. *et al.* (2019) The draft genomes of five agriculturally important African orphan crops. *GigaScience* 8, 152.

Doku, E.V. and Karikari, S.K. (1971) Operational selection in wild bambara groundnut. *Ghana Journal of Science* 11, 45–56.

Esan, V.I., Oke, G.O. and Ogunbode, T.O. (2023) Genetic variation and characterisation of Bambara groundnut (*Vigna subterranea* (L.) Verdc.) accessions under multi-environments considering yield and yield components performance. *Scientific Reports* 13(1), 1498. DOI: 10.1038/s41598-023-28794-8

FAO (2007) Data sheet *Vigna subterranea*. Ecocrop. Food and Agricultural Organization, Rome.

FAO (2022) *World Food and Agriculture – Statistical Yearbook 2022*. Rome. DOI: 10.4060/cc2211en

Frankel, O.H. (1984) Genetic perspective of germplasm conservation. In: Arber, W., Limensee, K., Peacock, W.J. and Stralinger, P. (eds) *Genetic Manipulations: Impact on Man and Society*. Cambridge University Press, Cambridge, UK, pp. 161–170.

Gbaguidi, A.A., Dansi, A., Dossou-Aminon, I., Gbemavo, D.S.J.C., Orobiyi, A., Sanoussi, F. and Yedomonhan, H. (2018) Agromorphological diversity of local bambara groundnut (*Vigna subterranea* (L.) Verdc.) collected in Benin. *Genetic Resources and Crop Evolution* 65, 1159–1171. DOI: 10.1007/s10722-017-0603-4

Goli, A.E. (1997) Bambara groundnut (*Vigna subterranea* (L) Verdc.). In: Heller, J., Begemann, F. and Mushonda, J. (eds) *Promoting the Conservation and Use of Underutilised and Neglected Crops* (Vol. 9, p. 167). International Plant Genetic Resources Institute, Rome.

Hepper, F.N. (1963) Plants of the 1957-58 West African expedition: II. The Bambara groundnut (*Voandzeia subterranea*) and Kersting's groundnut (*Kerstingiella geocarpa*) wild in West Africa. *Kew Bulletin* 16(3), 395–407.

Ho, W.K., Muchugi, A., Muthemba, S., Kariba, R., Mavenkeni, B.O. *et al.* (2016) Use of microsatellite markers for the assessment of bambara groundnut breeding system and varietal purity before genome sequencing. *Genome* 59, 427–431. DOI: 10.1139/gen-2016-0029

Ho, W.K., Chai, H.H., Kendabie, P., Ahmad, N.S., Jani, J. *et al.* (2017) Integrating genetic maps in bambara groundnut [*Vigna subterranea* (L) Verdc.] and their syntenic relationships among closely related legumes. *BMC Genomics* 18, 192.

Holm, J.M. and Marloth, B.W. (1940) The Bambara groundnut or njugo bean. Bulletin 215, South African Agricultural Technical Services, Government Printer, Pretoria.

Jorgensen, S.T., Aubanton, M., Harmonic, C., Dieryck, C., Jacobsen, S. *et al.* (2009) Identification of photo-period neutral lines of bambara groundnut (*Vigna subterranea*) from Tanzania. *IOP Conference Series: Earth and Environmental Science* 6, 2023.

Khan, F., Chai, H.H., Ajmera, I., Hodgman, C., Mayes, S. and Lu, C. (2017) A transcriptomic comparison of two bambara groundnut landraces under dehydration stress. *Genes* 8(4), 1–19. DOI: 10.3390/genes8040121

Khan, M.M.H., Rafii, M.Y., Ramlee, S. I., Jusoh, M. and Mamun, M.A. (2021) Genetic analysis and selection of Bambara groundnut (*Vigna subterranea* [L.] Verdc.) landraces for high yield revealed by qualitative and quantitative traits. *Scientific Reports* 11, 7597. DOI: 10.1038/s41598-021-87039-8

Khan, M.M.H., Rafii, M.Y., Ramlee, S.I., Jusoh, M., Oladosu, Y., Al Mamun, M. and Khaliqi, A. (2022) Unveiling genetic diversity, characterization, and selection of bambara groundnut (*Vigna subterranea* L. Verdc) genotypes reflecting yield and yield components in tropical Malaysia. *BioMed Research International* 2022, 6794475. DOI: 10.1155/2022/6794475

Khazaei, H., Street, K., Bari, A., Mackay, M. and Stoddard, F.L. (2013) The FIGS (Focused Identification of Germplasm Strategy) approach identifies traits related to drought adaptation in *Vicia faba* genetic resources. *PLoS ONE* 8, e63107.

Kunene, S., Odindo, A.O., Gerrano, A.S. and Mandizvo, T. (2022) Screening bambara groundnut (*Vigna subterranea* L. Verdc) genotypes for drought tolerance at the germination stage under simulated drought conditions. *Plants* 11, 3562. DOI: 10.3390/plants11243562

Linnemann, A.R. (1987) Bambara groundnut (*Vigna subterranea* (L.) Verdc.): a review. *Abstracts of Tropical Agriculture* 12(7), 9–25.

Linnemann, A.R. (1988) Cultivation of bambara groundnut (*Vigna subterranea* (L.) Verdc.) in Northern Nigeria. Report of a field study. Wageningen Agricultural University, Wageningen, the Netherlands.

Linneman, A.R. (1993) Phenological development in Bambara groundnut (*Vigna subterranea*) at constant exposure to photoperiods of 10-16h. *Annals of Botany* 71, 445–452.

Mohammed, M.S., Shimelis, H.A., Laing, M.D. and Usman, A. (2023) Progress and opportunities on Bambara groundnut (*Vigna subterranea* [L.] Verdc.): genetic improvement and product development. In: Farooq, M. and Siddique, K.H.M. (eds) *Neglected and Underutilised Crops*. Academic Press, Elsevier, London, pp. 617–645. DOI: 10.1016/B978-0-323-90537-4.00013-2

Mabhaudhi, T., Chibarabada, P., Chimonyo, V.G.P. and Modi, A.T. (2018) Modelling climate change impact: A case of bambara groundnut (*Vigna subterranea*). *Physics and Chemistry of the Earth, Parts A/B/C* 105, 25–31.

Mahmoodi, R., Dadpour, M.R., Hassani, D., Zeinalabedini, M., Vendramin, E. and Leslie, C.A. (2021) Composite core set construction and diversity analysis of Iranian walnut germplasm using molecular markers and phenotypic traits. *PLoS ONE* 16, e0248623.

Majola, N., Gerrano, A. and Shimelis, H. (2021) Bambara groundnut (*Vigna subterranea* [L.] Verdc.) production, utilisation and genetic improvement in Sub-Saharan Africa. *Agronomy* 11, 1345.

Manggung, R.E.R., Qadir, A. and Ilyas, S. (2016) Fenologi, Morfologi, dan Hasil Empat Aksesi Kacang Bambara (*Vigna subterranea* (L.) Verdc.). *Indonesian Journal of Agronomy* 44(1), 47–54. DOI: 10.24831/jai.v44i1.12492

Maphosa, Y., Jideani, V.A. and Maphosa, L. (2022a) Bambara groundnut production, grain composition and nutritional value: opportunities for improvements. *The Journal of Agricultural Science* 160, 448–458. DOI: 10.1017/S0021859622000521

Maphosa, Y., Jideani, V.A. and Ikhu-Omoregbe, D.I. (2022b) Novel *Vigna subterranea* (L.) Verdc soluble dietary fibre-starch nanocomposite: functional and antioxidant characteristics. *Food Technology and Biotechnology* 60(3), 361–374. DOI: 10.17113/ftb.60.03.22.7409

Massawe, F.J. (2000) Phenotypic and genetic diversity in bambara groundnut (*Vigna subterranea* (L.) Verdc) landraces. PhD thesis, University of Nottingham, UK.

Massawe, F., Dickinson, M., Roberts, J. and Azam-Ali, S. (2003a) Genetic diversity in bambara groundnut (*Vigna subterranea* (L.) Verdc) landraces revealed by AFLP markers. *Genome* 45, 1175–1180. DOI: 10.1139/g02-093

Massawe, F., Roberts, J., Azam-Ali, S. and Davey, M.R. (2003b) Genetic diversity in bambara groundnut (*Vigna subterranea* (L.) Verdc) landraces assessed by random amplified polymorphic DNA (RAPD) markers. *Genetic Resources and Crop Evolution* 50, 737–741. DOI: 10.1023/a:1025041301787

Massawe, F.J., Schenkel, W., Basu, S. and Temba, E.M. (2004) Artificial hybridisation in bambara groundnut (*Vigna subterranea* (L.) Verdc.). In: *Proceedings of the International Symposium on Bambara Groundnut*, 8–12 September 2003, Botswana College of Agriculture, Botswana.

Massawe, F.J., Mwale, S.S., Azam-Ali, S.N. and Roberts, J.A. (2005) Breeding in bambara groundnut (*Vigna subterranea* (L.) Verdc.): strategic considerations. *African Journal of Biotechnology* 4(6), 463–471.

Mayes, S., Ho, W.K., Chai, H.H., Gao, X., Kundy, A.C. *et al.* (2019) Bambara groundnut: an exemplar underutilised legume for resilience under climate change. *Planta* 250(3), 803–820. DOI: 1007/s00425-019-03191-6

Minnaar-Ontong, A., Gerrano, A.S. and Labuschagne, M.T. (2021) Assessment of genetic diversity and structure of Bambara groundnut [*Vigna subterranea* (L.) Verdc.] landraces in South Africa. *Scientific Reports* 11(1), 7408. DOI: 10.1038/s41598-021-86977-7

Molosiwa, O.O. (2012) Genetic diversity and population structure analysis of bambara groundnuts (*Vigna subterranea* (L.) Verdc.) landraces using morpho-agronomic characters and SSR markers. PhD thesis, University of Nottingham.

Molosiwa, O.O., Aliyu, S., Stadler, F., Mayes, K., Massawe, F., Kilian, A. and Mayes, S. (2015) SSR marker development, genetic diversity and population structure analysis of Bambara groundnut (*Vigna subterranea* (L.) Verdc.) landraces. *Genetic Resources and Crop Evolution* 62(7), 1225–1243. DOI: 10.1007/s10722-015-0226-6

Muhammad, F.F.A., Setiawan, A., Ilyas, S. and Ho, W.K. (2020) Genetic variability of Indonesian landraces of *Vigna subterranea*: Morphological characteristics and molecular analysis using SSR markers. *Biodiversitas* 21(9), 3929–3937. DOI: 10.13057/biodiv/d210902

Muhammad, I., Rafii, M.Y., Ramlee, S.I., Nazli, M.H., Harun, A.R. *et al.* (2022) Exploration of bambara groundnut (*Vigna subterranea* (L.) Verdc, an underutilized crop, to aid global food security: varietal improvement, genetic diversity and processing. *Agronomy* 10, 766.

Musango, R., Pasipanodya, J.T., Tamado, T., Mabasa, S. and Makaza, W. (2022) *Alectra vogelii*: a threat to bambara groundnut production under climate change: a review paper. *Journal of Agricultural Chemistry and Environment* 11, 83–105. DOI: 10.4236/jacen.2022.112006

Mwale, S.S., Azam-Ali, S.N., Massawe, F.J. and Roberts, J.A. (2004) Effect of soil moisture on growth and development of bambara groundnut (*Vigna subterranea* (L.) Verdc.). In: *Proceedings of the International Symposium on Bambara Groundnut*, 8–12 September 2003, Botswana College of Agriculture, Botswana.

Mwale, S.S., Azam-Ali, S.N. and Massawe, F.J. (2007) Growth and development of bambara groundnut (*Vigna subterranea*) in response to soil moisture: 1. Dry matter and yield. *European Journal of Agronomy* 26(4), 345–353.

Nair, R.M., Pujar, M., Cockel, C., Scheldeman, X., Vandelook, F. *et al.* (2023) *Global Strategy for the Conservation and Use of Vigna*. Global Crop Diversity Trust, Bonn, Germany. DOI: 10.5281/zenodo.7565174

Odesola, K.A., Olawuyi, O.J., Paliwal, R., Oyatomi, O.A. and Abberton, M.T. (2023) Genome-wide association analysis of phenotypic traits in Bambara groundnut under drought-stressed and non-stressed conditions based on DArTseq SNP. *Frontiers in Plant Science* 14, 1104417. DOI: 10.3389/fpls.2023.1104417

Odongo, F.O., Oyoo, M.E., Wasike, V., Owuoche, J.O., Karanja, L. and Korir, P. (2015) Genetic diversity of Bambara groundnut (*Vigna subterranea* (L.) Verdc.) landraces in Kenya using microsatellite markers. *African Journal of Biotechnology* 14(4), 283–291.

Ogwu, M.C., Ahana, C.M. and Osawaru, M.E. (2018) Sustainable food production in Nigeria: A case study for bambara groundnut (*Vigna subterranea* (L.) Verdc. Fabaceae). *Journal of Energy and Natural Resource Management (JENRM)* 1(1). DOI: 10.26796/jenrm.v1i1.125

Olukolu, B.A., Mayes, S., Stadler, F., Ng, N.Q., Fawole, I. *et al.* (2012) Genetic diversity in Bambara groundnut (*Vigna subterranea* (L.) Verdc.) as revealed by phenotypic descriptors and DArT marker analysis. *Genetic Resources and Crop Evolution* 59(2), 347–358. DOI: 10.1007/s10722-011-9686-5

Paliwal, R., Adegboyega, T.T., Abberton, M., Faloye, B. and Oyatomi, O. (2021) Potential of genomics for the improvement of underutilized legumes in sub-Saharan Africa. *Legume Science* 3(3), e69.

Pasipanodya, J.T., Horn, L.N., Achigan-Dako, E.G., Musango, R. and Sibiya, J. (2022) Utilisation of plant genetic resources of bambara groundnut conserved ex situ and genetic diversification of its primary genepool for semi-arid production. *Agriculture* 12, 492.

Pasquet, R.S. (2004) Bambara groundnut and cowpea gene pool organisation and domestication. In: *Conference Proceedings of the International Bambara Groundnut Symposium*, 5–12 September 2003, Gaborone, Botswana, pp. 265–274.

Pasquet, R.S., Schwedes, S. and Gepts, P. (1999) Isozyme diversity in Bambara groundnut. *Crop Science* 39(4), 1228–1236.

Rungnoi, O.-U., Suwanprasert, J., Somta, P. and Srinives, P. (2012) Molecular genetic diversity of bambara groundnut (*Vigna subterranea* L. Verdc.) revealed by RAPD and ISSR marker analysis. *SABRAO Journal of Breeding and Genetics* 44, 87–101.

Sesay, A., Edje, O.T. and Magagula, C.N. (2004) Working with farmers on the bambara groundnut research project in Swaziland. In: *Proceedings of the International Symposium on Bambara Groundnut*, 8–12 September 2003, Botswana College of Agriculture, Botswana.

Shareef, I., Sparkes, D. and Azam-Ali, S. (2014) Temperature and drought stress effects on growth and development of bambara groundnut (*Vigna subterranea* L.). *Experimental Agriculture* 50, 72–89. DOI: 10.1017/S0014479713000379

Singh, A.L. and Basu, M.S. (2005) Bambara groundnut: its physiology and introduction in India. In: Trivedi, P.C. (ed.) *Advances in Plant Physiology*. I.K. International Publishing House Pvt. Ltd. New Delhi, India, pp. 235–249.

Singrün, C. and Schenkel, W. (2004) Fingerprinting of bambara groundnut germplasm with molecular markers. In: *Proceedings of the International Symposium on Bambara Groundnut*, 8–12 September 2003, Botswana College of Agriculture, Botswana.

Smýkal, P., Coyne, C., Redden, R. and Maxted, N. (2013) *Genetic and Genomic Resources of Grain Legume Improvement*. Elsevier, Amsterdam, The Netherlands, pp. 41–80.

Snook, L.K., Dulloo, M.E., Jarvis, A., Scheldeman, X. and Kneller, M. (2011) Crop germplasm diversity: the role of gene bank collections in facilitating adaptation to climate change. *Crop Adaptation to Climate Change* 1, 495–506.

Somta, P., Chankaew, S., Rungnoi, O. and Srinives, P. (2011) Genetic diversity of the bambara groundnut (*Vigna subterranea* (L.) Verdc.) as assessed by SSR markers. *Genome* 54, 898–910. DOI: 10.1139/g11-056

Soumare, A., Diedhiou, A.G. and Kane, A. (2022) Bambara groundnut: a neglected and underutilised climate-resilient crop with great potential to alleviate food insecurity in sub-Saharan Africa. *Journal of Crop Improvement* 36(5), 747–767. DOI: 10.1080/15427528.2021.2000908

Tan, X.L., Azam-Ali, S., Goh, E.V., Mustafa, M., Chai, H.H. *et al.* (2020) Bambara groundnut: An underutilized leguminous crop for global food security and nutrition. *Frontiers in Nutrition* 7, 601496.

Wigglesworth, D.J. (1997) The potential for genetic improvement of bambara groundnut (*Vigna subterranea* L. Verdc.) in Botswana. *Proceedings of the International Symposium on Bambara Groundnut*, 23–25 July 1996, University of Nottingham, UK, pp. 181–191.

Yao, D.N., Kouassi, K.N., Erba, D., Scazzina, F., Pellegrini, N. and Casiraghi, M.C. (2015) Nutritive evaluation of the bambara groundnut Ci12 landrace [*Vigna subterranea* (L.) Verdc. (Fabaceae)] produced in Côte d'Ivoire. *International Journal of Molecular Sciences* 16(9), 21428–21441. DOI: 10.3390/ijms160921428

13 Winged Bean (*Psophocarpus tetragonolobus*)

Arpita Mahobia[1], Pravin Vishwanathrao Jadhav[1]*, Umesh Dnyaneshwar Shinde[1], Gopika Krishna Mote[1], Sagar Laxman Zanjal[1], Balwant Sayasrao Mundhe[1], Aditya Vishnudas Rathod[1], Elena Alexandrovna Torop[2], Sanjay Bapu Sakhare[1], Rameshwar Baliram Ghorade[1], Rahul Chandora[3] and Raviprakash Govindrao Dani[4]

[1]*Dr Panjabrao Deshmukh Krishi Vidyapeeth, Akola Maharashtra, India;* [2]*Voronezh State Agrarian University, Voronezh, St. Michurina, Russia;* [3]*ICAR-National Bureau of Plant Genetic Resources, Regional Station, Shimla, Himachal Pradesh, India;* [4]*Gen Scan Inc. Global Consultancy, Houston, Texas, USA and Namangan Engineering and Technology Institute, Namangan, Uzbekistan*

Abstract

Winged bean (*Psophocarpus tetragonolobus*) is an underutilized legume with great promise for widespread cultivation in the humid tropics, offering yields and nutritional value similar to soybean. It is a valuable substitute for soybeans in tropical regions worldwide, with nearly all parts of the plant being edible when properly prepared. Challenges arise, however, from the presence of antinutritional factors and the high costs associated with the indeterminate growth pattern of winged bean. Limited attention to its genomic study and germplasm characterization has hindered breeding efforts. In this chapter, we analyse the reasons for this and offer techniques for making better use of its genetic resources and allied *Psophocarpus* species in producing superior variations within the orphan legume. Employing genomics and breeding techniques, such as RNAi, marker assisted backcross breeding (MABB), transcriptomics, and CRISPR/Cas9, holds immense potential for developing disease-resistant, early maturing, high-yielding and determinate varieties of winged bean, as well as creating consumer-friendly value-added products from winged bean. A focused and organized research strategy is recommended to unlock the full potential of winged bean and improve global food security.

Keywords: Winged bean, underutilized legume, anti-nutritional factors, genetic resources, genomics, transcriptomics, CRISPR/Cas9

13.1 Introduction

During the course of the 20th century, there has been a significant reduction in the diversity of crops and species that humans consume despite the wide variety of crops that have been grown historically all over the world. Only three major cereal crops – rice, wheat, and maize – are extensively cultivated in monocultures and account for more than 50% of our food supply. Owing to

*Corresponding author: jpraveen26@yahoo.co.in

a lack of adequate agrodiversity, farming systems for food production are more susceptible to breaking down under more stressful conditions (such as drought, pests and diseases) (FAO 2010, 2015; Edreira *et al.*, 2011; Siebert *et al.*, 2017; Zampieri *et al.*, 2017). As a result, food production systems are less likely to produce enough food to meet demand.

In addition, a lack of genetic diversity makes it more challenging to use advantageous gene combinations in plant breeding, which hinders the creation of enhanced varieties of orphan crops such as winged bean. Agrobiodiversity, which is defined as the number of various crop species in a system, is also increasingly valued for its relationship to sustainability, socio-economic resilience, and improved human health (Chappell and LaValle, 2011; Adhikari *et al.*, 2014, 2017; Khoury *et al.*, 2014; Pellegrini and Tasciotti, 2014; Dwivedi *et al.*, 2017). Neglected and underutilized crop species have recently gained widespread attention due to their nutritional value in diet, ability to retain and promote agrobiodiversity, and withstand or resist abiotic and biotic stresses (FAO, 2010, 2015; Mayes *et al.*, 2012; Padulosi *et al.*, 2013; Chivenge *et al.*, 2015; Massawe *et al.*, 2016; Adhikari *et al.*, 2017).

The US National Academy of Sciences (NAS) formed a select committee in 1974 to carry out 'an extensive survey of underexploited tropical plants' as prospective crops for the future. Winged bean (*Psophocarpus tetragonolobus* L.), one of the 36 species they carefully evaluated, was picked out for special consideration and was promoted to the community of agricultural researchers (NAS, 1975). This underutilized leguminous species is grown mainly in hot, humid climates for its mature seeds, leaves and green pods. A tropical herbaceous legume plant, winged bean is known by many common names, including cigarillas, goa bean, four-angled bean, four-cornered bean, manila bean, princess bean, asparagus pea and dragon bean. Its genome is estimated to be 1.22 Gb in size and has a diploid structure (2n = 2x = 18) (Vatanparast *et al.*, 2016).

Compared to soybean, winged bean has a higher yield potential and greater nutritional value (Vatanparast *et al.*, 2016). It is a less popular tropical legume, cultivated mainly in South-east Asia and Malaysia (Lim, 2012). Due to its well-balanced amino acid make-up, excellent protein bioavailability and comparatively low amounts of antinutritional factors, winged bean

seeds are regarded as a good source of dietary protein (Wan Mohtar *et al.*, 2014).

Despite the nutritional potential of winged bean, minimal effort is invested into utilizing it because of its lengthy maturation period, indeterminate growth habit, low seed yield, need for a stake to support the vigorously growing vines, and presence of antinutritive factors such as chymotrypsin inhibitors, hemagglutinins and trypsin inhibitors (Lepcha *et al.*, 2017). In addition, there is limited information about the genetic diversity among the available winged bean germplasm (Laosatit *et al.*, 2022). Plant breeders need knowledge of genetic variability to create efficient methods and to provide details on heterotic pairings in crop development programmes. For their protection and, consequently, their use in breeding efforts, it is essential to comprehend the genetic diversity of the existing landraces from distinct geographic regions (Sriwichai, 2022). However, winged beans could be explored to be included in mainstream cultivation and consumption practices owing to their superior nutritional qualities (high protein and fatty acid content) and are, therefore, attracting academic and industry interest. Flowers, green leaves, tubers and green pods are examples of plant parts of this legume that can be eaten (Mohanty *et al.*, 2015). Understanding how the nutritional make-up of winged bean seeds can be used to replace soybean, which has similar properties, may be possible with this knowledge. Because soy proteins are widely used in food applications (Kojima *et al.*, 2010), they can be used as a benchmark when evaluating new protein materials (Firestone, 2009; Saadi *et al.*, 2012).

13.2 Origin, Distribution and Botanical Description

13.2.1 Origin

Winged bean is found in South-east Asia, some African nations, and the nations between India and Papua New Guinea (Fatihah *et al.*, 2012). According to morphology, it is primarily possible to determine the relationship between *P. tetragonolobus* and other *Psophocarpus* species. Comparing the genetic make-up of these plants will, however, be a more trustworthy method that can aid in determining the wild ancestor of this plant.

Various competing theories have been put forth to clarify the origin of winged bean. Four regions are recommended as likely centres of origin by Zeven and De Wet (1982): (i) the Indo-Malayan area, ascribed to its extensive cultivation history in eastern Assam, India; (ii) an Asiatic inception, proposing that winged bean potentially underwent domestication in this region from an unidentified and now extinct native Asian precursor; (iii) Papua New Guinea, supported by the considerable genetic diversity identified there; and (iv) an African hub, substantiated by the resemblances with African species. Notably, the African origin hypothesis has garnered substantial support owing to the close resemblance in shape between winged bean and *Psophocarpus grandifloras*, an African plant (Smartt, 1980).

This is supported by cytological and plant pathology data (Harder and Smartt, 1992). *P. tetragonolobus* (L.), a cultivated winged bean, has more chromosomes than other members of the Fabaceae family, and the presence of a fungus unique to the host provides further proof of its lineage. The ancestor of winged bean may have originated on the African side of the Indian Ocean, moved east as a wild plant, and subsequently experienced alterations because of farming.

13.2.2 Distribution

Winged bean exhibits significant morphological diversity, particularly in South-east Asia, Indonesia and Papua New Guinea (Khan, 1976; Harder and Smartt, 1992). Alternatively, it is probable that it had a wider geographic range and was first domesticated in the Indian subcontinent, or the islands of South-east Asia or Melanesia, given that winged bean is largely a crop of Asia and the Western Pacific with great antiquity in these regions. Geographically, the range of winged bean is bounded to the west by India and to the east by Papua New Guinea (Fig. 13.1). According to Hymowitz and Boyd (1977), winged bean is grown in the following countries: Sri Lanka, Indonesia, Bangladesh, Burma, Malaysia, Thailand and the Philippines. It has been introduced in several African nations, including Madagascar. The tuber, which is esteemed for its flavour and nutritional benefits, is widely farmed in Papua New Guinea's highlands and northern coastal regions (Khan *et al.*, 1977).

Compared to other significant staple crops, winged bean can withstand drought, flood, high temperatures, insect pests and diseases to a greater extent (Singh *et al.*, 2022). The diversification of global food systems with this vegetable could prove to be a valuable tool in battling extreme weather patterns or the effects of long-term climate change, in addition to addressing the problem of nutritional disparities that the world is now facing. Winged bean is typically grown in tropical regions. According to Latha *et al.* (2007), *P. tetragonolobus* can survive in various soil types and at elevations ranging from 0 to 2000 m. This is because symbiotic bacteria, primarily *Rhizobium* strains, can concentrate nitrogen that is transferred to plants through substances such as allantoic acid and allantois through nodules at the roots (Zahran, 1999).

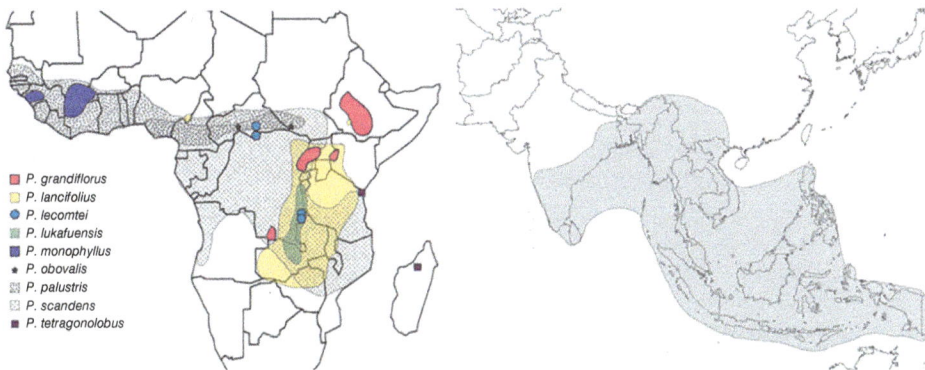

■ P. grandiflorus
☐ P. lancifolius
● P. lecomtei
▨ P. lukafuensis
■ P. monophyllus
• P. obovalis
▨ P. palustris
▨ P. scandens
■ P. tetragonolobus

Fig. 13.1. Distribution of winged bean in the world. (Figure used with permission.)

13.2.3 Botanical description

Psophocarpus tetragonolobus is an underutilized source of nutrition for the tropics. Since the leaves, blossoms, roots and pods can be consumed raw or cooked, *P. tetragonolobus* is sometimes referred to as 'poor man's food'. The fruits have anti-nociceptive, anti-inflammatory and antioxidant properties and are edible. It is a twining plant which climbs up to 3–5 m. It has three ovate to deltoid-shaped leaflets that make up its green trifoliate leaves. It blooms with 2.5–3.5 cm broad flowers that have colours from purple to red, white and blue. Its roots are tuberous, and each tuber is between 8 and 12 cm long and 2 and 4 cm wide. Most winged bean crops are self-pollinated. It is a short-day plant, and its peak flowering time is between 10 am and 12 pm. Excluding pollinators by caging winged bean plants led to a significant reduction in output and a delay in the development of fruits, an observation concluding that insect pollination aids in winged bean fruit setting (Singh *et al.*, 2022). The stigma is still responsive even after 34 h of flower blooming. Even before the flower opens, an anther dehisces. Cross-pollination has, however, been estimated to occur up to 7.6% of the time. Despite being a perennial twining plant, winged bean is normally planted as an annual, which has blue, bluish-white or purple flowers, and can be found alone or in groups of three on pseudoracemes. Pods can reach a length of 30–40 cm and hold 5–21 seeds. The pollen of winged bean is notably distinctive within the genus. Spheroidal and measuring between 43.4 and 49.9 cm (equatorial axis) and 42.3 and 51.6 cm (polar axis), pollen grains are round. The term 'winged bean' comes from the fact that it produces elongated pods (the fruit) with four sides, each of which has a wing. Each pod measures 3 cm in width and 15–30 cm in length.

13.3 Nutritional Significance

Winged bean has gained recognition as a crop with great potential for ensuring global food security in the decades to come. It is a notable source of carbohydrates, protein and B-complex vitamins. These beans also include vitamin B1(thiamine), B2 (riboflavin), B3 (niacin) and B6 (pyridoxine), among other B-complex vitamins

(Fig. 13.2). Winged bean is an excellent source of minerals as well. They contain concentrated amounts of some vital elements such as calcium, phosphorus, magnesium, iron, copper and manganese. The body uses manganese as a cofactor for superoxide dismutase, a potent antioxidant enzyme. The green leaves of winged bean are a great source of fibre, minerals, and vitamins A and C. The fresh leaves (100 g) also contain 8090 IU of vitamin A (270% of recommended dietary allowance) and 45 mg of vitamin C (75% of the recommended daily dose). Bean pods with young, fresh wings are one of the best sources of folic acid: 100 g of beans contains 66 g, or 16.5% of the recommended daily intake, of folates. Folate is one of the crucial building blocks for DNA synthesis and cell division, along with vitamin B12. Fresh winged bean has a considerable quantity of vitamin C, 18.3 mg per 100 g. Tender winged bean and immature pods have 49 kcal per 100 g, which is very low in terms of calorie count (Table 13.1).

13.3.1 Health benefits of winged bean

Processed winged bean can increase high-density lipoprotein (HDL), and antioxidant and enzymatic activity (Esan, 2020). The consumption of winged bean in the diet helps in building structural proteins (Esan, 2020). Processed winged bean decreased cholesterol and triglyceride levels and increased HDL relative to the standard control group. This reduces the risk of coronary heart disease (Esan, 2020). It supports the expression of the gene for ammonium elimination, which reduces creatinine levels and serves as a signal for creatinine clearance (Esan, 2020). The folate in winged bean can reduce the risk of birth problems because women who intake sufficient folate before, during and after conception significantly reduce the risk that their unborn child will have neural tube problems such as spina bifida (Greenberg *et al.*, 2011). Another important benefit of winged bean is that they are a rich source of calcium. This is especially important for women who want to maintain bone strength and prevent osteoporosis (Vannucci *et al.*, 2018). A healthy digestive system is also supported by these 'miracle' beans. They contain a variety of carbohydrates to feed and sustain the bacteria in our

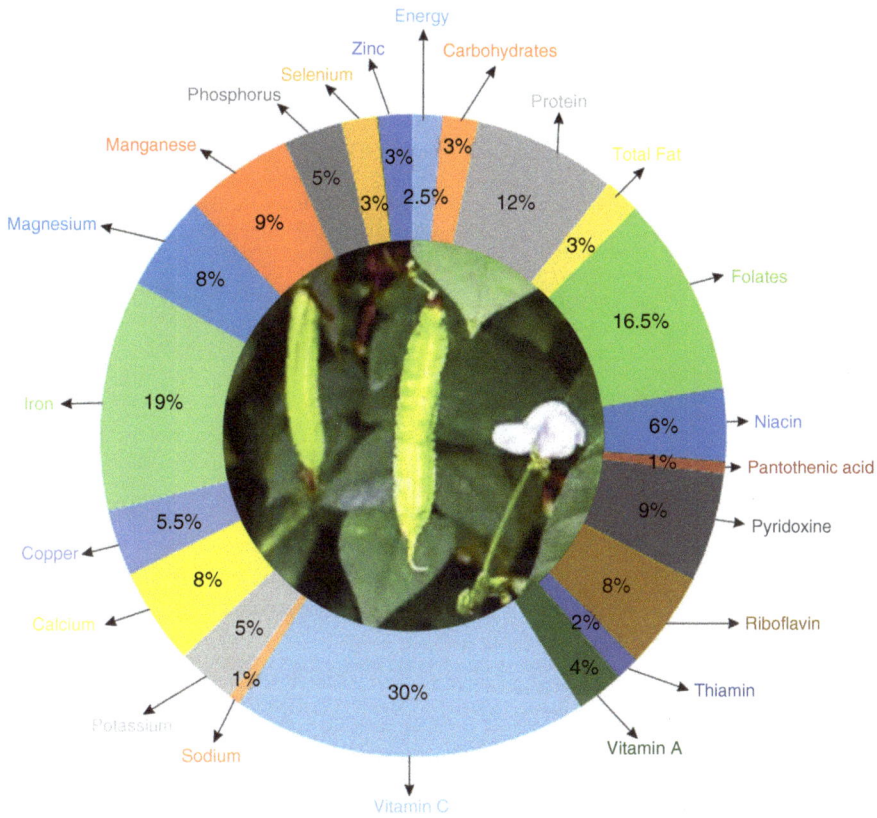

Fig. 13.2. Nutritive value of the immature fresh winged bean. The composition of nutrients is nutrition per 100 g of fresh green pods. (Author's own figure.)

guts that help in the digestion of food into the nutrients we need (Adegboyega *et al.*, 2019).

13.3.2 Antinutrient composition

Some of the antinutritional components identified in winged bean include phytic acid, phytates, flatulence factors, hydrogen cyanide, tannins, saponins, amylase inhibitors, hemagglutinins, phytates and phytic acid (Kortt *et al.*, 1984; Kantha *et al.*, 1986). The primary phytochemicals in *P. tetragonolobus* that are responsible for antioxidant and anti-inflammatory properties (Fig. 13.3) are listed in Table 13.2.

The majority of the trypsin inhibitors in winged bean are found in seeds, where their activity ranges from 11,300 to 74,700 IU/g, depending on the cultivar (Harding *et al.*, 1978;

King and Puwastien, 1987). Research on winged bean chymotrypsin inhibitors (WCIs), which are primarily found in tubers and seeds, has shown that 1 g of beans can inhibit between 30 and 48 mg of chymotrypsin. Their activity is greater than that of trypsin inhibitors (Kortt, 1984).

Numerous genes encode WCIs, which can be classified as Kunitz inhibitors KI (nine have been identified so far) and Bowman-brick-type inhibitors (three have been identified so far) (Kortt, 1984; Kortt *et al.*, 1989). Recently, the KI genes WCI-3b, WCI-2 and WCI-5 were discovered. WCI-5, which is solely expressed in the seeds, has been shown to have significant proteinase inhibitor activity whilst also preventing the growth of *Helicoverpa armigera* larvae (Telang *et al.*, 2008; Banerjee *et al.*, 2017).

It has been demonstrated that winged bean proteins psophocarpin B1, B2 and B3 resemble

Table 13.1. Nutritional status of winged bean (*Psophocarpus tetragonolobus*) per 100 g of green seeds.

Energy	49 kcal
Carbohydrates	4.31 g
Protein	6.95 g
Total fat	0.87 g
Cholesterol	0 mg
Vitamins	
Folates	66 µg
Niacin	0.900 mg
Pantothenic acid	0.059 mg
Pyridoxine	0.113 mg
Riboflavin	0.100 mg
Thiamin	0.140 mg
Vitamin A	128 IU
Vitamin C	18.3 mg
Electrolytes	
Sodium	4 mg
Potassium	240 mg
Minerals	
Calcium	84 mg
Copper	0.051 µg
Iron	1.5 mg
Magnesium	34 mg
Manganese	0.218 mg
Phosphorus	37 mg
Selenium	1.5 µg
Zinc	0.39 mg

Source: Singh *et al.* (2022), CC0 1.0 Universal (CC0 1.0); USDA National Nutrient Database.

the soybean lectins. In winged bean, hemagglutinin activity was shown to range from 77 to 154 hemagglutinin units/g of winged bean (Kortt, 1983). Winged bean has now been shown to have two hemagglutinins known as WBA-I and WBA-II that can agglutinate erythrocytes from various blood types (Srinivas *et al.*, 2000). However, heating destroys hemagglutinin, trypsin and chymotrypsin inhibitors; thus, only raw seeds will have such active levels of these antinutritive components. The activity of the hemagglutinin, trypsin and chymotrypsin inhibitors decreased after the flour made from seeds was briefly autoclaved; the activity of trypsin inhibitors decreased by 30–40%, that of chymotrypsin inhibitors by 15%, and that of hemagglutinins decreased by 75–96%. The activities of all three components were eliminated after the flour had been autoclaved for 30 minutes. However, a significant portion of the protein became insoluble. Lower than the levels of tannins in

other beans, the seeds also contain additional phenolic compounds and tannins in amounts ranging from 0.03 to 7.5 mg/g of beans (Tan *et al.*, 1983). They can precipitate with proteins and inadvertently block enzymes, reducing the quantity of protein readily available in the diet. Other antinutritional components known as phytates are found in seeds in concentrations of 6.1–7.5 mg/g, comparable to those found in soybeans. The level is not dangerous because these concentrations are not considered very high (Kantha *et al.*, 1986). Alkanes, another component, are found in the wax of mature leaves. This wax contains n-alkanes ranging from n-C16 to n-C20 and n-C22 to n-C35. The most abundant of these alkanes are n-C29 and n-C31, whereas n-C20 and n-C26 are the least abundant (Ray *et al.*, 2017).

13.4 Crop Gene Pool, Evolutionary Relationships and Systematics

The absence of wild *Psophocarpus* in Asia creates an enigmatic issue about the wild progenitor of winged bean, which leads to the possibility of the wild progenitor being extinct (Verdcourt and Halliday, 1978). Studies by various scientists have investigated the evolutionary relationships of winged bean (Yang *et al.*, 2018). Species analysis in the genus has been conducted on the basis of morphological phylogeny, and varied conclusions have been drawn (Bhadmus *et al.*, 2023). In particular, the position of *P. grandiflorus* found throughout East Africa, which is a wild species, is inconsistent. Some studies suggest that winged bean and *P. grandiflorus* are likely to be related (Verdcourt and Halliday, 1978; Smartt, 1980; Maxted, 1990; Harder and Smartt, 1992), whereas others place *P. grandiflorus* more distantly in the genus (Fatimah *et al.*, 2012). According to the most recent morphological investigation, winged bean is related to *Psophocarpus scandens* and *Psophocarpus palustris* is related to *P. scandens* (Fatimah *et al.*, 2012). *Erythrina* L. and *Otoptera* DC are relatives of *Psophocarpus*. Nine species of *Psophocarpus* have been classified as belonging to four subclades, i.e. subgenus *Psophocarpus* section *Psophocarpus*; subgenus *Psophocarpus* section *Vignopsis*; subgenus *Lophostigma*; and a new

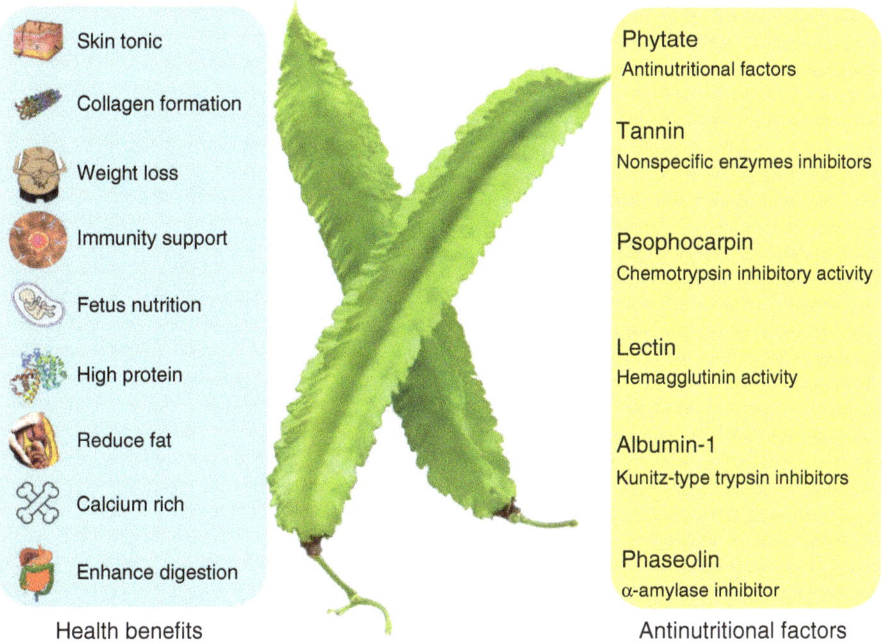
Skin tonic
Collagen formation
Weight loss
Immunity support
Fetus nutrition
High protein
Reduce fat
Calcium rich
Enhance digestion
Health benefits
Phytate
Antinutritional factors
Tannin
Nonspecific enzymes inhibitors
Psophocarpin
Chemotrypsin inhibitory activity
Lectin
Hemagglutinin activity
Albumin-1
Kunitz-type trypsin inhibitors
Phaseolin
α-amylase inhibitor
Antinutritional factors

Fig. 13.3. Health benefits and antinutritional factors of winged bean. (Author's own figure.)

Table 13.2. Principal metabolites controlling biological activity. (Author's own table.)

Metabolites	Activity	Plant part	Country	Reference
Phytate	Antinutritional factors, affinity for specific blood cell antigens	Seed	Malaysia	Kantha *et al.* (1986)
Tannin	Hemagglutinin activity and nonspecific enzyme inhibitors	Seed	Malaysia	Tan *et al.* (1983)
Psophocarpin	Chymotrypsin inhibitory activity	Pods and seed	India	Roy and Singh (1988)
Lectin	Hemagglutinin activity	Seed and roots	India	Shet and Madaiah (1987)
Albumin 1 (WBA-1)	Kunitz-type trypsin inhibitors	Seed	Malaysia	Kortt *et al.* (1989)
Phaseolin	α-Amylase inhibitor	Seed	Malaysia	Yao *et al.* (2016)

subgenus *Longipedunculares* (Fatihah *et al.*, 2012). Eight of the nine species are found only in West, Central and East Africa. The solitary cultivated member of the genus is found mainly in Asia, although it has been introduced to other tropical areas. The taxonomical hierarchy of *Psophocarpus* is well illustrated in Fig. 13.4. *Psophocarpus* is in the same family as soybean, i.e. Fabaceae with order Fabales and Class Rosids.

Understanding the controlling factors of gene expression during growth and differentiation is crucial for producing genetically modified cultivars with novel capacities. Significant findings have been obtained from preliminary studies on the structural and functional characterization of gene families linked to insect and disease resistance. The winged bean chitinase gene ChitWI, which confers antifungal properties in many crop plants, was discovered to be

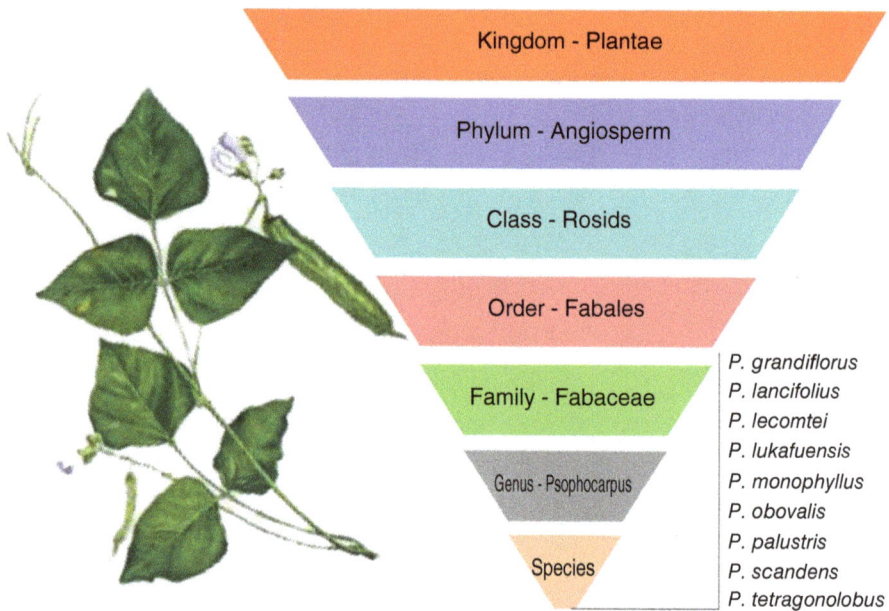

Fig. 13.4. Systematics of winged bean. (Author's own figure.)

strongly produced in roots (Esaka and Teramoto, 1998), indicating that it may be crucial for root development. A phylogeny of 48 sequences from the soybean trypsin inhibitor (STI) gene family discovered in the transcriptomes of Sri Lankan and Nigerian winged bean was described by Vatanparast *et al.* in 2016. According to their research, the winged bean lineage contains much more developed Kunitz-type inhibitor (KTI) genes than other legumes.

13.5 Germplasm Collection, Characterization, Evaluation and Conservation

Germplasm collection is paramount owing to its significant contributions to biodiversity conservation, agricultural sustainability, food security and scientific research. Winged bean has good genetic variability due to the occurrence of germplasm in different geographical locations, having the potential to form the foundation for both the understanding of genetic diversity and the development of improved crop varieties through targeted breeding efforts. These attempts are being hampered, however, by a lack of germplasm characterization. Winged bean is categorized as accessions in gene banks based on the source of germplasm, which includes Papua New Guinea (PNG/UPS accession), Sri Lankan (SL accession), Malaysian (M accession), Thailand (THAI accession), and International Institute of Tropical Agriculture (IITA; TPt accession). Three agroecological zones in Nigeria were used to evaluate 30 accessions of winged bean from the gene bank of the IITA, Ibadan, for seed morphometric and chosen agromorphological features. The acquired data were submitted to cluster analysis, correlation analysis, principal component (PC) analysis and analysis of variance. All evaluated attributes showed significant differences between the accessions. For all examined attributes, 88.2% of the variation between the accessions was explained by the top two PCs. Using the multi-spectral imaging technique, the accessions were divided into three clusters based on the agromorphological features (Bhadmus *et al.*, 2023). The cultivation of this crop is restricted to select places because it is underutilized, and people are unaware of its potential nutritional benefits. As a result, conserving winged bean is essential to save it from becoming extinct. After reviewing the relevance of winged bean, studying new dimensions to identify potential genes and proteins responsible for many significant qualities that may lead to effective crop improvement is crucial. Table 13.3 represents a list of commonly used varieties in India.

13.6 Use of Germplasm in Crop Improvement by the Conventional Approach

Erskine and Khan (1977) performed the first successful controlled crossovers, utilizing the technique for crossing devised by Erskine and Bala (1976). The authors used two crossings of three PNG pure lines (4F1054, UPS31 and 4F1061) to investigate the inheritance of five qualitative features. Purple was more prevalent than green in those crosses for stem colour, calyx colour, pod wing and pod specks. For the inheritance of each of these features, as well as the dominance of rectangular pod shape over flat pod shape, a single-gene method was postulated.

The first structured crossing programme to examine the inheritance of quantitative traits in winged bean was established using a partial diallel cross design between five well-researched PNG accessions (UPS89, UPS81, UPS66, UPS121 and UPS122), one Indonesian accession (UPS132) from Java, and one plant selection (relabelled as UPS132) from TPt1 (Anonymous, 1980; Erskine, 1981). In a different Port Moresby experiment, Erskine and Kesavan (1982) found that heterosis above parental means and narrow-sense heritability, as well as differences

in the number of pods per plant, were not significant. This was in a diallel analysis of the yield of green vegetable pods harvested every five days over a nine-week period. The authors suggested that environmental and non-additive gene components should take precedence in determining ultimate green pod yield. High effects of genotype × environment interaction have also been seen for several days to first blossom in a partial diallel comprising a similar set of eight parental lines in Perth, Western Australia. The data once again demonstrated the effect of day length on modulating days to first blossom and response variance to altering growth circumstances among genotypes. Unfortunately, the ambient parameters of soil temperature and air were not recorded, limiting the interpretation of these results.

Another key indicator of winged bean agronomic performance and manageability is plant growth habit. The primary characteristic was the number of primary branches in the first ten nodes, in addition to the days until the first flower that distinguished highland Papua New Guinea accessions from lowland South and South-east Asian accessions in a germplasm evaluation screen in Serdang (Malaysia) on 147 accessions from traditional growing centres of winged bean (Eagleton et al., 1985).

In one exploratory study, the role played by root-tuber heat-shock proteins and certain metabolites in the adaptation of winged bean (P. tetragonolobus (L.) DC.) to drought stress was studied (Lone et al., 2022).

Stephenson and colleagues in Waigani, a lowland area near Port Moresby, PNG, which is a suboptimal location for tuber production, used a tuber-rich PNG accession (UPS122) crossed

Table 13.3. Commonly used varieties of winged bean in India. (Author's own table.)

Varieties	Economic part	Seed yield (q/ha)	Specific features
AKWB-1	Green pods and seeds	10–12	It is a dual-purpose variety used as a vegetable and pulse.
IWB-1	Seeds	11–13	High-yielding, medium-duration variety and test weight is 36–38 g.
IWB-2	Green pods and seeds	13–14	This variety is amenable for seeds, green pods and fodder.
Chhattisgarh Pankhiya Sem-2	Green pods and seeds	10–12	Dual-purpose variety for backyard farming, used by tribal people of Chhattisgarh.
Revathy, PT-62, Pt-16, PT-49, PT-2	Green pods and seeds	10–15	High-yielding, comparatively free from diseases, most common in Kerala.

with a tuber-poor PNG accession (UPS31) (Stephenson *et al.*, 1979; Anonymous, 1980). They compared the F_1, F_2 and initial backcross generations (BC1 and BC2) in four replicated blocks. The plots were trellised and reproductively pruned to promote tuber bulking. The research revealed that, for root and tuber yield, the additive genetic variance was greater than the dominant or environmental variance (i.e. within the four replicate blocks).

Root and tuber yield, haulm yield and number of tubers per plant each had narrow-sense heritability of 0.45, 0.64 and 0.64, respectively. The F_1 outperformed the better parent UPS122 in haulm yield, which was closely linked with root and tuber production across all plots. To date, no similar study on the genetic control of tuber yield in central Myanmar, another traditional region for winged bean tuber cultivation, has been published in international literature. Researchers have also examined the genetic factors that influence nodulation and nitrogen-fixing. By dividing the variance among parental, F_1 and F_2 populations, Iruthayathas *et al.* (1985) investigated how much the genetic make-up of the host affected nodulation and nitrogen fixation. No non-allelic interaction was discovered, and additive effects accounted for most genetic variance despite nodulation having a low broad-sense heritability (0.25). The significant contextual influence on variance predicts, however, that in subsequent generations, the offspring of this hybrid may show genetic progress towards increased nodulation. Genetic studies of host tolerance to particular diseases that can restrict winged bean have also been noted in the literature. Sastrapradja and Lubis (Anonymous, 1981) used hybridization studies to show resistance to the false rust disease *Synchytrium psophocarpi* in some West Java accessions are controlled by two gene loci, where susceptibility alleles outnumber resistance alleles. West Javanese accessions were found to be resistant to *S. psophocarpi*, while Javanese germplasm was found to be resistant to the yellow mosaic virus, according to Thompson and Haryono (1979). Any controlled hybridization for germplasm enhancement in winged bean should consider the results of the controlled crossings performed by Jalani *et al.* (1983). They discovered evidence of maternal and paternal effects within each cross by evaluating six crosses in a full diallel design.

13.7 Integration of Genomic and Genetic Resources for Crop Improvement

To improve genetics for improved traits, it is crucial to understand the genetic diversity of a plant species. Furthermore, understanding how attributes differ across plants of various origins and comparing those variables to the genome and transcriptome could help with molecular breeding to produce winged bean varieties with desirable agronomic qualities (Schwembera *et al.*, 2017). Since its predecessor has not yet been discovered, the genetic resources for winged bean are limited. Early investigations on genetic diversity and population structure focused on phenotypic characteristics that were known to be regulated by a single gene (Erskine and Khan, 1977). Erskine and Khan (1980) investigated the organization of the winged bean population using three phenotypic markers gathered from several places in highland PNG. None the less, comparing its genome to that of other fully sequenced legumes can be intriguing. These legumes include *Glycine max*, *Cicer arietinum*, *Lotus japonicus*, *Cajanus cajan*, *Phaseolus vulgaris* and *Medicago truncatula* (Schwembera *et al.*, 2017).

The genomes of *Lupinus angustifolius*, *Trifolium pratense*, *Pisum sativum* and *Arachis hypogea* are currently the subject of additional studies (Galeano *et al.*, 2011). The affordable genomic tools that are now available, along with the recently published sequencing data, provide a way to study the genome of winged bean. According to various studies (Sastrapradja and Aminah-Lubis, 1975; Thompson and Haryono, 1979; Anonymous, 1981; Chandel *et al.*, 1984), PNG highland accessions and Indonesian and Burmese accessions seem to be more domesticated offspring of vegetable species that, for the most part, is a useful but easy to maintain, spreading vegetable crop that experiences little purpose-driven selection pressure. Such research has begun to investigate the diversity of winged bean and the population structure of germplasm collections based on phenotypic observations.

It was not until the development of molecular technology, however, that scientists could begin studying DNA polymorphisms without having to consider how a phenotype was recorded or how environmental factors might have affected it. In 2013, Mohanty *et al.* (2013) used 13 random amplification of polymorphic DNA (RAPD) and 7 inter-simple sequence repeat (ISSR) molecular markers to analyse the DNA of 24 accessions from four sources, including India, Nigeria and Thailand. This was the first study of its kind to be conducted on winged bean, and it took a while before it was published. Later, Chen *et al.* (2015) used a set of five ISSR markers to study a separate group of 45 winged bean accessions from eight nations (PNG, Indonesia, Costa Rica, Thailand, Colombia, Nigeria, Sri Lanka and China). Only 44 (65.7%) of the 67 identifiable marker fragments produced by the second group indicated polymorphism.

Despite the reduced level of polymorphism compared to the study by Mohanty *et al.* (2013) in the four groups identified by cluster analysis, there was inconsistency between genetic distance and the geographical origin of the accessions. Focusing on RNA rather than DNA is a distinct strategy for creating valuable genomic markers for genetic studies. For instance, genetic simple sequence repeat (SSR) has proved helpful in marker-assisted selection (MAS) in breeding as well as linkage analysis for genetic diversity studies (Varshney *et al.*, 2005). The winged bean accession TPt1 (Ibadan Local-1) was one of four 'orphan' (underutilized) crops whose transcriptomes were sequenced, assembled and annotated in the first attempt in this direction (Chapman, 2015).

SSR markers from two Sri Lankan accessions, one of which was obtained in the field and the other maintained at the USDA seed bank since 1984, were sequenced, annotated and compared in the second transcriptome analysis (Vatanparast *et al.*, 2016). The scientists discovered a substantial degree of similarity (fewer than 200 single nucleotide polymorphisms (SNPs)) between the two transcripts derived from these two Sri Lankan accessions. Transcriptomes from two further research efforts were published in 2017. Singh and colleagues compiled data from two accessions with different tannin concentrations that were held in India. The phenylpropanoid pathway candidate genes were isolated using expression analysis, and the differential expression levels were confirmed using real-time PCR (Singh *et al.*, 2017).

Later, to develop genic SSR markers, another study employed transcriptome information

collected from a Malaysian accession (Wong *et al.*, 2017). The authors then used this knowledge to validate 18 SSR markers in nine genotypes from five different countries (Malaysia, Bangladesh, Sri Lanka, Indonesia and PNG) (Wong *et al.*, 2017). Yang *et al.* (2018) used five of these markers to screen 53 accessions from the gene banks of the IITA, US Department of Agriculture (USDA) and National Agriculture and Food Research Organization (NARO, Japan). These mutations were infertile, although Class III chitinase, a gene with antifungal properties, was discovered to be expressed in the roots of winged bean plants (Esaka and Teramoto, 1998).

Unfortunately, this species did not receive the same level of attention as other economically significant Fabaceae species (such as soybean and peanut) in formulating breeding programmes to increase plant resistance to severe pests and diseases. It would be prudent to genotype deposited accessions and retrieve germplasm, independent of the declared geographical origin or assumptions based on phenotypic characterization, given the uncertainty that has emerged regarding the declared origins of winged bean accessions that are currently held in various gene banks around the world. This situation might benefit from using high-throughput genotyping-by-sequencing technology, which can provide a more detailed picture of the available germplasm. As a result, locations where more diversity could be mined would be more accurately identified, and truly diverse material would be more effectively chosen for phenotyping and breeding.

13.8 Roadmap of Advanced Approaches for Nutritional Improvement of Winged Bean

Several antinutritional factors are present in winged bean, which cause detrimental effects when digested. The antinutritional factors such as phytate, tannin, psophocarpin, lectin, albumin 1 (WBA-1) and phaseolin are present in winged bean. To take advantage of this multipotent nutrient-rich crop, there is a need to eliminate these antinutritional factors. To achieve this, different advanced strategies can be used. Here, we have discussed the roadmap to improve winged bean with advanced biotechnological tools. In crops, manipulating antinutritional features to increase nutrient bioavailability is a serious challenge because these qualities must be subdued without affecting productivity. Major antinutrients have been successfully reduced by these methods in crops, and the many techniques for changing their content are discussed here (Fig. 13.5).

13.8.1   RNA interference (RNAi)

A unique gene regulation mechanism called RNA silencing controls the number of transcripts by either inhibiting transcription (transcriptional gene silencing or TGS) or by causing sequence-specific RNA to degrade (post-transcriptional gene silencing (PTGS) or RNA interference (RNAi)). RNAi can be successfully used in winged bean to overcome various antinutritional factors. RNAi is successfully used in soybeans to reduce the phytate content by targeting the gene inositol polyphosphate 6-/3-/5-kinase gene with an inositol polyphosphate 6-/3-/5-kinase gene-specific RNAi construct. The produced putative transgenic populations were next subjected to genotypic, phenotypic and biochemical examinations, which showed extremely low phytic acid levels, a modest accumulation of inorganic phosphate and enhanced mineral content in several lines (Punjabi *et al.*, 2018). Phytic acid (myo-inositol1,2,3,4,5,6-hexakis-phosphate; InSP6) is the main phosphorus storage form in mature grain and legume seeds. It binds metallic cations such as calcium, zinc, magnesium and iron to produce a mixed salt known as phytate. Phytates thereby serve as antinutrients, making the minerals inaccessible. The main gene responsible for the production of phytate in legumes is called myo-inositol phosphate synthase (MIPS), and it is only found in a few important legumes such as soybean, green gram, black gram, cowpea and French bean. The entire sequence of the mRNA is necessary for techniques such as RNA-mediated gene silencing, which have the potential to lower the level of phytates in crops. To achieve this, in the case of dolichos bean, MIPS cDNA was extracted utilizing a one-step RT-PCR method and high-fidelity DNA polymerase from the growing seeds (Jagal *et al.*, 2020). This led to the cloning and sequence characterization of the dlMIPS gene. The 539 amino acids in the gene dlMIPS, which has a length of 1776 bp and an open reading frame of 1620 bp, are encoded (Jagal *et al.*,

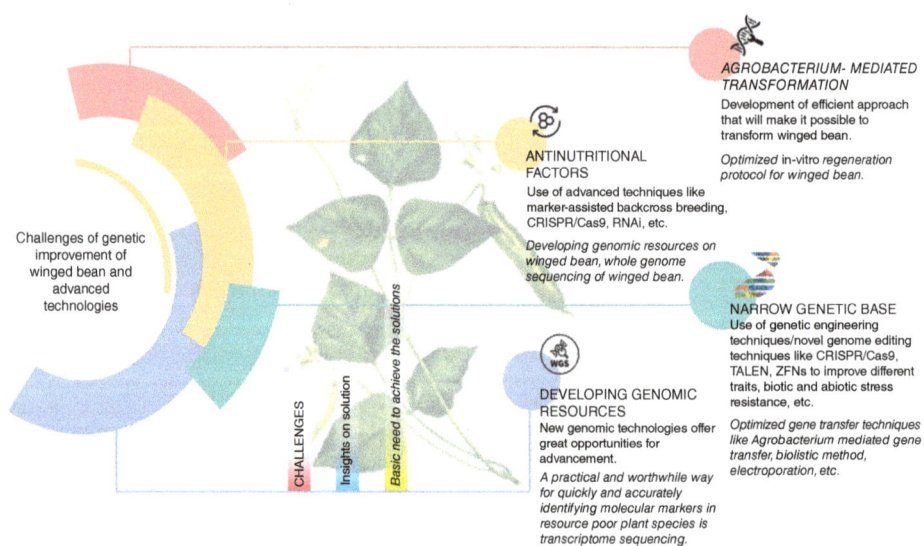

Fig. 13.5. Roadmap of advanced technologies for the genetic improvement of winged bean. (Author's own figure.)

2020). The dolichos bean is more closely related to the green gram, cowpea and French bean, according to phylogenetic studies of the sequence. Through homology modelling, the secondary and tertiary structures of the enzymes from cowpea, French bean and dolichos bean were predicted and compared. The active sites of these enzymes from these legumes were later predicted, and a comparative examination of these active sites revealed that different legumes have different ligand-binding residues (Jagal *et al.*, 2020). In order to precisely employ RNAi to quieten the responsible gene, an experiment such as that conducted in dolichos bean by Jagal *et al.* (2020) is required for winged bean to learn additional details about the gene responsible for phytate synthesis. Similarly, using RNAi to change MYB134 in poplar trees to reduce tannins increased the vulnerability of the plants to oxidative stress, highlighting the need for tissue-specific interventions to decrease antinutrients in crops (Gourlay *et al.*, 2020). A similar approach can be used for the reduction of tannin content of winged bean.

13.8.2 Marker-assisted backcross breeding

The creation of specific, quick and effective molecular markers for the rapid breeding of novel cultivars was made possible by advancements in genomic investigation and molecular biology (Ribaut *et al.*, 2002). In various crops, marker-assisted selection (MAS) is effective in introducing genes for resistance to disease and insects, and the list is growing with more and more successes each year in improving various traits. An antinutritional factor like Kunitz trypsin inhibitor (KTI) in winged bean must be inactivated before consumption. As in a study on the antinutritional factor KTI, pre-heat treatment is used for its elimination but this enhances processing costs and impacts the quality and solubility of the seed proteins (Maranna *et al.*, 2016). The genetic eradication of KTI is a crucial and practical strategy. In soybeans, the molecular marker-assisted backcross breeding (MABB) method is used successfully to remove KTI from two well-known soybean cultivars, DS9712 and DS9814, with the use of exotic germplasm line PI542044 as a donor for the *kti* allele, which prevents the generation of functional KTI peptides (Maranna *et al.*, 2016). Three closely related SSR markers were used for foreground selection of the *kti* allele, while 93 polymorphic SSR markers were used for the background selection. It was discovered that plants from the BC_1F_1 generation recovered 70.4–87.63% and 60.2–73.78%, respectively, of the recurrent parent genomes of DS9712 and DS9814. Similar results were seen in selected BC_2F_1 plants, which recovered their respective recurrent parent genomes by 93.01–98.92% and 83.3–91.67%, respectively. Recombinant selection was used to find plants with the least amount of linkage drag. The absence of KTI peptides in the seeds of the chosen plants (*ktikti*) was confirmed by biochemical investigation. In terms of yield and other attributes, the selected plants were phenotypically similar to the corresponding recurrent parent. With the use of the MABB technique, six KTI-free soybean breeding lines were developed (Maranna *et al.*, 2016). Developing a KTI-free winged bean cultivar can be accomplished using a similar method to eradicating KTI in soybeans.

13.8.3 Transcriptomics

The transcriptomics approach has been successfully used in winged bean. By the de novo transcriptome assembly and annotation of two Sri Lankan winged bean accessions, the genetic resources for this crop are strengthened. In this study, 12,956 SSRs were found, including 5190 high-confidence SNPs between Nigerian and Sri Lankan genotypes and 2594 repetitions for which primers were constructed. The transcriptome data sets produced here provide novel tools for gene discovery and marker creation in this orphan crop and are essential for upcoming plant breeding initiatives. Additionally, this study revealed the significance of the soybean trypsin inhibitor (STI) gene family in the context of related legumes and discovered evidence of the KTI gene family radiation in winged bean. Similar types of studies are expected to reveal more about different gene expressions in winged bean (Vatanparast *et al.*, 2016).

13.8.4 CRISPR/Cas9

There are many opportunities to modify the crop as desired with the use of the innovative genome editing technology CRISPR/Cas9. It will be interesting to use the CRISPR/Cas9 technology to develop an elite version of winged bean. In soybeans, with the use of CRISPR/Cas9 technology, the GmIPK1 gene is successfully edited to obtain a low-level phytic acid soybean cultivar. The GmIPK1 gene was disrupted using the CRISPR/Cas9 system, with sgRNAs 1 and 4 concentrating on the second and third exons, respectively. The PA content of soybean T2 seeds had dropped, but plant growth and seed development had not been impacted, according to the analysis of soybean GmIPK1 gene-edited lines (Song *et al.*, 2022). Future efforts should concentrate on lowering trypsin inhibitors only in seeds meant for human consumption. An alternative approach is provided by advanced gene-editing methods that target the two seed-specific KTI genes KTI1 and KTI3, which cause minor insertions and deletions in soybean open reading frames (Wang *et al.*, 2023). A non-transgenic, low TI soybean mutant in the William 82 cv. was created by CRISPR/Cas9. The easy molecular markers that are appropriate for high-throughput marker-aided selection are attached to the mutant gene alleles (Wang *et al.*, 2023). The CRISPR/Cas9 genome editing technique can also be used to target specific genes in winged bean that are involved in antinutritional factors, yield parameters, herbicide resistance, and biotic and abiotic stress tolerance. To harvest the best from the CRISPR/Cas9 genome editing technique in winged bean, there is a need to explore the genetics of winged bean further. The whole-genome sequencing of winged bean should be conducted to discover more about its genetic constitution. Genomic, proteomic, transcriptomic and metabolomic studies in winged bean would enable this crop to be explored further. Figure 13.5 indicates a roadmap of advanced technologies to overcome the challenges to the utilization of winged bean.

13.9 Erosion of Genetic Diversity from Traditional Areas and Limitations in Germplasm Use

The process of losing a section of a gene pool for a species in a particular area is known as genetic erosion. Although some conservationists use it to refer to the extinction of species, in this context, we use the phrase to refer to the loss of particular alleles or genotypes, subspecies or cultivars. Winged bean is an under-exploited crop; it has been grown traditionally in specific parts of the world. Because of limited knowledge about the nutritional value of this crop, however, it has not been fully utilized. Hence, international societies are now seeking to cultivate and utilize the potential benefits of the crop to reduce malnutrition and to find an alternative way to combat hunger.

Reduced cultivation, resulting in less availability of viable seeds, will pave the way for faster extinction of winged bean. By conserving and promoting the cultivation of winged bean, we contribute to the conservation of plant genetic diversity. Its nutrient content contributes to a well-rounded diet, especially in regions where access to diverse food sources is limited. This diversification enhances dietary diversity, reducing the risk of malnutrition and providing a safety net in case of crop failures. The ability of winged bean to grow in marginal conditions can help provide a consistent food supply even when other crops might struggle. Its market potential and versatility in culinary uses can contribute to rural livelihoods. This is important for maintaining resilient ecosystems and adapting to changing environmental conditions.

13.10 Cultivation Practices

Psophocarpus tetragonolobus develops slowly during the first six weeks, and fruit can be picked 10–12 weeks after planting. The first year of a plantation is when it generates the largest quantity of pods and seeds. This plant has a long lifespan, and research has shown that the second year of planting results in substantially superior tuber yields (369–392 g for each plant as opposed to 80–230 g in the first year) (Khan *et al.*, 1977).

13.10.1 Soil and climatic requirements

Winged bean needs sandy to heavy clay soils that are well-drained, rich in organic matter and have a pH of 4.3-7.5. They cannot survive in soils with a pH of less than 5.5 or in waterlogged conditions. The crop thrives up to an altitude of 2000 m (Singh *et al.*, 2022). Winged bean may grow best in hot, humid climates with annual rainfall ranging from 700 to 4100 mm and temperatures between 15.4 and 27.5°C (Singh *et al.*, 2022). Even though short days are common, temperatures above 32°C or below 18°C prevent flowering. Seeds must be soaked in water prior to planting and fungicides are frequently used. A very fine tilth is kept on the ground by ploughing it down 3 to 4 cm. In short-day circumstances, flowering is seen from mid-September through to October.

13.10.2 Method of sowing and seed rate

Winged bean is often sown in June or July, just before the monsoon season begins. The crop grown for tubers should be sown in August or September because early seeding promotes vigorous vegetative development and may inhibit tuberization. Stem cuttings may also be utilized in certain situations; however, seeds are typically the preferred method of propagation. A delivery of 15–20 kg of seed per hectare is the suggested quantity. However, due to their tough seed coat, seeds need to be pre-soaked for 1–2 days (Singh *et al.*, 2022). The seeds are sown at 3 to 4 cm depth, and they germinate 5–7 days after being sown. For commercial crops, the ideal spacing is 90 × 90 cm, while for seed crops, it is 45 × 45 cm (Singh *et al.*, 2022). In the case of dwarf types, a separation of 30 × 20 cm can be used (Singh *et al.*, 2022).

13.10.3 Integrated nutrient management

Winged bean does not need to be inoculated with bacteria that fix nitrogen because it is a leguminous plant with a high nodulation capacity, the ability to grow quickly and the ability to replenish the soil with nitrogen fixed from the atmosphere, especially in conditions where it is ploughed. Even though winged bean thrives in low-nutrient soils, tropical soils frequently lack nitrogen. In acidic soils, winged bean displays the same symptoms of aluminium toxicity as other legumes. In addition to 50, 80 and 50 kg/ha of N, P and K fertilizer, respectively, the crop requires 20 tons/ha of farmyard manure (Singh *et al.*, 2022). A split dose of P and K and a full dose of nitrogen are applied when seeding, and the remaining dose of nitrogen is top-dressed at 40–60 days after sowing (Singh *et al.*, 2022).

13.10.4 Weed control

Within a month of sowing, winged bean provides plant cover and grows quickly. One hand weeding at 15–20 days after sowing is required to control weeds during the early growth stage (Singh *et al.*, 2022). Staking is an essential technique for ensuring a respectable and high-quality crop because winged bean has unpredictable stem growth. Depending on how the crop is used and the available resources, different staking techniques are used. Higher seed production requires the support of the vines as well.

13.10.5 Plant protection

In India, there have been no significant reports of insect pests or diseases affecting winged bean. The significant fungal diseases are, however, false rust (*Synchytrium psophocarpi*) and leaf spot (*Pseudocercospora psophocarpi*). Winged bean is affected by five main fungal diseases, namely choanephora blight, collar rot disease, powdery mildew, dark leaf spot and false rust disease, caused by *Choanephora cucurbitarum*, *Fusarium semitectum*, *F. moniliforme*, *F. equiseti*, *Rhizoctonia solani*, *Macrophomina phaseolina*, *Erysiphe cichoracearum*, *Pseudocercospora psophocarpi* and *Synchytrium psophocarpi* (Bassal *et al.*, 2020). In Sri Lanka, winged bean is also susceptible to pulvinus rot disease (*Fusarium pallidoroseum*), leaf scorch disease (*Phoma sorghina*) and pod rot disease (*Botryodiplodia theobromae*) (Fortuner *et al.*, 1979).

Numerous nematodes have been found to negatively impact winged bean plants, with *Meloidogyne arenaria*, *M. incognita* and *M. javanica* infections being responsible for 70% of tuber loss (Reddy, 2015). Ring spot mosaic virus is known to cause 10–20% of yield loss and infects 9% of immature plants. In the Ivory Coast, winged bean can be prone to leaf curl disease and the recently discovered virus winged bean endornavirus (Fortuner *et al.*, 1979). Pest species that harm winged bean include members of Orthoptera, Coleoptera, Thysanoptera, Diptera, Hemiptera and Lepidoptera. The bean pod borer (*Maruca testulalis*), cotton boll worm (*Helicoverpa armigera*), *Mylabris afzelii* and *Mylabris pustulata*, as well as the winged bean blotch miner (*Leucoptera psophocarpella*), can harm either the leaves or the bean blooms (Khan, 1982). Fruit is also infected by cotton bollworm and bean pod borer (Khan, 1982). In order to reduce crop losses, appropriate plant protection measures may be adopted.

13.10.6   Harvesting and post-harvest management

The shoots and leaves are harvested while they are still tender. After seeding, the green pods can be picked 10 weeks later. Fresh pod and tuberous root yields vary from 5 to 10 t/ha, whereas seed yields are between 1 and 1.5 t/ha. To keep them fresh, winged bean can be placed in plastic bags and securely sealed at the neck. When stored at a temperature of 10°C and relative humidity of 90%, the shelf life of the pods can be extended to 4 weeks (Singh *et al.*, 2022).

13.11   Future Directions

Despite the most recent evolutionary analysis by Yang *et al.* (2018), further research is still needed to determine the origin of winged bean. A greater understanding of the taxonomy, origin and development of winged bean could be obtained through research efforts in this regard, which are based on bigger collections of farmed *P. tetragonolobus* and related *Psophocarpus* species. They could also enable the strategic utilization of the available resources for future advancement. Similar studies should be performed on germplasm (or subsets of it) that are currently held by various nations in institutions.

Such processes are crucial in the construction of mapping populations, linkage mapping, and genome-wide association studies for trait discovery, as shown in soybean, chickpea (Saeed and Darvishzadeh, 2017), mung bean (Noble

et al., 2018) and common bean (Tock *et al.*, 2017). This would enable us to evaluate the degree of adaptability and plasticity of winged bean by combining phenotypic and genetic data from various contexts. A crucial step in developing common descriptors to phenotype the examined germplasm for all types of screenings would be to update those used in earlier studies, which are currently available at: https://hdl.handle.net/10568/73415 (accessed 26 July 2024).

As with any other crop, winged bean is grown in a variety of environments, and its expansion and development have an impact on farming methods and expenses. Therefore, it makes sense to call for the creation of ideotypes that are appropriate for these situations through research that elucidates the relationship between plant architecture and yield as well as the best ways to balance these relationships (see Fig. 13.6 for potential breeding targets). For example, winged bean can be grown on a variety of supports (such as strings, trellises or single stalks), as well as directly on the ground with no support. Any desired modification to winged bean plant architecture and its potential effects on developmental processes would depend on the desired product or final farming purposes (for instance, for the production of pods, seeds or tubers), as well as the distinction between supported and unsupported cultivation methods. Additionally, it should be emphasized that individuals with longer growth cycles and longer flowering times are advantageous in some situations (for example, as a cover crop where biomass is the goal) as well as early maturing types, for instance, for seed production (Tanzi *et al.*, 2019).

Additionally, small-scale farmers and household users might prefer an extended flowering period since it guarantees a consistent supply of food and goods to sell in nearby marketplaces for an extended period of time. Unfortunately, no post-harvest study has been done to date; however, it may be useful in maintaining green pods and tubers for wider distribution areas. Potential breeding goals for winged bean are shown in Fig. 13.6.

The ability of winged bean to survive abiotic conditions such as drought, salt or waterlogging has perhaps been least explored previously. Karikari put up a list of traits for the potential identification of drought-resistant individuals in winged bean, which included the number and

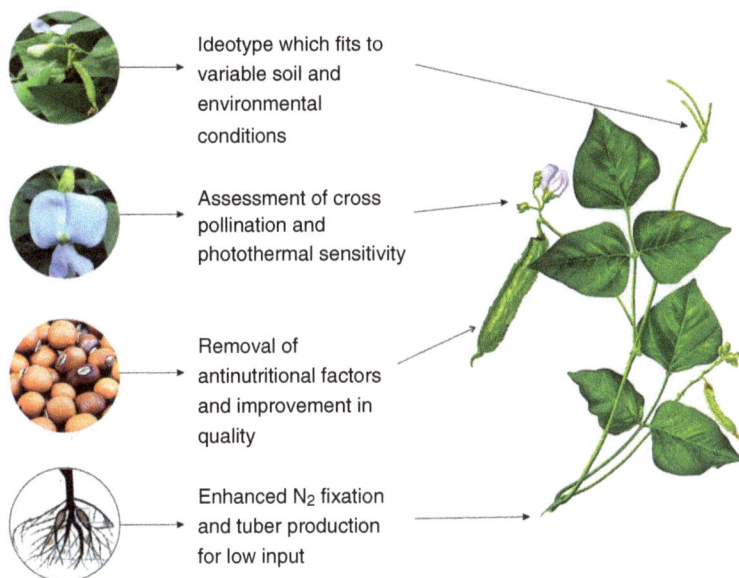

Fig. 13.6. Breeding targets for winged bean improvement. (Author's own figure.)

weight of nodules per plant and stem thickness (Karikari, 1978). Although winged bean had higher plant N and less foliage damage than soybeans, another study that examined the impact of salt stress on nitrogenase activity discovered that it was significantly lower in winged bean than in soybeans (Weil and Khalil, 1986). The researchers came to the conclusion that winged bean, which was only represented by one accession that is typically grown in Sri Lanka (SLS47), was comparable to soybean cv. 'Lee', which is the most salt-tolerant soybean cultivar and was used, among other things, for the discovery of the Ncl gene responsible for this trait (Weil and Khalil, 1986; Do *et al.*, 2016). Such results would motivate additional studies to examine the salt tolerance characteristics of winged bean. Along with drought resistance and photothermal sensitivity, the capacity of winged bean to withstand abiotic stresses will decide not only how widely it is adopted outside of its current production zones but also how it fits into low-input agricultural systems.

Lessons learnt from previous investigation endeavours are an essential component of the toolkit required to transform winged bean into a 'one-stalk supermarket'. Although the variety of winged bean allows breeders to enhance this crop in a variety of methods for economic interests, research should not overlook traditional patterns of consumption. Winged bean is more likely to be a horticultural crop first: the major aim for enhanced productivity and nutritional quality should be immature pods, tubers, green seeds and leaves. Focusing on improving cookability and reducing antinutritional elements could help the mature seed realize its full potential, as well as increase its economic value if it is used to produce value-added goods.

In order to create alternative food products with an array of nutritional profiles and flavours, such as tofu, tempeh and bean milk, food technologists have already started examine ways to adapt currently used processing methods for crops such as soybean (Kantha *et al.*, 1983; Omachi *et al.*, 1983; Prakash *et al.*, 1987; Yang and Tan, 2011). As a result, winged bean products would gain more traction and could be introduced to larger international markets. Reduced fibre content and pod wall conformation may aid the development of cultivars with increased palatability of vegetable pods. This would require careful consideration, however, because there is no certainty that lowering pod fibre content will improve eatable quality. The extent of seed loss caused by pod-shattering may be controlled by changing the pod wall (Koinange *et al.*, 1996). Unfortunately, there is no pertinent information accessible regarding winged bean production in the country.

It is difficult to obtain an accurate picture of the size of the cultivated area, production and market price because it is mainly grown on a small scale, in home gardens, and sold at local markets, a difficulty shared by other minor crop species (Khoury *et al.*, 2014). In a more detailed assessment of the diversity of national food production systems, it is more challenging to support government initiatives and research funding schemes that would improve diverse sources of food and income, including crops such as winged bean. However, because winged bean is a significant source of food and income in local markets, we anticipate that data will be collected on them as for other crops.

The evolution of orphan crops such as winged bean also depends on historical study, although this information is hard to obtain personally. To build on Graham Eagleton's work with wingedbeanflyer.org as a platform for disseminating research findings and interacting with farmers, researchers and commercial enterprises, Crops for the Future (CFF) is establishing wingedbeannetwork.org. The objective is also to promote relationships between the numerous actors involved in winged bean research and cultivation as well as the dissemination of recent research findings (Eagleton *et al.*, 2023).

Winged bean is a highly nutritious underused legume that could have some enhancement in terms of food use. This will aid in addressing the problems with nutrition and diet. In order to benefit from the nutritive significance of winged bean, research should focus on the current limitations that are creating a barrier to its utilization. Winged bean has some antinutritional components that need to be removed, including KTI, phytate, tannin, psophocarpin, lectin, albumin 1 (WBA-1), and phaseolin. Additionally, winged bean genetic resources are few, necessitating a thorough investigation of winged bean genetic resources using genomic, proteomic and transcriptomic research. Cutting-edge technologies such as CRISPR/Cas9, RNA interference, MABB, zinc-finger nucleases (ZFNs) and transcription activator-like effector nucleases (TALENs) and transcriptomics should be used to create elite cultivars that are free of antinutritional factors, are abiotic and biotic stress-resistant and have a higher yield to reap the nutritional benefits of this crop.

References

Adegboyega, T.T., Abberton, M.T., AbdelGadir, A.A.H., Dianda, M., Maziya-Dixon, B. *et al.* (2019) Nutrient and antinutrient composition of winged bean (*Psophocarpus tetragonolobus* (L.) DC.) seeds and tubers. *Journal of Food Quality* 2019, 3075208. DOI: 10.1155/2019/3075208

Adhikari, L., Hussain, A. and Rasul, G. (2017) Tapping the potential of neglected and underutilized food crops for sustainable nutrition security in the mountains of Pakistan and Nepal. *Sustainability* 9(2), 291. DOI: 10.3390/su9020291

Anonymous (1980) The winged bean flyer (3 No.1). International Documentation Centre for the Winged Bean Agricultural Information Bank of Asia, Laguna.

Anonymous (1981) The winged bean flyer (3 No.2). International Documentation Centre for the Winged Bean Agricultural Information Bank of Asia, Laguna.

Banerjee, S., Giri, A.P., Gupta, V.S. and Dutta, S.K. (2017) Structure-function relationship of a bio-pesticidal trypsin/chymotrypsin inhibitor from winged bean. *International Journal of Biological Macromolecules* 96, 532–537. DOI:10.1016/j.ijbiomac.2016.12.018

Bassal, H., Merah, O., Ali, A.M., Hijazi, A. and El Omar, F. (2020) *Psophocarpus tetragonolobus*: An underused species with multiple potential uses. *Plants* 9(12), 1730. DOI: 10.3390/plants9121730

Bhadmus, A., Abberton, M., Idehen, E., Ekanem, U., Paliwal, R. and Oyatomi, O. (2023) Genetic diversity assessment of winged bean [*Psophocarpus tetragonolobus* (L.) DC.] accessions using agronomic and seed morphometric traits. *Crops* 3(2), 170–183. DOI: 10.20944/preprints202303.0278.v1

Chandel, K.P.S., Pant, K.C. and Arora, R.K. (1984) Winged Bean in India. *NBPGR Sci. Monogr.* No. 8. National Bureau of Plant Genetics Resources, New Delhi, India.

Chapman, M.A. (2015) Transcriptome sequencing and marker development for four underutilized legumes. *Applications in Plant Sciences* 3(2), 1400111. DOI: 10.3732/apps.1400111

Chappell, M.J. and LaValle, L.A. (2011) Food security and biodiversity: can we have both? An agroecological analysis. *Agriculture and Human Values* 28, 3–26. DOI: 10.1007/s10460-009-9251-4

Chen, D., Yi, X., Yang, H., Zhou, H., Yu, Y., Tian, Y. and Lu, X. (2015) Genetic diversity evaluation of winged bean (*Psophocarpus tetragonolobus* (L.) DC.) using inter-simple sequence repeat (ISSR). *Genetic Resources and Crop Evolution* 62, 823–828. DOI: 10.1007/s10722-015-0261-3

Chivenge, P., Mabhaudhi, T., Modi, A.T. and Mafongoya, P. (2015) The potential role of neglected and underutilised crop species as future crops under water scarce conditions in Sub-Saharan Africa. *International Journal of Environmental Research and Public Health* 12(6), 5685–5711.DOI: 10.3390/ijerph120605685

Do, T.D., Chen, H., Hien, V.T.T., Hamwieh, A., Yamada, T. *et al.* (2016) Ncl synchronously regulates Na+, K+ and Cl– in soybean and greatly increases the grain yield in saline field conditions. *Scientific Reports* 6(1), 19147. DOI: 10.1038/srep19147

Dwivedi, S.L., Van Bueren, E.T.L., Ceccarelli, S., Grando, S., Upadhyaya, H.D. and Ortiz, R. (2017) Diversifying food systems in the pursuit of sustainable food production and healthy diets. *Trends in Plant Science* 22(10), 842–856. DOI: 10.1016/j.tplants.2017.06.011

Eagleton, G.E., Khan, T.N. and Erskine, W. (1985) Winged bean (*Psophocarpus tetragonolobus* (L.) DC.). In: Summerfield, R.J and Roberts, E.H. (eds) *Grain Legume Crops*, pp. 624–657. Collins, London.

Eagleton, G.E., Tanzi, A.S., Mayes, S., Massawe, F., Ho, W.K. *et al.* (2023) Winged bean (*Psophocarpus tetragonolobus* (L.) DC.). In: M. Farooq, M. and Siddique, K.H.M. (eds) *Neglected and Underutilized Crops.* Academic Press, Elsevier, London, pp. 437–486.

Edreira, J.R., Carpici, E.B., Sammarro, D. and Otegui, M.E. (2011) Heat stress effects around flowering on kernel set of temperate and tropical maize hybrids. *Field Crops Research* 123(2), 62–73. DOI: 10.1016/j.fcr.2011.04.015

Erskine, W. (1981) Inheritance of some economic characters in winged bean. In: *2nd International Seminar of Winged Bean, 19–23 January 1981*, Colombo, Sri Lanka.

Erskine, W. and Bala, A.A. (1976) Crossing technique in winged bean. *Tropical Grain Legume Bulletin* 6, 32–35.

Erskine, W. and Kesavan, V. (1982) Genetic variability in the green pod production of winged bean. *Journal of Horticultural Science* 57(2), 209–213. DOI: 10.1080/00221589.1982.11515042

Erskine, W. and Khan, T.N. (1977) Inheritance of pigmentation and pod shape in winged bean. *Euphytica* 26, 829–831. DOI: 10.1007/BF00021714

Erskine, W. and Khan, T.N. (1980) Variation within and between land races of winged bean (*Psophocarpus tetragonolobus* (L.) DC.). *Field Crops Research* 3, 359–364. DOI: 10.1016/0378-4290(80)90041-6

Esaka, M. and Teramoto, T. (1998) cDNA cloning, gene expression and secretion of chitinase in winged bean. *Plant and Cell Physiology* 39(3), 349–356. DOI: 10.1093/oxfordjournals.pcp.a029376

Esan, A.M., Charles, O.O., Samson, T.A., Folusho, E.B. and Kehinde, O.S. (2020) Biochemical and nutritional importance of winged bean (*Psophocarpus tetragonolobus* (L.)) on Wistar rats. *International Journal of the Science of Food and Agriculture* 4(2), 174–182. DOI: 10.26855/ijfsa.2020.06.009

FAO (2010) The second report on the state of the World's plant genetic resources for food and agriculture. Food and Agriculture Organization of the United Nations, Rome.

FAO (2011) The state of food insecurity: how does international price volatility affect domestic economies and food security? Food and Agriculture Organization of the United Nations, Rome.

FAO (2015) Coping with climate change—the roles of genetic resources for food and agriculture. Food and Agriculture Organization of the United Nations, Rome.

Fatihah, H.N., Maxted, N. and Arce, L.R. (2012) Cladistic analysis of *Psophocarpus* Neck. ex DC. (Leguminosae, Papilionoideae) based on morphological characters. *South African Journal of Botany* 83, 78–88. DOI: 10.1016/j.sajb.2012.07.010

Firestone, D. (2009) *AOCS Official Methods and Recommended Practices of the AOCS*. AOCS, Urbana, Illinois, USA.

Fortuner, R., Fauquet, C. and Lourd, M. (1979) Diseases of the winged bean in Ivory Coast. *Plant Disease Reporter* 63(3), 194–199.

Galeano, C.H., Fernandez, A.C., Franco-Herrera, N., Cichy, K.A., McClean, P.E., Vanderleyden, J. and Blair, M.W. (2011) Saturation of an intra-gene pool linkage map: towards a unified consensus linkage map for fine mapping and synteny analysis in common bean. *PLoS One* 6(12), 28135. DOI: 10.1371/journal.pone.0028135

Gourlay, G., Ma, D., Schmidt, A. and Constabel, C.P. (2020) MYB134-RNAi poplar plants show reduced tannin synthesis in leaves but not roots and increased susceptibility to oxidative stress. *Journal of Experimental Botany* 71(20), 6601–6611. DOI: 10.1093/jxb/eraa371

Greenberg, J.A., Bell, S.J., Guan, Y. and Yu, Y.H. (2011) Folic acid supplementation and pregnancy: more than just neural tube defect prevention. *Reviews in Obstetrics and Gynecology* 4(2), 52.

Habu, Y., Peyachoknagul, S., Umemoto, K., Sakata, Y. and Ohno, T. (1992) Structure and regulated expression of Kunitz chymotrypsin inhibitor genes in winged bean [*Psophocarpus tetragonolobus* (L.) DC.]. *The Journal of Biochemistry* 111(2), 249–258. DOI: 10.1093/oxfordjournals.jbchem.a123745

Harder, D.K. and Smartt, J. (1992) Further evidence on the origin of the cultivated winged bean, *Psophocarpus tetragonolobus* (L.) DC. (Fabaceae): Chromosome numbers and the presence of a host-specific fungus. *Economic Botany* 46, 187–191.

Harding, J., Martin, F.W. and Kleiman, R. (1978) Seed protein and oil yields of the winged bean, *Psophocarpus tetragonolobus*, in Puerto Rico. *Tropical Agriculture* 55.

Hymowitz, T. and Boyd, J. (1977) Origin, Ethnobotany and Agricultural Potential of the Winged Bean: *Psophocarpus tetragonolobus*. *Economic Botany* 31(2), 180–188. DOI: 10.1007/BF02866589

Iruthayathas, E.E., Vlassak, K. and Laeremans, R. (1985) Inheritance of nodulation and N2 fixation in winged beans. *Journal of Heredity* 76(4), 237–242. DOI: 10.1093/oxfordjournals.jhered.a110084

Jagal, K.S., Mathew, D., Shylaja, M.R., Francies, R.M. and Sujatha, R. (2020) Cloning and characterization of myo-inositol phosphate synthase gene (dlMIPS) and analysis of the putative structure of the enzyme responsible for the accumulation of anti-nutrient phytate in dolichos bean (*Dolichos lablab* L.). *Plant Physiology Reports* 25(2), 370–375. DOI: 10.1007/s40502-020-00507-7

Jalani, B.S., Benong, M. and Ismail, N. (1983) Evaluation of F1 hybrids on winged bean (*Psophocarpus tetragonolobus*). In: Yap, T.C., Graham, K.M. and Jalani, S. (eds.) *Proceedings of the Fourth International SABRAO Congress: Crop Improvement Research*. SABRAO, Malaysia, pp. 275–281. https://agris.fao.org/search (accessed 26 July 2024).

Kantha, S.S., Hettiarachchy, N.S. and Erdman, J.W. (1983) Laboratory scale production of winged bean curd. *Journal of Food Science* 48(2), 441–444. DOI: 10.1111/j.1365-2621.1983.tb10761.x

Kantha, S.S., Hettiarachchy, N.S. and Erdman Jr, J.W. (1986) Phytic acid, and selected minerals in winged bean flour. *Cereal Chemistry* 63(1), 9–13.

Karikari, S.K. (1972) Pollination requirements of winged beans (*Psophocarpus* et al.) in Ghana. *Ghana Journal of Agricultural Science* 5(3), 235–239.

Khalili, R.M.A., Shafekh, S.E., Norhayati, A.H., Fatahudin, I.M., Rahimah, R., Norkamaliah, H. and Azimah, A.N. (2013) Total phenolic content and in vitro antioxidant activity of winged bean (*Psophocarpus tetragonolobus*). *Pakistan Journal of Nutrition* 12(5), 416. DOI: 10.3923/pjn.2013.416.422

Khan, T.N. (1982) Winged bean production in the tropics. *FAO Plant Production and Protection Paper* 38, FAO, Rome.

Khan, T.N., Bohn, J.C. and Stephenson, R.A. (1977) Winged beans cultivation in Papua, New Guinea. *World Crops and Livestock* 29, 208–209.

Khoury, C.K., Bjorkman, A.D., Dempewolf, H., Ramirez-Villegas, J., Guarino, L. *et al.* (2014) Increasing homogeneity in global food supplies and the implications for food security. *Proceedings of the National Academy of Sciences* 111(11), 4001–4006. DOI: 10.1073/pnas.131349011

King, R.D. and Puwastien, P. (1987) Effects of germination on the proximate composition and nutritional quality of winged bean (*Psophocarpus tetragonolobus*) seeds. *Journal of Food Science* 52(1), 106–108. DOI: 10.1111/j.1365-2621.1987.tb13982.x

Koinange, E.M., Singh, S.P. and Gepts, P. (1996) Genetic control of the domestication syndrome in common bean. *Crop Science* 36, 1037–1045.

Kojima, M., Tachibana, N., Yamahira, T., Seino, S., Izumisawa, A. *et al.* (2010) Structured triacylglycerol containing behenic and oleic acids suppresses triacylglycerol absorption and prevents obesity in rats. *Lipids in Health and Disease* 9(1), 1–6. DOI: 10.1186/1476-511X-9-77

Kortt, A.A. (1983) Comparative studies on the storage proteins and anti-nutritional factors from seeds of *Psophocarpus tetragonolobus* (L.) DC from five Southeast Asian countries. *Plant Foods for Human Nutrition* 33, 29–40. DOI: 10.1007/BF01093735

Kortt, A.A. (1984) Purification and properties of the basic lectins from winged bean seed [*Psophocarpus tetragonolobus* (L.) DC]. *European Journal of Biochemistry* 138(3), 519–525. DOI: 10.1111/j.1432-1033.1984.tb07946.x

Kortt, A.A., Strike, P.M. and De Jersey, J. (1989) Amino acid sequence of a crystalline seed albumin (winged bean albumin-1) from *Psophocarpus tetragonolobus* (L.) DC: Sequence similarity with Kunitz-type seed inhibitors and 7S storage globulins. *European Journal of Biochemistry* 181(2), 403–408. DOI: 10.1111/j.1432-1033.1989.tb14739.x

Laosatit, K., Amkul, K., Chankaew, S. and Somta, P. (2022) Molecular genetic diversity of winged bean gene pool in Thailand assessed by SSR markers. *Horticultural Plant Journal* 8(1), 81-88.

Latha, L.Y., Sasidharan, S., Zuraini, Z., Suryani, S., Shirley, L., Sangetha, S. and Davaselvi, M. (2007) Antimicrobial activities and toxicity of crude extract of the *Psophocarpus tetragonolobus* pods. *African Journal of Traditional, Complementary and Alternative Medicines* 4(1), 59–63. DOI: 10.4314/ajtcam.v4i1.31195

Lepcha, P., Egan, A.N., Doyle, J.J. and Sathyanarayana, N. (2017) A review on current status and future prospects of winged bean (*Psophocarpus tetragonolobus*) in tropical agriculture. *Plant Foods for Human Nutrition* 72, 225–235. DOI: 10.1007/s11130-017-0627-0

Lim, T.K. (2012) *Edible Medicinal and Non-medicinal Plants* (Vol. 1). Springer, Dordrecht, The Netherlands, pp. 656–687.

Lone, R.A., Sarvendra, K., Singh, V., Bano, N., Bag, S.K., Mohanty, C.S. and Barik, S.K. (2022) Adaptation of winged bean (*Psophocarpus tetragonolobus* (L.) DC.) to drought stress is mediated by root-tuber heat-shock proteins and specific metabolites. *Current Plant Biology* 32, 100266.

Maranna, S., Verma, K., Talukdar, A., Lal, S.K., Kumar, A. and Mukherjee, K. (2016) Introgression of null allele of Kunitz trypsin inhibitor through marker-assisted backcross breeding in soybean (*Glycine max* L. Merr.). *BMC Genetics* 17(1), 1–9. DOI: 10.1186/s12863-016-0413-2

Massawe, F., Mayes, S. and Cheng, A. (2016) Crop diversity: an unexploited treasure trove for food security. *Trends in Plant Science* 21(5), 365–368. DOI: 10.1016/j.tplants.2016.02.006

Mayes, S., Massawe, F.J., Alderson, P.G., Roberts, J.A., Azam-Ali, S.N. and Hermann, M. (2012) The potential for underutilized crops to improve security of food production. *Journal of Experimental Botany* 63(3), 1075–1079. https://doi.org/10.1093/jxb/err396

Mohanty, C.S., Verma, S., Singh, V., Khan, S., Gaur, P. *et al.* (2013) Characterization of winged bean (*Psophocarpus tetragonolobus* (L.) DC.) based on molecular, chemical and physiological parameters. *American Journal of Molecular Biology* 3, 187–197. DOI: 10.4236/ajmb.2013.34025

Mohanty, C.S., Pradhan, R.C., Singh, V., Singh, N., Pattanayak, R. *et al.* (2015) Physicochemical analysis of *Psophocarpus tetragonolobus* (L.) DC seeds with fatty acids and total lipids compositions. *Journal of Food Science and Technology* 52, 3660–3670. DOI: 10.1007/s13197-014-1436-1

NAS (1975) *The winged bean: a high-protein crop for the humid tropics*. National Academy of Sciences, Washington, DC.

Noble, T.J., Tao, Y., Mace, E.S., Williams, B., Jordan, D.R., Douglas, C.A. and Mundree, S.G. (2018) Characterization of linkage disequilibrium and population structure in a mungbean diversity panel. *Frontiers in Plant Science* 8, 2102. DOI: 10.3389/fpls.2017.02102

Omachi, M., Ishak, E., Homma, S. and Fujimaki, M. (1983) Manufacturing of tofu from winged bean and the effect of presoaking and cooking. *Nippon Shokuhin Kogyo Gakkaishi* 30(4), 216–220. DOI: 10.3136/nskkk1962.30.216

Padulosi, S., Thompson, J. and Rudebjer, P.G. (2013) *Fighting poverty, hunger and malnutrition with neglected and underutilized species: needs, challenges and the way forward*. Bioversity International, Rome, Italy. DOI: 10.13140/RG.2.1.3494.3842

Pellegrini, L. and Tasciotti, L. (2014l) Crop diversification, dietary diversity and agricultural income: empirical evidence from eight developing countries. *Canadian Journal of Development Studies/Revue Canadienne D'études du Développement* 35(2), 211–227. DOI: 10.1080/02255189.2014.898580

Prakash, D., Misra, P.N. and Misra, P.S. (1987) Amino acid profile of winged bean (*Psophocarpus tetragonolobus* (L.) DC): a rich source of vegetable protein. *Plant Foods for Human Nutrition* 37, 261–264. DOI: 10.1007/BF01091791

Punjabi, M., Bharadvaja, N., Jolly, M., Dahuja, A. and Sachdev, A. (2018) Development and evaluation of low phytic acid soybean by siRNA triggered seed specific silencing of inositol polyphosphate 6-/3-/5-kinase gene. *Frontiers in Plant Science* 9, 804. DOI: 10.3389/fpls.2018.00804

Ray, S., Tah, J. and Sinhababu, A. (2017) Analysis of n-alkanes in leaf epicuticular wax of three cultivars of winged bean *Psophocarpus tetragonolobus*. *International Journal of ChemTech Research* 10, 178–185.

Reddy, P.P. (2015) *Plant Protection in Tropical Root and Tuber Crops* (No. 11591). Springer India, New Delhi, India.

Ribaut, J.M., Jiang, C. and Hoisington, D. (2002) Simulation experiments on efficiencies of gene introgression by backcrossing. *Crop Science* 42(2), 557–565. DOI: 10.2135/cropsci2002.5570

Roy, A. and Singh, M. (1988) Psophocarpin B1, a storage protein of Psophocarpus tetragonolobus, has chymotrypsin inhibitory activity. *Phytochemistry* 27(1), 31–34. DOI: 10.1016/0031-9422(88)80588-0

Saadi, S., Ariffin, A.A., Ghazali, H.M., Miskandar, M.S., Boo, H.C. and Abdulkarim, S.M. (2012) Application of differential scanning calorimetry (DSC), HPLC and pNMR for interpretation primary crystallisation caused by combined low and high melting TAGs. *Food Chemistry* 132(1), 603–612. DOI: 10.1016/j. foodchem.2011.10.095

Saeed, A. and Darvishzadeh, R. (2017) Association analysis of biotic and abiotic stresses resistance in chickpea (*Cicer* spp) using AFLP markers. *Biotechnology & Biotechnological Equipment* 31(4), 698–708. DOI: 10.1080/13102818.2017.1333455

Sastrapradja, S.E. and Aminah-Lubis, S.H. (1975) *Psophocarpus tetragonolobus* as a minor garden vegetable in Java. In: *Symposium on South East Asian Plant Genetic Resources*. Bogor, Indonesia.

Schwember, A.R., Carrasco, B. and Gepts, P. (2017) Unraveling agronomic and genetic aspects of runner bean (*Phaseolus coccineus* L.). *Field Crops Research* 206, 86–94. DOI: 10.1016/j.fcr.2017.02.020

Shet, M.S. and Madaiah, M. (1987) Distribution of lectin activity at different stages in the tissues of winged bean (*Psophocarpus tetragonolobus* (L.) DC). *Plant Science* 53(2), 161–165. DOI: 10.1016/0168-9452%2887%2990126-9

Siebert, S., Webber, H., Zhao, G. and Ewert, F. (2017) Heat stress is overestimated in climate impact studies for irrigated agriculture. *Environmental Research Letters*, 12(5), 054023. DOI: 10.1088/1748-9326/aa702f

Singh, P.K., Tiwari, J.K., Joshi, V., Lal, S.K., Selvakumar, R. *et al.* (2022) Winged bean–A nutritionally rich underutilized vegetable crop. *Indian Horticulture* 67(1). Available at: https://epubs.icar.org.in/index. php/IndHort/article/view/120978 (accessed 23 May 2024).

Singh, V., Goel, R., Pande, V., Asif, M.H. and Mohanty, C.S. (2017) De novo sequencing and comparative analysis of leaf transcriptomes of diverse condensed tannin-containing lines of underutilized *Psophocarpus tetragonolobus* (L.) DC. *Scientific Reports* 7(1), 44733. DOI: 10.1038/srep44733

Smartt, J. (1980) Some observations on the origin and evolution of the winged bean (*Psophocarpus tetragonolobus*). *Euphytica* 29, 121–123. DOI: 10.1007/BF00037256

Song, J.H., Shin, G., Kim, H.J., Lee, S.B., Moon, J.Y. *et al.* (2022) Mutation of GmIPK1 gene using CRISPR/ Cas9 reduced phytic acid content in soybean seeds. *International Journal of Molecular Sciences* 23(18), 10583. DOI: 10.3390/ijms231810583

Srinivas, V.R., Acharya, S., Rawat, S., Sharma, V. and Surolia, A. (2000) The primary structure of the acidic lectin from winged bean (*Psophocarpus tetragonolobus*): insights in carbohydrate recognition, adenine binding and quaternary association. *FEBS Letters* 474(1), 76–82. DOI: 10.1016/S0014-5793(00)01580-5

Sriwichai, S., Laosatit, K., Monkham, T., Sanitchon, J., Jogloy, S. and Chankaew, S. (2022) Genetic diversity of domestic (Thai) and imported winged bean [*Psophocarpus tetragonolobus* (L.) DC.] cultivars assessed by morphological traits and microsatellite markers. *Annals of Agricultural Sciences* 67(1), 34–41. DOI: 10.1016/j.aoas.2022.04.002

Stephenson, R.A., Kesavan, V., Claydon, A., Bala, A.A. and Kaiulo, J.V. (1979) Studies on tuber production in winged bean (*Psophocarpus tetragonolobus* (L.) DC.). In *Proceedings of the 5th International Symposium on Tropical Root and Tuber Crops*. Laguna, the Philippines.

Tan, N.H., Rahim, Z.H., Khor, H.T. and Wong, K.C. (1983) Winged bean (*Psophocarpus tetragonolobus*) tannin level, phytate content and hemagglutinating activity. *Journal of Agricultural and Food Chemistry* 31(4), 916–917. DOI: 10.1021/jf00118a063

Tanzi, A.S., Eagleton, G.E., Ho, W.K., Wong, Q.N., Mayes, S. and Massawe, F. (2019) Winged bean (*Psophocarpus tetragonolobus* (L.) DC.) for food and nutritional security: synthesis of past research and future direction. *Planta* 250, 911–931. DOI: 10.1007/s00425-019-03141-2

Telang, M.A., Giri, A.P., Pyati, P.S., Gupta, V.S., Tegeder, M. and Franceschi, V.R. (2009) Winged bean chymotrypsin inhibitors retard growth of *Helicoverpa armigera*. *Gene* 431(1-2), 80–85. DOI: 10.1016/j. gene.2008.10.026

Thompson, A.E. and Haryono, S.K. (1979) Sources of resistance to two important diseases of winged bean, *Psophocarpus tetragonolobus* (L.) DC. 1. *HortScience* 14(4), 532–533. DOI: 10.21273/HORTSCI.14.4.532

Tock, A.J., Fourie, D., Walley, P.G., Holub, E.B., Soler, A. *et al.* (2017) Genome-wide linkage and association mapping of halo blight resistance in common bean to race 6 of the globally important bacterial pathogen. *Frontiers in Plant Science* 8, 1170.

Tripathi, S., Tyagi, A.K., Bajaj, D., Upadhyaya, H.D., Srivastava, R. *et al.* (2016) Eco TILLING-based association mapping efficiently delineates functionally relevant natural allelic variants of candidate genes governing agronomic traits in chickpea. *Frontiers in Plant Science* 7, 450. DOI: 10.3389/fpls.2016.00450

Vannucci, L., Fossi, C., Quattrini, S., Guasti, L., Pampaloni, B. *et al.* (2018) Calcium intake in bone health: a focus on calcium-rich mineral waters. *Nutrients* 10(12), 1930. DOI: 10.3390/nu10121930

Varshney, R.K., Graner, A. and Sorrells, M.E. (2005) Genic microsatellite markers in plants: features and applications. *Trends Biotechnology* 23, 48–55. DOI: 10.1016/j.tibte ch.2004.11.005

Vatanparast, M., Shetty, P., Chopra, R., Doyle, J.J., Sathyanarayana, N. and Egan, A.N. (2016) Transcriptome sequencing and marker development in winged bean (*Psophocarpus tetragonolobus*; Leguminosae). *Scientific Reports* 6, 29070. DOI: 10.1038/srep29070

Verdcourt, B. and Halliday, P. (1978) A revision of *Psophocarpus* (Leguminosae-Papilionoideae-Phaseoleae). *Kew Bulletin* 33, 191–227. DOI: 10.2307/4109575

Wan Mohtar, W.A.A.Q.I., Hamid, A.A., Abd-Aziz, S., Muhamad, S.K.S. and Saari, N. (2014) Preparation of bioactive peptides with high angiotensin converting enzyme inhibitory activity from winged bean [*Psophocarpus tetragonolobus* (L.) DC.] seed. *Journal of Food Science and Technology* 51, 3658–3668. DOI: 10.1007/s13197-012-0919-1

Wang, Z., Shea, Z., Rosso, L., Shang, C., Li, J., Bewick, P., Li, Q., Zhao, B. and Zhang, B. (2023) Development of new mutant alleles and markers for KTI1 and KTI3 via CRISPR/Cas9-mediated mutagenesis to reduce trypsin inhibitor content and activity in soybean seeds. *Frontiers in Plant Science* 14, 1111680. DOI: 10.3389/fpls.2023.1111680

Weil, R.R. and Belmont, G.S. (1991) Dry matter and nitrogen accumulation and partitioning in field grown winged bean. *Experimental Agriculture* 27(3), 323–328. DOI: 10.1017/S0014479700019049

Weil, R.R. and Khalil, N.A. (1986) Salinity tolerance of winged bean as compared to that of soybean. *Agronomy Journal* 78, 67. DOI: 10.2134/agronj1986.00021962007800010015x

Wong, Q.N., Tanzi, A.S., Ho, W.K., Malla, S., Blythe, M. *et al.* (2017) Development of gene-based SSR markers in winged bean (*Psophocarpus tetragonolobus* (L.) DC.) for diversity assessment. *Genes* 8(3), 100. DOI: 10.3390/genes8030100

Yang, J. and Tan, H. (2011) Study on winged bean milk. In: *International Conference on New Technology of Agricultural, 2011*, IEEE, pp. 814–817. DOI: 10.1109/ICAE.2011.5943916

Yang, S., Grall, A. and Chapman, M.A. (2018) Origin and diversification of winged bean (*Psophocarpus tetragonolobus* (L.) DC.), a multipurpose underutilized legume. *American Journal of Botany* 105(5), 888–897. DOI: 10.1002/ajb2.1093

Yao, Y., Hu, Y., Zhu, Y., Gao, Y. and Ren, G. (2016) Comparisons of phaseolin type and α-amylase inhibitor in common bean (*Phaseolus vulgaris* L.) in China. *The Crop Journal* 4(1), 68–72. DOI: 10.1016/j.cj.2015.09.002

Zahran, H.H. (1999) Rhizobium-legume symbiosis and nitrogen fixation under severe conditions and in an arid climate. *Microbiology and Molecular Biology Reviews* 63(4), 968–989. DOI: 10.1128/mmbr.63.4.968-989.1999

Zampieri, M., Ceglar, A., Dentener, F. and Toreti, A. (2017) Wheat yield loss attributable to heat waves, drought and water excess at the global, national and subnational scales. *Environmental Research Letters* 12(6), 064008. DOI: 10.1088/1748-9326/aa723b

Zeven, A.C. and De Wet, J.M. (1982) *Dictionary of Cultivated Plants and their Regions of Diversity: Excluding Most Ornamentals, Forest Trees and Lower Plants*. Pudoc, Centre for Agricultural Publishing and Documentation, Wageningen, the Netherlands.

14 Moth Bean (*Vigna aconitifolia* (Jacq.) Maréchal)

Ramavtar Sharma[1]*, Sushil Kumar[2]*, Hans Raj Mahla[1], Khushwant B. Choudhary[1] and Vikas Khandelwal[3]

[1]*Division of Plant Improvement and Pest Management, ICAR-Central Arid Zone Research Institute (CAZRI), Jodhpur, India;* [2]*Department of Agricultural Biotechnology, Anand Agricultural University (AAU), Anand, India;* [3]*ICAR-All India Coordinated Research Project on Pearl Millet, Mandor, Jodhpur, India*

Abstract

Moth bean (*Vigna aconitifolia* (Jacq.) Maréchal), an annual summer crop, has been of substantial importance for human survival in drier northwest regions of India. Being a short-day, long-duration crop with a trailing growth habit, it suited traditional agriculture, taking care of both food and fodder requirements. The greater phenotypic plasticity and deep root system ensure production under scanty and erratic precipitation. It survives long spells of drought and high temperatures by ameliorating phenological developments. However, poor yield potential, trailing growth habit, susceptibility to bird damage and sensitivity to humidity have confined this to drier areas of the Thar Desert. Initially identified varieties represented long-duration spreading genotypes (maturing in 90-120 days) predominantly accessible in the germplasm collections. The development of short-duration, compact varieties through mutagenesis favoured advanced agricultural practices, lifestyle and changed rainfall patterns. The unsuccessful artificial crossing efforts attributed mainly to small flower size and paucity of diversity for yield and desirable agronomic traits favoured the use of mutation breeding techniques for genetic improvement. This crop needs inheritance studies, and most of the information generated pertains to germplasm studies. However, scientific interventions initiated in molecular genetics can help elucidate the complex genetic architecture of moth bean. It is worth mentioning that amongst the cultivated pulses, the moth bean provides an efficient system for in vitro manipulations, enabling gene expression studies and crop improvement.

Keywords: Breeding, diversity, moth bean, mutation, regeneration

14.1 Introduction

Moth bean (*Vigna aconitifolia* (Jacq.) Maréchal; 2n = 22), belonging to the Fabaceae (Leguminaceae) family, holds significant agricultural value in arid regions (Srivastava and Soni, 1995). Referred to locally as mat or moth bean, this species exhibits exceptional drought resistance, thriving even in adverse conditions. It distinguishes itself as a robust pulse crop within India's rainfed ecosystems. Particularly well-suited to regions with limited soil moisture, this annual leguminous plant adeptly copes with evaporative stress. This resilience can be attributed to its genetic composition and inherent protective mechanisms, enabling rapid adaptation to the

---

*Corresponding authors: ramavtar.cazri@gmail.com; sushil254386@gmail.com

© CAB International 2024. *Potential Pulses: Genetic and Genomic Resources*
(eds R. Chandora, T. Basavaraja and A. Pratap)
DOI: 10.1079/9781800624658.0014

unpredictable desert climate marked by scarce moisture and nutrients in the soil. The herbaceous, creeping growth of moth bean, reaching 40–50 cm, is accompanied by a deep and rapidly expanding root system. This contributes to its capacity to endure temperatures as high as 40–50°C in scorching fields. Its deep roots (reaching down to about 1.5 m depth in loose soils), coupled with a short life cycle and impressive drought tolerance, significantly enhance its ability to acclimate to the harsh, arid environment. Moreover, the expansive canopy and trailing growth habit of the plant are pivotal in preserving soil moisture and regulating temperature. In the driest conditions, moth bean emerges as a vital choice for mixed cropping, frequently paired with pearl millet (Kandpal *et al.*, 2009). This positions it as a cornerstone within resource-scarce, arid agricultural systems, where the juxtaposition of low moisture levels and high temperatures is a characteristic occurrence. This phenomenon is particularly prevalent in desert dunes and degraded lands, solidifying the importance of moth bean as an adaptable plant.

Moth bean's origin in India is evident through its abundant presence in both wild and cultivated forms. Its cultivation spans from the Himalayas in the north to Sri Lanka in the south, encompassing regions such as India, Pakistan and Myanmar (Singh and Singh, 2013). On a global scale, India holds the primary role in moth bean cultivation and production. While moth bean flourishes across the Indian subcontinent and East Asia, especially in Thailand, its cultivation has extended to drier areas in unconventional countries such as China, Africa and the southern USA. In India, Rajasthan, Maharashtra and Gujarat have emerged as key moth bean producers. Rajasthan, covering 86% of the country's area of moth bean, leads in cultivation, with a notable 2.77 lakh tonnes production between 2012 and 2015. Rajasthan's prominence continues with a cultivation area of 13.87 lakh ha and a yield of 4.34 lakh tonnes, boasting a productivity of 310 kg ha⁻¹ (Sahoo *et al.*, 2019).

Moth bean is extensively cultivated for its tender pods and mature seeds, which are consumed worldwide, especially in underdeveloped countries, owing to their substantial nutrient content, protein richness and mineral abundance (Vijendra *et al.*, 2016). The seeds, a noteworthy source of albumin and glutamine protein, are particularly abundant in amino acids, including lysine and leucine (Saxena, 2004). This versatile pulse serves various roles: as food, feed, fodder, green manure and pasture. Its lush green stems and leaves provide protein-rich and palatable fodder for cattle. This makes moth bean a nutritional resource for both humans and animals in arid regions. The legume is a key ingredient in diverse confections such as papad, bhujia, mangori, vada and kheech (Rajora *et al.*, 2009). Moreover, mature seeds of moth bean form the foundation for delectable snacks, fostering the growth of a seed-based export market and generating employment within agro-based sectors.

Despite the potential of moth bean as a versatile resource for food, fodder, feed and various industries, coupled with its impressive drought tolerance, nitrogen-fixation ability, adaptability to diverse cropping systems, and suitability for less fertile soils, there has been limited focus on genetic-level improvements for this legume (Kumar, 2002a). The paucity of superior varieties exacerbates the situation. Furthermore, moth bean cultivation spans a mere 13.50 lakh ha, which needs to be improved to meet the growing demand for high-quality seeds. The crop's subpar response to nitrogen-containing fertilizers and excessive rainfall poses challenges, emphasiszng the importance of optimal genotypes by specific precipitation patterns. The absence of high-yielding, early-maturing varieties, along with the prevalence of insect pests in rainfed areas, hinders adequate yields of this versatile crop.

To ensure enhanced seed yield stability, revising breeding methodologies is imperative, aiming to develop appropriate crop architecture paired with stress resistance. However, the existing morpho-phenotypic diversity in moth germplasm proves inadequate for breeding high-yield, stress-tolerant genotypes. The limited intrinsic variability for yield and stress tolerance restricts the expansion of the genetic base for such traits. Consequently, in conjunction with conventional approaches such as selection, hybridization, and mutation breeding, incorporating non-conventional crop improvements strategies such as DNA markers, tissue culture and genetic engineering is crucial. This holistic approach will be pivotal in unlocking the full potential of moth bean and addressing the challenges posed by its cultivation and utilization.

14.2   Botanical Description

The moth bean plant features a compact structure, with short internodes and an angular stem that grows up to 30 cm. The densely branched plant can span up to 150 cm and bear pods on peduncles that vary in size. Trifoliated leaves vary in size and shape, exhibiting deep to shallow lobes that transition from hairy when young to glabrescent as they mature (Figs 14.1 and 14.2). The raceme inflorescence appears on a 5–6 cm axillary peduncle adorned with yellowish hairs (Brink and Jansen, 2006). Bisexual papilionaceous flowers boast a campanulated calyx of 4–5 mm and display various shades of yellow due to its subgenus *Ceratotropis* classification (Boudoin and Marechal, 1988). The 5–5.5 mm yellowish corolla has a sickle-shaped keel, and ten stamens are arranged with nine united and one free. The superior ovary is sessile, accompanied by an incurved style. Fruits are cylindrical pods (2–4 × 0.5 cm) with a curved, short-hair-covered beak, turning brown or pale grey upon maturity, typically containing 5–8 seeds. These 4.5 × 3 mm seeds are variously coloured from whitish green to yellow or brown, often mottled with black. Majumdar (2011) observed epigeal germination as the predominant pattern in moth bean.

14.3   Cytology

The structural and morphological characteristics of chromosomes are pivotal in unravelling the origins, evolution and classification of taxa, as well as in gauging genetic relatedness across diverse genomes (Stace, 2000; Kumar and Rao, 2002). However, *Vigna* species, including moth bean, face challenges in karyological studies due to their remarkably diminutive chromosome size (1.6–3.7 µm) (She *et al.*, 2015). Inter-species cross-ability within *Vigna* is limited, and this scarcity has been explored from cytological and evolutionary perspectives (Ojomo and Chheda, 1971; Smartt, 1985; Kumar *et al.*, 2011; Sehrawat and Yadav, 2014; Aremu *et al.*, 2016). Cytogenetic insights can be harnessed to surmount interspecific crossing barriers, thereby unlocking the potential of genetic resources within moth bean's wild relatives. Moth bean chromosomes exhibit binding sites for chromomycin A3 (CMA), suggesting GC-rich regions at the terminal short arm(s) (Shamurailatpam *et al.*, 2015). The co-presence of CPD (combined propidium iodide and DAPI (4′,6-diamidino-2-phenylindole)), DAPI+ and post-FISH (fluorescent *in situ* hybridization) DAPI+ bands within pericentromeric areas implies heterochromatin housing GC- and AT-rich sequences

Fig. 14.1. Deep and shallow leaved spreading moth bean genotypes with long peduncle. Pods on branches damaged by soil moisture. (Author's own figure.)

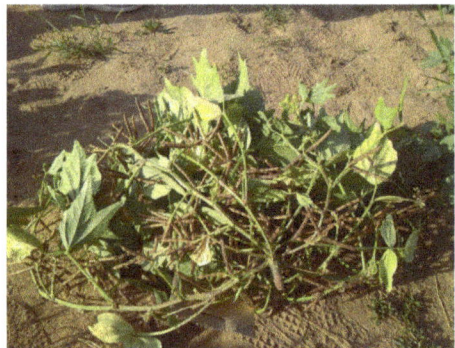

Fig. 14.2. Heavy-bearing moth bean plant with short peduncle. (Author's own figure.)

and other repeat types (She *et al.*, 2015). Sub-telocentric chromosomes were reported by Sinha and Roy (1979) and Shamurailatpam *et al.* (2015). According to Bhatnagar *et al.* (1974), the karyotype composition is characterized by one elongated sub-median centromere measuring between 2.7 and 3.5 μm, along with five moderately sized sub-median centromeres ranging from 1.96 to 2.6 μm. Additionally, there is one medium-sized centromere located at the median position, as well as four small centromeres measuring less than 1.95 μm, also positioned at the median. According to She *et al.* (2015), it is 9m+2 μm. The karyotypes adhere to Stebbins' category 2B (1971) and have a haploid complement length of 42.53+3.93 μm. Arm ratios span between 1.08 and 1.60, while relative lengths of chromosomes range from 6.43 to 13.01. Remarkably, the longest chromosome is more than 2.1 × the shortest (Shamurailatpam *et al.*, 2015; She *et al.*, 2015). In moth bean, the 45S rDNA site spans the entire short arm. Unusually, the single 5S locus resides centromerically on pair 11, contrasting to other Asiatic *Vigna*, possibly indicating an inversion involving an interstitial locus.

To enhance our understanding of global macrosynteny and karyoevolutionary mechanisms within *Vigna* and *Phaseolus*, Oliveira *et al.* (2020) employed a comparative BAC (bacterial artificial chromosome)–FISH mapping approach. This approach was applied to moth bean (subgenus *Ceratotropis*), a species lacking a sequenced genome, and the results were compared to the findings from cowpea (subgenus *Vigna*) and common bean (*Phaseolus vulgaris* (Pv)). The study involved the hybridization of 17 Pv BACs, eight Vu (*V. unguiculata* [Vu, subgenus Vigna]) BACs, and 5S and 35S rDNA probes *in situ* onto the 11 chromosome pairs of Vac (*V. aconitifolia* [Vac, subgenus Ceratotropis]).

Significantly, five chromosomes (6, 7, 9, 10 and 11) demonstrated high levels of synteny and collinearity within *V. aconitifolia*. Furthermore, the research revealed macrosynteny disruptions of the translocation type, specifically encompassing chromosomes 1 and 8 of *V. aconitifolia*. This comprehensive approach not only sheds light on the intricate relationships between different species within the *Vigna* and *Phaseolus* genera but also contributes to our broader comprehension of chromosomal evolution and synteny patterns in these legume plants.

14.4 Genetic Improvement

14.4.1 Conventional breeding interventions

In India, research efforts concerning arid legumes like guar, moth bean, cowpea and horse gram received a significant boost when they were integrated into the 'All India Coordinated Research Project on Guar Improvement' in 1992, following the recommendations of the Rao Committee. The project's name was later changed to 'All India Coordinated Research Project (AICRP) on Arid Legumes', and it is operated by ICAR – National Bureau of Plant Genetic Resources (NBPGR); however, a committee chaired by Dr P.N. Bahl, Ex. ADG, ICAR, in 1995, proposed relocating its headquarters to Rajasthan Agricultural University, Bikaner (Rajasthan), owing to its prominent contributions to the cultivation and production of these crops. Following the Jain Committee's suggestions in 1999, the AICRP on Arid Legumes became part of the National Network Project on Arid Legumes, with its base moving to ICAR – Central Arid Zone Research Institute (CAZRI), Jodhpur. More recently, to streamline pulse research, it relocated to ICAR – Indian Institute of Pulses Research (IIPR), Kanpur, in 2015, operating as the 'All India Network Project on Arid Legumes'.

14.4.2 Germplasm, varieties and variability

Through dedicated and intensive collection endeavours conducted by NBPGR, a remarkable assembly of 3422 moth bean accessions has been amassed from 1971 up to the present day. The collections were mainly collected from the states of Haryana, Gujarat, Madhya Pradesh and Rajasthan based on growth habit, leaf location, pod and seed colour. Exotic collections (41) were also acquired from Sri Lanka, Thailand, Akola, Mexico, the USA, the USSR and Taiwan. Furthermore, various institutions across the globe also maintain smaller moth bean collections. Notably, the University of Georgia and the United States Department of Agriculture (USDA) curate 56 accessions, the N.I. Vavilov All

Russian Institute of Plant Genetic Resources (VIR) manages 48, the Australian Grain Gene Bank safeguards 35, and AVRDC – The World Vegetable Centre holds 26. Additionally, CIAT in Columbia oversees eight accessions, and the Leibniz Institute of Plant Genetics and Crop Plant Research preserves six, collectively contributing to the conservation and study of moth bean diversity internationally (Kanishka *et al.*, 2023). At the forefront of conservation efforts, NBPGR's Regional Station in Jodhpur, Rajasthan, diligently maintains active collections. Additionally, IIPR, Kanpur, along with its coordinating centres outlined by Asthana (1998), curate working collections encompassing diverse *Vigna* species. These dedicated endeavours contribute significantly to the preservation and study of the rich genetic diversity within the genus *Vigna*.

Between 1991 and 1998, the majority of moth bean germplasm underwent evaluation in successive batches, unveiling significant diversity across an array of morpho-physiological traits. These traits encompassed days to flowering (ranging from 32.0 to 84), days to maturity (spanning 57 to 105), plant height (ranging between 11.68 and 49.30 cm), branches per plant (1.58 to 16), clusters per plant (varying from 9.00 to 73.67), pod length (2.00 to 5.30 cm), pods per cluster (ranging from 1.49 to 7.67), seeds per pod (fluctuating between 2.00 and 10.00), and seed index (ranging from 1.50 to 4.60 g) (Dwivedi and Bhatnagar, 2002). Notably, this assessment encompassed a broad spectrum of traits, including yield and growth characteristics, which exhibited remarkable diversity. The extent of variation in various studies has been elaborated upon in a recent review paper authored by Kanishka *et al.* (2023).

The journey of genetic improvement in moth bean commenced well before the establishment of the AICRP on Arid Legumes in 1992. Initially, the focus was on releasing dual-purpose varieties (such as Type 1, Type 2 and Baleshwar 12) with extended growth periods (100–120 days) to cater to fodder and grain needs. These early varieties, derived from landrace selections, were characterized by their spreading growth habit, forming a mat on the ground. Subsequent selection efforts honed in on developing varieties such as Jadia, Jwala and IPCMO-912, which exhibited shorter maturity periods of 80–90 days.

Notably, these varieties demonstrated robust performance in regions with normal rainfall distributions in the range 250–400 mm across 4–5 rain events (Sharma *et al.*, 2003a).

The primary objectives of genetic improvement centred on addressing primitive traits intrinsic to the moth bean. Challenges encompassed its highly indeterminate growth pattern (yielding either large vine or small bushy types depending on the environment), substantial flower and pod drop, pod-shattering tendencies, and sensitivity to light and temperature variations. While unfavourable cultivation conditions might have inadvertently preserved these survival-suited primitive attributes, they hindered yield enhancement through selection methods. Conventional hybridization was also impeded owing to the intricate nature of the crop's small flowers, diminutive anthers and inadequate pod setting (Sharma and Majumdar, 1982), further decelerating genetic progress.

In a fortunate turn, progress was achieved by discovering dwarf mutants that were early in maturing (60–70 days). These mutants, released as varieties in 1990, emerged as revolutionary genotypes. SK Rajasthan Agricultural University, Bikaner, and CAZRI, Jodhpur, emerged as significant contributors to this advancement, presenting varieties such as RMO-40, FMM-96, RMO-225, RMO-257, RMO-435, CAZRI moth-1 and CAZRI moth-2 (Sharma *et al.*, 2021). In recent years, the development of short-duration varieties has addressed the evolving climatic dynamics, offering solutions to cope with changing weather patterns. This steady progression in genetic improvement endeavours signifies the commitment to enhancing the agricultural performance and adaptability to the shifting environment of moth bean (Fig. 14.3).

Significant diversity has been documented and preserved within the moth bean crop, encompassing yield and yield-contributing characteristics. Numerous studies have reported this variability, rendering the initial selection efforts effective (Bhavsar and Birari, 1991; Dwivedi *et al.*, 1999; Dwivedi and Bhatnagar, 2002; Garg *et al.*, 2003; Kakani *et al.*, 2003; Solanki *et al.*, 2003; Om Vir and Singh, 2015). Daulay *et al.* (2000) further explored this diversity, revealing a wide array of morpho-physiological traits in addition to yield-related traits. These traits

Fig. 14.3. A set of early maturing mutants in the field. (Author's own figure.)

spanned leaves, branches, flowering and maturity periods, cluster per plant (CP), cluster per peduncle (CPed), peduncle length (PedL), pods per cluster (PC) and plant (PP), seeds per pod (SP), as well as pod and seed size, shape and colour. Moreover, these studies examined factors such as pod and seed weight per plant, along with 100-seed weight. Sihag *et al.* (2004) and Patel *et al.* (2008) found significant diversity in branches per plant, followed by grain yield per plant (GYP), CP and PP. They reported substantial heritability for GYP, days to 50% flowering (DFF), days to maturity (DM), plant height (PH), branches per plant, CP, PP and harvest index. Yaqoob *et al.* (2005) observed a broad range for PH (45–147 cm), PP (14–410), PC (2.3–4.0), PL (2.1–3.8 cm), SP (5.8–7.2), YMV incidence (scored 1 to 3) and grain yield (3.25–33.40 g). Rajora *et al.* (2012) noted significant genotype differences in days to flowering, PH, PedL, DM and GYP. Conversely, except for 100-seed weight, Yogeesh *et al.* (2016) observed substantial variations in yield and its related traits. Patil *et al.* (2007) determined that biomass per plant, PP and 100-seed weight serve as reliable predictors of seed yield, facilitating direct selection for improved yield. Notably, PP, SP and DFF were crucial for superior genotypes (Bhavsar and Birani, 1989; Bangar *et al.*, 2008). In the study by Rajora *et al.* (2012), the emphasis was placed on choosing longer pods, and shorter leaf stalks and promoting early maturity as strategies to achieve higher yields. Bhandari (1991) identified the direct positive influence of straw yield, PP, 1000-seed weight, SP and plant habit on seed yield. Other studies revealed key contributors to yield, including days to flowering, biomass, branches per plant, CP, PP, SP and seed weight (Vijaya Kumar *et al.*, 2016; Yogeesh *et al.*, 2016). Stability analysis indicated significant linear and non-linear genotype × environment interactions (Jindal and Vir, 1983; Henry, 2001; Kumar, 2001; Henry and Kackar, 2002a, b).

14.4.3 Distant hybridization

Among the domesticated Asian *Vigna* species, the moth bean is particularly captivating owing to its remarkable resilience against drought and high temperatures (Tomooka *et al.*, 2014). The precise description of the wild ancestor of moth bean remains elusive, yet it is believed to have originated in India (Arora and Nayar, 1984). More attention should be given to the study of interspecific hybridization between moth and other related species. An initial attempt for interspecific hybridization between moth and *V. trilobata*, as conducted by Chavan *et al.* (1966), yielded eight seeds, of which only one successfully reached the flowering stage, accompanied by a pollen sterility rate of 90.86%. Subsequently, Biswas and Dana (1976) managed to acquire a limited number of pods, and by employing polyploidization techniques, they could restore pollen fertility to around 89.7%. The combination of Seredex-B and GA3 was identified as effective in enhancing pod formation in inter- and intra-specific hybrids. Singh (1990) attempted wide hybridization involving `many cultivated and wild *Vigna* species, including *V. aconitifolia*. His work encompassed discussions on achieved successes, encountered barriers and proposed strategies to overcome them. Moth bean has been employed as the seed parent in hybridizations with *V. trilobata*. Conversely, *V. trilobata* has proven effective as a pollen parent in crosses with *V. mungo*, *V. radiata* and *V. aconitifolia*. None the less, attempts to achieve successful results in the reverse crosses have been unsuccessful, as documented by Bisht *et al.* (2005). However, concrete outcomes from inter-specific hybridization endeavours in moth bean are yet to be reported.

14.4.4   Mutation breeding

Research focused on enhancing moth bean crop varieties has been limited; however, in 1983, a mutation breeding programme was initiated for moth bean. This programme aimed to isolate mutants that exhibit a high yield potential and early maturation and simultaneously possess tolerance to drought conditions.

Mutation breeding methods, particularly utilizing gamma rays, have yielded remarkable successes (Kumar, 1996). Henry and Daulay (1983) harvested 25 mutants with better pod counts per plant in the M_2 and M_5 generations compared to the parent plants. These mutants demonstrated enhanced yield potential, early maturity and heightened drought tolerance. This study highlighted the potential for introducing variability in pods per plant and growth patterns through induced mutations. However, radiation treatment in both M_1 and M_2 generations led to reduced germination, survival rates and plant height, coupled with delayed flowering (Kothekar, 1989). A mutagen-treated population displayed better variability than controls in the M_2 and M_4 generations, affecting traits such as flowering initiation, maturity period, branches per plant, clusters per leaf axil, PedL, PC and seed yield. Among the treatments applied to moth bean, specifically 20 kR + ethyl methane sulfonate (EMS; 0.4%), 30 kR + EMS (0.6%), 40 kR + EMS (0.6%) and 40 kR combined with Jadia and Jwala varieties, emerged as the utmost potent. These treatments demonstrated a significant capacity to induce substantial economic variability across various traits in moth bean, as highlighted in the study by Sharma *et al.* (2003a, b). Sharma *et al.* (2002) conducted a meticulous assessment of 305 moth bean mutants, revealing the highest prevalence in the Jwala variety (46.7%), particularly at a 20 kR (34.4%) dose. Notably, increased variability

beyond controls was observed in M_2 and M_3 generations for traits such as flowering, maturity period, branches per plant, cluster per leaf axil, and seed yield per plant, as documented in the studies by Sharma *et al.* (2003b) and Singh and Kumar (2005). In a related effort, Kumar (2002b) successfully generated unbranched mutants from the branched early maturing genotype RMO-40 through the application of 30 kR gamma radiation to seeds. The application of chemical mutagens such as sodium azide (SA), hydroxylamine (HA) and EMS treatments induced variability in various traits, as demonstrated in studies by Mahna *et al.* (1993) and Sharma *et al.* (2003a, b). In their research, Mahna *et al.* (1993) successfully identified morphological and chlorophyll mutants, some of which also affected nodulation in moth bean. They proposed that single-gene changes were responsible for the observed variations in plant height (tall and dwarf), fasciation, excessive branching, reduced nodulation and chlorophyll content. Furthermore, their study revealed that the mutagenic effectiveness was most pronounced with a sodium azide concentration of 0.001 M and a hydroxylamine concentration of 0.010 M. The response of four moth bean varieties, namely RMO-40, RMO-257, Jwala and CZM-1, was investigated in terms of their reactions to EMS, methyl methanesulfonate (MMS), SA, and HA treatments. The study focused on assessing induced biological damage and the emergence of polygenic variability, as well as conducting a comparative analysis of mutagenic effectiveness and efficiency among these mutagens. The evaluation of the mutagenic progeny was undertaken to assess their yield and yield-related traits in both the M_2 and M_3 generations. The treatments displayed decreasing effectiveness in the order of MMS > SA > EMS > HA. In this study, the moth bean RMO-40 demonstrated the highest mutagenic effectiveness at 1.0 mM SA. Jangid *et al.* (2010) introduced microwave-oven-generated microwaves to induce mutagenesis in moth bean. They treated aseptically cultured seedlings, observing improved callusing and shoot emergence within 5 seconds of exposure. Prolonged treatment, however, delayed callusing. Random amplified polymorphic DNA (RAPD) analysis of regenerated plantlets revealed significant variations, showcasing microwave-based mutagenesis as a novel method for genetic diversity enhancement. Khadke and Kothekar (2018) conducted a study to investigate the impact of EMS and SA on the trypsin inhibitor content of moth bean. Through electrophoresis analysis, seven viable and 78 micro-mutants were identified, each exhibiting between three and seven iso-inhibitors of trypsin. Notably, specific viable mutants displayed substantial variations in their trypsin inhibitor (TI) profiles compared to the control group. Remarkably, these mutants exhibited a reduction in TI content by approximately 25–45%. Moreover, a marked distinction in seed protein content was observed between the micro-mutants and viable mutants. This finding underscores the significant variations arising from the induced mutations in trypsin inhibitor and seed protein content among the studied moth bean mutants.

14.4.5 Biotechnological interventions

Tissue culture

IN VITRO REGENERATION.   Tissue culture technologies have been extensively explored to enhance crop quality and yield, encompassing various crops, including legumes. In legumes, *in vitro* regeneration is often challenging, except for a notable exception found in moth bean, as demonstrated by Jain and Chopra in 1988. While several legume species have exhibited the potential for *de novo* regeneration, their impact on crop improvement remains limited, as highlighted by Mohan and Krishnamurthy in 2003.

In the case of moth bean, remarkable progress has been made in plantlet regeneration from various tissues, including roots (Eapen and Gill, 1986), hypocotyls (Bhargava *et al.*, 1983; Eapen *et al.*, 1986; Gill *et al.*, 1986, 1987), cotyledons (Bhargava *et al.*, 1983; Godbole *et al.*, 1984; Eapen *et al.*, 1986; Gill *et al.*, 1986), and leaves (Eapen *et al.*, 1986; Bhargava and Chandra, 1989). Additionally, callus cultures have successfully induced plantlets (Bhargava and Chandra, 1983; Godbole *et al.*, 1984; Krishnamurthy *et al.*, 1984).

For instance, Bhargava and Chandra (1983) initiated callus cultures in moth bean by transferring root and hypocotyl explants (2–5 mm) from 3–4-day-old seedlings onto MS medium containing K (1.0 µM) and 2,4-D (0.9 µM).

Within 5–7 days, both explants produced soft, friable, light cream-yellowish callus. The development of shoot buds from callus varied significantly, however, depending on the growth medium and hormone combinations. Notably, the researchers did not observe the induction of roots in the differentiated shoot buds.

Subsequently, in 1984, Godbole and colleagues fine-tuned the regeneration environments to obtain whole plants from shoot apices and cotyledonary explants of the PLMO-13 moth bean variety. These regenerated plantlets were successfully transferred to field conditions. Notably, the explants utilized by Godbole *et al.* (1984) were derived from plantlets cultured on a B5BA medium. Interestingly, explants from pre-conditioned seedlings demonstrated the ability to develop into plants regardless of the growth medium (14S or B5) and the presence of hormones. This suggests that pre-conditioning may stimulate the explants, possibly influenced by endogenous plant growth regulators within the moth bean. Multiple shoots were generated through callus differentiation from shoot apices and cotyledons, and they were rooted on filter paper bridges in a specialized rooting medium. Remarkably, the survival and rooting success rate of these shoots reached 60–70%, with robust roots emerging within just 21 days. These well-rooted plants were effectively transitioned to pots containing a vermiculite and soil mixture for a 14-day hardening.

Gill *et al.* (1986) achieved plant regeneration from both hypocotyl and stem explants of moth bean without using phytohormones in the basal medium. These explants exhibited callus formation and the development of embryoids/shoot buds on hormone-free basal medium. In contrast, cotyledon and leaf explants from the same study did not show successful differentiation. They highlighted the importance of factors such as pre-conditioning germinating seeds with hormones and the size and positioning of tissues on the growth medium in influencing explant regeneration capacity. The resulting plantlets, complete with roots, were successfully grown to maturity in the field using the same medium. Eapen *et al.* (1986) further demonstrated that a combination of auxins (2,4D or IAA) and cytokinins (BA or Kn) in MS medium induced callus formation in hypocotyl, stem, leaf and cotyledon explants while also showing that callus induction and regeneration from hypocotyl and stem explants could occur on basal media without phytohormones.

Eapen and Gill (1986) were the pioneers in demonstrating the possibility of directly inducing plantlet differentiation from cultured root explants of moth bean seedlings. In line with their previous findings, a medium devoid of plant growth regulators could initiate plantlet formation. However, including cytokinins such as BA, Z, Kn and 2,i-P significantly enhanced both the frequency of plant regeneration and the average number of shoot buds per culture. Seeking to boost totipotency for accelerated regeneration and stimulate shoot bud development, Gill *et al.* (1987) exposed 6–7-day-old hypocotyl explants to electrical pulses (a single pulse of 58 Ω) at room temperature. The results demonstrated that electrically treated hypocotyls exhibited a marked increase in the stimulation of shoot bud formation compared to control explants.

Jain and Chopra (1988) undertook an extensive evaluation of 196 moth bean genotypes on MS media supplemented with phytohormones, using various explants like hypocotyls, epicotyls, cotyledons and leaf tissues. Their findings revealed that the capacity for regeneration is fundamentally influenced by genetic factors, as regeneration frequencies ranged from 5% to 26%. In a parallel study, Bhargava and Chandra (1989) highlighted the significant impact of hormonal ratios, hormone types, and the age of donor plants on leaf-based regeneration. Similar to the observations made by Jain and Chopra (1988), they noted that genotype significantly influences regeneration frequency. Still, the frequency can be enhanced by adjusting phytohormone concentrations in the growth medium. None the less, regardless of genotype, it was evident that regeneration frequency declined with an increase in the age of donor plants.

Many of the above studies employed various combinations of auxins and cytokinins to facilitate callus formation, differentiation and regeneration in moth bean. Surprisingly, none of these investigations delved into using gibberellic acid (GA_3), flurprimidol or triazoles such as paclobutrazol and uniconazole in the context of *in vitro* moth bean culture.

To explore the influence of the potent triazoles paclobutrazol and uniconazole on growth and regeneration within moth bean callus cultures, Sankhla *et al.* (1991) conducted *in vitro* culturing of hypocotyls. Their research revealed that triazoles had a detrimental effect on both callus formation and regeneration efficiency in moth bean. Furthermore, Gehlot *et al.* (1989) introduced paclobutrazol (1.7 to 6.8 µM) into the regeneration medium, which resulted in a notable reduction in root and shoot differentiation *in vitro*. However, it was observed that GA_3, when used in isolation, had a limited impact on callus growth. Remarkably, it counteracted the adverse effects of triazoles, as reported by Sankhla *et al.* (1991).

Flurprimidol, known for its role as a gibberellin biosynthesis inhibitor affecting organ formation and somatic embryogenesis in tissue cultures, was also examined by Upadhyaya *et al.* (1989). In the case of moth bean *in vitro* growth, metabolism and regeneration, their findings pointed to a negative influence of flurprimidol. This underscores the notion that the gibberellin status of tissue maintained *in vitro* can significantly impact the regeneration process.

The above studies collectively provide insights into the factors influencing tissue culture and regeneration in *Vigna aconitifolia*, offering valuable knowledge for improving propagation and crop improvement strategies for moth bean.

Protoplast culture

Protoplasts serve as an excellent means to induce genetic variation in *Vigna* species. Arya *et al.* (1990) reported the successful regeneration of both protoplasts and calli, followed by the development of plants from these protoplasts, which were subsequently transferred to field conditions. These derived plants exhibited notable variations in seed, pod and plant characteristics. In a related study, Shekhawat and Galston (1983) recovered mesophyll protoplasts with driselase, a cellulolytic enzyme derived from the fungus *Irpex lacteus* (Basidiomycete). Their experiments showed that cellulase and meicelase, individually or in combination, were ineffective for this purpose. The isolated protoplasts had diameters ranging from 25 to 45 µm. Protoplast culturing was conducted using hanging drops on the inner surface of Petri dish lids. Notably, the formation of cell plates and the initial division were observed after 60–70 hours.

By the fifth day of culture, approximately 50% of the cells had undergone their first division, and this percentage increased to 60–70% by the sixth day. The researchers achieved a higher cell division and plating efficiency in protoplasts by supplementing the culture medium with glutamine and asparagine. Subsequently, the growth of green meristematic zones on the protoplasts derived on GS-3 medium, along with the successive development of embryoids bearing two cotyledons from these green islands on GS-4 medium (auxin absent and little cytokinins). These findings strongly support the idea that the protoplasts underwent regeneration into plants through somatic embryogenesis.

In contrast to the approach taken by Shekhawat and Galston (1983), Krishnamurthy *et al.* (1984) employed cellulases and a macerozyme mixture for the isolation of protoplasts. These isolated protoplasts were subsequently embedded in agar or cultured in a liquid medium. The emergence of callusing became evident on the agar surface after approximately 7–8 weeks of protoplast culture. Notably, the plating efficiency in this study ranged from 20% to 25%, a notable improvement compared to the previously reported 10% by Shekhawat and Galston in 1983. Krishnamurthy *et al.* (1984) also observed that the induction of multiple shoot bud formation in the protoplast-derived callus was more successful when using an MS basal medium without plant growth regulators compared to other combinations of MS with added hormones. Arya *et al.* (1990), like Shekhawat and Galston (1983), utilized driselase to isolate mesophyll protoplasts with diameters ranging from 25 to 50 μm sourced from leaf tissue of 21-day-old plants. Their culturing methodology remained consistent with Shekhawat and Galston's approach. Arya *et al.* (1990) achieved the successful development of plantlets derived from these protoplasts, nurturing them to maturity. The cultivated plantlets exhibited variations in characteristics such as seed germination, pod length, plant height and the ratio of seed weight to pod weight. These findings indicated that protoplasts from highly homozygous legumes could serve as a source of genetic variation, offering potential opportunities for further exploitation in moth bean improvement programmes. This research thus contributes to our understanding of genetic diversity enhancement in *Vigna aconitifolia* and its implications for crop improvement.

In vitro *flowering*

Flowering, marking the shift from vegetative to reproductive stages, is orchestrated by complex genetic pathways influenced by internal and external cues (Kazan and Lyons, 2016). A significant advancement in crop genetic enhancement has emerged by manipulating flower evocation and induction, allowing for the remarkable induction of *in vitro* flowering. This technique expedites reproductive development, with potential implications for crop improvement programmes. *In vitro* flowering, fruiting and seed set accomplishments across various species offer a strategy to hasten improvement efforts while evading natural field-based selection pressures. Recognizing these advantages, Saxena *et al.* (2005) embarked on investigating the feasibility of manipulating *in vitro* flowering in moth bean. Their approach involved introducing de-rooted 7–8-day-old seedlings onto an MS medium with N6-benzylaminopurine (BAP) and Kn (ranging from 0.1 to 3.0 mg/l) alone and in combination with indole acetic acid (IAA) (0.1–1.0 mg/l). A notable achievement of up to 46.66% flowering was recorded using 1 mg/l of Kn. Although BAP initially hindered flowering, the negative effect was mitigated by adding 1 mg/l of IAA. This innovative methodology holds the promise of advancing crop development through controlled *in vitro* flowering, bypassing traditional constraints and maximizing genetic potential.

14.5 Molecular Biology: DNA Markers, Transcriptomics and Genetic Transformation

14.5.1 Marker study for diversity analysis

The effectiveness of any breeding strategy hinges upon genetic diversity and its accurate evaluation. Traditionally, plant breeders have employed morphological traits (both quantitative and qualitative) to assess diversity and selection. However, these traits often fail to indicate genetic variation precisely owing to their susceptibility to environmental influences, leading to selection inefficiencies. DNA markers have emerged as a more suitable tool for evaluating genetic diversity, confirming cultivar/hybrid purity, identifying varieties/genotypes, and selecting diverse genotypes for hybridization. Furthermore, molecular markers facilitate the tracking of genes of interest before phenotypic expression (Lübberstedt and Varshney, 2013). Although RAPD, inter-simple sequence repeat (ISSR) and amplified fragment length polymorphism (AFLP) have been utilized for moth bean phylogenetic studies, information regarding the extent and pattern of genetic variability, particularly at the DNA level, remains insufficient.

Although variability in moth germplasm is apparent at the morpho-agronomic level, a comprehensive investigation of genetic diversity using DNA techniques remains incomplete. Despite the limited reproducibility of RAPD markers, they have found extensive use in assessing genetic diversity, germplasm characterization, cultivar identification, genetic purity testing and gene tagging (Table 14.1). Kaga *et al.* (1996) examined genetic variation across five species, including the moth bean, within the *Vigna* subgenus *Ceratotropis* using RAPD analysis. They observed greater inter-specific differences among moth, green gram and black gram than in others. Moitra and Mandal (2003) initially attempted DNA fingerprinting of seven moth bean accessions using the M13 forward primer as a RAPD marker, revealing noticeable polymorphism among the genotypes. Saxena *et al.* (2006) utilized RAPD primers to evaluate induced and natural genetic diversity in moth bean. Their study involved 12 accessions, including released varieties, mutant lines and breeding materials. Among 72 primers, 208 bands (average of 4.0 per primer) were generated; 171 were scorable and repeatable, with 79 exhibiting polymorphisms. Genotype-specific amplicons were more common in mutants. Genetic similarity coefficients ranged from 0.93 to 0.76, with a mean of 0.83 ± 0.0058, reflecting diversity variations across accessions and regions.

Imran (2011) evaluated 31 moth bean accessions using seven RAPD primers. Jaccard's similarity coefficient was between 0.20 and 1.00 (average 0.55), dividing genotypes into three subgroups. Mamta (2012) analysed ten

Vigna aconitifolia genotypes via RAPD, with five drought-tolerant (Jwala, PLMO-7, IC-39749, PLMO-2 and IC-39754) and five drought-susceptible (RMO-40, IC-140678, IC-20992, IC-39848 and IC-472239) genotypes. Among 40, 20 primers produced 101 bands, of which 86 were polymorphic. The genotypes formed five clusters: RMO-40, PLMO-2 and IC-140678 in one, IC-20992, IC-39848 and IC-472239 in another, IC-39754 and IC-39749 in the third, and PLMO-7 and Jwala separately, matching their physiological drought-resistance status.

AFLP analysis of diploid species within *Vigna* subgenus *Ceratotropis* indicated proximity between sections *Ceratotropis* (black gram and green gram) and *Aconitifoliae* (moth bean) (Tomooka *et al.*, 2002). Wang *et al.* (2008) employed gene-specific markers and sequencing to evaluate phylogenetic relationships and genetic variability in USDA *Vigna* germplasm, including moth bean. Among three examined moth bean accessions, two (PI 165479 and PI 372355) closely grouped with minimal genetic variation (genetic distance = 0.001, 66% bootstrap value), distinct from PI 164419 (genetic distance = 0.027). This unveiled some genetic diversity within the moth bean, positioning it closer to the black gram, a *Ceratotropis* section member.

While the wild moth bean variant was recognized in India (Arora and Nayar, 1984), no living samples exist in gene banks (Goel *et al.*, 2002). This scarcity of preserved samples results in a need for more information regarding the identity and beneficial traits of the wild form. In a study by Takahashi *et al.* (2016), DNA sequencing of ribosomal RNA-internal transcribed spacer and atpB-rbcL was performed on 71 *Vigna* genus accessions, including six moth bean accessions. Molecular analysis revealed the longest atpB-rbcL length (699–700 bp) in moth bean, akin to findings by Doi *et al.* (2002). Phylogenetic trees distinguished the subgenus *Ceratotropis* as a distinct group from subgenera *Vigna* and *Plectrotropis*. This study unveiled the wild ancestor of moth bean within gene bank collections, represented by two Tamil Nadu accessions (ID-5 and ID-6) and one Andhra Pradesh accession (ID-4), pinpointing south-eastern India as the primary habitat for the wild moth bean.

Imran (2011) studied the cross-species amplification of SSR markers developed in adzuki bean (Wang *et al.*, 2004) and green gram

Table 14.1. DNA marker studies in *Vigna aconitifolia* (moth bean) for diversity analysis. (Author's own table.)

Details of plant material	Country	Type of marker	Number and list of primer	Reference
Accessions: 23 of 5 species in the subgenus *Ceratotropis*, genus *Vigna*	Japan	RAPD	24 random decamers	Kaga *et al.* (1996)
Accessions: 7 moth bean	India	RAPD	M13 forward primer	Moitra and Mandal (2003)
Accessions: 12 moth bean	India	RAPD	72 random decamer primers (OPA, OPB, OPC and OPG series)	Saxena *et al.* (2006)
Accessions: 31 moth bean	India	RAPD and SSR	7 RAPD primers and 14 SSR primers of adzuki bean, and 12 SSR primers of green gram	Imran (2011)
Genotypes: 10 *V. aconitifolia* genotypes	India	RAPD	40 RAPD primers	Mamta (2012)
Germplasm: 18 out of the 21 species in the subgenus *Ceratotropis*	Japan	AFLP	–	Tomooka *et al.* (2002)
Accessions: 48 representing 12 *Vigna* species	USA	Gene-derived markers	30 gene-derived markers from legumes	Wang *et al.* (2008)
29 *Vigna* species, selected from five of the nine subgenera, and nine species of *Phaseolus*	India	Internal transcribed spacer (ITS) sequences	ITS region of the 18S-26S nuclear ribosomal DNA repeat	Goel *et al.* (2002)
71 accessions of the genus *Vigna*, consisting of 28 species and three subgenera (*Ceratotropis*, *Plectrotropis* and *Vigna*)	Japan	DNA sequence-based markers	Nuclear rDNA-ITS and chloroplast atpB-rbcL spacer regions	Takahashi *et al.* (2016)
Genus *Vigna* subgenus *Ceratotropis*	Japan	ITS and atpB-rbcL intergenic spacer	Sequence data from the ribosomal DNA ITS and atpB-rbcL intergenic spacer of chloroplast DNA regions	Doi *et al.* (2002)
Five populations of wild adzuki bean	Japan	SSR	50 primers pair from an (AG)(n)-enriched library for adzuki bean	Wang *et al.* (2004)
Accessions: 17 mung bean	Thailand	SSR	192 SSRs	Tangphatsornruang *et al.* (2009)

(Tangaphatsornruang *et al.*, 2009). Fourteen and 12 SSR primers of adzuki bean and green gram were chosen for amplification in moth. Of 26, four markers, namely CEDG-06, CEDG-264, VR-274 and VR-304, failed to amplify in moth.

The remaining markers successfully exhibited amplification in the range of its original species. Out of 14 adzuki bean SSRs, 85% (12 of 14) displayed successful amplification. Similarly, from the 12 mung bean primers, 83% (10 out of 12) exhibited amplification in moth. None of the primers was found to be polymorphic in moth bean, which suggests a conservation of the repeat motif in moth bean.

Chaitieng *et al.* (2006) demonstrated successful amplification of *V. angularis* SSR markers in *V. aconitifolia*; of 205 SSRs, 141 (68.8%) effectively amplified in moth. In cross-species amplification, Somta *et al.* (2009) found that 68 out of 85 (80%) green gram genic SSR primers amplified well in moth bean. These findings underscore the close phylogenetic relationship within Asian *Vigna* species. While inter-specific hybrids between green gram and moth bean exist, these SSRs could identify such hybrids. The high amplification success from green gram and adzuki bean SSRs highlights genome homology among *Vigna* species, which is valuable for genetics and genomics studies. None the less, limited polymorphism restricts broader applications.

14.5.2 Transcriptomics studies

Heat stress inflicts permanent harm on plant functionality and development. The combination of elevated temperatures and drought not only leads to reduced yields but also adversely impacts the quality of the produce (Mullarkey and Jones, 2000; Chen *et al.*, 2006). Moth bean, a resilient crop thriving in the arid Thar Desert, employs a multitude of structural and functional mechanisms to evade drought stress induced by high temperatures and sporadic rainfall.

In recent years, functional genomics approaches, as demonstrated by Rampuria *et al.* (2012), have played a crucial role in unravelling the stress-responsive mechanisms in plants. However, compared to other significant pulse crops, research aimed at identifying genes that confer tolerance to abiotic stresses in moth bean remains scarce. Genomics-based studies, such as those conducted on chickpea and cowpea (Gazendam and Oelofse, 2007; Deokar *et al.*, 2011), have led to the identification and characterization of candidate genes responsible for withstanding heat and drought. While substantial progress has been made in understanding stress tolerance in various crops, exploring similar mechanisms in moth bean lags behind. Expanding our knowledge in this area can enhance the resilience and productivity of this vital crop in the face of challenging environmental conditions. Some information related to the identification of expressed sequence tags (ESTs), markers and genes has been obtained in moth bean using the suppression subtractive hybridization (SSH) technique by Rampuria *et al.* (2012) and Gurjar *et al.* (2014), using differential display reverse transcriptase (DDRT)-PCR by Soni (2010) and transcriptome study by Tiwari *et al.* (2018b).

Rampuria *et al.* (2012) utilized an SSH approach to create a cDNA library to identify genes that are induced early in response to high temperatures in moth bean. To isolate RNA for the generation of the SSH library, they exposed 15-day-old seedlings of RMA-40 to a brief temperature treatment of 42°C for just 5 minutes. Following rigorous bioinformatics analysis, they successfully assembled a set of 488 unigenes, comprising 114 contigs and 374 singletons, from the alignment of 738 ESTs. Among these unigenes, 206 ESTs (28%) coded for unknown proteins, and notably, 160 ESTs (14%) were identified as novel to moth bean. During a similarity search, top Blast hits found matches with soybean proteins. Remarkably, 578 ESTs displayed substantial similarity during BLASTX analysis against the nr-database. Extensive analysis identified gene ontology annotations for 479 sequences (65%), including 339 sequences annotated with 165 EC codes, which were mapped to 68 distinct KEGG pathways. Significantly, several ESTs were discovered to have pivotal functions in the biosynthesis of crucial amino acids, including serine (JK266289), proline (JK266411), leucine (JK265743) and methionine (JK265944). These amino acids serve as precursors for the production of various specific metabolites and are directly involved in the osmotic regulation of cells. Rampuria *et al.* (2012) further highlighted the up-regulation of RNA-binding zinc finger CCCH domain-containing protein (JK265825, InterPro: COIL, TMHMM) under stress conditions, suggesting its likely involvement in the stress response. Additionally, they observed the upregulation of the AP2/ERF domain-containing protein (JK266265) in moth subjected to high temperatures, which correlated with the development of drought stress alongside heat stress. Their study notably unveiled the participation of around 20 distinct signalling genes and 16 diverse transcription factors linked to heat stress in moth bean. This offers fresh insights into the responses of the moth bean plant to environmental challenges. Rampuria *et al.* (2012) swiftly constructed an SSH library in genotype RMO-40 by subjecting it to a mere 5 minutes of heat stress, resulting in the identification of myriad signalling genes, including transcription factors and chaperones. These genes set in motion a signalling cascade in response to stress conditions. Seeking to delve deeper into the mechanisms underlying heat tolerance, Gurjar *et al.* (2014) embarked on another SSH experiment to identify genes exhibiting overexpression after a 30-minute heat stress period. Employing the same genotype, RMO-40, Gurjar *et al.* (2014) focused on unravelling heat-related ESTs in moth through SSH.

For the RNA extraction used in the SSH analysis, they subjected two-week-old seedlings to an elevated temperature of 42°C for 30 minutes. This experiment yielded 300 clones from the SSH library, with 200 high-quality ESTs >100 base pairs generated. Subsequent sequence alignment produced 125 unigenes, comprising 33 contigs and 92 singletons. The majority of Blast hits exhibited a strong resemblance to genes in *Glycine max* and *Populus trichocarpa*. Among the differentially expressed ESTs, a fascinating 21% segment comprised stress-related proteins. These included ubiquitin (JK087319), ubiquitin-conjugating enzyme (JK087278), ubiquitin ligase-like protein (JK087321), glutathione S-transferase (JK087205), peptidylprolyl isomerase (JK087268), leucine zipper protein (JK087310), CC-NBS-LRR resistance protein (JK087265) and putative Ca^{2+} endoplasmic reticulum ATPase (JK087187). Notably, the upregulation of the leucine zipper protein proposes the contribution of the abscisic acid biosynthesis pathway in thermotolerance. Furthermore, approximately 35% of the ESTs were annotated as heat-shock proteins (HSP), shedding light on their critical role during heat stress in the plant. This research represents a significant step toward unravelling the molecular mechanisms governing heat tolerance in moth bean.

In a meticulous effort to uncover drought-related genes in moth bean, Soni (2010) initiated a comprehensive screening process involving a set of 87 genotypes, which comprised 12 distinct varieties and 75 mutant lines. The objective was to assess desiccation tolerance across these genotypes at both the germination and seedling stages. Various concentrations of polyethylene glycol (PEG)-6000 and alcohol were employed for this purpose. Ultimately, among these genotypes, RMO-435 exhibited notable drought tolerance based on physiological parameters and was subsequently chosen for transcriptomic analysis through DDRT-PCR. A total of 27 primer combinations, consisting of three anchored and nine arbitrary primers, were utilized in the DDRT-PCR procedure. Remarkably, three of these primer combinations (P1/T8, P3/T1, P3/T3) failed to produce any amplification bands. However, the remaining 24 primer combinations collectively generated 95 bands, averaging approximately 4.04 bands per primer combination. The majority of amplicons fell below the 1-kb threshold, although some approached a size of 1.5 kb.

Following amplification, 40 clones were sequenced and subsequently matched with ESTs from various plant genera. Intriguingly, 19.35% of these differentially expressed genes exhibited matches with *Arabidopsis thaliana* clones, with an additional 9.68% corresponding to *Glycine max*. Notably, there was a notable prevalence of putative heat shock proteins (7) and regulatory proteins (10) among the matched genes. These 40 differentially expressed genes were categorized into six distinct classes based on their presumed functions, encompassing cell signalling-related proteins, transporter proteins, HSPs, regulatory proteins, photosynthetic proteins and miscellaneous proteins.

Furthermore, a BLAST analysis unveiled the similarity of these genes with a range of previously identified sequences, such as transposase proteins from *Vibrio* spp., transmembrane channel proteins from *Cicer arietinum*, heat shock proteins from *Arabidopsis thaliana*, malate dehydrogenase from *Citrullus lanatus*, and RO-TAMASE FKBP 1 (ROF1) with FK506 binding, calmodulin-binding, and peptidyl-prolyl cis-trans isomerase properties from *A. thaliana*. This comprehensive study represents a significant stride in unravelling the molecular underpinnings of drought tolerance in moth bean. Transcriptome analysis has become a valuable tool in recent years for studying gene expression, identifying novel transcripts and pinpointing differentially expressed genes. Tiwari *et al.* (2018b) conducted a comprehensive study in which they evaluated 240 moth genotypes, representing cultivars from various regions of moth bean cultivation in India, for the identification of a moisture stress-tolerant genotype named Marumoth followed by a transcriptome study of Marumoth. For transcriptome analysis, they isolated total RNA from the leaves of both controls. They stressed samples and generated a substantial number of reads on both the Roche 454 and the ABI Solid platform for both samples. Their analysis identified 1287 SSRs and 5606 transcripts associated with 179 pathways. Additionally, they observed the upregulation and downregulation of ten and 41 transcripts, respectively, under water stress conditions.

Transcription factors (TFs) are critical for the adaptability of a plant to abiotic stress. In this study, bHLH-type proteins, which act as transcriptional regulators, were abundant during stress conditions, suggesting their role in stress tolerance in moth bean. Other TFs, such as ethylene responsive factor, NAC and WRKY, were also observed to play a role in moisture-deficient conditions, confirming their regulatory functions in stress responses. Furthermore, catalase activity, associated with stress response, increased significantly in the leaves of drought-stressed plants, in line with the upregulation of catalase transcripts observed in the study.

In an investigation, Tiwari *et al.* (2018a) conducted quantitative PCR (qPCR) experiments to examine the expression profiles of three selected putative annotated stress-induced genes, namely, HSP, dehydration responsive protein (DRP), and DNA-3-methyladenine glycosylase 1 (Tag1), in both control and stressed samples. The RT-PCR analysis revealed varying gene expression levels in response to heat stress treatments across different moth bean genotypes. Notably, genotypes such as Jwala, RMO 40, Marumoth, Jadia, and IC 36157 exhibited substantial upregulation of HSP, DRP and Tag1 genes. This heightened gene expression suggests that these particular genotypes possess a strengthened capacity to withstand high-temperature stress. As a result, these genotypes are promising candidates for breeding programmes to develop improved heat-tolerant moth varieties.

In a recent study conducted by Suranjika *et al.* (2022), a *de novo* transcriptome assembly was performed in moth (genotype RMO-435), encompassing six different tissue types representing various developmental stages. These tissues included leaves, roots, flowers, pods and seed tissue at early and late developmental stages. The RNA-seq was performed in three replications on the Illumina NextSeq platform and yielded an impressive 520 million high-quality paired-end reads for each tissue. Subsequently, rigorous quality filtering retained approximately 494 million clean reads, constituting 95% of the raw data, which were utilized to construct the transcriptome of *V. aconitifolia*. The assembly process resulted in 150,938 unigenes, characterized by an average length of 937.78 base pairs, with an N50 value of 1227 base pairs. The biggest unigene exceeded a substantial length of 16 kb. Predominantly, the unigenes fell within the 500–1000 base pair range. Notably, approximately 79.9% of these unigenes were successfully annotated in public databases, with 12,839 exhibiting significant matches in the KEGG database. Differential expression analysis revealed that the majority of unigenes exhibited significant changes in expression levels during the late stages of seed development when compared to leaves. An in-depth *in silico* expression analysis unveiled 35,685 differentially expressed unigenes across various tissues relative to leaves. Furthermore, the study identified 74,082 unigenes as transcription factors and categorized them into 58 different families. The bHLH family of transcription factors was the most abundant, with 7539 members, followed by MYB-related (5268), NAC (5055) and ERF (3933). Additionally, 12,096 SSRs were identified within the genic regions of moth bean. Digital expression analysis uncovered tissue-specific gene activities, subsequently validated using qPCR analysis. Notably, the qPCR study demonstrated significant up-regulation of Hsp70 (TRINITY_DN56042_c11_g1_i1.p1) in the root and late seed tissues of moth bean.

14.5.3 QTL mapping and GWAS

Somta *et al.* (2018) mapped quantitative trait loci (QTL) using an F_2 population consisting of 188 plants of moth bean. This population was created through a cross between a resistant genotype, TN67 (♂ parent), and a susceptible genotype, IPCMO056 (♀ parent), to investigate seed resistance against the weevil (*Callosobruchus chinensis*). The segregation pattern indicated that a single dominant gene governs the resistance to weevil in TN76. Two significant QTLs for resistance were identified: the primary QTL named qVacBrc2.1 and a modifying QTL called qVacBrc5.1. qVacBrc2.1 was mapped on linkage group 2, located between the SSR, namely markers CEDG261 and DMB-SSR160. It explained 50.41–64.23% of the variance in resistance-related traits, varying depending on the specific trait and population. The study by Somta *et al.* (2018) recommended the use of markers CEDG261 and DMB-SSR160 in marker-assisted selection for weevil resistance in *V. aconitifolia*.

Oliveira *et al.* (2020) introduced the initial physical map of *V. aconitifolia* and conducted a comparative analysis of its chromosomes with *Vigna* and *Phaseolus* species. This comparison revealed a significant level of genome synteny between *V. aconitifolia* and related pulse crops such as *Vigna radiata*, *V. angularis*, *V. umbellata*, *V. unguiculata* and *Phaseolus vulgaris*. As demonstrated by Yundaeng *et al.* (2019), this genomic similarity suggests that genomic resources established for related species will be valuable to improve moth bean. A recent study by Yundaeng *et al.* (2019) has reported the mapping of domestication-related traits (DRTs) in moth bean, representing a significant advancement in our understanding of this crop. Utilizing a comprehensive set of 1644 SSR markers sourced from various *Vigna* species, including *V. angularis*, *V. radiata*, *V. unguiculata* and *Phaseolus vulgaris*, this study marked the first genetic mapping effort for DRTs in moth bean. In this research, an F_2 population comprising 188 plants, resulting from a cross between TN67 and the susceptible genotype, i.e. IPCMO056, served as the experimental basis. The investigation successfully identified 50 QTLs and three genes associated with 20 different DRTs. These QTLs varied in number per DRT, ranging from one to eight, with a predominant presence in linkage groups (LGs) 1, 2, 4, 7 and 10. These LGs played pivotal roles in the genetic control of crucial DRTs, including seed dormancy, pod shattering and seed size. Notably, seven major QTLs, accounting for 20% of the phenotypic variance (PVE), were identified on LGs 1, 4 and 7. Furthermore, for organ size attributes such as seed size, pod size, leaf size and stem thickness, QTLs were distributed across LGs 1, 2, 3, 4, 7, 8, 9 and 10 (Yundaeng *et al.*, 2019), underscoring the multifaceted genetic control of these essential traits in moth bean.

A comprehensive comparative genomic analysis of QTLs associated with DRTs in moth has unveiled intriguing insights. Notably, some of these QTLs exhibit correspondence with previously identified QTLs linked to equivalent traits in other *Vigna* crops, as documented in studies by Isemura *et al.* (2007, 2010, 2012) and Kongjaimun *et al.* (2012). For instance, moth bean features two QTLs responsible for pod dehiscence, one on LG 1 and the other on LG 7. Among these, the QTL located on LG 1 bears semblance to those discovered in green gram and cowpea, while the QTL residing on LG 7 aligns with findings from other *Vigna* species. However, it is important to note that there are also significant QTLs with large effects specific to moth bean, underscoring the involvement of additional genes or genomic regions. This observation aligns with the unique adaptation of moth bean to arid climates, where it thrives when other *Vigna* species struggle to grow efficiently (Yundaeng *et al.*, 2019).

Among the myriad DRTs in moth, one trait of particular interest to researchers is seed size, given its notably smaller dimensions compared to other *Vigna* species. The small seed size in moth bean may be linked to genetic mutations that could have led to significant changes in seed size, similar to other *Vigna* species; however, these mutations were not utilized during the domestication of moth bean. This distinctive characteristic encourages further exploration and investigation by researchers in the field. The genome-wide association study (GWAS) approach has been employed to investigate the genetic underpinnings of intricate traits in model legume crops. Despite the moth bean's commendable nutritional profile, there is a conspicuous absence of association studies aimed at unravelling the genetic variability and architectural intricacies of desirable traits. The dearth of comprehensive genomic resources has significantly curtailed investigations into genetic diversity and gene mapping endeavours in this crop. There has been limited utilization of the SSR marker system to explore the genetic underpinnings of domestication-related traits within the F_2 mapping population of moth bean.

A study by Yadav *et al.* (2023) aimed to explore genetic diversity, population structure and marker-trait associations for flowering traits in moth bean, utilizing a diversity panel of 428 accessions. Through genotyping by sequencing, 9078 high-quality single nucleotide polymorphisms (SNPs) were identified. Population analysis classified moth bean accessions into two subpopulations, with higher variability observed in accessions from the north-western region of India. Analysis of molecular variance indicated significant variation within individuals and among individuals, emphasizing the genetic diversity within the species. Marker-trait association analysis using multiple models identified 29 genomic regions associated with days to 50% flowering, demonstrating consistency across different models. Four of these regions showed significant phenotypic effects. Additionally, genetic relationships among *Vigna* species were examined, with the localization of moth bean SNPs suggesting a close relationship with *Vigna mungo*. These findings highlight the centre of diversity in the north-western regions of India and provide potential genomic regions and candidate genes associated with flowering traits, offering valuable insights for breeding programmes to develop early-maturity moth bean varieties.

14.6   Genetic Transformation

Despite substantial political and regulatory challenges, genetically modified crops have witnessed rapid adoption as a groundbreaking agricultural innovation. Legumes have, however, been traditionally considered 'recalcitrant' for genetic transformation studies, with the primary hurdle being the regeneration of transformed plantlets (Somers *et al.*, 2003). Genetic transformation has been accomplished in major pulse crops such as chickpea, red gram, green pea, green gram, *Phaseolus* spp, *Vicia* spp., etc. However, reports on transgenic moth bean remain limited. Despite the emergence of transgenics in the 1980s and the initial success of developing a transgenic legume crop, the moth bean, in the same decade (Eapen *et al.*, 1987; Kohler *et al.*, 1987a, b), growth in this area has not been as significant as in cereals.

Transformation of *V. aconitifolia* has been achieved using *Agrobacterium*-mediated techniques (Eapen *et al.*, 1987; Lee *et al.*, 1993; Malabadi and Nataraja, 2007; Kumar *et al.*, 2008,

2010) and direct DNA transfer through protoplast treatment with PEG and electroporation (Kohler *et al.*, 1987a, b; Bhargava and Smigocki, 1994; Kamble *et al.*, 2003). Various explants, including protoplasts (Kohler *et al.*, 1987a, b; Eapen *et al.*, 1987), hypocotyl (Lee *et al.*, 1993), cotyledonary nodes (Kumar *et al.*, 2010) and leaves (Kumar *et al.*, 2008), have been employed for transformation. Kohler *et al.* (1987a) successfully transformed isolated protoplasts from primary plants by facilitating DNA uptake induced by PEG, leading to the development of plantlets. Subsequently, direct gene-gun-mediated transformation of ungerminated mature embryos was successfully accomplished (Bhargava and Smigocki, 1994). Recent advancements in moth bean transformation techniques include co-cultivation with *Agrobacterium* (Kumar *et al.*, 2008) and vacuum infiltration of cotyledonary nodes with *Agrobacterium* (Kumar *et al.*, 2010).

14.7   Conclusion and Future Perspectives

Moth bean, a crop cultivated primarily on marginal lands, has received limited attention in terms of scientific interventions. The absence of high-yielding, early-maturing, upright varieties, coupled with challenges related to fungal and bacterial diseases, insect pests and bird damage in rainfed cultivation, imposes significant constraints on achieving adequate production of this versatile crop. Consequently, there is a pressing need to formulate breeding strategies aimed at developing an appropriate plant architecture while enhancing tolerance to both abiotic and biotic stresses, thereby ensuring a higher level of moth bean production.

At the morphological level, it is evident that the existing variability within moth bean germplasm is insufficient to confer the necessary stress tolerance and support sustained improvements in yield. Thus, in addition to traditional breeding methods such as selection, hybridization, and mutation breeding, there is an urgent imperative to explore non-conventional genetic improvement approaches, including molecular markers, tissue culture, and genetic transformation technologies. These advanced techniques hold the potential to revolutionize moth bean cultivation and address the existing limitations, ultimately leading to increased production levels.

The limited scientific intervention for moth bean extends to areas such as constructing linkage maps and identifying QTLs and GWAS. The development of linkage maps, specific to moth bean, can play a crucial role in understanding the genetic basis of various traits, including stress tolerance and yield-related characteristics. Likewise, QTL analysis and GWAS offer promising avenues for pinpointing key genetic markers associated with desirable traits, facilitating more precise breeding efforts. As we delve into the realm of molecular genetics, these techniques can help elucidate the complex genetic architecture of moth bean, enabling breeders to target specific regions of the genome associated with stress tolerance and productivity. By integrating various advanced genetic approaches with traditional breeding methods, one can unlock the full potential of moth bean, ushering in a new era of enhanced production and resilience on marginal lands.

References

Aremu, B.R., Adegbite, A.E. and Babalola, O.O. (2016) Cytological studies of some *Vigna* (L.) Walp. species. *Journal of Life Sciences* 8, 33–38.

Arora, R.K. and Nayar, E.R. (1984) Wild relatives of crop plants in India, NBPGR Sci. Monograph No. 7. National Bureau of Plant Genetic Resources, New Delhi.

Arya, I.D., Arya, S., Rao, D.V. and Shekhawat, N.S. (1990) Variation amongst protoplast-derived mothbean (*Vigna aconitifolia*) plants. *Euphytica* 47, 33–38.

Asthana, A.N. (1998) Pulse crops research in India. *The Indian Journal of Agricultural Sciences* 68 (8), 448–452.

Bangar, N.D., Lakra, A. and Chavan, B.H. (2008) Correlation and path coefficient analysis in moth bean. *Journal of Maharashtra Agricultural Universities* 33, 164–166.

Baudoin, J.P. and Marechal, R. (1988) Taxonomy and evolution of the genus *Vigna*. In: Shanmugasundaram, S. and McLean, B.T. (eds) *Mungbean Proceedings of Second International Symposium*, Asian Vegetable Research and Development Centre, Taiwan, pp. 2–12.

Bhandari, M.M. (1991) Association analysis in moth bean (*Vigna aconitifolia* Jacq. Marechal). *International Journal of Tropical Agriculture* 9, 38–41.

Bhargava, S. and Chandra, N. (1983) In vitro differentiation in callus cultures of moth bean (*Vigna aconitifolia* Jacq. Marechal). *Plant Cell Reports* 2, 47–50.

Bhargava, S. and Chandra, N. (1989) Factors affecting regeneration from leaf explants of moth bean, *Vigna aconitifolia* (Jacq.) Marechal. *Indian Journal of Experimental Biology* 27, 55–57.

Bhargava, S. and Smigocki (1994) Transformation of tropical grain legumes using particle bombardments. *Current Science* 66, 439-42.

Bhargava, S., Upadhyay, S., Garg, K. and Chandra, N. (1983) Differentiation of shoot buds in hypocotyl explants and callus cultures of some legumes. In: Sen, S.K. and Giles, K.L. (eds) *Plant Cell Cultures in Crop Improvement*. Plenum Press, New York, pp. 431–433.

Bhatnagar, C.P., Chandola, R.P., Saxena, D.K. and Sethi, S. (1974) Cytotaxonomic studies on genus *Phaseolus*. In: *Proceeding of 2nd General Congress of SABRAO*. New Delhi, pp. 800–804.

Bhavsar, V.V. and Birari, S.P. (1989) Variability, correlation and path analysis in moth bean. *Journal of Maharashtra Agricultural Universities* 14, 148–150.

Bhavsar, V.V. and Birari, S.P. (1991) Genetic divergence in moth bean, *Vigna aconitifolia* (Jacq.) Marechal. *Biovigyanam* 17, 118–120.

Bisht, I.S., Bhat, K.V., Lakhanpaul, S., Latha, M., Jayan, P.K. *et al*. (2005) Diversity and genetic resources of wild *Vigna* species in India. *Genetic Resources and Crop Evolution* 52, 53–68.

Biswas, M.R. and Dana, S. (1976) *Phaseolus aconitifolia* x *P. trilobata* cross. *Indian Journal of Genetics and Plant Breeding* 36, 125–131.

Brink, M. and Jansen, P.C.M. (2006) *Vigna aconitifolia* (Jacq.) Marechal. In: Brink, M. and Belay, G. (eds) *Plant Resources of Tropical Africa 1: Cereals and pulses*. PROTA Foundation, Wageningen, Netherlands/Backhuys Publishers, Leiden, Netherlands/CTA, Wageningen, Netherlands, pp. 201.

Chaitieng, B., Kaga, A., Tomooka, N., Isemura, T., Kuroda, Y. *et al*. (2006) Development of a black gram [*Vigna mungo* (L.) Hepper] linkage map and its comparison with an azuki bean [*Vigna angularis* (Willd.) Ohwi and Ohashi] linkage map. *Theoretical and Applied Genetics* 113, 1261–1269.

Chavan, V.M., Patel, G.D. and Bhaskar, D.G. (1966) Improvement of cultivated phaseolus species need for interspecific hybridization. *Indian Journal of Genetics and Plant Breeding* 26A, 125–131.

Chen, J., Burke, J.J., Velten, J. and Xin, Z. (2006) FtsH11 protease plays a critical role in Arabidopsis thermotolerance. *The Plant Journal* 48, 73– 84.

Daulay, H.S., Henry, A. and Bhati, T.K. (2000) Screening of suitable genotypes of cluster-bean, moth bean and green gram in legume based intercropping with pearl millet. *Current Agriculture* 24, 85–88.

Deokar, A.A., Kondawar, V., Jain, P.K., Karuppayil, M., Raju, N.L. *et al.* (2011) Comparative analysis of expressed sequence tags (ESTs) between drought-tolerant and susceptible genotypes of chickpea under terminal drought stress. *BMC Plant Biology* 11, 70.

Doi, K., Kaga, A., Tomooka, N. and Vaughan, D.A. (2002) Molecular phylogeny of genus *Vigna* sub genus *Ceratotropis* based on rDNA ITS and atpB-rbcL intergenic spacer region of cpDNA sequences. *Genetica* 114, 129–145.

Dwivedi, N.K. and Bhatnagar, N. (2002) Genetic resources. In: Kumar, D. and Singh, N.B. (eds) *Mothbean in India.* Scientific Publishers, Jodhpur, India, pp. 80–92.

Dwivedi, N.K., Bhandari, D.C., Bhatnagar, N., Dabas, B.S. and Chandel, K.P.S. (1999) Conservation of genetic diversity of arid legumes. In: Faroda, A.S., Joshi, N.L., Kathju, S. and Amal Kar (eds) *Recent Advances in Management of Arid Ecosystem.* Proceedings of a Symposium held in India. Jodhpur, India, pp. 49–56.

Eapen, S. and Gill, R. (1986) Regeneration of plants from cultured root explants of moth bean (*Vigna aconitifolia* L. Jacq. Marechal). *Theoretical and Applied Genetics* 72, 384–387.

Eapen, S., Gill, R. and Rao, P.S. (1986) Tissue culture studies in mothbean. *Current Science* 55, 707–709.

Eapen, S., Kohler, F., Gerdemann, M. and Schieder, O. (1987) Cultivar dependence of transformation rates in mothbean after co-cultivation of protoplasts with *Agrobacterium tumefaciens*. *Theoretical and Applied Genetics* 75, 207–210.

Garg, D.K., Kakani, R.K., Sharma, R.C. and Gupta, P.C. (2003) Genetic variability and character association in moth bean under hyper arid region. In: Henry, A., Kumar, D. and Singh, N.B. (eds) *Advances in Arid Legumes Research.* Scientific Publishers, Jodhpur, India, pp. 93–97.

Gazendam, I. and Oelofse, D. (2007) Isolation of cowpea genes conferring drought tolerance: Construction of a cDNA drought expression library. *Water SA* 33, 387–391.

Gehlot, H.S., Upadhyaya, A., Davis, T.D. and Sankhla, N. (1989) Growth and organogenesis in moth bean callus as affected by paclobutrazol. *Plant and Cell Physiology* 30, 933–936.

Gill, R., Eapen, S. and Rao, P.S. (1986) Tissue culture studies in mothbean –factors influencing plant regeneration from seedling explants of different cultivars. *Proceedings of the Indian Academy of Science (Plant Science)* 96, 55–61.

Gill, R., Mishra, K.P. and Rao, P.S. (1987) Stimulation of shoot regeneration of *Vigna aconitifolia* by electrical control. *Annals of Botany* 60, 399–403.

Godbole, D.A., Kunachgi, M.N., Potdar, U.A., Krishnamurthy, K.V. and Mascarenhas, A.F. (1984) Studies on drought resistant legume: the moth bean, *Vigna aconitifolia* (Jacq.) Marechal. II. Morphogenetic studies. *Plant Cell Reports* 3, 75–78.

Goel, S., Raina, S.N. and Ogihara, Y. (2002) Molecular evolution and phylogenetic implications of internal transcribed spacer sequences of nuclear ribosomal DNA in the *Phaseolus-Vigna* complex. *Molecular Phylogenetics and Evolution* 22, 1–19.

Gurjar, K., Rampuria, S., Joshi, U., Palit, P., Bhatt, R. *et al.* (2014) Identification of heat-related ESTs in moth bean through suppression subtraction hybridization. *Applied Biochemistry and Biotechnology* 1173, 2116–2128.

Henry, A. (2001) Genotype x environment interaction for seed yield in moth bean germplasm. *Current Agriculture* 25, 85–89.

Henry, A. and Daulay, H.S. (1983) Genotype x environment interactions of induced mutants in mothbean. *Madras Agricultural Journal* 70, 705–708.

Henry, A. and Kackar, N.L. (2002a) Adaptability and relative performance of moth bean varieties. *Current Agriculture* 26, 67–69.

Henry, A. and Kackar, N.L. (2002b) Phenotypic stability in moth bean. *Current Agriculture* 26, 70–73.

Imran, M. (2011) Identification of transferable SSR markers of major pulses in Moth bean [*Vigna aconitifolia* (Jacq.) Marechal]. MSc Thesis. Swami Keshwanand Rajasthan Agricultural University, Bikaner, India.

Isemura, T., Kaga, A. and Konishi, S. (2007) Genome dissection of traits related to domestication in azuki bean (*Vigna angularis*) and their comparison with other warm season legumes. *Annals of Botany* 100, 1053–1071.

Isemura, T., Kaga, A., Tomooka, N., Shimizu, T. and Vaughan, D.A. (2010) The genetics of domestication of rice bean, *Vigna umbellata*. *Annals of Botany* 106, 927–944.

Isemura, T., Kaga, A., Tabata, S., Somta, P. and Srinives, P. (2012) Construction of a genetic linkage map and genetic analysis of domestication related traits in mungbean (*Vigna radiata*). *PLoS ONE* 7(8), e41304. DOI: 10.1371/journal.pone.004130.

Jain, J. and Chopra, V.L. (1988) Genotypic differences in response to regeneration of in vitro cultures of moth bean *Vigna aconitifolia* (Jacq.) Marechal. *Indian Journal of Experimental Biology* 26, 654–656.

Jangid, R.K., Sharma, R., Sudarsan, Y., Eapen, S., Singh, G. *et al.* (2010) Microwave treatment induced mutations and altered gene expression in *Vigna aconitifolia*. *Biologia Plantarum* 54, 703–706.

Jindal, S.K. and Vir, S. (1983) Adaptability in moth bean (*Vigna aconitifolia* (Jacq.) Marechal) for seed yield in rainfed conditions. *Annals of Arid Zone* 22, 83–86.

Kaga, A., Tomooka, N., Egawa, Y., Hosaka, K. and Kamijima, D. (1996) Species relationship in the sub-genus *Ceratotropis* (genus *Vigna*) revealed by RAPD analysis. *Euphytica* 88, 17–24.

Kakani, R.K., Garg, D.K., Sharma, R.C. and Narayan, S. (2003) Evaluation of germplasm lines of moth bean. In: Henry, A., Kumar, D. and Singh, N.B. (eds) *Advances in Arid Legumes Research*. Scientific Publishers, Jodhpur, India, pp. 107–111.

Kamble, S., Misra, H.S., Mahajan, S.K. and Eapen, S. (2003) A protocol for efficient biolistic transformation of moth bean *Vigna aconitifolia* L. Jacq. Marechal. *Plant Molecular Biology Reporter* 21, 457–458.

Kandpal, B.K., Mertia, R.S., Kumawat, R.N. and Mahajan, S.S. (2009) Performance of moth bean + pearl millet in intercropping systems as influenced by delayed sowing in extreme arid environment. In: Kumar, D., Henry, A. and Vittal, K.P.R. (eds) *Legumes in Dry Areas*. Scientific Publisher, Jodhpur, India, pp. 239–248.

Kanishka, R., Gayacharan, C., Basavaraja, T., Chandora, R. and Rana, J.C. (2023) Moth bean (*Vigna aconitifolia*): A minor legume with major potential to address global agricultural challenges. *Frontiers in Plant Science* 14, 1179547. DOI: 10.3389/fpls.2023.1179547

Kazan, K. and Lyons, R. (2016) The link between flowering time and stress tolerance. *Journal of Experimental Botany* 67, 47–60.

Khadke, S. and Kothekar, V. (2018) Effect of EMS and SA on trypsin inhibitor content in Moth bean (*Vigna aconitifolia* (Jacq.) Marechal), In: *International Symposium on Plant Mutation Breeding and Biotechnology, No. IAEA-CN-263-197*. International Atomic Energy Agency (IAEA), Austria.

Kohler, F., Golz, C., Eapen, S., Kohn, H. and Schieder, O. (1987a) Stable transformation of moth bean (*Vigna aconitifolia*) via direct gene transfer. *Plant Cell Reports* 6, 313–317.

Kohler, F., Golz, C., Eapen, S. and Schieder, O. (1987b) Influence of plant cultivar and plasmid DNA on transformation rates in tobacco and mothbean. *Plant Science* 53, 87–91.

Kongjaimun, A., Kaga, A., Tomooka, N., Somta, P. and Vaughan, D.A. *et al.* (2012) The genetics of domestication of yardlong bean, *Vigna unguiculata* (L.) Walp. ssp. *unguiculata* cv.-gr. Sesquipedalis. *Annals of Botany* 109, 1185–1200.

Kothekar, V.S. (1989) Differential radiation sensitivity in moth bean. *Current Science* 58, 758–760.

Krishnamurthy, K.V., Godbole, D.A. and Mascarenhas, A.F. (1984) Studies on a drought resistant legume: The moth bean, *Vigna aconitifolia* Marechal I. Protoplast culture and organogenesis. *Plant Cell Reports* 3, 30–32.

Kumar, A. and Rao, S.R. (2002) Cytological investigations in some important tree species of Rajasthan I. Karyomorphological studies in some species of *Anogeissus* (DC.) Guill., Perr. & A. Rich. *Silvae Genetica* 51, 100–104.

Kumar, D. (1996) Variability studies in induced mutants of moth bean on rainfed arid lands. *Annals of Arid Zone* 35, 125–128.

Kumar, D. (2001) Phenotypic stability of quantitative traits in moth bean. *Indian Journal of Pulses Research* 14, 116–118.

Kumar, D. (2002a) New concepts of Moth bean improvement. In: Kumar, D. and Singh N.B. (eds) *Moth Bean in India*. Scientific Publishers, Jodhpur, India, pp. 31–48.

Kumar, D. (2002b) *Production Technology for Moth Bean in India*. Indian Council of Agricultural Research, Central Arid Zone Research Institute, Jodhpur, pp. 8.

Kumar, S., Saxena, S.N., Jat, R.S. and Sharma, R. (2008) *Agrobacterium tumefaciens* mediated genetic transformation of moth bean (*Vigna aconitifolia* (Jacq) Marechal). *Indian Journal of Genetics and Plant Breeding* 68, 327–329.

Kumar, S., Singh, G., Jat, R.S., Saxena, S.N. and Sharma, R. (2010) Transformation of moth bean (*Vigna aconotifolia*) cotyledonary nodes through vacuum infiltration of *Agrobacterium tumefaciens*. *Green Farming* 1(6), 551–554.

Kumar, S., Imtiaz, M., Gupta, S. and Pratap, A. (2011) Distant hybridization and alien gene introgression. In: Pratap, A. and Kumar, J. (eds) *Biology and Breeding of Food Legumes*. CABI, Oxfordshire, UK, pp. 81–110.

Lee, N.G., Stein, B.H., Suzuki and Verma, D.P.S. (1993) Expression of antisense nodulin-35S RNA in *Vigna aconitifolia* transgenic root nodules retards peroxisome development and affects nitrogen availability to the plant. *The Plant Journal* 3, 599–606.

Lübberstedt, T. and Varshney, R.K. (2013) *Diagnostics in Plant Breeding*. Springer Science, Dordrecht, the Netherlands. DOI: 10.1007/978-94-007-5687-8__1

Mahna, S.K., Parvateesam, M. and Garg, R. (1993) Genetic effect of sodium azide and hydroxylamine on moth bean (*Vigna aconitifolia*). *Indian Journal of Agricultural Sciences* 63, 551–555.

Majumdar, D.K. (2011) *Pulse Crop Production: Principles and Technologies*. Prentice-Hall of India Pvt. Ltd, India, pp. 281.

Malabadi, R.B. and Nataraja, K. (2007) *Agrobacterium tumefaciens* mediated genetic transformation of *Vigna aconitifolia* and stable transmission of the genes to somatic seedlings. *International Journal of Agricultural Research* 2, 450–458.

Mamta (2012) Studies on diversity of *Vigna aconitifolia* (Jacq.) Marechal through RAPD analysis. MSc Thesis. Swami Keshwanand Rajasthan Agricultural University, Bikaner, India.

Mohan, M.L. and Krishnamurthy, K.V. (2003) *In vitro* morphogenesis in grain legumes: An overview. In: Jaiwal, P.K. and Singh, R.P. (eds) *Improvement Strategies for Leguminosae Biotechnology*. Kluwer Academic Publishers, UK, pp. 23–63.

Moitra, N. and Mandal, N. (2003) Random amplified polymorphic DNA (RAPD) assay in moth bean [*Vigna aconitifolia* (Jacq.) Marechal]. *Journal of Interacademicia* 7, 480-483.

Mullarkey, M. and Jones, P. (2000) Isolation and analysis of thermo-tolerant mutant in wheat. *Journal of Experimental Botany* 51, 139–146.

Ojomo, O.A. and Chheda, H.R. (1971) Mitotic events and other disorders induced in cowpeas (*Vigna unguiculata*) Walp. by ionizing radiation. *Radiation Botany* 11, 375–381.

Oliveira, A.R., Martins, L.V., Bustamante, F.D.O., Amatriaín, M.M., Close, T. *et al.* (2020) Breaks of macrosynteny and collinearity among moth bean (*Vigna aconitifolia*), cowpea (*Vigna unguiculata*), and common bean (*Phaseolus vulgaris*). *Chromosome Research* 28, 293–306. DOI: 10.1007/s10577-020-09635-0

Om, Vir and Singh, A.K. (2015) Moth bean [*Vigna aconitifolia* (Jacq.) Marechal] germplasm: Evaluation for genetic variability and inter characters relationship in hot arid climate of western Rajasthan, India. *Legume Research* 38, 748–752.

Patel, J.D., Desai, N.C., Intwala, C.G. and Kodappully, V.C. (2008) Genetic variability, correlation and path analysis in moth bean. *Journal of Food Legumes* 21, 158–160.

Patil, S.C., Patil, V.P., Patil, H.E. and Pawar, S.V. (2007) Correlation and path coefficient studies in moth bean. *International Journal of Plant Sciences* 2, 141–144.

Rajora, M.P., Patidar, M. and Mahajan, S.S. (2009) Effect of pre-sowing seed treatments on seed germination, plant establishment and yield of moth bean (*Vigna aconitifolia*). In: Kumar, D., Henry, A. and Vittal, K.P.R. (eds) *Legumes in Dry Areas*. Scientific Publishers, Jodhpur, India, pp. 218–224.

Rajora, M.P., Bhatt, R.K. and Ram, R. (2012) Analysis for seed yield and its components in moth-bean (*Vigna aconitifolia*) under hot arid conditions. *Madras Agricultural Journal* 99, 21–25.

Rampuria, S., Joshi, U., Paramita, P., Amit, A.D., Meghwal, R.R. *et al.* (2012) Construction and analysis of an SSH cDNA library of early heat-induced genes of *Vigna aconitifolia* variety RMO-40. *Genome* 55, 783–796.

Sahoo, S., Kumar, S., Sanadya, Kumari, N. and Bhagawati, B. (2019) Estimation of the various genetic variability parameters for seed yield and its component traits in Mothbean germplasm [*Vigna aconitifolia* (Jacq.) Marechal]. *Journal of Pharmacognosy and Phytochemistry* 8, 49–52.

Sankhla, N., Davis, T.D., Gehlot, H.S., Upadhyaya, A., Sankhla, A. *et al.* (1991) Growth and organogenesis in moth bean callus cultures as influenced by triazole growth regulators and gibberellic acid. *Journal of Plant Growth Regulation* 10, 41–45.

Saxena, S.N. (2004) Physio-molecular analysis of somaclonal variation in cultures of moth bean [*Vigna aconitifolia* (Jacq.)] Marechal.] derived from different level of cytokinins. PhD Thesis. Rajasthan Agricultural University, Bikaner.

Saxena, S.N., Rohit, K. and Sharma, R. (2005) Prospects of manipulating in vitro flowering in moth bean for crop improvement. *Journal of Arid Legumes* 2, 210–214.

Saxena, S.N., Sharma, R. and Mahla, H.R. (2006) Analysis of induced and naturally available genetic diversity in moth bean (*Vigna aconitifolia* Jacq. Merachal) using RAPD markers. *Journal of Arid Legumes* 3, 60–64.

Sehrawat, N. and Yadav, M. (2014) Screening and cross-compatibility of various *Vigna* species for yellow mosaic virus resistance. *Journal of Innovative Biology* 1, 31–34.

Shamurailatpam, A., Madhavan, L., Yadav, S.R., Bhat, K.V. and Rao, S.R. (2015) Heterochromatin distribution and comparative karyo-morphological studies in *Vigna umbellata* Thunberg, 1969 and *V. aconitifolia* Jacquin, 1969 (Fabaceae) accessions. *Comparative Cytogenetics* 9, 119–132.

Sharma, R.C. and Majumdar, J.R. (1982) Studies of flower shedding behaviour in moth bean. *Scientific Culture* 48, 65–66.

Sharma, R.C., Kakani, R.K., Solanki, S.S. and Garg, D.K. (2002) Genetic divergence in moth bean. In: *National Symposium on Arid Legumes for Food, Nutrition, Security and Promotion of Trade.* Indian Society of Arid Legumes, CAZRI, Jodhpur, pp. 28.

Sharma, R.C., Kakani, R.K., Solanki, S.S. and Garg, D.K. (2003a) Genetic divergence in moth bean. In: Henry, A., Kumar, D. and Singh, N.B. (eds) *Advances in Arid Legumes Research.* Scientific Publishers, Jodhpur, India, pp. 104–106.

Sharma, R.C., Joshi, P. and Kakani, R.K. (2003b) Mutation breeding in moth bean. In: Henry, A., Kumar, D. and Singh, N.B. (eds) *Advances in Arid Legumes Research.* Scientific Publishers, Jodhpur, India, pp. 88–92.

Sharma, R., Kumar, S., Mahla, H., Khandelwal, V., Roy, P.K. *et al.* (2021) Moth Bean. In: Pratap, A. and Gupta, S. (eds) *The Beans and the Peas from Orphan to Mainstream Crops.* Elsevier, Amsterdam, pp. 67–88.

She, C.W., Jiang, X., Ou, L.J., Liu, J., Long, K.L., *et al.* (2015) Molecular cytogenetic characterization and phylogenetic analysis of the seven cultivated *Vigna* species (Fabaceae). *Plant Biology* 17, 268–280.

Shekhawat, N.S. and Galston, A.W. (1983) Isolation, culture and regeneration of moth bean (*Vigna aconitifolia*) leaf protoplasts. *Plant Science Letters* 32, 43–51.

Sihag, S.K., Khatri, R.S. and Joshi, U.N. (2004) Genetic variability and heritability for grain yield and its components in mothbean [*Vigna aconitifolia* (Jacq.) Marechal]. *Annals of Biology* 20, 219–222.

Singh, D.P. (1990) Distant hybridization in genus *Vigna* - a review. *Indian Journal of Genetics and Plant Breeding* 50, 268–276.

Singh, I.S. and Singh, M. (2013) Asian *Vigna*. In: Singh, I.S., Upadhyaya, H.D. and Singh, M. (eds) *Genetic and Genomic Resources of Grain Legume Improvement.* Elsevier, the Netherlands, pp. 237–267.

Singh, M. and Kumar, D. (2005) Mutants with high productivity induced in moth bean. *Indian Journal of Pulses Research* 18, 244-245.

Sinha, S.S.N. and Roy, H. (1979) Cytological studies in the genus *Phaseolus* I. Mitotic analysis in fourteen species. *Cytologia* 44, 191–199.

Smartt, J. (1985) Evolution of grain legumes. III. Pulses in the genus *Vigna*. *Experimental Agriculture* 21, 87–100.

Solanki, S.S., Sharma, R.C., Kakani, R.K. and Garg, D.K. (2003) Estimation of genetic parameters in moth bean. *Advances in Arid Legumes Research.* Scientific Publishers (India), Jodhpur, pp. 84–87.

Somers, D.A., Somac, D.A. and Olhoft, P.M. (2003) Recent advances in legume transformation. *Plant Physiology* 131, 892–899.

Somta, P., Seehalak, W. and Srinive, P. (2009) Development, characterization and cross-species amplification of mungbean (*Vigna radiata*) genic microsatellite markers. *Conservation Genetics* 10, 1939–1943.

Somta, P., Jomsangawong, A., Yundaeng, C., Yuan, X., Chen, J. *et al.* (2018) Genetic dissection of azuki bean weevil (*Callosobruchus chinensis* L.) resistance in moth bean (*Vigna aconitifolia* [Jaqc.] Maréchal). *Genes* 9, 555. DOI: 10.3390/genes9110555

Soni, P. (2010) Identification of drought related genes in moth bean through differential display. PhD Thesis. S.K. Rajasthan Agricultural University, Bikaner, India.

Srivastava, J.P. and Soni, C.P. (1995) Physiological basis of yield in moth bean [*Vigna aconitifolia* (Jacq.) Marechal] under rainfed conditions. *Journal of Agronomy and Crop Science* 174(2), 73–77.

Stace, C.A. (2000) Cytology and cytogenetics as a fundamental taxonomic resource for the 20th and 21st centuries. *Taxon* 49, 451–477.

Suranjika, S., Pradhan, S., Nayak, S.S. and Parida, A. (2022) De novo transcriptome assembly and analysis of gene expression in different tissues of moth bean (*Vigna aconitifolia*) (Jacq.) Marechal. *BMC Plant Biology* 22(1), 1–15. DOI: 10.1186/s12870-022-03583.

Takahashi, Y., Somta, P., Muto, C., Iseki, K., Naito, K. *et al.* (2016) Novel genetic resources in the genus *Vigna* unveiled from gene bank accessions. *PLoS ONE* 11(1): e0147568. DOI: 10.1371/journal. pone.0147568.

Tangphatsornruang, S., Somta, P., Uthaipaisanwong, P., Chanprasert, J., Sangsrakru, D. *et al.* (2009) Characterization of microsatellites and gene contents from genome shotgun sequences of mungbean (*Vigna radiata* (L.) Wilczek). *BMC Plant Biology* 9, 137.

Tiwari, B., Kalim, S., Bangar, P., Kumari, R., Kumar, S. *et al.* (2018a) Physiological, biochemical, and molecular responses of thermotolerance in moth bean (*Vigna aconitifolia* (Jacq.) Marechal). *Turkish Journal of Agriculture and Forestry* 42, 176–184.

Tiwari, B., Kalim, S., Tyagi, N., Kumari, R., Bangar, P. *et al.* (2018b) Identification of genes associated with stress tolerance in moth bean [*Vigna aconitifolia* (Jacq.) Marechal], a stress hardy crop. *Physiology and Molecular Biology of Plants* 24, 551–561.

Tomooka, N., Yoon, M.S., Doi, K., Kaga, A. and Vaughan, D.A. (2002) AFLP analysis of diploid species in the genus *Vigna* subgenus *Ceratotropis*. *Genetic Resources and Crop Evolution* 49, 521–530.

Tomooka, N., Naito, K., Kaga, A., Sakai, H., Isemura, T., Ogiso-Tanaka, E. *et al.* (2014) Evolution, domestication and neo-domestication of the genus *Vigna*. *Plant Genetic Resources* 12 (S1), S168–S171. DOI: 10.1017/S1479262114000483

Upadhyaya, A., Gehlot, H.S., Davis, T.D., Sankhla, N., Sankhla, A. *et al.* (1989) *In vitro* growth, metabolism and regeneration of moth bean callus as influenced by flurprimidol. *Biochemie und Physiologie der Pflanzen* 185, 245–252.

Vijaya Kumar, A.G., ShrutiKoraddi, D.T., Kallesh, S.T., Hundekar, I. and Boodi, H. (2016) Genetic variability studies in mothbean [*Vigna aconitifolia* (Jacq.) Marechal] in an semi-arid environment of North Karnataka. *The Bioscon* 11, 1017–1021.

Vijendra, P.D., Huchappa, K.M., Lingappa, R., Basappa, G., Jayanna, S.G. and Kumar, V. (2016) Physiological and biochemical changes in moth bean (*Vigna aconitifolia* L.) under cadmium stress. *Journal of Botany* 2016, 6403938. DOI: 10.1155/2016/6403938

Wang, M.L., Barkley, N., Gillaspie, G. and Pederson, G.A. (2008) Phylogenetic relationships and genetic diversity of the USDA *Vigna* germplasm collection revealed by gene-derived markers and sequencing. *Genetics Research* 90, 461–480.

Wang, X.W., Kaga, A., Tomooka, N. and Vaughan, D.A. (2004) The development of SSR markers by a new method in plants and their application to gene flow studies in azuki bean (*Vigna angularis* (Willd.) Ohwi and Ohashi). *Theoretical and Applied Genetics* 109, 352–360.

Yadav, A.K., Singh, C.K., Kalia, R.K., Mittal, S., Wankhede, D.P. *et al.* (2023) Genetic diversity, population structure, and genome-wide association study for the flowering trait in a diverse panel of 428 moth bean (*Vigna aconitifolia*) accessions using genotyping by sequencing. *BMC Plant Biology* 23, 228.

Yaqoob, M., Najibullah and Khaliq, P. (2005) Mothbean germplasm evaluation for yield and other important traits. *Indus Journal of Plant Sciences* 4, 241–248.

Yogeesh, L.N., Viswanatha, K.P., Madhusudhan, R. and Gangaprasad, S. (2016) Morphological diversity and genetic variability in moth bean. *International Journal of Science, Environment and Technology* 5, 1912–1924.

Yundaeng, C., Somta, P., Amkul, K., Kongjaimun, A., Kaga, A. *et al.* (2019) Construction of genetic linkage map and genome dissection of domestication-related traits of moth bean (*Vigna aconitifolia*), a legume crop of arid areas. *Molecular Genetics and Genomics* 294, 621–635.

15 Faba Bean (*Vicia faba* L.)

Lynn Abou-Khater, Rind Balech and Fouad Maalouf*

*International Center for Agricultural Research in the
Dry Areas (ICARDA), Beirut, Lebanon*

Abstract

Faba bean (*Vicia faba* L.) is an important cool season food legume, rich in protein and essential micronutrients. It plays a critical role in improving land structure and contributing to wild pollinator maintenance. It was domesticated in the Fertile Crescent of the Near East and spread to more than 60 countries with diverse agroecology around the world. Despite its nutritional value and adaptability, faba bean cultivation faces challenges, including biotic and abiotic stresses, reducing cultivation areas in various countries. Conservation efforts, notably by organizations like the ICARDA Gene Bank, have preserved diverse genetic resources, ensuring the genetic resilience of this valuable crop. These genetic repositories are invaluable for breeding programmes, enabling the development of high-yielding and stress-resistant varieties. Despite the historical limited funding provided to faba bean, the research on crop improvement and agronomy led to a substantial increase in its yield. Moreover, to achieve higher productivity in the farming field, modern tools were developed to ensure higher genetic gains and faster release of cultivars. These include speed breeding with four generations per year, and the discoveries of single nucleotide polymorphism markers which are key achievements towards the development of marker-assisted selection techniques.

15.1 Introduction

Faba bean (*Vicia faba* L.), also known as broad bean, field bean, fava bean, horse bean or Windsor bean, is a crop of versatile uses and numerous benefits. It plays a substantial role in human food and animal feed and delivers multiple services for the ecosystem that align with sustainability principles. Faba bean is loaded with protein and energy and is a good source of fibre, vitamins, minerals and key essential amino acids (Crépon *et al.*, 2010; Singh *et al.*, 2014; Koivunen *et al.*, 2016). It is considered as the 'poor man's meat' in low-income countries where people cannot afford animal proteins. Faba bean is consumed green and dry; the green immature seeds and pods are consumed as vegetables, whilst mature and dry seeds are an essential ingredient in many dishes in the Middle East, Mediterranean region, Ethiopia, Sudan and China. Beyond its nutritional value, numerous noteworthy health-enhancing properties of faba bean have been reported (Martineau-Côté *et al.*, 2022).

In addition, faba bean provides livestock farmers with the required amount of starch and protein. However, its inclusion in livestock diets was hindered by the presence of antinutritional components, namely tannins, vicines and

*Corresponding author: f.maalouf@cgiar.org

convicines, though numerous processing methods can be used to reduce their amount. Moreover, the faba bean is an excellent atmospheric nitrogen fixer. Its capacity to fix up to 90% of its nitrogen requirements makes it highly suitable for low-input cropping systems, resulting in a substantial reduction of greenhouse gas emissions and its inclusion in the cropping system aids in improving the properties of the soil, breaking the cycle of pests and diseases and in diversifying the entire cropping system (Jensen *et al.*, 2010). Also, faba bean plays an essential role in conserving wild pollinators that are facing numerous threats (Sentil *et al.*, 2021).

Despite the diverse benefits of faba bean, it is still neglected and underutilized in many regions of the world owing to a range of factors from the lack of extension and poor cultural practices to the abiotic and biotic stresses that impact faba bean yields and quality specifications desired by the end market, in addition to the low investments in faba bean research and genetic improvement, resulting in neglect and abandonment of the crop by the scientific community. The projected population growth and climate change are intensifying the pressure on food production, leading to more environmental degradation and greenhouse gas emissions. Moreover, the demand for food is going beyond quantity as people are favouring healthier food choices. Faba bean properties align well with the shifting trends towards healthy eating and environmentally friendly behaviours. Further research and efforts are needed, however, by scientists and agricultural policy makers to increase faba bean production to meet future demand and to help solidify its place in the growing market.

15.2   Origin, Botanical Description and Distribution

Faba bean domestication began around the 10th millennium BC in the Near East (south-western Asia) (Cubero, 1974), and afterwards, its cultivation spread to Europe through Anatolia, to the west through the African Mediterranean coast, to Ethiopia through the Nile, and to India through Mesopotamia (Cubero, 1974; Cubero and Nadal, 2005). Subsequently, during the 10th century, the crop reached China; during the 16th century, it reached America; later, during the 18th century, it reached Australia and South Africa (Cubero, 1974). The faba bean wild progenitor was presumed to be extinct until 14,000-year-old seeds offering hints about the time and place that the faba bean plant grew naturally were discovered in the Mount Carmel region (Caracuta *et al.*, 2015).

Faba bean is a partially allogamous species that belongs to the Fabaceae family and the vetches *Vicia* genus, even though some scientists treat it as a separate monotypic genus, *Faba* (Miller, 1754). Due to the extreme variation in seed size, faba bean was classified initially based on the seed size into four subspecies, namely: major (large seeded), equina (medium seeded), minor (small seeded) and paucijuga (small seeded) (Muratova, 1931), but later, because of the high cross-compatibility between their different seeds, Cubero (1974) considered them as four botanical groups of the same species resulting from cycles of selection over thousands of years on growth habit and seed size.

Faba bean holds a significant position among cool season food legumes, ranking fourth in global production. Currently, faba bean cultivation has spread to more than 70 countries worldwide, with production averaging 6 and 1.7 million tons harvested from 2.7 and 0.2 million ha with an average yield of 2.2 and 6.2 tons/ha for dry and green faba bean, respectively, during 2021 (FAOSTAT, 2023). China predominates as the largest producer of dry faba bean, followed by Ethiopia and the UK. For green faba bean pods and grains, Algeria holds the top position, followed by Egypt (Fig. 15.1) and China (FAOSTAT, 2023). Prior to this year, the data on green faba bean were not included in FAOSTAT. Still, with the inclusion of these data, stakeholders can now have a more precise understanding of the global production and acreage trends of faba bean. While dry seeds are the main form of consumption, there has been a rising interest in faba bean green pods and grains in many countries around the world, such as Columbia and Egypt, for example, where their production has surpassed that of dry seeds in recent years.

Globally, the acreage dedicated to dry faba bean cultivation experienced a decline between 1961 and 1990; however, since then, it has reached a stable level. Despite this stabilization, production has substantially increased, as depicted in Fig. 15.2 (FAOSTAT, 2023). In contrast, green faba bean has experienced a modest increase in acreage since 1961 (Fig. 15.3). However, this crop's production has significantly increased, particularly after the year 2000 (Fig. 15.3). The increase in faba bean production serves as evidence of the positive impact of breeding and agronomic achievements on global faba bean cultivation. As breeding techniques continue to advance and agronomic practices evolve, the production of faba bean is expected to further increase, meeting the increased demand and interest in this crop.

Fig. 15.1. Faba bean plants of improved cultivar in middle Egypt. (Author's own figure.)

15.3 Faba Bean Gene Pools and Evolution

Faba bean belongs to the *Vicia* genus, with more than 140 species, located in Europe, Asia and North America (Kupicha, 1981; Allkin *et al.*, 1986), while its major diversity is found in West Asia. The *Vicia* species of this genus are widely adapted to different harsh environments, while cultivated faba bean is adapted to high rainfall and irrigated agriculture and responds to inputs. The faba bean products are mainly dry and green seeds for human consumption and animal feeding. Other species such as *Vicia sativa*, *V. ervilia*, *V. articulata*, *V. narbonensis*, *V. villosa*, *V. benghalensis* and *V. pannonica* are cultivated for their straw and seeds for animal feeding. The *Vicia* species considered as in the most common neighbourhood of *V. faba* are *Vicia peregrina* and *Vicia michauxii* (Van de Ven *et al.*, 1993). These related vetches have seven chromosomes (Sitte *et al.*, 2002). *Vicia narbonensis* is considered the nearest relative to *Vicia faba* (ven and Hopf, 1973). However, several attempts to cross cultivated faba bean that has 2n = 12 chromosomes with different wild species such as *Vicia narbonensis* L. that have 2n = 14 chromosomes (Zohary and Hopf, 1973; Hajjar and Hodgkin, 2007) have failed, although they are closely related in terms of taxonomy.

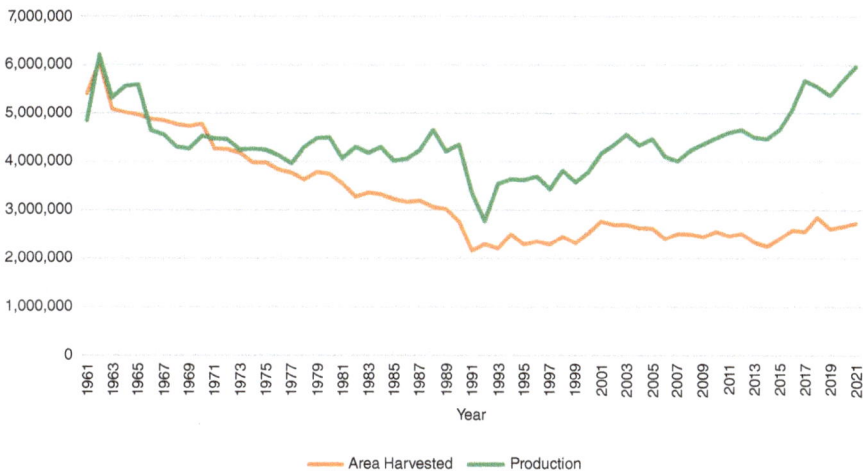

Fig. 15.2. Evolution trends of the global production (tons) of dry faba bean between 1961 and 2021. (Author's own figure.)

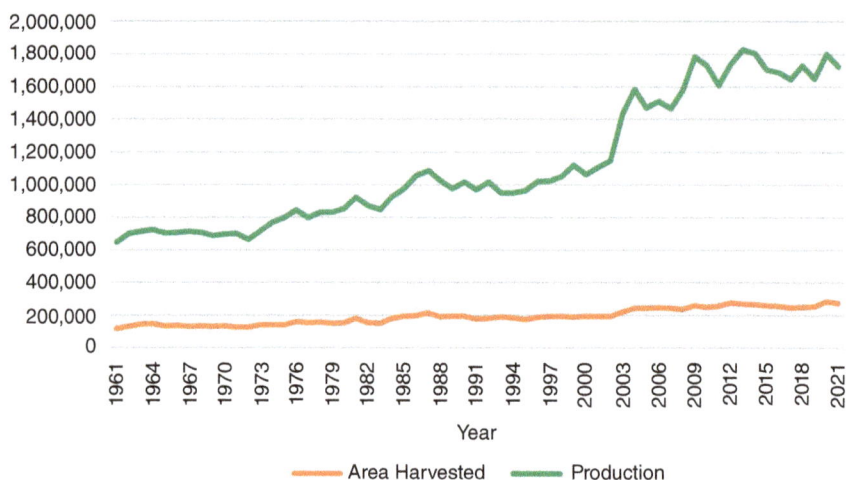

Fig. 15.3. Evolution trends of the global area harvested (ha) of green faba bean between 1961 and 2021. (Author's own figure.)

Faba bean intraspecific diversity is mainly described by seed size (Fig. 15.4). Persoon (1807) described the small-seeded group as *V. faba minor* (roundish seeds of up to 0.6 or 0.7 g weight per seed), the medium-large-seeded group as *V. faba equina* (single seed weight less than about 1 g) and *V. faba major* with its impressive, large, and flat seeds weighing more than 1 g per seed. Muratova (1931), in addition to the three subspecies mentioned above, has one more subspecies, namely *V. faba paucijuga*. These groups present a wide range of variability in cultivated faba bean for diverse traits, including seed size, number of seeds per pod and pod length (Maalouf *et al.*, 2013). The major types were common in the South Mediterranean countries and China; *equina* types are grown throughout the Nile Valley and Ethiopia, North Africa, and Australia, and the small-seeded *minor* is found in North Europe. The *minor* and *major* types are common in Europe, while *paucijuga* types are common in Central and South Asia and Europe.

The faba bean is considered a unique monospecies (Duc *et al.*, 2010). To present, no closest living direct wild ancestor of the faba bean has been found. Caracuta *et al.* (2016) considered the possible areas where the wild progenitor of faba bean grew. In addition, Caracuta demonstrated that the wild-type faba bean would grow in a habitat that today resembled Mount Carmel

14,000 years ago. This important finding is essential to trace back the evolution of faba bean crops because it provides the scientific community with clear evidence of the initiation of the domestication of this crop. Based on the early developed system by Harlan and de Wet (1971), three tiers of gene pools were recognized. For different crops, genetic resources in the second tiers were classified, such as for wheat, lentil and chickpea. Since the diverse botanical groups of faba bean are easily crossable (Cubero, 1973), these groups are classified into the first tiers. This assumption is also confirmed by Maalouf *et al.* (2022) by using single nucleotide polymorphism (SNP) markers to evaluate diverse collections. They confirmed that a wide range of genetic variability is observed in faba bean germplasm. Still, only one large heterogeneous group was classified, endorsing Cubero's observation of a lack of genetic or sterility barriers between different subspecies.

15.4   Germplasm Collection and Conservation

15.4.1   *Ex-situ* conservation

According to the global trust strategy (published in https://www.croptrust.org/), the top ten gene

Fig. 15.4. Variability in seed shape and size of faba bean. HSW, 100-seed weight. (Author's own figure.)

banks that conserve faba bean and related *Vicia* species are presented in Table 15.1. ICARDA safeguards the largest collection of faba bean worldwide (>32% of the total world collection).

The faba bean global collection conserved at ICARDA is collected from 71 countries with a high percentage of unique accessions. The collection held at ICARDA also conserves more than 6000 accessions of other *Vicia* species, including about 3000 accessions of wild species of *Vicia*. Collections with a high percentage of wild relatives, landraces and materials originally collected by ICARDA are most likely to encompass unique accessions prioritized in a rational global system.

15.4.2 Germplasm maintenance

Faba bean is an entomophilous and partially allogamous species. The value of outcrossing varied from practically near zero to 0.7 (Link *et al.*, 1994; Maalouf *et al.*, 1999; Suso *et al.*, 1999, 2001; Gasim *et al.*, 2004) among cultivars and environmental conditions (Link *et al.*, 1994;

Maalouf *et al.*, 1999; Suso *et al.*, 2001). The faba bean germplasm maintenance requires preventing the effect of insect pollinators. Insect-proof cages protecting faba bean plants are one of the most efficient techniques applied in faba bean to multiply plant genetic resources. However, besides being an expensive system to prevent intercrossing in this crop, this technique has limited capacity and increases the inbreeding depression of faba bean, affecting the yield potential of different cultivars (Drayner *et al.*, 1956). It is recommended for use only when small samples are obtained from a breeding programme or any seed collection (Hawtin and Omar, 1980).

Managing a large number of accessions for large increases is difficult and very costly to handle under insect-proof cages, in addition to the negative impact of cages on the level of inbreeding depression. Other techniques have been developed to reduce germplasm mixtures resulting from pollination by reducing the geneflow between plots; for example, in a combination of isolation by 3-m distance and pollination barriers, *Brassica napus* L. and *Triticale* reduce by more than 95% the intercrossing between plots (Robertson

Table 15.1. Number of cultivated faba bean accessions and related *Vicia* species conserved *ex-situ* in gene banks with more than 1000 accessions. (Author's own table.)

Country gene bank	Institute	Number of accessions (*V. faba*)	Number of accessions wild relatives/ *Vicia* species
Global	International Centre for Agricultural Research in Dry Areas (ICARDA	10,034	6148
China	Institute of Crop Science (CAAS)	5,229	–
Australia	Australian Temperate Field Crops Collection, Victoria	2,665	2
Germany	GenBank, Leibniz Institute of Plant Genetics and Crop Plant Research (IPK), Gatersleben	1,925	1106
Italy	Istituto di Genetica Vegetale (IGV), CNR, Bari	1,875	69
Ecuador	Instituto de Ciencias Naturales Universidad Central del Ecuador (ICN)	1,650	–
Russian Federation	N.I. Vavilov All Russian Scientific Research Institute of Plant Industry (VIR)	1,441	–
Ethiopia	Biodiversity Conservation and Research Institute (BCRI)	1,208	–
Spain	Centro de Recursos Fitogenéticos (CRF), INIA, Madrid	1,179	1
	Centro de Investigación y Desarrollo Agrario Alameda del Obispo	1,098	–
France	Station d'Amélioration des Plantes (INRA)	1,057	–

and Cardona, 1986). Cultivating the barrier separating the faba bean reduces the gene flow between plots rather than leaving the separated area with no cultivation flow depending on the genotype and the location (Suso *et al.*, 2006). These authors also considered that a *V. narbonensis* zone discourages pollinators from leaving the plots (bee emigration) at a rate higher than other trap crops. At ICARDA, the plant genetic resources are maintained under insect-proof cages to maintain the identity of collections, while improved lines are multiplied in isolated fields in open fields using *Brassica* or *Vicia narbonensis*.

15.5 Germplasm Characterization and Diversity

As a stand-alone cultivated species, faba bean possesses a wide spectrum of variation for different agro-morphological traits such as plant architecture, leaf size and shape, seed colour,

seed shape and size (generally measured by 100-seed weight). There is a more than ten-fold range of seed size among the faba bean accessions (20–250 g/100 seeds) at ICARDA (Terbol, Lebanon). The larger seed is considered a response of the faba bean genome to human selection because all the archaeological specimens dated prior to late Pre-Pottery Neolithic B (PPNB) are small-seeded type (Tanno and Willcov, 2006; Caracuta *et al.*, 2016). The medium-sized type was found in the Iberian Peninsula, both in Portugal and Spain, and Central Europe in 5000 BP, and the flattened larger types were not known before 1500 BP (Ladizinski, 1998). The research evaluated the variability within and among entries in the world collections of faba bean begun by Abdalla *et al.* (1976). Before the advent of molecular marker technology, investigations focused on morphological characteristics, protein-based markers (seed storage protein and isozyme) and geographical patterns of the variation of selected traits in the context of cultivar diversification. Since the mid-1990s, the application of various types of DNA-based markers offered a great opportunity to assess the genetic diversity of faba bean germplasm. The marker systems used in faba bean included restriction fragment length polymorphism (RFLP) (Torres *et al.*, 1993), random amplified polymorphic DNA (RAPD) (Torres *et al.*, 1993), long terminal repeat (LTR) sequence retrotransposon-base markers or sequence-specific amplification polymorphism (SSAP) (Sanz *et al.*, 2007), inter-simple sequence repeat (ISSR) (Terzopoulos and Bebeli, 2008; Wang *et al.*, 2012), amplified fragment length polymorphism (AFLP) (Zeid *et al.*, 2003; Zong *et al.*, 2009), target region amplification polymorphism (TRAP) (Kwon *et al.*, 2010), simple sequence repeat (SSR) (Oliveira *et al.*, 2016; Göl *et al.*, 2017), and single nucleotide polymorphism (SNP) markers generated by hybridization of PCR amplified site allele-specific sequences to bead chips (Kaur *et al.*, 2014; Abou Khater *et al.*, 2021; Maalouf *et al.*, 2022). These studies generated information of critical importance in managing and ensuring the long-term success of faba bean improvement.

15.6 Crop Improvement by Conventional Approaches

15.6.1 Breeding methods

Being a partially allogamous plant with an outcrossing rate of approximately 0.5 in several environments, faba bean is, in theory, amenable to different breeding methods. Mass selection, recurrent selection techniques, and traditional cultivar development techniques such as phenotype-based pedigree selection, single-seed descent line breeding, or the creation of synthetic cultivars are the main conventional techniques used to create improved faba bean cultivars. Parental lines are selected for these approaches based on their phenotypic characteristics and pedigree.

The effectiveness of the various breeding approaches was assessed by Gharzeddin *et al.* (2019) using faba bean pure lines obtained by the pedigree method, bulk lines obtained through recurrent selection, and synthetic varieties. The synthetic and recurrent lines demonstrated superior performance in terms of grain yield and biological yield, showing greater genetic gains

per year than the inbred lines. While the exploitation of heterosis has led to yield increases in open-pollinated populations (Link *et al.*, 1994), it is not yet feasible at a commercial level due to the lack of stable cytoplasmic male sterility (Palmer *et al.*, 2011) that is very crucial for unlocking the benefits of heterosis in faba bean. Recently, the speed breeding (SB) approach proved that it offers great potential to increase the rate of the genetic gains in crops such as faba bean. SB can be achieved using different methods, one of which involves extending the daily duration the plants are exposed to light, combined with early seed harvest, to cycle quickly from seed to seed, thereby reducing the generation times for some long-day or day-neutral crops (Ghosh *et al.*, 2018). Recently, generation advancement under SB was conducted at ICARDA, confirming the possibility of developing four generations per year. SB was applied successfully to faba bean by completing four crop cycles per year (Table 15.2). ICARDA also developed a shuttle breeding strategy in faba bean to increase generation turnover by developing two generations per year and screening under target environments.

15.6.2 Breeding for resistance to biotic stresses

Numerous biotic stresses, including foliar diseases, insects, viruses and parasitic weeds, can lead to significant yield losses if not managed properly. Breeding is widely regarded as the most effective means of disease control; however, many pests or diseases are not the main goals of breeding programmes because they are of local or minor significance. The high level of genetic diversity in faba bean accessions made it possible to develop resistant lines through several cycles of single plant selection, with self-pollination. Advances in pest and disease resistance in faba bean were recently reviewed by Rubiales and Khazaei (2022) and by Adhikari *et al.* (2021).

Foliar diseases

Globally, *Ascochyta* blight (*Ascochyta fabae*), chocolate spot (*Botrytis fabae*) and rust (*Uromyces viciae-fabae*) are the most severe and devastating diseases of faba bean. The development of resistance to these diseases is somehow complicated

Table 15.2. Time in days to crop phenology cycle under speed breeding confirming four cycles per year. (Author's own table.)

Traits	Mean	Ranges
Days to flowering	34±0.7	31–37
Days to podding	48±0.6	44–49
Days to maturity	94±0.3	92–94

due to their pathogen variability (Gorfu, 1996; Herath, 1997; Stoddard *et al.*, 1999; Ijaz *et al.*, 2021).

Sources for resistance to *Ascochyta* blight and chocolate spot were identified by ICARDA (Robertson, 1984; Hanounik and Robertson, 1989) and were used to develop lines with a good level of resistance to these diseases through transferring resistance or through direct selection in collaboration with the National Agricultural Research systems. The lines L-831818, V-26 and V-958 had stable resistance to *Ascochyta* blight across environments. Atienza *et al.* (2016) considered line 29H one of the most resistant to *Ascochyta* blight. Finally, Ascot and Farah are two resistant varieties available for use in Australia (Kaur *et al.*, 2014), and recently, PBA Amberley, PBA Bendoc, PBA Samia and Nura were reported as resistant to two different pathotypes (Adhikari *et al.*, 2022).

Despite the difficulty of breeding for a high level of chocolate spot resistance, because it is believed to be a multigenic disease (Beyene *et al.*, 2016), breeding efforts led to the release of the following chocolate spot resistant lines: Moti, Gebelcho, Gora, Obsie, Walki, Gora, Icarus, Amberley and PBA in Ethiopia and Australia (Temesgen *et al.*, 2015; Maalouf *et al.*, 2019); however, the wide adaptation of these varieties is yet to be evaluated.

Several sources of resistance to rust have been reported (Sillero *et al.*, 2010; Rubiales *et al.*, 2013; Adhikari *et al.*, 2016; Ijaz and Adhikari, 2016) but their resistance is incomplete and is described as 'slow rusting'. In Australia, the three varieties PBA Warda, PBA Nasma and PBA Nanu with moderate resistance to rust were released recently (Adhikari *et al.*, 2022). Hypersensitive rust resistance resulting from the death of some hypersensitive cells and the formation of others but with reduced disease severity is also a type of incomplete rust resistance (Sillero *et al.*, 2000; Adhikari *et al.*, 2016).

In addition to the foliar diseases mentioned above, faba bean gall, a relatively new disease caused by *Physioderma viciae*, has been threatening faba bean production in Japan, China and Ethiopia during the past ten years. Efforts to address this constraint led to the identification of several resistant lines, namely Dosha, Tumsa, Walki, Hachalu, Degaga and Nc 58 (Yitayih and Azmeraw, 2017; Alehegn *et al.*, 2018), but no genetic studies have been reported so far. The development of faba bean cultivars with multiple disease resistance should be a high priority for faba bean breeders around the globe, but this does not seem to be the case because only a few studies have reported the identification of sources for multiple disease resistance in faba bean (Villegas-Fernández *et al.*, 2011; Maalouf *et al.*, 2016).

Soil-borne diseases

Besides the mentioned foliar diseases, root rot diseases have been reported to cause severe yield losses in high rainfall and irrigated areas where waterlogging occurs (Maalouf *et al.*, 2021; Paul and Gupta, 2021). Different agents cause root rot diseases, but *Fusarium solani* (black root rot) is the most common. Root rot management is challenging since pathogens are found close to the rhizosphere and can persist for a long time by developing resistant structures. Varietal resistance is an essential and cost-effective component in the control package proposed for effectively managing the disease (Habtegebriel and Boydom, 2016). Resistant and moderately resistant varieties to black root rot are available (Wayu and Wolki), and sources of resistance were identified and are considered for further improvement (Habtegebriel and Boydom, 2016; Mahmoud and Abd El-Fatah, 2020).

Insects

Sitona weevil (*Sitona lineatus* L.), black bean aphid (*Aphis fabae* Scopoli), and faba bean stem borer (*Lixus algirus*) are only a few of the insects that seriously harm faba bean production by feeding on leaves, stems, roots and flowers of the plants or by transmitting viruses (Mwanauta *et al.*, 2015; Carrillo-Perdomo *et al.*, 2019). Sources of resistance to borer weevil and seed weevil were identified and will be integrated into breeding programmes to develop insect-resistant cultivars (Carrillo-Perdomo *et al.*, 2019; Ait Taadaouit *et al.*, 2021).

Viruses

More than 50 viruses can infect faba bean through seed or insect vectors (Makkouk *et al.*, 2012; Hema *et al.*, 2014), but the most common are broad bean mottle virus (BBMV), broad bean stain virus (BBSV), bean leaf roll virus (BLRV), bean yellow mosaic virus (BYMV), alfalfa mosaic virus (AMV) and faba bean necrotic yellows virus (FBNYV). Viruses are often not host-specific, which allows them to spread easily from one species to another. Prevention is the key to a virus-free crop because, once infected, the cure is impossible, which highlights the importance of developing virus-resistant cultivars. To this end, resistance to BLRV and BYMV was reported (Robertson and Saxena, 1993; Makkouk and Kumari, 1995; Makkouk *et al.*, 2002), and 27 resistant genotypes to FBYNV were identified recently at ICARDA (Kumari, 2018). Also, resistance to BLRV of the two Australian cultivars PBA Nasma and PBA Nanu has been reported (Adhikari *et al.*, 2021).

Weeds

Annual and parasitic weeds are considered one of the major biotic constraints for faba bean production, but parasitic weeds are the most dangerous.

Broomrapes (*Orobanche crenata* and *Orobanche foetida*) are the most devastating root parasitic weeds that infect faba bean fields, causing significant yield losses in the Mediterranean basin, North Africa and the Near East and they are expanding to Ethiopia and Sudan (Abebe *et al.*, 2013; Rubiales *et al.*, 2016). Faba bean is a poor weed competitor so the control of weeds in faba bean is crucial and challenging because none of the proposed management techniques have shown satisfactory results. Hence, breeding for broomrape resistance has proven to be the most promising method, especially because, after some *Orobanche*-resistant cultivars were released, faba bean regained some farming areas in the infected regions. Numerous lines with *Orobanche* resistance have been identified, leading to the release of *Orobanche*-resistant cultivars such as Giza 402,

Cairo 843, Misr 1 and Misr 3 in Egypt, Najah and Chourouk in Tunisia, and Baraka and Hashbengie in Ethiopia (Cubero, 1994; Maalouf *et al.*, 2019). There is no complete resistance to *Orobanche* among the lines identified so far; however, the cultivar Baraca and many accessions, including Quijote, Navio, 674/154/85 L3-4, or 402/29/84, V-1268, V-1302, V-1301, V-268, V-231, V-319 and V-1272, were reported to be resistant to both *Orobanche* species in numerous sites (Kharrat and Halila, 1994; Abbes *et al.*, 2007; Fernández-Aparicio *et al.*, 2012; Rubiales *et al.*, 2014). Because complete weed resistance is not yet available, herbicide application is therefore still regarded as the most efficient method to control parasitic weeds. The negative effects that might result from the application of these chemicals and their limitations has encouraged scientists to identify herbicide-resistant lines of faba bean. To this end, PBA Bendoc, a variety resistant to imidazolinone, was released in Australia (Adhikari *et al.*, 2022), and lines resistant to glyphosate, metribuzin and imazethapyr were identified at ICARDA (Abou Khater *et al.*, 2021).

15.6.3   Breeding for tolerance to abiotic stresses

Genetic improvement of abiotic stress tolerance is one of the main objectives of the faba bean breeding programme around the world. Numerous abiotic stresses affect faba bean production on a regional level, such as waterlogging, soil acidity and soil salinity, but heat, frost, and drought or waterlogging are the ones of global importance. The high level of diversity in faba bean accessions facilitated the identification and development of tolerant lines.

Heat tolerance

Heat stress during the reproductive stage of the faba bean affects pollen growth and viability, resulting in significant yield loss (Bishop *et al.*, 2016). Therefore, when evaluated under heat conditions, pollen viability can be an indicator of the tolerance level of faba bean genotypes (Maalouf *et al.*, 2019). Many studies reported the identification of heat-tolerant faba bean accessions, namely C.52/1/1/1, IG11743,

IG11843, VF351, VF420, VF522, VF626 and INRA1631 (Abdelmula and Abuanja, 2007; Maalouf et al., 2019). In addition, the high tolerance to the heat of two Sudanese cultivars, Ed-Damer and Hudeiba-93, was re-confirmed recently (Maalouf et al., 2019).

Cold tolerance

As faba bean cultivation is expanding to northern latitudes such as in North America and North Europe, and with the many advantages that winter faba bean offers over the spring type (Link et al., 2010), frost tolerance is becoming a pressing task for breeders and scientists. Breeding for cold tolerance that is based on non-destructive methods led to the identification of several frost-tolerant cultivars and improved lines, such as Arthur, Diver, Gladice, Côte d'Or, Hiverna, ILB3187, ILB2999, ILB14 and ILB345 (Link et al., 2010; Inci and Toker, 2011; Landry et al., 2015; Ali et al., 2016). The ability of some genotypes to tolerate frost at the seedling stage does not mean that they will be able to tolerate it at the vegetative stage (Link et al., 2010). To this end, genotypes with tolerance to frost at the vegetative stage, such as 11NF010a-2, PBA Warda, PBA Nasma and PBA Nanu, were identified in Australia (Alharbi, 2018).

Drought tolerance

Faba bean is sensitive to rainfall anomalies, especially when they occur late at the productive stage and when the crop is grown under rain-fed conditions (Siddique et al., 2001). The identification of early maturing genotypes is a strategy adopted by breeders to ensure the drought escape of the crop. Although several studies highlighted the presence of genotypic variation for drought tolerance (Abdelmula et al., 1999; Amede et al., 1999; Link et al., 1999; Khazaei et al., 2013; Maalouf et al., 2015; Muktadir et al., 2021), and several physiological parameters such as carbon isotope discrimination, leaf temperature, photosynthetic rate, transpiration rate, stomatal conductance, and the spectral indices structure-insensitive pigment index and normalized pheophytinization index (Khan et al., 2010; Maalouf et al., 2015; Mansour et al., 2021) can be used in screening drought tolerance physiological traits associated with drought, the progress toward

developing drought tolerant cultivars is very slow. The high and stable yield of Flip06-010FB across environments led to its classification as drought tolerant (Maalouf et al., 2015) and Nubaria-2, Nubaria-3, Misr-1, Sakha-3, Giza-3-Improved, Giza-843 and GZA were reported as drought tolerant by Mansour et al. in 2021.

Waterlogging tolerance

Faba bean production suffers from waterlogging in many high-rainfall and irrigated areas such as the vertisol area in Ethiopia (Keneni et al., 2001) and the Nile River basin in Sudan and Egypt. Waterlogging can lead to severe growth and yield reduction and can escalate foliar diseases (Pampana et al., 2016; Tadele, 2020). To address this notable limiting factor, the faba bean research programme in Ethiopia released six faba bean varieties (Selale, Wayu, Wolki, Hachalu, Didia and Ashebeka) that have high yield potential under waterlogging conditions (Tadele, 2020).

Soil salinity

The deleterious effects of soil salinity make it one of the most severe problems facing semi-arid agriculture systems. Approximately 20% of agricultural land in the world and 35% of agricultural land in Egypt are affected by salt stress (Yamaguchi and Blumwald, 2005; Omar et al., 2021).

Soil salinity is the most serious problem facing semi-arid agriculture systems because it restricts the ability of the plant to access water and nutrients. Saline soil restricts the ability of faba bean plants to access water and nutrients, leading to severe yield losses (Farooq et al., 2017). The practices applied to reduce the detrimental effects of soil salinity do not seem very effective as they are applied at post emergence stages and faba bean seed germination is highly affected by saline soil. Hence, growing faba bean cultivars with tolerance to soil salinity is the best approach. The cultivars Giza429, Giza843, Misr1, ILB-4347, Sakha, Misr-3, Flip12501FB, Sel.TH/4002/09, Sel.TH/4199/09 and VF112 were reported as tolerant to soil salinity and thus represented a step toward the development of salt stress tolerance (del Pilar Cordovilla et al., 1995; Abdelhamid et al., 2010; Afzal et al., 2022; Maalouf et al., 2022).

Soil acidity

Soil acidity, associated with low nutrient availability, is one of the major limiting factors for faba bean production in Ethiopian highlands (Fageria *et al.*, 2012; Keneni *et al.*, 2016). The high cost of managing acidic soil, which is not affordable for poor Ethiopian farmers, made breeding for tolerance to acidic soils one of the major breeding objectives in Ethiopia. Successful screening results were reported by Tolessa *et al.* (2015), Belachew and Stoddard (2017) and more recently by Tadele and Shiferaw (2022), where GLA 1103, CS20DK, Obse, Wolki, Didi'a, NC 58, Kassa and Dosha were identified as tolerant to acidic soils.

15.7   Erosion of Genetic Diversity

The genetic diversity of any crop changes over time owing to various biotic and abiotic factors, which lead to changes in genetic diversity through mutation, natural and artificial selection, genetic drift, and changes in gene flow. Breeding plays a substantial role in the choice of superior genotypes by narrowing down the variability, which significantly affects the gene and genotypic frequencies of the population (Ray *et al.*, 2015). Several factors affect the genetic diversity of faba bean, which can be summarized here.

First, being a partially allogamous crop, different pollination systems can affect seed production patterns. Faba bean is visited by a significant number of bees, bumble bees, honeybees, and wild bees (Bond and Kirby 1999, 2001; Pierre *et al.*, 1999). It might be opportune to consider that animal-based pollination from wild species fosters reproductive potential and genetic resilience in many ecosystems. Pollinators help seed and pod set and, even more importantly, contribute to the escape from stresses. For instance, the mean disease severity is lower, and the number of resistant lines selected is higher under open-pollinated conditions than in self-pollinated conditions. Pollinators play a positive role in maintaining the diversity of faba bean crops.

The second point is the worldwide reduction of the cultivated area of faba bean due to several abiotic and abiotic stresses as shown by the data compiled from FAO up to 1992 (Fig. 15.2), listing the global annual harvested area, yield, and production of faba bean and showing a reduction of 50% of the overall area since 1961. This reduction induced the loss of some landraces, which in turn could be reflected in the change or loss of alleles because of reduced population size and shrinking in a number of distinct habitats or environments.

In Morocco, the area allocated to faba bean has been reduced by 50% following infestation by the *Orobanche* parasitic weed, which has forced farmers to replace the dominant susceptible landraces with improved developed cultivars. FBNYV also devastated the faba bean crop in 1992, which has led to a drastic reduction in area and, therefore, the disappearance of traditional landraces and cultivars (Katul *et al.*, 1993; Makkouk *et al.*, 1994). Similarly, due to recurrent drought in Yunnan province, the area of faba bean has been reduced significantly; meaning significant losses of endemic landraces grown in this area. Artificial selection through deployment and selection of desirable traits led to the replacement of old landraces with new resistant/tolerant cultivars or by other species. In addition to improved agricultural practices, the observed increase in average yield could result from the increased adoption of modern varieties and traditional landrace replacement; this could be another indicator of the genetic erosion of this crop. In Egypt, more than 25 varieties have been released since 1980, with a 30% adoption rate. In China, the cultivar Yundou147, released from a K0285 × ILB8047 cross, is estimated to account for more than 30% of the faba bean acreage in Yunnan province. Several varieties replacing the old landraces in different regions have been released by various Chinese academies.

Surveys undertaken by ICARDA within the dry-land agrobiodiversity project including four countries of the Fertile Crescent – Jordan, Lebanon, Syria and the Palestinian Authority – showed that the landraces of several field crops (cereals and food legumes) were replaced by introduced fruit tree species, such as apples, cherries and olives (Mazid *et al.*, 2014).

The consumer preference in North Africa has changed from total consumption of dry faba bean to the inclusion of green faba bean in the diet, especially in the case of Egypt, which imposed in the case of Egypt and Sudan the replacement of equina types by major types to address the consumer preferences. These also contributed to the erosion of valuable landraces of equina types in areas where they are traditionally grown. In addition, the living wild relatives of faba bean crops are not yet identified and may be eroded totally because of urbanization and cultivation of food trees in the high Abreus Mountain where fossil forms of this wild type were discovered. The only information available is on fossil specimens attributed to the first parent of faba bean cultivated varieties (Caracuta *et al.*, 2016).

15.8 Genomic and Genetic Resources Integration in Crop Improvement

Despite the advantages that molecular markers offer in improving the productivity and accuracy of classical plant breeding and despite the moderate level of resistance in selected faba bean germplasm for the majority of stresses, little effort has been put into developing reliable DNA markers that breeders can use in many aspects such as genetic maps, quantitative trait loci (QTL) mapping, marker-assisted breeding, and genetic diversity studies as part of their breeding strategies. The delay in the assembly of the faba bean reference genome has greatly complicated these studies; however, the high-throughput approaches such as transcriptome analysis (Ocaña *et al.*, 2015) and the development of genetic maps using different types of molecular markers (reviewed by Khazaei *et al.*, 2021) paved the way for the better exploitation of the crop through the discrimination of varying faba bean landraces based on their geographical origin, ecological behaviours and response to diverse biotic/abiotic stresses. For instance, in terms of abiotic stresses, SNPs/QTLs associated with heat tolerance, drought tolerance and soil salinity tolerance were recently identified (Asif and Paul, 2021; Maalouf *et al.*, 2022; Gutiérrez *et al.*, 2023) (Table 15.3). Also, Sallam *et al.* (2022) identified nine novel SNPs and genomic regions controlling root traits under frost stress. For biotic stresses, many molecular studies resulted in the identification of genes, QTLs and/or markers associated with resistance to *Ascochyta*

blight (Avila *et al.*, 2004; Díaz-Ruiz *et al.*, 2009a; Atienza *et al.*, 2016; Ocaña-Moral *et al.*, 2017; Faridi *et al.*, 2021; Gutiérrez and Torres, 2021) and chocolate spot (Gela *et al.*, 2022) (Table 15.3). Some of the single genes controlling hypersensitive rust resistance were mapped, including Uvf-1, Uvf-2 and Uvf-3 (Avila *et al.*, 2003; Ijaz *et al.*, 2021) (Table 15.3). Moreover, QTLs/genes associated with resistance to *Orobanche crenata* and *Orobanche foetida* in faba bean were identified and validated (Díaz-Ruiz *et al.*, 2009b; Gutiérrez *et al.*, 2013; Ocaña-Moral *et al.*, 2017; Gutiérrez and Torres, 2021) and a more saturated map with reduced QTL intervals was proposed recently by Gutiérrez and Torres (2021). Also, Abou-Khater *et al.* (2022) identified SNPs associated with herbicide tolerance (Table 15.3). Concerning the antinutritional factors of faba bean, high-throughput, low-cost Kompetitive allele specific PCR (KASP) markers for low vicine-convicine concentration and for low tannin content were developed (Khazaei *et al.*, 2017; Zanotto *et al.*, 2020) (Table 15.3).

Recent faba bean genetic studies reported identified QTL markers associated with key agronomic and seed-related traits (Li *et al.*, 2023; Skovbjerg *et al.*, 2023; Zhao *et al.*, 2023). While most published faba bean genetic studies were based on biparental F_2 and recombinant inbred line (RIL) populations lacking precision and having low genetic diversity (Scott *et al.*, 2020), Webb *et al.* published in 2016 the first faba bean genetic map constructed from six F_2 and RIL populations and produced a map of 687 markers in six linkage groups corresponding to the six haploid chromosomes. In 2020, Carrillo-Perdomo *et al.* constructed the most saturated genetic map using three mapping populations and encompassing 1728 SNP markers distributed in six linkage groups. In addition to linkage mapping, genome-wide association studies (GWAS) are frequently performed now to maximize the allele frequency and genetic diversity that is lacking in biparental populations, resulting in higher mapping resolution (Faridi *et al.*, 2021; Abou-Khater *et al.*, 2022; Maalouf *et al.*, 2022; Gutiérrez *et al.*, 2023). It was not until early this year that the giant genome of faba bean was assembled (Jayakodi *et al.*, 2023), enabling breeders and geneticists to accelerate the improvement of sustainable faba bean production in the different production areas.

Table 15.3. Details of some recent genetic/genomic studies on faba bean traits of interest. (Author's own table.)

Trait of interest	Population	Marker/loci	Number of markers/loci	Reference
Heat tolerance	Diverse	SNP	11	Maalouf *et al.* (2022)
Cold tolerance		SNP	9	Sallam *et al.* (2022)
Drought tolerance	Diverse	SNP	29	Gutiérrez *et al.* (2023)
Soil salinity tolerance	RIL	QTL	7	Asif and Paul (2021)
Ascochyta blight	RIL	QTL	3	Gutiérrez and Torres (2021)
resistance	Diverse	AFLP, SNP	9 AFLP, 3 SNPs	Faridi *et al.* (2021)
Chocolate spot resistance	RIL	QTL	5	Gela *et al.* (2022)
Rust resistance	RIL	KASP	2	Ijaz *et al.* (2021)
Broomrape resistance	RIL			Gutiérrez and Torres (2021)
Herbicide tolerance	Diverse	SNP	14	Abou-Khater *et al.* (2022)
Low vicine-convicine		KASP	1	Khazaei *et al.* (2017)
Low tannin		KASP	1	Zanotto *et al.* (2020)
Agronomic traits	F_2	QTL	98	Li *et al.* (2023)

15.9 Cultivation Practices

15.9.1 Agronomy

Faba bean is a cool season crop that thrives in a wide range of climatic conditions. Under Mediterranean climates, with mild winters and dry and warm summers, faba beans are typically sown in autumn and harvested in late spring to early summer. In cooler regions, faba bean are usually sown in spring to prevent frost damage (Sallam *et al.*, 2015); however, it is worth noting that some cultivars with winter hardiness can be sown in autumn. In subtropical environments, faba bean can be grown during the two distinct rainy seasons that spans from February to June/July and from June to October/November. For optimal faba bean production, the growing season should ideally have moderate temperatures ranging from 18 to 27°C (Duke, 1981) with no excessive heat. Faba bean grows best on a deep, well-drained loam soil with a pH range of 6.5–9.0 (Jensen *et al.*, 2010). Faba requires an average of well-distributed rainfall of 300 mm throughout its growth cycle. It grows well without irrigation unless grown in a dry and hot climatic zone where rainfall is scarce. Supplemental irrigation is crucial in these conditions to avoid yield penalties.

Faba bean production and productivity can be significantly influenced by the presence of foliar and soil-borne diseases, which are prevalent in many countries. Growers employ various disease management strategies, including the use of resistant cultivars, cultural practices, fungicides and the integration of multiple practices, to minimize disease outbreaks and maximize yield potential.

Although nitrogen fertilization is not necessary for faba bean plants, applying nitrogen fertilizer at a rate of 20 kg/ha and phosphorus fertilizer at a rate of 40 kg/ha can enhance the nodulation process, especially because the biological nitrogen fixation process is known to be energetically demanding.

15.9.2 Disease management

The key to effective disease management is the application of an integrated approach that involves the use of disease-resistant cultivars along with other practices such as ploughing, adjusting the sowing dates, sowing density, crop rotation, intercropping with cereals, wide row spacing, use of certified seeds, weed management, straw burning when not used for animal feed, broad bed and furrow (BBF) and cultivation in the case of root rot disease (Keneni *et al.*, 2001; Davidson and Kimber, 2007; El-Mougy and Abdel-Kader, 2009; Stoddard *et al.*, 2010; Ahmed *et al.*, 2016) as well as the application of fungicides (Table 15.4) whether for seed treatments or foliar spray.

15.9.3 Insect management

Applying an effective integrated pest management programme that includes genetic resistance

Table 15.4. Fungicides are used to manage diseases in faba bean. (Author's own table.)

Disease	Seed treatment	Foliar spray
Ascochyta blight	Thiabendazole, benomyl	Chlorothalonil, mancozeb, azoxystrobin
Chocolate spot	*Rhizobium*	Mancozeb, carbendazim
		boron, calcium, molybdenum
Rust	NA	Mancozeb, dichlofluanid, difenoconazole, epoxiconazole, chlorothalonil, carbendazim, tebuconazole, flusilazole
Faba bean gall	Carbendazim, thiram, triadimefon	Triadimefon, metalaxyl, difenoconazole
Root rot disease	Thiamethoxam, mefenoxam, difenoconazole, carbathiin, thiram	NA

sanitation, insecticide application and biological control is essential to ensure healthy plants; however, among the mentioned techniques, scientists agreed that insecticide application is the most efficient method to control insects in faba bean fields and to control storage pests. Although numerous families of insecticides can be applied, their application should be used to delay and prevent insecticide resistance. An example of a possible biological control method is the use of *Lysiphlebus fabarum* Marshall (Hymenoptera) as a parasitoid of black bean aphid (Mahmoudi *et al.*, 2010). As for sanitation, treating faba bean seeds with thiamethoxam before planting might be helpful in preventing insects.

15.9.4 Virus management

Different methods are used to manage viruses, beginning with avoiding aphids, their most important vectors. Because aphids are polyphagous, however, it makes them hard to control. As is the case for disease management, combining genetic resistance with pesticide application along with other control measures is reported to reduce virus epidemics in faba bean (Makkouk *et al.*, 2003; Thresh, 2003). Of these control measures, we cite rouging, adjusting planting date, planting density and row spacing, planting non-host plants as borders, removing crop residues, minimum tillage and intercropping with non-hosts. The most popular insecticides used to control aphids are imidacloprid, acetamiprid and thiamethoxam.

15.9.5 Weed management

Weeds (parasitic and annual) in faba bean fields cannot be managed properly without the application of an integrated approach that combines the integration of partially resistant cultivars along with cultural practices, biological control and the application of chemical herbicides (Rubiales *et al.*, 2006, 2018; Abang *et al.*, 2007; Fernández-Aparicio *et al.*, 2016). Cultural practices include hand weeding, no-tillage farming, late planting, crop rotation with spring crops, intercropping, solarization and prevention of parasitic weed seed movement from one field to another through irrigation, animals or farming tools (Sauerborn *et al.*, 1989; López-Bellido *et al.*, 2009; Rubiales *et al.*, 2012; Karkanis *et al.*, 2016; Mishra *et al.*, 2016; Abbes *et al.*, 2019). As for biological control, scientists benefit from the allelopathic effect of some plant species as a bioherbicide for controlling annual and parasitic weeds (Evidente *et al.*, 2007; Alsaadawi *et al.*, 2017; Cimmino *et al.*, 2018; Messiha *et al.*, 2018; El-Dabaa *et al.*, 2019; Ahmed *et al.*, 2020). Nevertheless, the application of herbicides is still the most common method used to manage parasitic weeds in conventional agriculture (Joel *et al.*, 2007); glyphosate, imidazolinones and sulfonylureas and metribuzin are used to control parasitic weeds whilst pendimethalin, clomazone, bentazon, quizalofop-ethyl, and propaquizafop are commonly used for annual weeds (Karkanis *et al.*, 2018).

15.9.6 Soil acidity/salinity

The application of lime is the best way to remediate acid soil. A recent study conducted by Yacob

(2022) revealed that the application of 117 kg/ha of lime and 46 kg/ha of phosphorus resulted in a significant yield improvement of faba bean. Calcium-based products are used as liming material, such as calcium carbonate ($CaCO_3$), calcium hydroxide ($Ca(OH)_2$) and calcium oxide (CaO). To mitigate the effects of soil salinity, many techniques have been recommended, including leaching, treating salty drainage water, applying biostimulants such as biochar and jasmonic acid (El Nahhas et al., 2021), melatonin, exopolysaccharide-producing bacteria (Abd El-Ghany and Attia, 2020), and salicylic acid (Ahmad et al., 2018) and biofertilizer like the soil yeast *Rhodotorula glutinis* (Gaballah and Gomaa, 2004) and *Rhizobium leguminosarum* bv. *viciae* (El-Akhdar, 2020) and the foliar application of silicon (Hellal et al., 2012), as well as treating the seeds with gamma radiation prior to sowing (El-Awadi et al., 2017).

15.10 Conclusions

Faba bean is one of the promising cool-season food legumes in many production areas. The crop suffers from various biotic and abiotic stresses; however, efforts have been made by faba bean breeders around the globe to develop varieties adapted to heat and drought stress and resistant to key important diseases. Owing to the limited funding for faba bean, the breeding programmes are based mostly on traditional field screening. However, recent advancements in modern technologies, including speed breeding, whole-genome sequencing of faba bean and identifying crucial markers linked to economic traits, mark significant progress in developing new cultivars. The modern speed breeding method allows the development of at least four generations per year and is considered a pioneer innovation that can lead to a significant increase in genetic gains. The application of association studies will be useful for trait mapping, genetic diversity and linkage disequilibrium studies or map-based cloning. This method allows the application of the molecular-assisted selection technique and genomic-assisted breeding for major economic traits. Identifying markers associated with economically important traits is becoming a reality in faba bean, which will allow the screening of substantial population sizes in early generations, ensuring a higher genetic rate.

References

Abang, M.M., Bayaa, B., Abu-Irmaileh, B. and Yahyaoui, A. (2007) A participatory farming system approach for sustainable broomrape (Orobanche spp.) management in the Near East and North Africa. *Crop Protection* 26, 1723–1732.

Abbes, Z., Kharrat, M., Delavault, P., Simier, P. and Chaïbi, W. (2007) Field evaluation of the resistance of some faba bean (*Vicia faba* L.) genotypes to the parasitic weed *Orobanche foetida* Poiret. *Crop Protection* 26, 1777–1784.

Abbes, Z., Trabelsi, I., Kharrat, M. and Amri, M. (2019) Intercropping with fenugreek (*Trigonella foenum-graecum*) enhanced seed yield and reduced *Orobanche foetida* infestation in faba bean (*Vicia faba*). *Biological Agriculture & Horticulture* 35, 238–247.

Abd El-Ghany, M.F. and Attia, M. (2020) Effect of exopolysaccharide-producing bacteria and melatonin on faba bean production in saline and non-saline soil. *Agronomy* 10, 316.

Abdalla, M.M.F., Morad, M.M. and Roushdi, M. (1976) Some quality characteristics of selections of *Vicia faba* L. and their bearing upon field bean breeding. *Zeitschrift für Pflanzenzüchtung* 77, 72–79.

Abdelhamid, M.T., Shokr, M.M. and Bekheta, M.A. (2010) Growth, root characteristics, and leaf nutrients accumulation of four faba bean (*Vicia faba* L.) cultivars differing in their broomrape tolerance and the soil properties in relation to salinity. *Communications in Soil Science and Plant Analysis* 41, 2713–2728.

Abdelmula, A.A. and Abuanja, I.K. (2007) Genotypic responses, yield stability, and association between characters among some of Sudanese faba bean (*Vicia faba* L.) genotypes under heat stress. In: *Proceedings of Conference on International Agricultural Research for Development*. University of Kassel Witzenhausen and University of Gottingen, Germany, pp. 9–11.

Abdelmula, A.A., Link, W., Kittlitz, E.V. and Stelling, D. (1999). Heterosis and inheritance of drought tolerance in faba bean, *Vicia faba* L. *Plant Breeding* 118, 485–490.

Abebe, T., Meles, K., Nega, Y., Beyene, H. and Kebede, A. (2013) Interaction between broomrape (*Orobanche crenata*) and resistance faba bean genotypes (*Vicia faba* L.) in Tigray region of Ethiopia. *Canadian Journal of Plant Protection* 1, 104–109.

Abou-Khater, L., Maalouf, F., Patil, S.B., Balech, R., Nacouzi, D., Rubiales, D. and Kumar, S. (2021) Identification of tolerance to metribuzin and imazethapyr herbicides in faba bean. *Crop Science* 61, 2593–2611.

Abou-Khater, L., Maalouf, F., Jighly, A., Alsamman, A.M., Rubiales, D. *et al.* (2022) Genomic regions associated with herbicide tolerance in a worldwide faba bean (Vicia faba L.) collection. *Scientific Reports* 12, 158.

Adhikari, K.N., Zhang, P., Sadeque, A., Hoxha, S. and Trethowan, R. (2016) Single independent genes confer resistance to faba bean rust (*Uromyces viciae-fabae*) in the current Australian cultivar Doza and a central European line Ac1655. *Crop and Pasture Science* 67, 649–654.

Adhikari, K.N., Khazaei, H., Ghaouti, L., Maalouf, F., Vandenberg, A., Link, W. and O'Sullivan, D.M. (2021) Conventional and molecular breeding tools for accelerating genetic gain in faba bean (*Vicia faba* L.). *Frontiers in Plant Science* 12, 744259.

Adhikari, K.N., Khater, L.A. and Maalouf, F. (2022) Application of genetic, genomic strategies to address the biotic stresses in faba bean. In: *Genomic Designing for Biotic Stress Resistant Pulse Crops.* Springer International Publishing, Cham, Switzerland, pp. 353–380.

Afzal, M., Alghamdi, S.S., Migdadi, H.H., El-Harty, E. and Al-Faifi, S.A. (2022) Agronomical and physiological responses of faba bean genotypes to salt stress. *Agriculture* 12, 235.

Ahmad, P., Alyemeni, M.N., Ahanger, M.A., Egamberdieva, D., Wijaya, L. and Alam, P. (2018) Salicylic acid (SA) induced alterations in growth, biochemical attributes and antioxidant enzyme activity in faba bean (Vicia faba L.) seedlings under NaCl toxicity. *Russian Journal of Plant Physiology* 65, 104–114.

Ahmed, A.F. and Fakkar, A.A. (2020) Effect of humic acid and powders of some plants producing allelopathic compounds on soil properties and productivity of faba bean. *Asian Journal of Research and Review in Agriculture* 2, 52–69.

Ahmed, S., Abang, M.M. and Maalouf, F. (2016) Integrated management of Ascochyta blight (*Didymella fabae*) on faba bean under Mediterranean conditions. *Crop Protection* 81, 65–69.

Ait Taadaouit, N., El Fakhouri, K., Sabraoui, A., Maalouf, F., Rohi, L. and El Bouhssini, M. (2021) First sources of resistance in faba bean (*Vicia faba* L.) to the stem borer weevil, *Lixus algirus* L. (Coleoptera: Curculionidae). *Phytoparasitica* 49, 349–356.

Alehegn, M., Tiru, M. and Taddess, M. (2018) Screening of faba bean (*Vicia faba*) varieties against faba bean gall diseases (*Olpidium viciae*) in east Gojjam zone, Ethiopia. *Journal of Biology, Agriculture and Healthcare* 8, 50–55.

Alharbi, N. (2018) Factors of yield determination on faba bean (*Vicia faba* L.) as influenced by varying sowing times. Doctoral thesis. University of Sydney, Sydney, Australia.

Ali, M.B., Welna, G.C., Sallam, A., Martsch, R., Balko, C. *et al.* (2016) Association analyses to genetically improve drought and freezing tolerance of faba bean (*Vicia faba* L.). *Crop Science* 56, 1036–1048.

Allkin, R., Macfarfarlane, T.D., White, R.J., Bisby, F.A. and Adey, M.E. (1986) Names and synonyms of species and subspecies in the Vicieae: issue 3. Vicieae database project. Experimental Taxonomic Information Products Publication, University of Southampton, Southampton, UK.

Alsaadawi, I.S., Tawfiq, A.A. and Malih, H.M. (2017) Effect of allelopathic sorghum mulch on growth and yield of faba bean (*Vicia faba*) and companion weeds. *Journal of Plant Protection* 12, 123–127.

Amede, T., Kittlitz, E.V. and Schubert, S. (1999) Differential drought responses of faba bean (*Vicia faba* L.) inbred lines. *Journal of Agronomy and Crop Science* 183, 35–45.

Asif, M.A. and Paull, J.G. (2021) An approach to detecting quantitative trait loci and candidate genes associated with salinity tolerance in faba bean (*Vicia faba*). *Plant Breeding* 140, 643–653.

Atienza, S.G., Palomino, C., Gutiérrez, N., Alfaro, C.M., Rubiales, D., Torres, A.M. and Ávila, C.M. (2016) QTLs for ascochyta blight resistance in faba bean (*Vicia faba* L.): validation in field and controlled conditions. *Crop and Pasture Science* 67, 216–224.

Avila, C.M., Sillero, J.C., Rubiales, D., Moreno, M.T. and Torres, A.M. (2003) Identification of RAPD markers linked to the Uvf-1 gene conferring hypersensitive resistance against rust (*Uromyces viciae-fabae*) in *Vicia faba* L. *Theoretical and Applied Genetics* 107, 353–358.

Avila, C.M., Satovic, Z., Sillero, J.C., Rubiales, D., Moreno, M.T. and Torres, A.M. (2004) Isolate and organ-specific QTLs for ascochyta blight resistance in faba bean (*Vicia faba* L). *Theoretical and Applied Genetics* 108, 1071–1078.

Belachew, K.Y. and Stoddard, F.L. (2017) Screening of faba bean (*Vicia faba* L.) accessions to acidity and aluminium stresses. *PeerJ* 5, e2963. DOI: 10.7717/peerj.2963

Beyene, A.T., Derera, J., Sibiya, J. and Fikre, A. (2016) Gene action determining grain yield and chocolate spot (*Botrytis fabae*) resistance in faba bean. *Euphytica* 207, 293–304.

Bishop, J., Potts, S.G. and Jones, H.E. (2016) Susceptibility of faba bean (*Vicia faba* L.) to heat stress during floral development and anthesis. *Journal of Agronomy and Crop Science* 202, 508–517.

Bond, D.A. and Kirby, E.J.M. (1999) *Anthophora plumipes* (Hymenoptera: Anthophoridae) as a pollinator of broad bean (*Vicia faba major*). *Journal of Apicultural Research* 38, 199–203.

Bond, D.A. and Kirby, E.J.M. (2001) Further observations of *Anthophora plumipes* visiting autumn-sown broad bean (*Vicia faba major*) in the United Kingdom. *Journal of Apicultural Research* 40, 113–114.

Caracuta, V., Barzilai, O., Khalaily, H., Milevski, I., Paz, Y. *et al.* (2015) The onset of faba bean farming in the Southern Levant. *Scientific Reports* 5, 14370.

Caracuta, V., Weinstein-Evron, M., Kaufman, D., Yeshurun, R., Silvent, J. and Boaretto, E. (2016) 14,000-year-old seeds indicate the Levantine origin of the lost progenitor of faba bean. *Scientific Reports* 6, 37399.

Carrillo-Perdomo, E., Raffiot, B., Ollivier, D., Deulvot, C., Magnin-Robert, J.B., Tayeh, N. and Marget, P. (2019) Identification of novel sources of resistance to seed weevils (*Bruchus* spp.) in a faba bean germplasm collection. *Frontiers in Plant Science* 9, 1914.

Carrillo-Perdomo, E., Vidal, A., Kreplak, J., Duborjal, H., Leveugle, M. *et al.* (2020) Development of new genetic resources for faba bean (*Vicia faba* L.) breeding through the discovery of gene-based SNP markers and the construction of a high-density consensus map. *Scientific Reports* 10, 6790.

Cimmino, A., Masi, M., Rubiales, D., Evidente, A. and Fernández-Aparicio, M. (2018) Allelopathy for parasitic plant management. *Natural Product Communications* 13, 289–294. DOI: 10.1177/1934578X1801300307

Crépon, K., Marget, P., Peyronnet, C., Carrouée, B., Arese, P. and Duc, G. (2010) Nutritional value of faba bean (*Vicia faba* L.) seeds for feed and food. *Field Crops Research* 115, 329–339.

Cubero, J.I. (1973) Evolutionary trends in *Vicia faba* L. *Theoretical and Applied Genetics (Theoretische und angewandte Genetik)* 43, 59–65.

Cubero, J.I. (1974) On the evolution of *Vicia faba* L. *Theoretical and Applied Genetics* 45, 47–51.

Cubero, J.I. (1994) Breeding work in Spain for *Orobanche* resistance in faba bean and sunflower. In: *Biology and Management of Orobanche. Proceedings of the Third International Workshop on Orobanche and Related Striga Research.* Royal Tropical Institute, Amsterdam, The Netherlands, pp. 465–473.

Cubero, J.I. and Nadal, S. (2005) Faba bean (*Vicia faba* L.). In: *Genetic Resources, Chromosome Engineering and Crop Improvement Grain Legumes. Series II-Grain Legumes.* CRC, Boca Raton, Florida, pp. 163–186.

Davidson, J.A., and Kimber, R.B. (2007) Integrated disease management of ascochyta blight in pulse crops. *Ascochyta Blights of Grain Legumes* 119, 99–110.

del Pilar Cordovilla, M., Ligero, F. and Lluch, C. (1995) Influence of host genotypes on growth, symbiotic performance, and nitrogen assimilation in faba bean (*Vicia faba* L.) under salt stress. *Plant and Soil* 172, 289–297.

Díaz-Ruiz, R., Satovic, Z., Avila, C.M., Alfaro, C.M., Gutiérrez, M.V., Torres, A.M. and Román, B. (2009a) Confirmation of QTLs controlling *Ascochyta fabae* resistance in different generations of faba bean (*Vicia faba* L.). *Crop and Pasture Science* 60, 353–361.

Díaz-Ruiz, R., Torres, A., Gutierrez, M.V., Rubiales, D., Cubero, J.I. *et al.* (2009b) Mapping of quantitative trait loci controlling *Orobanche foetida* Poir. resistance in faba bean (*Vicia faba* L.). *African Journal of Biotechnology* 8, 2718–2724.

Drayner, J.M. (1956) Regulation of outbreeding in field beans (*Vicia faba*). *Nature* 177, 489–490.

Duc, G., Bao, S., Baum, M., Redden, B., Sadiki, M. *et al.* (2010) Diversity maintenance and use of *Vicia faba* L. genetic resources. *Field Crops Research* 115, 270–278.

Duke, J.A. (1981) Legume species. In: *Handbook of Legumes of World Economic Importance.* Springer, Boston, Massachusetts, pp. 5–310.

El Nahhas, N., AlKahtani, M.D., Abdelaal, K.A., Al Husnain, L., Al Gwaiz, H.I. *et al.* (2021) Biochar and jasmonic acid application attenuates antioxidative systems and improves growth, physiology, nutrient uptake, and productivity of faba bean (*Vicia faba* L.) irrigated with saline water. *Plant Physiology and Biochemistry* 166, 807–817.

El-Akhdar, I. (2020) Improvement of faba bean production in salt affected soils by *Rhizobium leguminosarum* bv. *viciae* inoculation and phosphorus application. *Microbiology Research Journal International* 29, 1–15.

El-Awadi, M.E., Sadak, M., Dawood, M.G., Khater, M.A. and Elashtokhy, M.M.A. (2017) Amelioration the adverse effects of salinity stress by using γ-radiation in faba bean plants. *Bulletin of the National Research Centre* 41, 293–310.

El-Dabaa, M.T., Abd-El-Khair, H. and E-Nagdi, W.M.A. (2019) Field application of Clethodim herbicide combined with Trichoderma spp. for controlling weeds, root knot nematodes and Rhizoctonia root rot disease in two faba bean cultivars. *Journal of Plant Protection Research* 59, 255–264.

El-Mougy, N.S. and Abdel-Kader, M.M. (2009) Seed and soil treatments as integrated control measure against faba bean root rot pathogens. *Plant Pathology Bulletin* 18, 75–87.

Evidente, A., Fernández-Aparicio, M., Andolfi, A., Rubiales, D. and Motta, A. (2007) Trigoxazonane, a mono-substituted trioxazonane from *Trigonella foenum-graecum* root exudate, inhibits *Orobanche crenata* seed germination. *Phytochemistry* 68, 2487–2492.

Fageria, N.K., Baligar, V.C., Melo, L.C. and De Oliveira, J.P. (2012) Differential soil acidity tolerance of dry bean genotypes. *Communications in Soil Science and Plant Analysis* 43, 1523–1531.

FAOSTAT (2023) Food and Agriculture Organization of The United Nations. Available at: https://www.fao.org/faostat/en/#data/QCL (accessed 15 July 2022).

Faridi, R., Koopman, B., Schierholt, A., Ali, M.B., Apel, S. and Link, W. (2021) Genetic study of the resistance of faba bean (*Vicia faba*) against the fungus *Ascochyta fabae* through a genome-wide association analysis. *Plant Breeding* 140, 442–452.

Farooq, M., Gogoi, N., Hussain, M., Barthakur, S., Paul, S. *et al.* (2017) Effects, tolerance mechanisms and management of salt stress in grain legumes. *Plant Physiology and Biochemistry* 118, 199–217.

Fernández-Aparicio, M., Moral, A., Kharrat, M. and Rubiales, D. (2012) Resistance against broomrapes (*Orobanche* and *Phelipanche* spp.) in faba bean (*Vicia faba*) based in low induction of broomrape seed germination. *Euphytica* 186, 897–905.

Fernández-Aparicio, M., Flores, F. and Rubiales, D. (2016) The effect of *Orobanche crenata* infection severity in faba bean, field pea, and grass pea productivity. *Frontiers in Plant Science* 7, 1409.

Gaballah, M.S. and Gomaa, A.M. (2004) Performance of faba bean varieties grown under salinity. *Journal of Applied Sciences* 4, 93–99.

Gasim, S., Abel, S. and Link, W. (2004) Extent, variation and breeding impact of natural cross-fertilisation in German winter faba beans using hilum colour as marker. *Euphytica* 136, 193–200.

Gela, T.S., Bruce, M., Chang, W., Stoddard, F.L., Schulman, A.H., Vandenberg, A. and Khazaei, H. (2022) Genomic regions associated with chocolate spot (*Botrytis fabae* Sard.) resistance in faba bean (*Vicia faba* L.). *Molecular Breeding* 42, 35. DOI: 10.1007/s11032-022-01307-7

Gharzeddin, K., Maalouf, F., Khoury, B., Abou Khater, L., Christmann, S. and Jamal El Dine, N.A. (2019) Efficiency of different breeding strategies in improving the faba bean productivity for sustainable agriculture. *Euphytica* 215, 1–15.

Ghosh, S., Watson, A., Gonzalez-Navarro, O.E., Ramirez-Gonzalez, R.H., Yanes, L., Mendoza-Suárez, M. *et al.* (2018) Speed breeding in growth chambers and glasshouses for crop breeding and model plant research. *Nature Protocols* 13, 2944–2963.

Göl, Ş., Doğanlar, S. and Frary, A. (2017) Relationship between geographical origin, seed size and genetic diversity in faba bean (*Vicia faba* L.) as revealed by SSR markers. *Molecular Genetics and Genomics* 292, 991–999.

Gorfu, D. (1996) Morphological, cultural and pathogenic variability among nine isolates of *Botrytis fabae* from Ethiopia. *Fabis Newsletter* (38/39), 37–41.

Gutiérrez, N. and Torres, A.M. (2021) QTL dissection and mining of candidate genes for *Ascochyta fabae* and *Orobanche crenata* resistance in faba bean (*Vicia faba* L.). *BMC Plant Biology* 21, 1–12.

Gutiérrez, N., Palomino, C., Satovic, Z., Ruiz-Rodríguez, M.D., Vitale, S. *et al.* (2013) QTLs for *Orobanche* spp. resistance in faba bean: identification and validation across different environments. *Molecular Breeding* 32, 909–922.

Gutiérrez, N., Pégard, M., Balko, C. and Torres, A.M. (2023) Genome-wide association analysis for drought tolerance and associated traits in faba bean (*Vicia faba* L.). *Frontiers in Plant Science* 14, 1091875.

Habtegebriel, B. and Boydom, A. (2016) Integrated management of faba bean black root rot (*Fusarium solani*) through varietal resistance, drainage and adjustment of planting time. *Journal of Plant Pathology and Microbiology* 7, 1–4.

Hajjar, S.B. and Robertson, L.D. (1989) Resistance in *Vicia faba* germplasm to blight caused by Ascochyta fabae. *Plant Disease* 73, 202–205.

Harlan, J.R. and de Wet, J.M. (1971) Toward a rational classification of cultivated plants. *Taxon* 20, 509–517.

Hawtin, G.C. and Omar, M.A. (1980) International faba bean germplasm collections at ICARDA. *FABIS Newsletter-Faba Bean Information Service (ICARDA)* 2, 20–22.

Hellal, F.A., Abdelhameid, M., Abo-Basha, D.M. and Zewainy, R.M. (2012) Alleviation of the adverse effects of soil salinity stress by foliar application of silicon on faba bean (*Vicia faba* L.). *Journal of Applied Sciences Research*, August, 4428–4433.

Hema, M., Sreenivasulu, P., Patil, B.L., Kumar, P.L. and Reddy, D.V. (2014) Tropical food legumes: virus diseases of economic importance and their control. *Advances in Virus Research* 90, 431–505.

Herath, I.H.B. (1997) Genetic studies on faba bean rust and on rust resistance of faba bean. Doctoral thesis. University of Sydney, Sydney, Australia.

Ijaz, U., Sudheesh, S., Kaur, S., Sadeque, A., Bariana, H., Bansal, U. and Adhikari, K. (2021) Mapping of two new rust resistance genes Uvf-2 and Uvf-3 in faba bean. *Agronomy* 11, 1370.

Inci, N.E. and Toker, C. (2011) Screening and selection of faba beans (*Vicia faba* L.) for cold tolerance and comparison to wild relatives. *Genetic Resources and Crop Evolution* 58, 1169–1175.

Jayakodi, M., Golicz, A.A., Kreplak, J., Fechete, L.I., Angra, D. *et al.* (2023) The giant diploid faba genome unlocks variation in a global protein crop. *Nature* 615, 652–659.

Jensen, E.S., Peoples, M.B. and Hauggaard-Nielsen, H. (2010) Faba bean in cropping systems. *Field Crops Research* 115, 203–216.

Joel, D.M., Hershenhorn, J., Eizenberg, H., Aly, R., Ejeta, G. *et al.* (2007) Biology and management of weedy root parasites. *Horticultural Reviews* 33, 267–349.

Karkanis, A., Ntatsi, G., Kontopoulou, C.K., Pristeri, A., Bilalis, D. and Savvas, D. (2016) Field pea in European cropping systems: Adaptability, biological nitrogen fixation and cultivation practices. *Notulae Botanicae Horti Agrobotanici Cluj-Napoca* 44, 325–336.

Karkanis, A., Ntatsi, G., Lepse, L., Fernández, J.A., Vågen, I.M. *et al.* (2018) Faba bean cultivation–revealing novel managing practices for more sustainable and competitive European cropping systems. *Frontiers in Plant Science* 9, 1115.

Katul, U., Vetten, H.J., Maiss, E., Makkouk, K.M., Lesemann, D.E. and Casper, R. (1993) Characterisation and serology of virus-like particles associated with faba bean necrotic yellows. *Annals of Applied Biology* 123, 629–647.

Kaur, S., Kimber, R.B., Cogan, N.O., Materne, M., Forster, J.W. and Paull, J.G. (2014) SNP discovery and high-density genetic mapping in faba bean (*Vicia faba* L.) permits identification of QTLs for ascochyta blight resistance. *Plant Science* 217, 47–55.

Keneni, G., Asmamaw, B. and Jarso, M. (2001) Efficiency of drained selection environments for improving grain yield in faba bean under undrained target environments on vertisol. *Euphytica* 122, 279–285.

Keneni, G., Fikre, A. and Eshete, M. (2016) Reflections on highland pulses improvement research in Ethiopia. *Ethiopian Journal of Agricultural Sciences* Special 17–50.

Khan, H.R., Paull, J.G., Siddique, K.H.M. and Stoddard, F.L. (2010) Faba bean breeding for drought-affected environments: A physiological and agronomic perspective. *Field Crops Research* 115, 279–286.

Kharrat, M. and Halila, M.H. (1994) *Orobanche* species on faba bean (*Vicia faba* L.) in Tunisia: problems and management. In: *Biology and management of Orobanche. Proceedings of the Third International Workshop on Orobanche and Related Striga Research.* Royal Tropical Institute, Amsterdam, The Netherlands, pp. 639–643.

Khazaei, H., Street, K., Santanen, A., Bari, A. and Stoddard, F.L. (2013) Do faba bean (*Vicia faba* L.) accessions from environments with contrasting seasonal moisture availabilities differ in stomatal characteristics and related traits? *Genetic Resources and Crop Evolution* 60, 2343–2357.

Khazaei, H., Purves, R.W., Song, M., Stonehouse, R., Bett, K.E., Stoddard, F.L. and Vandenberg, A. (2017) Development and validation of a robust, breeder-friendly molecular marker for the vc-locus in faba bean. *Molecular Breeding* 37, 1–6.

Khazaei, H., O'Sullivan, D.M., Stoddard, F.L., Adhikari, K.N., Paull, J.G. *et al.* (2021) Recent advances in faba bean genetic and genomic tools for crop improvement. *Legume Science* 3, e75.

Koivunen, E., Partanen, K., Perttilä, S., Palander, S., Tuunainen, P. and Valaja, J. (2016) Digestibility and energy value of pea (*Pisum sativum* L.), faba bean (*Vicia faba* L.) and blue lupin (narrow-leaf) (*Lupinus angustifolius*) seeds in broilers. *Animal Feed Science and Technology* 218, 120–127.

Kumari, S. (2018) Screening faba bean (*Vicia faba* L.) germplasm for resistance to persistently aphid transmitted viruses. In: *Proceedings of Seventh International Food Legumes Research Conference.* Marrakesh, Morocco, p. 100.

Kupicha, F.K. (1981) Vicieae. *Advances in Legume Systematics* 1, 377–381.

Kwon, S.J., Hu, J. and Coyne, C.J. (2010) Genetic diversity and relationship among faba bean (*Vicia faba* L.) germplasm entries as revealed by TRAP markers. *Plant Genetic Resources* 8, 204–213.

Ladizinski, G. (1998) *Plant Evolution Under Domestication.* Kluwer Academic, Dordrecht, The Netherlands.

Landry, E.J., Lafferty, J.E., Coyne, C.J., Pan, W.L. and Hu, J. (2015) Registration of four winter-hardy faba bean germplasm lines for use in winter pulse and cover crop development. *Journal of Plant Registrations* 9, 367–370.

Li, M.W., He, Y.H., Rong, L.I.U., Guan, L.I., Dong, W.A.N.G. *et al.* (2023) Construction of SNP genetic map based on targeted next-generation sequencing and QTL mapping of vital agronomic traits in faba bean (*Vicia faba* L.). *Journal of Integrative Agriculture* 22, 2095–3119. DOI: 10.1016/j.jia.2023.01.003

Link, W., Ederer, W., Metz, P., Buiel, H. and Melchinger, A.E. (1994) Genotypic and environmental variation for degree of cross-fertilisation in faba bean. *Crop Science* 34, 960–964.

Link, W., Abdelmula, A.A., Kittlitz, E.V., Bruns, S., Riemer, H. and Stelling, D. (1999) Genotypic variation for drought tolerance in *Vicia faba*. *Plant Breeding* 118, 477–484.

Link, W., Balko, C. and Stoddard, F.L. (2010) Winter hardiness in faba bean: Physiology and breeding. *Field Crops Research* 115, 287–296.

López-Bellido, R.J., Benítez-Vega, J. and López-Bellido, L. (2009) No-tillage improves broomrape control with glyphosate in faba bean. *Agronomy Journal* 101, 1394–1399.

Maalouf, F.S., Suso, M.J. and Moreno, M.T. (1999) Choice of methods and indices for identifying the best parentals for synthetic varieties in faba bean. *Agronomie* 19, 705–712.

Maalouf, F., Mohammed, N., Aladdin, H., Ahmed, A., Xuxiao, Z., Shiying, B. and Tao, Y. (2013) Faba bean. In: Singh, M., Upadhyaya, H.D. and Bisht, I.S. (eds) *Genetic and Genomic Resources of Grain Legume Improvement.* Elsevier, Amsterdam, The Netherlands, pp. 113–136.

Maalouf, F., Nachit, M., Ghanem, M.E. and Singh, M. (2015) Evaluation of faba bean breeding lines for spectral indices, yield traits and yield stability under diverse environments. *Crop and Pasture Science* 66, 1012–1023.

Maalouf, F., Ahmed, S., Shaaban, K., Bassam, B., Nawar, F., Singh, M. and Amri, A. (2016) New faba bean germplasm with multiple resistances to Ascochyta blight, chocolate spot and rust diseases. *Euphytica* 211, 157–167.

Maalouf, F., Hu, J., O'Sullivan, D.M., Zong, X., Hamwieh, A., Kumar, S. and Baum, M. (2019) Breeding and genomics status in faba bean (*Vicia faba*). *Plant Breeding* 138, 465–473.

Maalouf, F., Ahmed, S. and Bishaw, Z. (2021) Faba bean. In: Pratap, A. and Gupta, S. (eds) *The Beans and the Peas.* Woodhead Publishing, Philadelphia, Pennsylvania, pp. 105–131.

Maalouf, F., Abou-Khater, L., Babiker, Z., Jighly, A., Alsamman, A.M. *et al.* (2022) Genetic dissection of heat stress tolerance in faba bean (*Vicia faba* L.) using GWAS. *Plants* 11, 1108.

Mahmoud, A.F. and Abd El-Fatah, B.E. (2020) Genetic diversity studies and identification of molecular and biochemical markers associated with fusarium wilt resistance in cultivated faba bean (*Vicia faba*). *The Plant Pathology Journal* 36, 11–28.

Mahmoudi, M., Sahragard, A. and Sendi, J.J. (2010) Foraging efficiency of *Lysiphlebus fabarum* Marshall (Hymenoptera: Aphidiidae) parasitising the black bean aphid, *Aphis fabae* Scopoli (Hemiptera: Aphididae), under laboratory conditions. *Journal of Asia-Pacific Entomology* 13, 111–116.

Makkouk, K.M. and Kumari, S.G. (1995) Transmission of broad bean stain comovirus and broad bean mottle bromovirus by weevils in Syria. *Journal of Plant Diseases and Protection* 102, 136–139.

Makkouk, K.M., Rizkallah, L., Madkour, M., El-Sherbeeny, M., Kumari, S.G., Amriti, A.W. and Solh, M.B. (1994) Survey of faba bean (*Vicia faba* L.) for viruses in Egypt. *Phytopathologia Mediterranea* 33, 207–211.

Makkouk, K.M., Kumari, S.G. and van Leur, J.A. (2002) Screening and selection of faba bean (*Vicia faba* L.) germplasm resistant to Bean leafroll virus. *Australian Journal of Agricultural Research* 53, 1077–1082.

Makkouk, K.M., Hamed, A.A., Hussein, M. and Kumari, S.G. (2003) First report of Faba bean necrotic yellows virus (FBNYV) infecting chickpea (*Cicer arietinum*) and faba bean (*Vicia faba*) crops in Sudan. *Plant Pathology* 52, 412.

Makkouk, K., Pappu, H. and Kumari, S.G. (2012) Virus diseases of peas, beans, and faba bean in the Mediterranean region. *Advances in Virus Research* 84, 367–402.

Mansour, E., Desoky, E.S.M., Ali, M.M., Abdul-Hamid, M.I., Ullah, H., Attia, A. and Datta, A. (2021) Identifying drought-tolerant genotypes of faba bean and their agro-physiological responses to different water regimes in an arid Mediterranean environment. *Agricultural Water Management* 247, 106754.

Martineau-Côté, D., Achouri, A., Karboune, S. and L'Hocine, L. (2022) Faba bean: an untapped source of quality plant proteins and bioactives. *Nutrients* 14, 1541. DOI: 10.3390/nu14081541

Mazid, A., Shideed, K. and Amri, A. (2014) Assessment of on-farm conservation of dry-land agrobiodiversity and its impact on rural livelihoods in the Fertile Crescent. *Renewable Agriculture and Food Systems* 29, 366–377.

Messiha, N.K., El-Dabaa, M.A.T., El-Masry, R.R. and Ahmed, S.A.A. (2018) The allelopathic influence of *Sinapis alba* seed powder (white mustard) on the growth and yield of *Vicia faba* (faba bean) infected with *Orobanche crenata* (broomrape). *Middle East Journal of Applied Sciences* 8, 418–425.

Miller, P. (1754) *The Gardeners Dictionary* (4th edn). John and James, Rivington, London.

Mishra, J.S., Rao, A.N., Singh, V.P. and Kumar, R. (2016) Weed management in major crops. *Weed Science and Management* 9, 169–190.

Muktadir, M.A., Adhikari, K.N., Ahmad, N. and Merchant, A. (2021) Chemical composition and reproductive functionality of contrasting faba bean genotypes in response to water deficit. *Physiologia Plantarum* 172, 540–551.

Muratova, V.S. (1931) *Boby (Vicia faba L.).* Izd. Instituta Rastenievodstva, Leningrad.

Mwanauta, R.W., Mtei, K.M. and Ndakidemi, P.A. (2015) Potential of controlling common bean insect pests (bean stem maggot (*Ophiomyia phaseoli*), Ootheca (*Ootheca bennigseni*) and Aphids (*Aphis fabae*)) using agronomic, biological and botanical practices in field. *Agricultural Sciences* 6, 489.

Ocaña, S., Seoane, P., Bautista, R., Palomino, C., Claros, G.M., Torres, A.M. and Madrid, E. (2015) Large-scale transcriptome analysis in faba bean (*Vicia faba* L.) under *Ascochyta fabae* infection. *PLoS One* 10, e0135143.

Ocaña-Moral, S., Gutiérrez, N., Torres, A.M. and Madrid, E. (2017) Saturation mapping of regions determining resistance to Ascochyta blight and broomrape in faba bean using transcriptome-based SNP genotyping. *Theoretical and Applied Genetics* 130, 2271–2282.

Oliveira, H.R., Tomás, D., Silva, M., Lopes, S., Viegas, W. and Veloso, M.M. (2016) Genetic diversity and population structure in *Vicia faba* L. landraces and wild related species assessed by nuclear SSRs. *PloS One* 11, e0154801.

Omar, M.E.D.M., Moussa, A.M.A. and Hinkelmann, R. (2021) Impacts of climate change on water quantity, water salinity, food security, and socioeconomy in Egypt. *Water Science and Engineering* 14, 17–27.

Palmer, R.G., Gai, J., Dalvi, V.A. and Suso, M.J. (2011) Male sterility and hybrid production technology. In: Pratap, A. and Kumar, J. (eds) *Biology and Breeding of Food Legumes.* CABI, Wallingford, UK, pp. 193–207. DOI: 10.1079/9781845937669.0193

Pampana, S., Masoni, A. and Arduini, I. (2016) Response of cool-season grain legumes to waterlogging at flowering. *Canadian Journal of Plant Science* 96, 597–603.

Paul, S.K. and Gupta, D.R. (2021) Faba bean (*Vicia faba* L.), A promising grain legume crop of Bangladesh: A review. *Agricultural Reviews* 42, 292–299.

Persoon, C.H. (1807) Neottia. In: Persoon, C.H. (ed.) *Synopsis Plantarum seu Enchiridium Botanicum.* Treuel & Wurtz, Paris, France, pp. 510–511.

Pierre, J., Suso, M.J., Moreno, M.T., Esnault, R. and Le Guen, J. (1999) Diversite et efficacite de l'entomofaune pollinisatrice (Hymenoptera: Apidae) de la feverole (*Vicia fabae* L.) sur deux sites, en France et en Espagne. *Annales-Societe Entomologique De France* 35, 312–318.

Ray, H., Bock, C. and Georges, F. (2015) Faba bean: transcriptome analysis from etiolated seedling and developing seed coat of key cultivars for synthesis of proanthocyanidins, phytate, raffinose family oligosaccharides, vicine, and convicine. *The Plant Genome* 8. DOI: 10.3835/plantgenome2014.07.0028

Robertson, L.D. (1984) A note on the ILB source of *Botrytis fabae* resistance. In: Chapman, G.P. and Tarawali, S.A. (eds) *Proceedings of a Seminar in the EEC Programme of Coordination of Research on Plant Productivity.* Springer, Dordrecht, The Netherlands, p. 79.

Robertson, L.D. and Cardona, C. (1986) Studies on bee activity and outcrossing in increase plots of *Vicia faba* L. *Field Crops Research* 15, 157–164.

Robertson, L.D. and Saxena, M.C. (1993) Problems and prospects of stress resistance breeding in faba bean. In: Singh, K.B. and Saxena, M.C. (eds) *Breeding for Stress Tolerance in Cool-season Food Legumes.* Wiley, Chichester, UK, pp. 37–50.

Rubiales, D. and Fernández-Aparicio, M. (2012) Innovations in parasitic weeds management in legume crops. A review. *Agronomy for Sustainable Development* 32, 433–449.

Rubiales, D. and Khazaei, H. (2022) Advances in disease and pest resistance in faba bean. *Theoretical and Applied Genetics* 135, 3735–3756.

Rubiales, D., Pérez-de-Luque, A., Fernández-Aparico, M., Sillero, J.C., Román, B. *et al.* (2006) Screening techniques and sources of resistance against parasitic weeds in grain legumes. *Euphytica* 147, 187–199.

Rubiales, D., Rojas-Molina, M.M. and Sillero, J.C. (2013) Identification of pre-and posthaustorial resistance to rust (*Uromyces viciae-fabae*) in lentil (*Lens culinaris*) germplasm. *Plant Breeding* 132, 676–680.

Rubiales, D., Flores, F., Emeran, A.A., Kharrat, M., Amri, M., Rojas-Molina, M.M. and Sillero, J.C. (2014) Identification and multi-environment validation of resistance against broomrapes (*Orobanche crenata* and *Orobanche foetida*) in faba bean (*Vicia faba*). *Field Crops Research* 166, 58–65.

Rubiales, D., Rojas-Molina, M.M. and Sillero, J.C. (2016) Characterisation of resistance mechanisms in faba bean (*Vicia faba*) against broomrape species (*Orobanche* and *Phelipanche* spp.). *Frontiers in Plant Science* 7, 1747.

Rubiales, D., Fernández-Aparicio, M., Vurro, M. and Eizenberg, H. (2018) Advances in parasitic weed research. *Frontiers in Plant Science* 9, 236.

Sallam, A., Martsch, R. and Moursi, Y.S. (2015) Genetic variation in morpho-physiological traits associated with frost tolerance in faba bean (*Vicia faba* L.). *Euphytica* 205, 395–408.

Sallam, A., Moursi, Y.S., Martsch, R. and Eltaher, S. (2022) Genome-wide association mapping for root traits associated with frost tolerance in faba beans using KASP-SNP markers. *Frontiers in Genetics* 13, 907267.

Sanz, A.M., Gonzalez, S.G., Syed, N.H., Suso, M.J., Saldaña, C.C. and Flavell, A.J. (2007) Genetic diversity analysis in *Vicia* species using retrotransposon-based SSAP markers. *Molecular Genetics and Genomics* 278, 433–441.

Sauerborn, J., Saxena, M.C. and Meyer, A. (1989) Broomrape control in faba bean (*Vicia faba* L.) with glyphosate and imazaquin. *Weed Research* 29, 97–102.

Scott, M.F., Ladejobi, O., Amer, S., Bentley, A.R., Biernaskie, J., Boden, S.A., Clark, M., Dell'Acqua, M., Dixon, L.E., Filippi, C.V. and Fradgley, N. (2020) Multi-parent populations in crops: a toolbox integrating genomics and genetic mapping with breeding. *Heredity* 125, 396–416.

Sentil, A., Lhomme, P., Michez, D., Reverté, S., Rasmont, P. and Christmann, S. (2021) "Farming with Alternative Pollinators" approach increases pollinator abundance and diversity in faba bean fields. *Journal of Insect Conservation* 26, 401–414.

Siddique, K.H.M., Regan, K.L., Tennant, D. and Thomson, B.D. (2001) Water use and water use efficiency of cool season grain legumes in low rainfall Mediterranean-type environments. *European Journal of Agronomy* 15, 267–280.

Sillero, J.C., Moreno, M.T. and Rubiales, D. (2000) Characterisation of new sources of resistance to *Uromyces viciae-fabae* in a germplasm collection of *Vicia faba*. *Plant Pathology* 49, 389–395.

Sillero, J.C., Villegas-Fernández, A.M., Thomas, J., Rojas-Molina, M.M., Emeran, A.A. *et al.* (2010) Faba bean breeding for disease resistance. *Field Crops Research* 115, 297–307.

Singh, A.K., Bhardwaj, R. and Singh, I.S. (2014) Assessment of nutritional quality of developed faba bean (*Vicia faba* L.) lines. *Journal of AgriSearch* 1, 96–101.

Sitte, P., Weiler, E., Kadereit, H.J., Bresinsky, A. and Körner, C. (2002) *Strasburger — Lehrbuch der Botanik*, 35. Auflage, Spektrum Akademischer Verlag, Heidelberg, Germany, pp. 339–351.

Skovbjerg, C.K., Angra, D., Robertson-Shersby-Harvie, T., Kreplak, J., Keeble-Gagnère, G. *et al.* (2023) Genetic analysis of global faba bean diversity, agronomic traits and selection signatures. *Theoretical and Applied Genetics* 136, 114.

Stoddard, F.L., Kohpina, S. and Knight, R. (1999) Variability of *Ascochyta fabae* in South Australia. *Australian Journal of Agricultural Research* 50, 1475–1481.

Stoddard, F.L., Nicholas, A.H., Rubiales, D., Thomas, J. and Villegas-Fernández, A.M. (2010) Integrated pest management in faba bean. *Field Crops Research* 115, 308–318.

Suso, M.J., Moreno, M.T. and Melchinger, A.E. (1999) Variation in outcrossing rate and genetic structure on six cultivars of *Vicia faba* L. as affected by geographic location and year. *Plant Breeding* 118, 347–350.

Suso, M.J., Pierre, J., Moreno, M.T., Esnault, R. and Le Guen, J. (2001) Variation in outcrossing levels in faba bean cultivars: role of ecological factors. *The Journal of Agricultural Science* 136, 399–405.

Suso, M.J., Gilsanz, S., Duc, G., Marget, P. and Moreno, M.T. (2006) Germplasm management of faba bean (*Vicia faba* L.): Monitoring intercrossing between accessions with inter-plot barriers. *Genetic Resources and Crop Evolution* 53, 1427–1439.

Tadele, M. (2020) Impact assessment of abiotic production constrains of Faba Bean (*Vicia faba* L.) and breeding mechanisms for acid soil, drought and waterlogging affected environments. *Journal of Biology, Agriculture and Healthcare* 10, 49–56.

Tadele, M. and Shiferaw, T. (2022) Reaction of elite faba bean genotypes for soil acidity stress. *International Journal of Horticulture, Agriculture and Food Science* 6, 5–18.

Tanno, K.I. and Willcox, G. (2006) The origins of cultivation of *Cicer arietinum* L. and *Vicia faba* L.: early finds from Tell el-Kerkh, north-west Syria, late 10th millennium BP. *Vegetation History and Archaeobotany* 15, 197–204.

Temesgen, T., Keneni, G., Sefera, T. and Jarso, M. (2015) Yield stability and relationships among stability parameters in faba bean (*Vicia faba* L.) genotypes. *The Crop Journal* 3, 258–268.

Terzopoulos, P.J. and Bebeli, P.J. (2008) Genetic diversity analysis of Mediterranean faba bean (*Vicia faba* L.) with ISSR markers. *Field Crops Research* 108, 39–44.

Thresh, J.M. (2003) Control of plant virus diseases in sub-Saharan Africa: the possibility and feasibility of an integrated approach. *African Crop Science Journal* 11, 199–223.

Tolessa, T.T., Keneni, G. and Mohammad, H. (2015) Genetic progresses from over three decades of faba bean ('*Vicia faba*' L.) breeding in Ethiopia. *Australian Journal of Crop Science* 9, 41–48.

Torres, A.M., Weeden, N.F. and Martin, A. (1993) Linkage among isozyme, RFLP and RAPD markers in *Vicia faba*. *Theoretical and Applied Genetics* 85, 937–945.

Van de Ven, W.T.G., Duncan, N., Ramsay, G., Phillips, M., Powell, W. and Waugh, R. (1993) Taxonomic relationships between *V. faba* and its relatives based on nuclear and mitochondrial RFLPs and PCR analysis. *Theoretical and Applied Genetics* 86, 71–80.

Villegas-Fernández, A.M., Sillero, J.C., Emeran, A.A., Flores, F. and Rubiales, D. (2011) Multiple-disease resistance in *Vicia faba*: multi-environment field testing for identification of combined resistance to rust and chocolate spot. *Field Crops Research* 124, 59–65.

Wang, H.F., Zong, X.X., Guan, J.P., Yang, T., Sun, X.L., Ma, Y. and Redden, R. (2012) Genetic diversity and relationship of global faba bean (*Vicia faba* L.) germplasm revealed by ISSR markers. *Theoretical and Applied Genetics* 124, 789–797.

Webb, A., Cottage, A., Wood, T., Khamassi, K., Hobbs, D. *et al.* (2016) A SNP-based consensus genetic map for synteny-based trait targeting in faba bean (*Vicia faba* L.). *Plant Biotechnology Journal* 14, 177–185.

Yacob, A. (2022) Response of faba bean and acid soil properties to lime and P fertilizer application in Hagereselam District Sidama Ethiopia. *Plant* 10, 76–80.

Yamaguchi, T. and Blumwald, E. (2005) Developing salt-tolerant crop plants: challenges and opportunities. *Trends in Plant Science* 10, 615–620.

Yitayih, G. and Azmeraw, Y. (2017) Adaptation of faba bean varieties for yield, for yield components and against faba bean gall (*Olpidium viciae* Kusano) disease in South Gondar, Ethiopia. *The Crop Journal* 5, 560–566.

Zanotto, S., Vandenberg, A. and Khazaei, H. (2020) Development and validation of a robust KASP marker for zt2 locus in faba bean (*Vicia faba*). *Plant Breeding* 139, 375–380.

Zeid, M., Schön, C.C. and Link, W. (2003) Genetic diversity in recent elite faba bean lines using AFLP markers. *Theoretical and Applied Genetics* 107, 1304–1314.

Zhao, N., Xue, D., Miao, Y., Wang, Y., Zhou, E. *et al.* (2023) Construction of a high-density genetic map for faba bean (*Vicia faba* L.) and quantitative trait loci mapping of seed-related traits. *Frontiers in Plant Science* 14, 1201103.

Zohary, D. and Hopf, M. (1973) Domestication of pulses in the Old World: legumes were companions of wheat and barley when agriculture began in the Near East. *Science* 182, 887–894.

Zong, X., Liu, X., Guan, J., Wang, S., Liu, Q., Paull, J.G. and Redden, R. (2009) Molecular variation among Chinese and global winter faba bean germplasm. *Theoretical and Applied Genetics* 118, 971–978.

16 Policies and Incentives to Promote Pulses: Indian Perspective

A. Amarender Reddy¹, Shravani Sanyal¹ and Aditya Pratap²*

¹*School of Crop Health Policy Support Research, ICAR-National Institute of Biotic Stress Management (ICAR-NIBSM), Raipur, India;* ²*All India Coordinated Research Project on Kharif Pulses, ICAR-Indian Institute of Pulses Research, Kanpur, India*

Abstract

This chapter evaluates the historical trends in the production and productivity of Indian pulses, including a comparison of relative costs and returns associated with alternative crops. Additionally, it conducts a social cost–benefit analysis, emphasizing the need to incentivize pulses by recognizing their nitrogen (N)-fixation capabilities and lowering greenhouse gas emissions. The chapter estimates the non-market externalities, particularly N-fixation, and advocates for equivalent incentives by the government to foster both the production and consumption of pulses. It also examines the pivotal role of pulses in achieving balanced nutrition for humans and considers consumer preferences for specific pulse traits. Domestic marketing channels and the value chain are reviewed to enhance competitiveness, focusing on government interventions to strengthen the pulse sector. Furthermore, the chapter explores trends in import and export scenarios, addressing future challenges in making pulses affordable for lower-income consumers without compromising the profitability of pulse farmers. A dedicated section discusses government policies, existing support, and their impact on pulse production and availability. Nutritional aspects of pulses are also considered, contemplating potential distribution through government initiatives targeting impoverished populations. Policy alternatives are thoroughly examined for their historical influence on pulse production, considering factors such as Minimum Support Price, procurement and fertilizer subsidies. The chapter also evaluates existing demand–supply disparities and advocates for the replacement of ad hoc trade policies with long-term trade policies to tackle production uncertainties effectively. Finally, it proposes incentive strategies rooted in the ecological contributions of pulses. The chapter also proposes using advanced statistical methods for accurate production estimates for planning. Recommendations include establishing long-term government contracts for pulse import/export to ensure stable domestic prices, with the overarching goal of maintaining consumer affordability in the face of production shocks, particularly for commodities such as arhar (pigeon pea), mungbean and urdbean where global trade is thin.

Keywords: Benefit–cost, global trade, pulses, policy, subsidies, value chain

16.1 Introduction

Pulses constitute a diverse group of 12 grain legume crops, including chickpea, arhar (pigeon pea), mungbean, urdbean, lentil, among others. These are the dried, edible seeds derived from various leguminous plants. It is essential to note that although peas and beans fall under the legume category, they are commonly classified as vegetables rather than pulses. Pulses boast a

---

*Corresponding author: adityapratapgarg@gmail.com

DOI: 10.1079/9781800624658.0016

high protein and carbohydrate content, making them valuable sources of nutrients (Ali *et al.*, 2000; Materne and Reddy, 2007; Rao *et al.*, 2010; Singh and Pratap, 2016; Joshi and Rao, 2017; Mishra *et al.*, 2023). Additionally, they are typically low in fat, distinguishing them from other legumes such as soybean and groundnuts, which have higher fat content. The nutritional richness and adaptability of pulses contribute significantly to global food systems, particularly in countries like India. Pulses play a crucial role in both Indian cuisine and the nation's agricultural economy. A traditional Indian platter incorporates pulses alongside staples such as wheat or rice, offering a high-protein source. This is particularly significant as the majority of the Indian population is vegetarian and pulses provide essential amino acids necessary for proper growth, development and overall health.

India holds the position of being the world's leading producer, importer and consumer of pulses, contributing to 24% of global production and occupying one-third of the global cultivation area (Reddy, 2013; Singh *et al.*, 2022). Despite this, pulse production has struggled to match the pace of population growth, leading to increased relative prices, making them unaffordable for poor households and a decline in per capita consumption below the Indian Council of Medical Research (ICMR) recommendations. Despite technological advancements in pulse production in recent years, continuous neglect and insufficient investment in pulse research and development have contributed to this challenge. Urgent measures are needed to enhance pulse productivity, especially of arhar, mungbean and urdbean, through cost-saving technologies to prevent production costs from rising and prices from exceeding the affordability of impoverished consumers. Unfortunately, India still relies on significant imports to the tune of 2–3 million tonnes of pulses annually. Even after this, the per capita availability of pulses in the country remains low and has, in fact, decreased over time. Protein-energy malnutrition (PEM), reflected by indicators such as being underweight, stunting and wasting, is a major concern in the Indian population, with high prevalence rates among children under the age of five (Bhutia, 2014).

The decline in the relative profitability of pulses compared to other crops, such as cereals, has been a significant factor contributing to their slower production growth. Lower yield potential and input subsidies like water and fertilizer, primarily benefiting cereal crops, along with guaranteed procurement of paddy and wheat at minimum support price (MSP), have enhanced the attractiveness of paddy and wheat at the cost of pulses, pushing pulses to marginal lands, which reduced relative yields, and enhanced costs and hence prices to consumers. To boost both the production and consumption of pulses, there is an urgent need to reevaluate the incentive system in Indian agriculture in favour of pulses. Furthermore, the contribution of pulses to ecology and soil health by enhancing soil fertility through nitrogen fixation, reducing the need for fertilizers, and generating fewer greenhouse gas emissions compared to cereal crops needs to be internalized in price policy. Unfortunately, these ecological contributions are often overlooked when designing incentives for pulse crops. Recognizing the impact of subsidies on relative farm costs and revenue, on natural resource uses and the environment is becoming increasingly crucial. A comprehensive evaluation of pulses versus non-pulse crops should consider these distortions and account for sustainability contributions. In addressing the intricate policy challenges surrounding pulse crops, this chapter aims to assess their current status and provide policy recommendations for the sustainable development of both production and consumption in India. The chapter is organized into sections to cover comprehensively various aspects of the policy issues in pulses sector.

16.2 Trends in Production and Yields

There has been a significant increase in pulses production, especially from 2016–17 onwards. The pulses production in 2015–16 was recorded at just 16.3 mt, which increased by 7 mt in the next year, and almost the same level was maintained until 2019–20 followed by a record production of 27.3 mt in the year 2021–22. Most of the growth in pulses production has been contributed by chickpea, where the production increased progressively from 7.1 mt in 2015–16 to 13.5 mt in 2021–22 (Table 16.1). Likewise, mungbean also saw an impressive growth of >182% in the past 1.5 decades (Pratap, 2022).

Table 16.1 Pulses production in million tonnes. Source: Agriculture at a Glance (2023). (Author's own table.)

Crop	Arhar	Gram	Urdbean	Mungbean	Lentil	Other pulses	Total pulses
2014–15	2.8	7.3	2	1.4	1	2.6	17.2
2015–16	2.6	7.1	1.9	1.6	1	2.2	16.3
2016–17	4.9	9.4	2.8	2.2	1.2	2.7	23.1
2017–18	4.3	11.4	3.5	2	1.6	2.6	25.4
2018–19	3.3	9.9	3.1	2.5	1.2	2.1	22.1
2019–20	3.9	11.1	2.1	2.5	1.1	2.4	23.0
2020–21	4.3	11.9	2.2	3.1	1.5	2.4	25.5
2021–22	4.2	13.5	2.8	3.2	1.3	2.3	27.3
2022–23	3.3	12.3	2.6	3.7	1.6	2.6	26.1

The steep rise in production of pulses in India is attributed mainly to two factors: one is the increase in MSP since 2016–17 and the other is technological progress in major pulses, which revolutionized pulses production, especially in states such as Andhra Pradesh and Karnataka with the spread of heat tolerant varieties of chickpea. Interestingly, there has been a slow yet significant increase in the production, especially from rabi pulses, both in chickpea and lentil. Besides, an upward trend has also been observed in irrigated mungbean and urdbean, where yields are higher than the average for pulses.

The production of arhar has varied between 2.6 and 4.9 million tonnes in the last decade. The production of urdbean was recorded at 2 mt in the year 2014–15, while the peak production of urdbean was recorded in the year 2017–18 at 3.5 million tonnes. Mungbean production ranged between 1.4 mt and 3.7 mt with the peak production recorded in the year 2022–23. Lentil production was recorded between 1 mt and 1.6 mt and maximum production was recorded in 2017–18 and 2022–23. The production of other pulses was recorded in the range of 2.1 to 2.7 mt, with the minimum in the year 2018–19 and the maximum in the year 2016–17. The production of total pulses was recorded in the range of 16.3 to 27.3 mt, minimum in the year 2015–16 and the maximum in the year 2022–23.

A discernible distinction has surfaced in the yield levels of kharif and rabi pulses. Important kharif pulses, such as pigeon pea or arhar, exhibit consistently low yields owing to factors such as reliance on predominantly rainfed conditions, uncertain rainfall, and limited input use. In contrast, rabi pulse crops such as chickpea are experiencing an increase in yields owing to the expansion of crop area under irrigated conditions, assured and controlled irrigation, and the adoption of improved crop management practices such as the judicious use of fertilizers and pesticides (Fig. 16.1). Moreover, the use of optimal doses of fertilizers is pursued with the anticipation of guaranteed yields and prices.

16.3   Geographical Spread of Crops

In this section, we have divided the major pulse growing states in India into high, medium and low productivity states based on their average yield being more than 1t/ha, between 1 and 0.5 t/ha and less than 0.5 t/ha (Table 16.2). On the basis of this categorization, three states are in the high productivity category for chickpea (Maharashtra, MP and Rajasthan), two for arhar (Karnataka and Maharashtra) and in one each for urdbean (MP) and mungbean (Rajasthan). For lentil, no state is categorized as highly productive. Again, four states for chickpea are considered as medium productivity, two each in case of mungbean and lentil, and one for urdbean. It is high time to shift low productivity states to medium or high productive states, and medium productive states to high productive states. In high and medium productive states, the area for pulses can be expanded so that the pulse crops become competitive in both domestic and international markets.

Figure 16.2 illustrates the distribution of districts based on yield levels for both chickpea and arhar. Notably, there has been a substantial rise in the number of districts with yields exceeding 1 ton/acre from 21% in 2001 to 54% in 2017 for chickpea. In contrast, the corresponding figures for arhar-producing districts show an

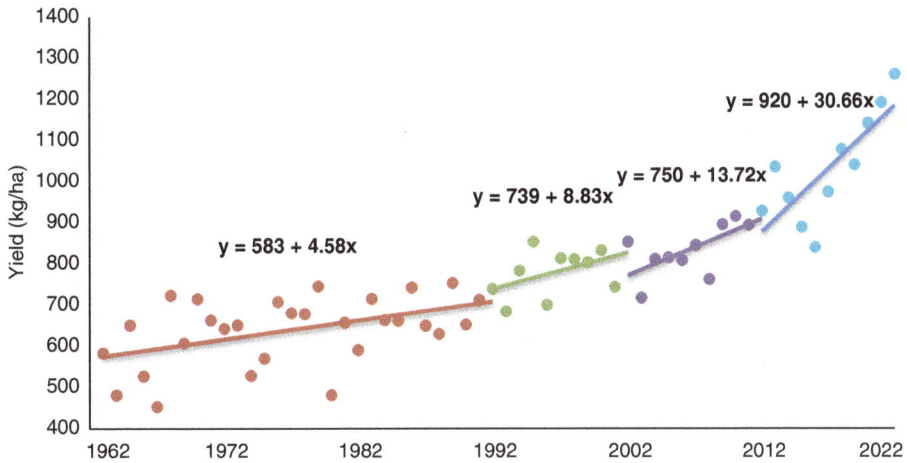

Fig. 16.1. Chickpea yield (kg/ha). Source: Agriculture at a Glance (2023). (Author's own figure.)

Table 16.2. Grouping of states based on productivity of different pulse crops. (Author's own table.)

State category	Arhar	Chickpea	Urdbean	Mungbean	Lentil
High productive (> 1 t/ha)	Karnataka, Maharashtra	Maharashtra, MP, Rajasthan	MP	Rajasthan	–
Medium productive (1–0.5 t/ha)	–	Gujarat, Karnataka, UP, AP	UP	MP, Karnataka	MP, UP
Low productive (>0.5 t/ha)	Telangana, UP, Gujarat, Jharkhand, AP, MP, Odisha, Chhattisgarh	Chhattisgarh, Jharkhand, Telangana	Rajasthan, TN, AP, Maharashtra, Gujarat, Karnataka, Chhattisgarh, Odisha	Odisha, Bihar, TN, Gujarat, AP, UP, Telangana	WB, Bihar, Jharkhand, Assam, Rajasthan, Chhattisgarh, Uttarakhand, Odisha

Source: Directorate of Economic and Statistics (2023b).

increase from 16% to 37%. It is concerning that approximately 23% of arhar districts still experience yields below 0.5 t/ha, highlighting a critical situation that requires improvement.

16.4 Costs and Returns

This section conducts a comparative analysis of the costs and benefits associated with pulse crops and their competing crops during both kharif and rabi seasons. The study focuses on two major pulse-growing states, Maharashtra and Andhra Pradesh, utilizing the cost of cultivation scheme provided by the Government of India (Directorate of Economic and Statistics, 2023c). The analysis employs the cost concept of 'Cost A2 plus Family Labour' to compare costs and benefits. The findings reveal that, across all states, the costs incurred in cultivating pulse crops are consistently lower when compared to their competing crops. Although the returns are slightly lower as well, the cost–benefit analysis demonstrates a favourable ratio, ranging from 1.3 to 1.6 for most pulse crops. This suggests higher profitability despite the lower costs, emphasizing the comparative advantage of pulses for resource-poor farmers (Table 16.3).

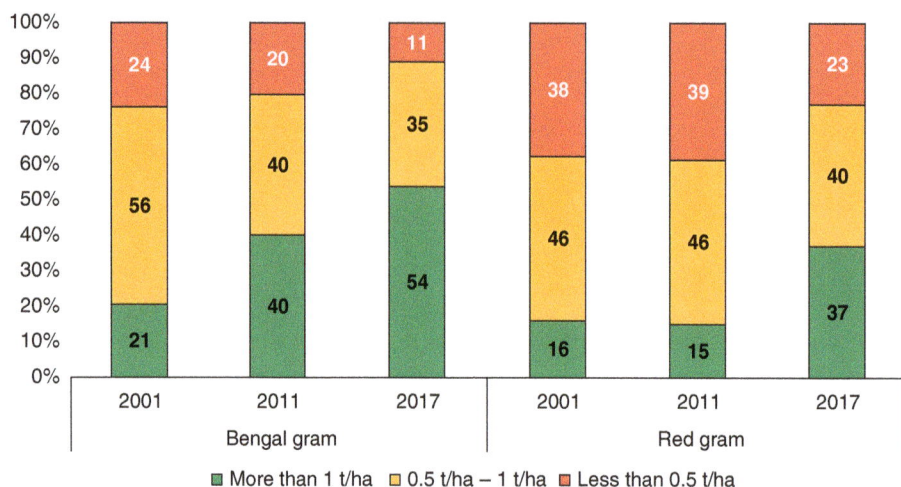

Fig. 16.2. Distribution of districts by yield levels (%). Source: ICRISAT (2023). (Author's own figure.)

Table 16.3 provides insights into the cost of cultivation and production of pulses in Andhra Pradesh and Maharashtra for the year 2020–2021. In Andhra Pradesh, human labour accounted for 26.4% of the total cost (chickpea) to 45.7% (urdbean), animal labour varied from 1.3% (mungbean and urdbean) to 7.2% (arhar), and machine labour ranged from 9.8% (cotton) to 29.1% (arhar) across different crops. The maximum and minimum costs were observed in cotton (Rs 72,765/ha) and arhar (Rs 28,340/ha). Among these crops, arhar reported the minimum yield (6.4 q/ha), while maize recorded the maximum yield (71.9 q/ha). Gross revenue varied from Rs 41,905 (arhar) to Rs 126,311 (maize). The net returns over costs were highest in maize (Rs 62,711/ha) and lowest in mungbean (Rs 12,457/ha).

In Maharashtra, the pattern of costs was also similar. The seed expenses ranged from 3.5% of the total cost (arhar) to 12.7% (chickpea), fertilizer and manure differed from 7.2% (urdbean) to 13.6% (cotton), insecticides varied from 2% (mungbean) to 5.8% (cotton), and irrigation cost ranged from 0.5% (mungbean) to 3.1% (maize). The maximum and minimum costs were reported in cotton (Rs 76,896/ha) and arhar (Rs 38,142/ha), respectively. Across these crops, mungbean and urdbean recorded the minimum yield (8.1 q/ha), while maize exhibited the maximum yield (57 q/ha). The minimum

and maximum gross revenue was observed in mungbean (Rs 49,343/ha) and maize (Rs 124,581/ha), respectively, while the net returns over the costs were highest in maize (Rs 57,612/ha) and lowest in cotton (Rs 10,393/ha). The source for the data is the Commission for Agricultural Costs and Prices (CACP, 2023), various reports.

Figure 16.3 illustrates the yield trends of cotton and arhar in Maharashtra from 2000 to 2022. The cotton yield fluctuated between 7.58 q/ha to 15.34 q/ha, while the arhar yield ranged from 1.96 q/ha to 5.39 q/ha during the same period. The consistently low and stagnant yields of arhar pose a significant challenge that needs to be addressed for a potential revolution in pulses, similar to the one observed in chickpea during the past two decades. In contrast to arhar, cotton yield has shown a steady increase over the years. This rise is primarily attributed to the introduction of *Bt* cotton in 2002, with its widespread adoption reaching nearly 100%. Examining arhar yields, there is a notable variability. The arhar yield initially experienced a slight increase and then started to decline until 2009, reaching its lowest point at 1.96 q/ha in that year. Subsequently, arhar yield witnessed an upswing from 2011 to 2014. After a decrease in 2015, there was another increase until 2017, with the highest recorded yield at 5.39 q/ha. Post-2017, apart from the years 2018 and

Table 16.3. Cost of cultivation and cost of production of pulses and other crops (2020–2021). (Author's own table.)

Cost item	Andhra Pradesh						Maharashtra					
	Chickpea	Arhar	Mungbean	Urdbean	Cotton	Maize	Chickpea	Arhar	Mungbean	Urdbean	Cotton	Maize
Human labour (%)	26.4	28.2	30.4	45.7	45.0	32.1	35.9	48.6	48.0	49.4	48.4	41.4
Animal labour (%)	6.0	7.2	1.3	1.3	5.7	1.7	4.1	7.9	6.1	14.6	11.9	5.7
Machine labour (%)	28.5	29.1	20.7	19.8	9.8	20.2	25.3	19.9	25.7	15.2	10.1	21.6
Seed (%)	16.6	5.6	9.6	11.5	7.1	12.9	12.7	3.5	5.5	5.7	4.7	8.1
Fertilizer (%)	12.5	16.1	0.8	4.1	11.9	17.1	9.2	8.5	8.8	7.2	13.6	13.5
Insecticides (%)	6.5	9.4	24.6	10.8	8.5	8.8	2.6	7.0	2.0	3.1	5.8	2.5
Irrigation charges (%)	0.0	0.1	1.3	0.7	0.2	3.7	6.1	0.9	0.5	1.5	1.9	3.1
Crop insurance (%)	0.0	0.0	0.0	0.0	0.0	0.0	0.0	0.1	0.0	0.0	0.1	0.0
Other costs (%)	3.6	4.3	11.3	6.0	11.8	3.7	4.1	3.6	3.3	3.2	3.4	4.1
A2+FL (Rs/ha)	37,228	28,340	31,836	33,094	72,765	63,600	38,142	74,670	42,699	38,753	76,896	66,969
Yield (Qt/ha)	18.2	6.4	7.0	10.9	16.6	71.9	13.2	19.9	8.1	8.1	12.7	57.0
A2+FL (Rs/qtl)	2,034	4,410	4,525	3,039	4,382	884	2,834	3,662	5,226	4,765	5,970	949
Gross revenue (GR) (Rs/ha)	87,019	41,905	44,293	72,056	117,247	126,311	63,489	120,976	49,343	51,203	87,289	124,581
GR cost A2+FL (Rs/ha)	49,790	13,565	12,457	38,962	44,482	62,711	25,347	46,305	6,644	12,450	10,393	57,612
B–C ratio	2.3	1.5	1.4	2.2	1.6	2.0	1.7	1.6	1.2	1.3	1.1	1.9

Source: Directorate of Economic and Statistics, 2023c.

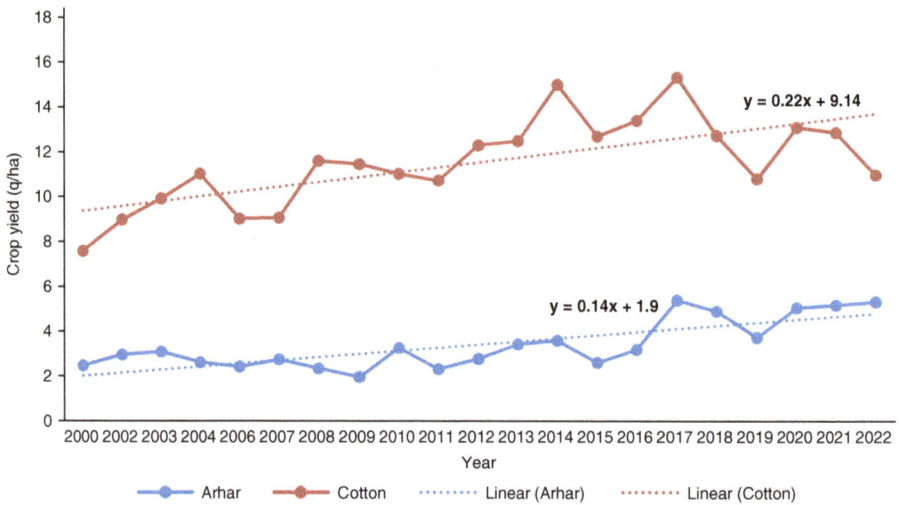

Fig. 16.3. Crop yield, Maharashtra (q/ha). Source: CACP (2023) Commission for Agricultural Costs and Prices, various reports. (Author's own figure.)

2019, the arhar yield exhibited a generally increasing trend until 2022. Addressing the challenges associated with arhar yields remains crucial for sustaining and improving pulse production in the region.

As a soft fibre crop, cotton is highly susceptible to insect pest attacks, particularly from pests such as bollworms, aphids and thrips. The tender leaves, buds and bolls of the cotton plant attract these pests. In contrast, arhar has tougher foliage, making it comparatively less appealing to common pests. The extensive use of pesticides in cotton cultivation has led to the emergence of pesticide-resistant insect populations over time. The climatic conditions favourable for cotton, such as warm and moist environments, also promote the proliferation of various insect species. On the contrary, arhar, with less intensive pesticide use, does not face this issue to the same extent (Fig. 16.4). Arhar, being more adaptable to a range of climatic conditions, does not create a uniformly favourable environment for pests. While cotton breeding has historically prioritized yield and fibre quality (except for *Bt* cotton), arhar breeding placed more emphasis on pest and disease resistance. Arhar plants possess natural defence mechanisms such as tougher leaves and the presence of certain chemicals that act as deterrents to pests. Cotton is often cultivated in large monocultures, where vast areas are dedicated to a single crop. This monoculture practice can lead to an increase in specific pests targeting cotton as they find an abundant food source. In contrast, arhar is often grown in mixed cropping systems, contributing to pest control by diversifying the insect population and reducing the prevalence of crop-specific pests. As a result of all these multiplicities of issues, the use of insecticides is much lower in leguminous crops such as arhar compared to cotton, making the cost of cultivation lower for arhar.

The application of farmyard manure (FYM) in arhar cultivation has remained relatively stable at 1 t/ha (Fig. 16.5). Arhar, being a leguminous crop, can fix atmospheric nitrogen in its root nodules, reducing the need for an exogenous application of manure. In contrast, the use of manure in cotton cultivation in Maharashtra has continuously declined from 2000 to 2021. This decline coincides with a significant increase in the use of chemical fertilizers during the same period. The reduction in manure usage is attributed to factors such as the need to maintain a large livestock population for manure production, the labour-intensive and costly nature of manure preparation, management, and storage, especially in the changing landscape of agriculture. Additionally, the bulkiness and non-marketability of manure pose obstacles to developing a robust supply chain, contributing to scarcity and rising

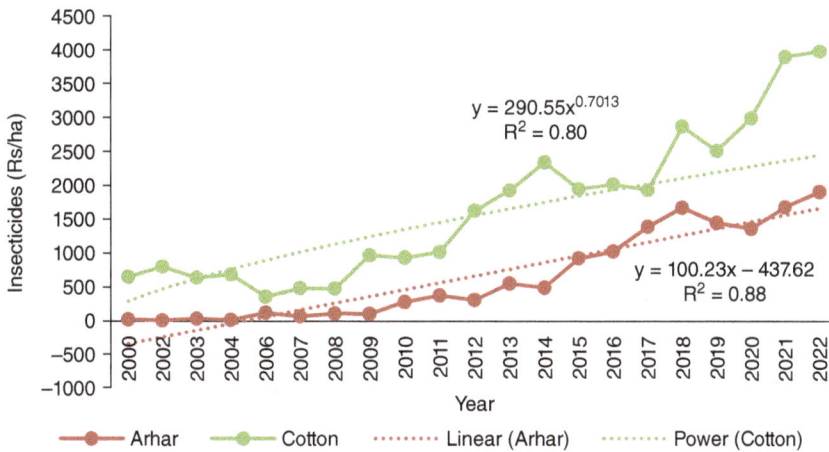

Fig. 16.4. Use of insecticides (Rs/ha) in Maharastra. Source: CACP (2023) Commission for Agricultural Costs and prices, various reports. (Author's own figure.)

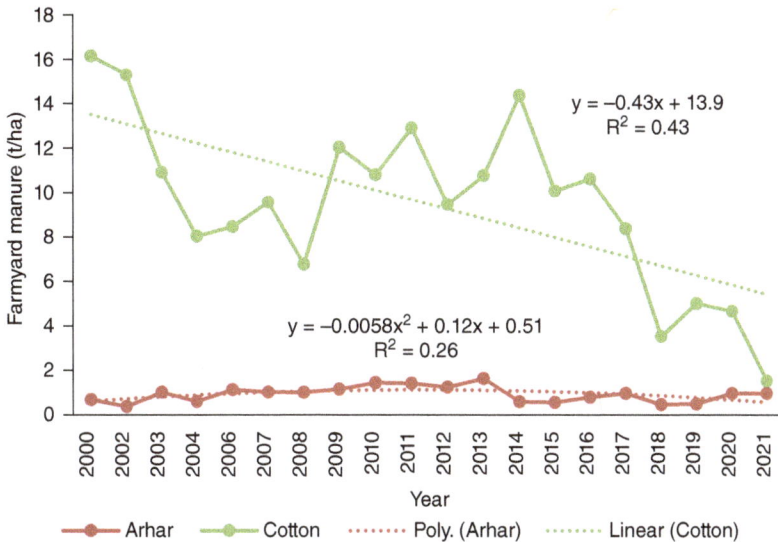

Fig. 16.5. Use of farmyard manure (t/ha) in Maharastra. Source: CACP (2023) Commission for Agricultural Costs and prices, various reports. (Author's own figure.)

prices. Ready-to-use chemical fertilizers, in contrast, are more accessible and affordable to farmers, because they do not require the labour-intensive processes associated with manure. This shift in fertilizer-use patterns reflects modern agriculture's changing dynamics and challenges.

The use of fertilizers in arhar cultivation has been comparatively low, especially in the initial decade from 2000 to 2010, where it remained minimal at 10 kg/ha. This average is for the pooled data of farmers who used fertilizer versus those who did not apply it (Fig. 16.6). This is because arhar is a leguminous crop benefitting from N fixation. The lesser use of fertilizers in arhar was also due to it being grown mainly by poor farmers, who do not have enough financial liquidity to purchase fertilizers. In the subsequent decade, however, there was a gradual increase in fertilizer consumption, reaching a peak of 50 kg/ha in the year 2022. The rise in fertilizer

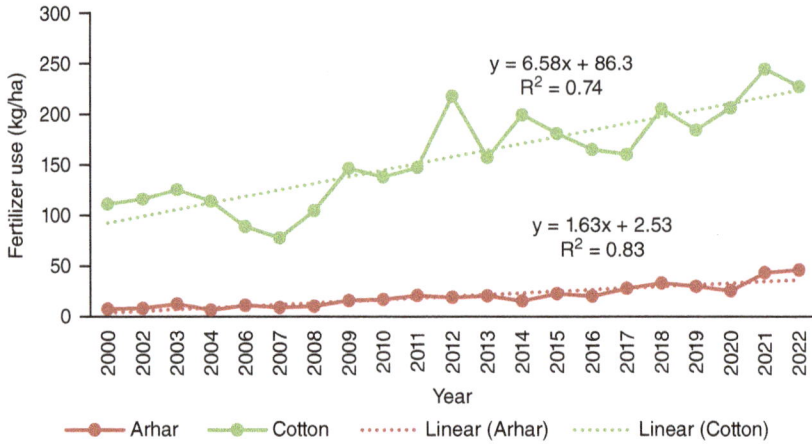

Fig. 16.6. Fertilizer use (kg/ha) in Maharashtra. Source: CACP (2023) Commission for Agricultural Costs and Prices. (Author's own figure.)

consumption after 2016 may be due to rising output prices (government increased MSP for all pulses after 2016–17). In contrast, fertilizer use in cotton experienced a steep increase from 2006 to 2012 from the higher base, coinciding with the rapid spread of *Bt* cotton, which reached almost 100% adoption during that period. Subsequently, the rate of increase in fertilizer consumption slowed down. Market dynamics, such as changes in cotton and input prices, played a significant role in influencing fertilizer use. High cotton prices in 2009 motivated farmers to increase fertilizer use to maximize yields, while lower prices or higher input costs in 2006 resulted in reduced fertilizer application. In the years with a higher incidence of pests and diseases, farmers tended to use more fertilizers to compensate for yield losses caused by pest attacks. The continuous decline in soil fertility caused by the growth of cotton as a monoculture also contributed to higher fertilizer use in later periods to offset lower soil fertility. These factors highlight the complex interplay of agronomic, economic and environmental factors influencing fertilizer use in both arhar and cotton cultivation.

16.5 Subsidies and Externalities

During the 1960s, when India was facing severe food grain shortages, the government took long-lasting measures to secure the country's food. These involved heavily incentivizing paddy and wheat cultivation through MSP and fertilizer subsidies (Saini and Gulati, 2017). Policies providing subsidies for irrigation and chemical fertilizers disproportionately increased the profitability of paddy and wheat compared to pulses. While these subsidies benefitted paddy and wheat farmers, they reduced the relative profitability of pulses, leading to a significant shift in cultivated areas from pulses to paddy and wheat. However, starting from 2002, a new shift has been observed in dryland areas, with farmers moving from pulses to cotton cultivation. The major driving force behind this shift is the adoption of *Bt* cotton, which is resistant to bollworms, and the procurement of cotton at MSP by the Cotton Corporation of India (CCI). The decline in the profitability of pulse crops has prompted farmers to shift pulses to marginal lands with lower fertility. It is important to note that these incentives and subsidy costs are borne by the entire society but are ignored while calculating private profits to farmers. Therefore, there is a need to assess the social cost–benefits by including subsidy costs and nitrogen (N)-fixation benefits of pulses to the society. The private cost–benefits and profits are typically assessed based on market prices of inputs and outputs without considering social cost–benefits. Social benefits of pulses, such as enhancing soil fertility by introducing N during the growth phase, reducing reliance on fertilizers, ameliorating soil physical properties, and improving overall soil suitability for agriculture, can only be assessed through a social cost–benefit analysis.

The subsequent section explores the net income from pulses and competing crops at prevailing market prices, followed by incorporating social benefits and costs, including subsidies, N fixation, and the net impact of greenhouse gas (GHG) emissions. Despite generating less revenue at market prices, pulses demonstrate higher profitability at the societal level compared to competing crops. Additionally, the cultivation of pulses contributes significantly fewer GHG emissions than other crops, making pulses environmentally favorable. By internalizing subsidies and acknowledging the multifaceted contributions of pulses, this section establishes a scientific framework for the social benefit–cost assessment of pulses compared to competing crops. The goal is to advocate for government incentives that align with the non-market benefits provided by pulses.

This is carried out in two steps:

1. Comparison of Net Returns. This encompasses evaluating the net return from cultivating pulses and other crops under current input prices. Assess the net return from pulses and alternative crops when input prices are considered without subsidies.

2. Environmental and Natural Resource Valuation. This involves calculating and comparing the revenue generated from cultivating pulses with that of alternative crops after incorporating environmental and natural resource valuations (detailed methodology in Kumar *et al.*, 2017; Hemming *et al.*, 2018; Akber *et al.*, 2022; Nuthalapati *et al.*, 2022).

Through these steps, the section aims to provide a more nuanced understanding of the overall benefits of pulse cultivation, both in economic terms and in terms of environmental sustainability, and advocate for policy incentives that align with these social benefits. This analysis focuses on the states of Maharashtra and Andhra Pradesh, utilizing data primarily sourced from the Ministry of Agriculture, Government of India's Cost of Cultivation Reports for the three most recent years, concluding with 2021 (Directorate of Economic and Statistics (2023a, b, c). Data on input subsidies were also obtained from the Central Statistical Office and National Account Statistics of the Government of India.

Steps in the measurement of subsidies:

1. Cost–benefit analysis at market prices for pulses and competing crops.

2. Measurement of subsidies (fertilizer + water); subsidies are costs to the government.

3. Measurement of externalities (CO_2 + nitrogen fixation); negative externalities are costs to society and positive externalities are benefits to society.

16.6 Social Cost–Benefit Analysis

For the assessment of social benefits, subsidies for crop-specific inputs, including fertilizers, irrigation, and electricity supplies, are deducted from the net income at market prices. For urea, where the actual cost for the government for supply is Rs 2200/45 kg, the maximum retail price (MRP) is Rs 242/45 kg bag (effective from 1 March 2018), the imputed subsidy amounts to Rs 43.5/kg of urea, resulting in Rs 94.5/kg of nitrogen. The total irrigation and electricity subsidies are distributed among the selected crops based on the proportion of irrigated area under these crops in the state. The value of N fixation by pulse crops are valued at the prevailing N price of Rs 106/kg of N. The monetary value assigned to GHG emissions by both pulses and non-pulse crops are measured in terms of CO_2 kg equivalent, is determined to be Rs 750 (US$10/t) (Ali and Gupta, 2012; Chand *et al.*, 2015). For Maharashtra, red gram and its competing crop, cotton, are considered, while in Andhra Pradesh, mungbean, urdbean and their competing crop, maize, have been selected for evaluation.

Figure 16.7 provides a comprehensive overview of the private costs and benefits to farmers, as well as social benefits after adjusting for subsidies and externalities in Maharashtra for pulses (red gram) and their competitive crop (cotton). Key focus areas include gross returns, private costs, private profits, total subsidies, externalities and social benefits. For red gram, the gross returns are reported at Rs 95,000, indicating the total revenue generated from the cultivation. In contrast, cotton shows comparatively lower gross returns at Rs 80,000. The Cost A1, which includes all paid-out costs, is Rs 49,000 for red gram and Rs 53,000 for cotton. Private profits, calculated as gross returns minus Cost A1, amount to

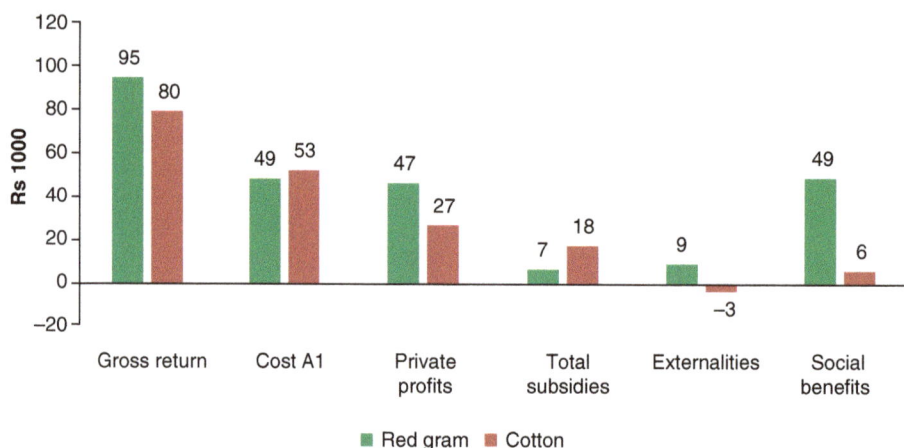

Fig. 16.7. Private profits and social cost–benefit analysis of pulse and non-pulse crops in Maharashtra. Source: Authors' calculations from Directorate of Economic and Statistics (2023c). (Author's own figure.)

Rs 47,000 for red gram and Rs 27,000 for cotton. This reflects the income retained by farmers after deducting all the costs associated with the cultivation. When accounting for subsidies and externalities, social benefits are significantly higher for red gram at Rs 49,000 compared to cotton, which stands at Rs 6000. Social benefits represent the broader societal gains, taking into consideration the impact of subsidies on costs and the positive externalities associated with pulse cultivation. This analysis highlights the potential social benefits of red gram cultivation, considering not only the economic returns to farmers but also the broader positive impacts on society, including environmental and sustainability factors.

From Figure 16.8, it is evident that maize exhibits higher gross returns at Rs 117,000, surpassing urdbean at Rs 63,000 and mungbean at Rs 43,000. Maize cultivation incurs higher costs at Rs 49,000, in contrast to urdbean at Rs 25,000 and mungbean at Rs 21,000. Private profits are notably elevated in maize (Rs 68,000) compared to urdbean (Rs 38,000) and mungbean (Rs 23,000). Total subsidies are greater for maize (Rs 18,000) compared to both mungbean and urdbean (Rs 1000). Generally, maize cultivation is recognized as a high-input and high-output system, involving high use of fertilizers, manure, human and animal labour, and associated costs. Maize cultivation demands more resources, including machines, labour, pesticide spray and higher irrigation costs, when juxtaposed with the cultivation of mungbean and urdbean. The data indicate that pulse cultivation embodies a more environmentally friendly approach, with the value of N-fixation and GHG emissions at Rs 10,000 in urdbean and Rs 5000 in mungbean, in contrast to negative Rs 3000 in maize. Furthermore, pulse cultivation is characterized by a higher degree of labour intensity, suggesting a potential reduction in the environmental footprint linked to machinery. This aligns with more sustainable and environmentally friendly farming practices, emphasizing the positive environmental attributes of pulses cultivation. Despite a substantial gap between maize and pulse crops in private profits, the social benefits are nearly equal at Rs 47,000 for both maize and urdbean, and the gap between maize and mungbean has diminished. This highlights the comparable social benefits of cultivating both maize and pulse crops, underscoring the potential role of pulses in fostering sustainable and eco-friendly cropping systems. Given that pulses externalities to environment are positive, while for competing crops it is negative, incentives need to be designed so that the pulses farmers get at least monetary benefits equivalent to their positive externalities.

16.7   Contribution of Pulses to Human Nutrition

Preferences for different pulse crops and varieties differ markedly across states. Pulses boast substantial protein and fibre content (Table 16.4),

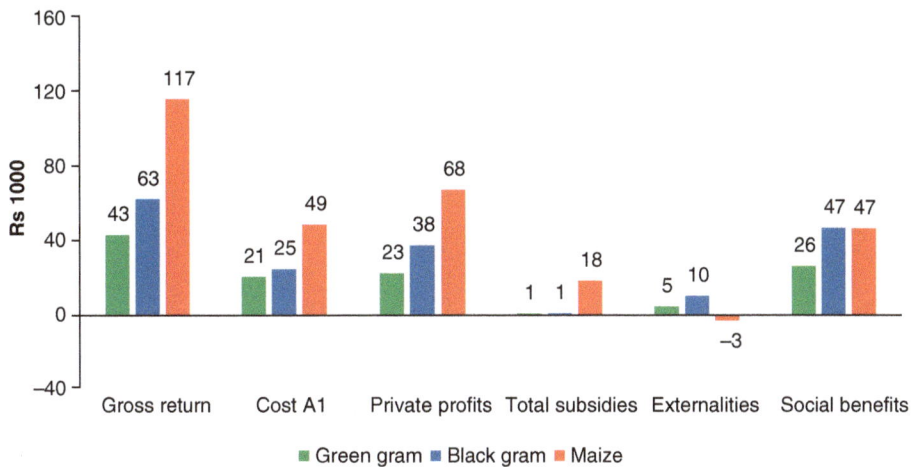

Fig. 16.8. Private and social cost–benefit analysis of pulse and non-pulse crops in Andhra Pradesh. Source: Authors' calculations from data taken from Directorate of Economic and Statistics (2023c). (Author's own figure.)

Table 16.4. Nutrition value of different pulses (100 g). (Author's own table.)

Crops	Protein (g)	Carbohydrates (g)	Fibre (g)	Fat (g)	Reference
Lentil	23–32	46	12	0.8–2.0	Hedley, 2001
Arhar	19.5–22.9	44.3	10	1.3–3.8	Hedley, 2001
Chickpea	15.5–28.2	44.4	9	3.1–7.0	Hedley, 2001
Mungbean	21–31.3	8.9–12.9	–	1.2–1.6	Anwar et al. 2007
Peas	18.3–31	45	12	0.6–5.5	Hedley, 2001
Common bean	20.9–27.8	41.5	5	0.9–2.4	Hedley, 2001

coupled with a lower glycaemic index (Singh and Pratap, 2018). The fibre within pulses acts to decelerate the digestion and absorption of carbohydrates, resulting in a gradual and consistent release of glucose into the bloodstream. This attribute can be advantageous for individuals seeking to regulate blood sugar levels, particularly those with diabetes or individuals focused on blood sugar control.

The energy content across different pulse crops ranges from 81 to 116 kcalories per 100 g. Lentils exhibit the highest calorie count at 116, while peas have the lowest at 81 calories. In general, pulses contain about 20 g protein per 100 g of grain, with slight variation among them, with slightly higher in lentils and mungbeans. Lentils and arhar also lead in carbohydrates with about 45 g, while mungbeans have the least at about 10 g. Regarding fibre, lentils and peas record the highest at 12 g, followed by arhar (10 g) and chickpeas (9 g),

with the common bean the least at 5.0 g. The fat content varies from 0.6 to 7.0 g, with chickpeas having the highest fat content at 3.1 to 7.0 g, and lentils, mungbeans, and peas having less than 2 g.

India witnessed a significant increase in pulses production from 2021 to 2023, with harvests totalling 25.5 mt in 2021, 27.3 mt in 2022 and 26.1 mt in 2023 (Table 16.5). However, despite the bumper harvests, India remained a net importer during these years, with imports ranging from 1.7 to 2.5 million tons annually. The domestic supply, comprising domestic production and net imports, fluctuated between 27 and 28 million tons during this period. Of the total domestic supply, approximately 80% was utilized for food consumption, 13–14% for feed, 3.3–3.6% for seed, and the remaining 3–3.3% accounted for losses (Parappurathu et al., 2014; Nuthalapati et al., 2022). The increased production post 2016–2017, coupled with higher imports, resulted

Table 16.5. Food balance sheet of pulses (1000 tons). (Author's own table.)

	2021	2022	2023
Production	25,500	27,300	26,100
Import quantity	1710	2570	2370
Export quantity	230	310	670
Stock variation	−500	1151	−600
Domestic supply quantity	27,480	28,409	28,400
Food (% of supply)	79.2	79.8	80.1
Feed (%)	14.6	13.7	13.0
Seed (%)	3.3	3.4	3.6
Losses (%)	2.9	3.2	3.3
Food supply quantity (kg/capita/yr)	15.3	16.0	16.0
Food supply quantity (g/capita/day)	41.9	43.8	43.8
Food supply (kcal/capita/day)	138.7	144.4	145.4
Protein supply quantity (g/capita/day)	9.0	9.4	9.4
Fat supply quantity (g/capita/day)	1.7	1.8	1.7

Calculated from Directorate of Economic and Statistics (2023b) data using Sheehy *et al.* (2019) and FAO methodology.

in an augmented net availability for food consumption from 2021 to 2023. The availability of pulses for food purposes increased to 15.3 kg per capita per year in 2021, rising to 16 kg per capita per year in 2022 and 2023. This equates to the utilization of pulses for food at 41 to 43 g per capita per day, which still falls short of the Indian Council of Medical Research (ICMR) recommendation (National Family Health Survey, 2007, 2017; NSSO, 2013; Hall *et al.*, 2017; Kumar *et al.*, 2017; Deepti *et al.*, 2023). On average, about 9 g protein per day are derived from pulses, reflecting their crucial role in providing protein nutrition to the population. Despite the increased availability, the consumption figures highlight the challenge of meeting the recommended daily intake, emphasizing the need for continued efforts to enhance pulse production and distribution for better nutritional outcomes.

The per capita pulse consumption, after showing an upward trend from 1988 to 2000, showed a declining trend thereafter (Fig. 16.9). However, between 2010 and 2013, there was an increasing trend (NSSO, 2013). One of the striking observations is that the rural households consumed less pulses than the urban consumers in all the years from 1988 to 2013. Per capita consumption in rural households increased from 22 g/capita/day in 1988 to 28 g/capita/day in 2000 while it decreased again to 22 g/capita/day in 2010 and increased to 26 g/capita/day in 2013. Evidently, the per capita daily consumption of the urban population was always 4–5 g higher than

the rural population (Deepti *et al.*, 2023), but the per capita consumption both in rural and urban areas is less than the ICMR recommended consumption. Low production and limited availability of pulses in the global market has limited their consumption, unlike the edible oils, where although domestic production is limited, India imports vast quantities and now average consumption is above ICMR recommendations. From 2010 to 2013, there was a slight increase in pulse consumption, mainly driven by the increased domestic production from 15 to 19 mt and increased net imports from 2 to 4 mt.

16.8 Average Consumption of Pulses

Overall, pulses consumption across all states in India has been recorded below the Indian Council of Agricultural Research (ICAR) recommended level of 80 g per capita per day. Among the states, Punjab exhibited the highest consumption of pulses at 30.6 g per capita per day, followed by Madhya Pradesh (30.03 g), Uttar Pradesh (29.73 g), Bihar (26.83 g), and the lowest in Rajasthan (19.87 g) (Fig. 16.4). In Punjab, chickpea constituted the largest portion of consumption at 14.6 g, followed by mungbean (5.2 g), lentil (3.7 g), and urdbean (2.6 g). Consumption of arhar, peas, and lathyrus (*Khesari*) was negligible. Chickpea was predominantly consumed as split (5.73 g), followed by whole

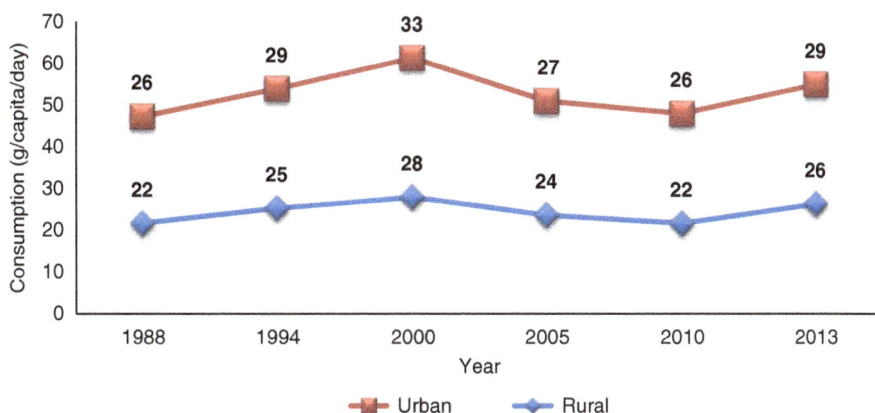

Fig. 16.9. Pulses consumption g/capita/day (rural and urban). Source: NSSO consumption survey, (2013). (Author's own figure.)

grain (5.33 g) and *besan* (powder of dehusked cotyledons) (3.5 g). In Rajasthan, chickpea again represented the highest consumption (8.3 g per capita per day), followed by mung bean (6.1 g), lentil (1.47 g), and arhar (1.27 g). The primary forms of chickpea consumption were split (4 g) and *besan* (3.83 g). In Madhya Pradesh, arhar had the highest consumption (13.73 g), followed by chickpea (7.3 g), mungbean (3.87 g), urdbean (2.1 g), and lentil (1.8 g). Chickpea was primarily consumed as gram-split and *besan*. Bihar exhibited the highest consumption of lentil (9.63 g), followed by chickpea (8.67 g), arhar (3.77 gr), and mung bean (2.6 g). In Uttar Pradesh, arhar had the highest consumption (10.27 g), followed by chickpea (5.6 g), peas (5.73 g), urdbean (3 g), and lentil (2.9 g). Chickpea was mostly consumed as *besan*, followed by gram-split and gram-whole.

16.8.1   Consumption habits of households

We also explored the consumption patterns of the households during the reference month, revealing significant variations in the consumption of different pulses across states (Table 16.6). In Punjab and Rajasthan, the majority consumed mung bean and chickpea (split), while chickpea (whole) was consumed more in Punjab. Furthermore, besan was more commonly consumed by households in Punjab and Rajasthan. Arhar consumption was most prevalent in Madhya Pradesh

(89%), followed by Uttar Pradesh (81.3%), Bihar (37.7%), Rajasthan (26.3%), and least in Punjab (10.8%). Interestingly, about 10% of households in Madhya Pradesh, approximately 4.3% in Uttar Pradesh and 4.4% in Bihar reported consuming homegrown arhar during the reference month. Overall, the figures indicate that, although pulses consumption increased post-2017, the regional/local preferences are widely varied, which need to be considered when making dietary recommendations.

16.9   Domestic Pulses Markets

One of the main concerns about pulse consumption is increased prices, which makes them unaffordable to poor consumers (Saini and Gulati, 2017). This issue can be addressed through strengthening the pulse value chain from farmers to end consumers. This section considers domestic marketing channels and price trends across different pulse value chain participants.

16.9.1   Marketing channels

The marketing channel refers to a set of intermediaries or 'middlemen' that help move a product from the farmer to the end consumer. It is the path or route that a product takes from production

Table 16.6. Percentage of households reporting consumption of pulses in a reference month (2012–2013). (Author's own table.)

Crop	Top consuming states	Least consuming states
Arhar	MP (93.8), Gujarat (91.8), TN (90.2), Maharashtra (89.7), AP (89.0)	HP (13.8), Arunachal Pradesh (14.7), Punjab (22.2), Haryana (35.2), Rajasthan (39.2)
Mungbean	Haryana (84.5), Chandigarh (85.8), Punjab (83.4), Rajasthan (83.0), Gujarat (81.6),	Chhattisgarh (15.1), Jharkhand (25.8), Arunachal Pradesh (27.6), AP (46.4)
Urdbean	TN (85.8), Kerala (73.7), Karnataka (68.2), HP (66.7), AP (61.0)	Arunachal Pradesh (1.3), Bihar (1.9), Jharkhand (4.3), Chhattisgarh (24.2), MP (24.5)
Chickpea: whole	Punjab (78.5), Kerala (71.6), HP (59.1), Gujarat (43.4), Jharkhand (40.5)	Arunachal Pradesh (4.6), AP (5.6), Rajasthan (7.2), MP (7.6), Odisha (9.5)
Chickpea: split	Haryana (79.9), Punjab (79.6), HP (78.1), Chandigarh (69.3), Rajasthan (68.8)	Kerala (4.2), Arunachal Pradesh (8.7), Chhattisgarh (18.5), Jharkhand (35.4), MP (38.1)
Chickpea products	TN (58.7), Jharkhand (20.8), AP (7.9), Bihar (7.7), Karnataka (6.2)	MP (0.2), Rajasthan (0.3), Arunachal Pradesh (0.4), Chhattisgarh (0.5), Odisha (0.6)
Besan	Haryana (78.6), Gujarat (69.0), Rajasthan (65.8), Punjab (64.3), MP (58.7)	Kerala (0.30), AP (0.62), TN (1.27), Odisha (2.81), Jharkhand (3.09)
Lathyrus	Chhattisgarh (11.0), Bihar (02.1), Kerala (0.12), MP (0.07), TN (0.04)	Odisha (1.0), Jharkhand (1.0), AP (1.0), Arunachal Pradesh (2.0), Punjab (3.0)
Lentil	Arunachal Pradesh (79.6), Punjab (74.4), Jharkhand (66.4), HP (56.0), Rajasthan (49.1)	TN (2.0), Kerala (3.1), AP (8.6), Chhattisgarh (18.6), MP (23.4)
Peas	Kerala (30.7), Karnataka (29.4), Maharashtra (16.3), Chhattisgarh (5.3), TN (8.0)	Rajasthan (1.0), Punjab (1.2), Haryana (1.4), MP (2.6), HP (3.1)
Other pulse products	Gujarat (21.3), Kerala (19.2), TN (16.9), Punjab (16.4), HP (16.2)	Jharkhand (3.3), Bihar (3.3), AP (3.5), MP (3.6), Arunachal Pradesh (4.5)
Other pulses	HP (68.2), Punjab (47.3), Karnataka (45.8), TN (27.8), Kerala (26.2)	Jharkhand (1.2), MP (2.1), Bihar (3.4), Odisha (3.8), AP (4.0)
Pulses and pulse products	Bihar (98.6), MP (96.2), Arunachal Pradesh (95.5), Punjab (95.3)	Karnataka (85.7), HP (86.8), AP (89.5), TN (90.7), Kerala (90.7)

Source: NSSO (2013).

to the final consumer (Reddy and Charyulu, 2016). The purpose of a marketing channel is to deliver the product efficiently and effectively to the target market. Some of the major marketing channels in pulses are given below. Almost >95% of the pulses produced in India go through private channels.

    Private channels:

Producer – *Dal* Miller – Consumer

Producer – Village Trader – *Dal* Miller – Wholesaler – Retailer – Consumer

Producer – *Dal* Miller – Retailer – Consumer

Producer – Wholesaler – *Dal* Miller – Retailer – Consumer

Producer – Wholesaler – *Dal* Miller – Wholesaler – Retailer – Consumer

Producer – Wholesaler – Retailer – Consumer (for whole mungbean/chickpea)

Producer – Commission Agent – *Dal* Miller – Wholesaler – Retailer – Consumer

Institutional channels:

Producer – Procuring Agency – *Dal* Miller – Consumer

Producer – Procuring Agency – *Dal* Miller – Wholesaler – Retailer – Consumer

Producer – Procuring Agency – *Dal* Miller – Retailer – Consumer

    For imported pulses, the channel will be: Exporting country – Importing country – Private / Government Agencies – Wholesaler / Cooperative Societies – Retailer / Village Societies – Consumer.

Typically, a higher share of the consumer rupee goes to the farmers when there are fewer intermediaries involved (Figs 16.10 and 16.11). The primary aim of the recent initiatives such as Farmer Producers' Organizations (FPOs) is to diminish the number of intermediaries in the value chain. FPOs take on certain value-added activities, such as *dal* milling, flour mills, packaging and branding, to enhance the farmers' portion of the consumer rupee. The Indian Government and ICAR – Indian Institute of Pulses Research are encouraging the FPOs and farmers to purchase *dal* mills, which are not only efficient with higher turnover ratio but also report lesser post-harvest losses (Kumar *et al.*, 2019). India generally imports pulses whenever there is a shortage and domestic prices are higher. Although import prices are lower, arrival costs are

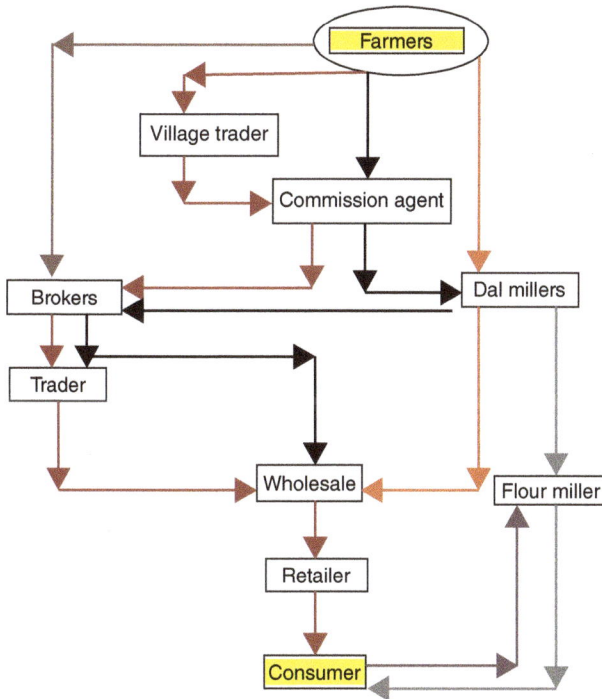

Fig. 16.10. Sale price of pulses at different nodes of the value chain in pulses. (Author's own figure.)

Fig. 16.11. The sale price of pulses at different nodes of the value chain in pulses (pigeon pea as an example). (Author's own figure.)

higher because of customs compliance and logistics and transport expenses.

16.9.2   Input markets (seed)

The value chain starts with seed. Procurement
of quality seed is an important aspect in increasing farmers' incomes. It was observed that about
32% of the pulse farmers were using their own
farm-saved seeds, 59% purchased it from local
markets and about 3% purchased from input
dealers. The role of cooperatives and government departments was meagre, which needs to
be increased to maintain the quality of the seeds
(Table 16.7).

16.9.3   Output markets

In the context of pulse markets, government
intervention is primarily limited to the announce-
ment of the MSP before the sowing season and the
procurement of the harvest during the harvest
time. Initially, there was only a nominal increase
in MSP for pulses until 2008–09 (Fig. 16.11).
However, during the period from 2001 to 2008,
consumption demand witnessed growth, leading to increased retail price inflation for pulses
due to the heightened demand and a shortage
of supply. In response to this situation, the government made the decision to increase the relative
MSP for pulses compared to competing crops as
an incentive for farmers to grow more pulses.
Notably, a significant increase in MSP occurred
after the 2016–17 period as a response to a spike

in retail prices of some of the pulses, which
reached as high as Rs 250/kg in major metro-
politan cities like Delhi. This adjustment in MSP
reflects a strategic measure taken to address the
shortage of pulses and ensure fair returns to
pulse-growing farmers. However, the procurement at MSP by the government procurement
agencies for pulses remains low (about 6% of
total production) compared to more than 35% in
paddy and wheat (Fig. 16.12).

Surprisingly, only 52.8% of pulse farmers
engaged themselves in market sales, indicating
that nearly half (47.2%) of them are subsistence
farmers, not actively participating in the market
(Table 16.8). Specifically, the data reveals that
only 44.8% of mungbean farmers and 63.8% of
chickpea farmers sold their produce in the
market. Sales to the government procurement
agencies remained minimal, accounting for <1–
2% of farmers. The key marketing channels for
pulse farmers are transactions with local private
traders (80.9% of the farmers who sold), farmer–
mandi transactions (10.1%), farmer-input dealers (1%), and farmer-private processors (2.3%).
Notably, mungbean, urdbean and lentil were
primarily marketed through local private traders,
while arhar was traded through *mandis*, and lentil was predominantly marketed through private
processors.

It was observed that the awareness of MSP
among the pulse farmers was comparatively
lower than that among the paddy and wheat
farmers (Table 16.9). A high per cent of paddy
farmers (53%) were familiar with MSP, followed
by wheat (37%), arhar (33%), and chickpea
farmers (29%). On an average, only 26% of pulse
farmers were aware of MSP. Moreover, a mere

Table 16.7. Source of seed for pulse growing farmers (per cent). (Author's own table.)

Crop	Arhar	Urdbean	Mungbean	Chickpea	Lentil	All pulses
Own farm	43.1	27.0	20.8	28.5	42.9	31.9
Local market	47.8	66.4	62.5	62.4	54.5	59.0
APMC market	1.2	0.9	1.6	1.2	0.0	0.8
Input dealer	3.8	1.3	2.8	2.7	0.1	3.2
Cooperative	1.4	2.8	1.6	0.7	0.0	0.9
Government agencies	1.4	0.4	0.6	0.4	0.2	1.0
Others	1.2	1.2	10.0	3.8	2.3	3.1
All	100	100	100	100	100	100

APMC, Agricultural Produce Market Committee. NSSO (2019) NSS REPORT NO. 587: Situation Assessment of
Agricultural Households and Land and Livestock Holdings of Households in Rural India, 2019.

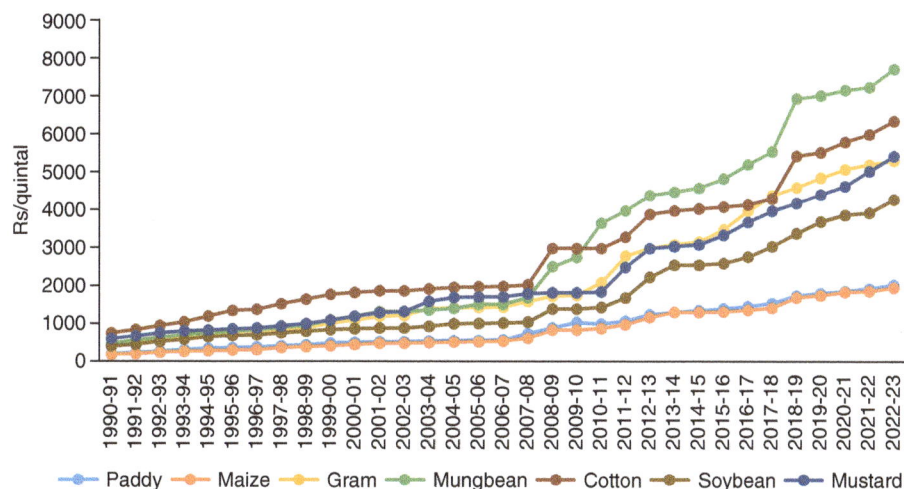

Fig. 16.12. Minimum Support Price for pulses and competing crops. Source: RBI (2024). (Author's own figure.)

Table 16.8. Distribution of farmers (%) reporting sale of crops by major disposal agencies, 2019. (Author's own table.)

Crop	Arhar	Urdbean	Mungbean	Chickpea	Lentil	All
Farmers reporting sale (%)	60.4	50.5	44.8	63.2	47.8	52.8
Sold to:						
Local market	76.0	87.4	87.4	79.7	87.5	80.9
APMC market	12.4	6.5	7.1	9.2	2.5	10.1
Input dealer	1.7	0.6	2.7	0.8	0.0	1.0
Cooperative	0.2	0.6	0.4	1.5	2.0	0.7
Govt. agency	1.2	2.5	0.3	1.9	0.0	1.3
FPO	0.3	0.0	0.0	0.1	0.0	0.1
Private processors	4.3	0.4	0.7	3.1	6.1	2.3
Contract farming	0.0	0.0	0.0	0.0	0.0	0.0
Others	3.9	2.0	1.4	3.7	1.8	3.6
All	100	100	100	100	100	100

NSS REPORT NO. 587: Situation Assessment of Agricultural Households and Land and Livestock Holdings of Households in Rural India, 2019.

17% of pulse farmers were cognizant of the existence of government procurement agencies, and only 3.8% sold their produce to these agencies. Approximately 6.5% of pulse output, on average, had been sold to the procurement agencies at MSP. One contributing factor to the limited awareness among pulse farmers is the historical trend where pulse prices have consistently remained above MSP (Table 16.9). However, in recent years, particularly during the harvest period, an oversupply has led to market prices falling below MSP. This shift prompts farmers to be more willing to sell to procurement

agencies at MSP. Unfortunately, the scattered and thin distribution of pulses production, coupled with the government's low priority of procuring pulses at villages and block levels, has resulted in the lack of procurement centres. This lack of access to government procurement centres led to over dependence of farmers on local traders. To tackle this issue, incorporating pulses into the Public Distribution System (PDS) could generate a consistent demand for procurement from government agencies. If government procurement at MSP constitutes a minimum of 10-15% of total production, it has the potential to positively

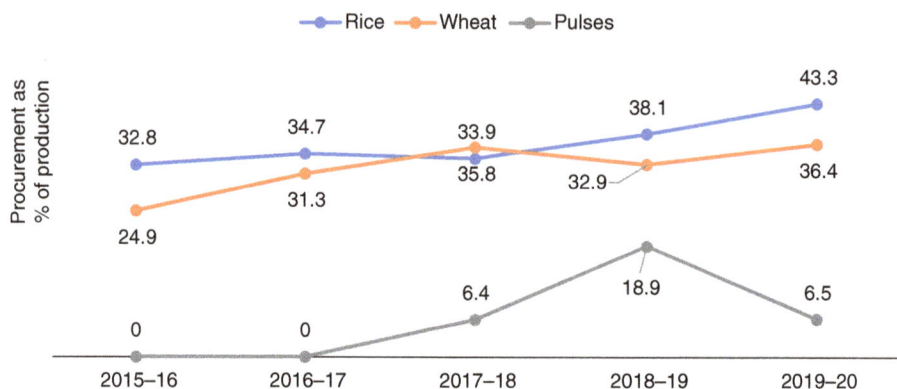

Fig. 16.13. Procurement of rice, wheat and pulses at MSP (% of production). Source: Indian Parliament (2023). (Author's own figure.)

Table 16.9. Farmers (%) reporting sale of crops while having awareness about MSP for selected crops, 2019. (Author's own table.)

Crop	Aware of MSP	Aware of procurement agency under MSP	Sold to procurement agency	% of output sold under MSP
Paddy	52.8	40.3	18.5	24.7
Wheat	37.1	27.2	9.7	20.8
Urdbean	20.9	15.7	5.4	11.5
Chickpea	29.0	18.8	5.5	8.2
Lentil	24.2	13.8	1.8	2.6
Arhar	32.6	20.5	2.2	2.1
Mungbean	24.9	17.1	4.0	8.1
All pulses	26.3	17.2	3.8	6.5

NSS REPORT NO. 587: Situation Assessment of Agricultural Households and Land and Livestock Holdings of Households in Rural India, 2019.

influence open market prices, ensuring they remain higher than the procurement prices.

Figure 16.14 indicates the average monthly retail price of different pulses (Rs/kg) in the years 2014 to 2023. The retail price of arhar varied from Rs 93 to Rs 106 in January to December. The maximum retail price of urdbean was 103 Rs/kg in November and the minimum retail price was 94 Rs/kg in March. The maximum and minimum retail prices of mungbean were 95 Rs/kg and 92 Rs/kg in November and February, respectively. The retail price of lentil was a maximum of 80 Rs/kg in October and November, while it was a minimum of 75 Rs/kg in February and March. The analysis reveals that the retail price of most of the pulses remained at a maximum in November and a minimum in March. Hence there is a need for procuring and storing pulses in the months of March–April and selling them in October–November to smooth the seasonal price changes and reduce the price inflation led by a shortage of pulses in November.

16.9.4 Long-term price cycle

Apart from the seasonal variations, there are long-term fluctuations also in pulse prices, as illustrated for arhar (Fig. 16.15). Arhar prices witnessed an increase from 2014 to 2015, followed by a decline until 2018, and then a prolonged upward trajectory. During the periods of rising prices, the disparity between wholesale and retail prices diminished. Conversely, when prices were on a downward trend, the gap widened.

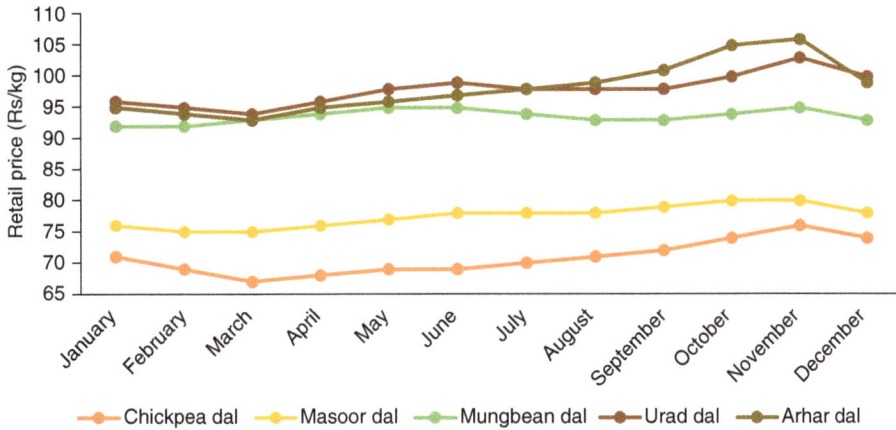

Fig. 16.14. Retail price (Rs/kg) average from 2014 to 2023. Source: Department of Consumer Affairs (2024). (Author's own figure.)

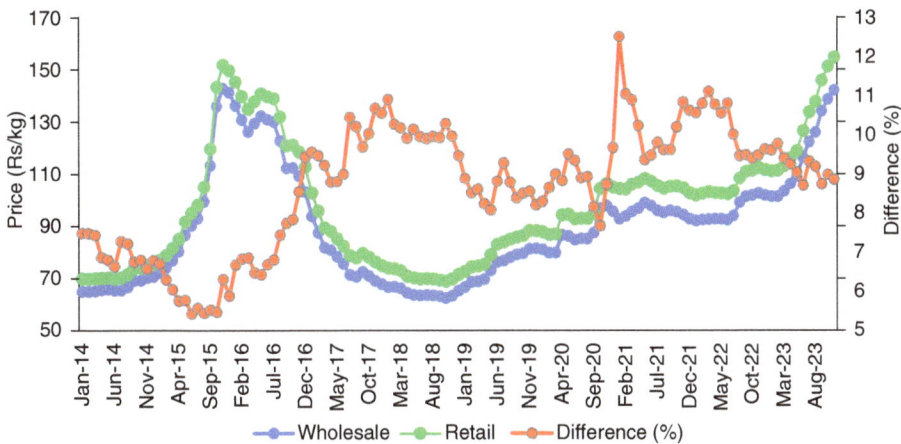

Fig. 16.15. Arhar dal: wholesale and retail prices (Rs/kg). Source: Department of Consumer Affairs (2024). (Author's own figure.)

This suggests that retailers and traders absorbed some of the wholesale price rises and passed them on to consumers only partially during price increases. This also indicates that traders and retailers are not engaging in speculative hoarding but are contributing to the smoothing of prices.

16.10   Global Trade in Pulses

As shown in Table 16.10, the highest recorded pulse imports occurred in the year 2016–17,

totaling 6.07 mt, followed by 2017–18 with 5.53 mt. Subsequently, a declining trend is evident as the domestic production kept on increasing. Peas constitute a disproportionate share of the imports, followed by chickpea, lentil, mungbean and arhar. Peas saw the highest import in 2016–17, reaching 3.17 mt, and the lowest in 2020–21 at 0.05 mt. Notably, no peas imports were recorded in 2021–22 and 2022–23. The highest import of chickpeas was in 2016–17 at 1.08 mt and the lowest in 2022–23 at 0.06 mt. India imports peas (dry) because of their lower prices in the international markets, which are

Table 16.10. Trends in imports and exports of pulses (million tons). (Author's own table)

Year	Imports						Exports					
	Pea	Chickpea	Mungbean/urdbean	Lentil	Arhar	Total pulses	Pea	Chickpea	Mungbean/urdbean	Lentil	Arhar	Total pulses
2010–11	1.50	0.10	0.43	0.16	0.00	2.20	0.00	0.20	0.00	0.00	0.00	0.21
2011–12	2.04	0.21	0.43	0.12	0.00	2.79	0.00	0.17	0.00	0.00	0.00	0.17
2012–13	1.37	0.70	0.64	0.51	0.47	3.68	0.00	0.19	0.00	0.00	0.00	0.20
2013–14	1.33	0.28	0.62	0.71	0.58	3.51	0.00	0.33	0.00	0.00	0.00	0.34
2014–15	1.95	0.42	0.62	0.82	0.46	4.27	0.00	0.19	0.00	0.01	0.00	0.21
2015–16	2.25	1.03	0.58	0.13	0.70	4.69	0.01	0.22	0.01	0.01	0.00	0.25
2016–17	3.17	1.08	0.57	0.83	0.41	6.07	0.01	0.09	0.01	0.02	0.01	0.13
2017–18	2.88	0.98	0.35	0.80	0.53	5.53	0.00	0.13	0.02	0.01	0.01	0.17
2018–19	0.85	0.19	0.57	0.25	0.45	2.31	0.00	0.23	0.02	0.15	0.01	0.41
2019–20	0.67	0.37	0.38	0.85	0.44	2.72	0.00	0.13	0.02	0.02	0.01	0.19
2020–21	0.05	0.29	0.42	0.11	0.84	1.71	0.01	0.16	0.03	0.02	0.02	0.23
2021–22	0.00	0.20	0.81	0.67	0.89	2.57	0.06	0.12	0.08	0.02	0.04	0.31
2022–23	0.00	0.06	0.56	0.86	0.89	2.37	0.16	0.35	0.04	0.09	0.03	0.67

Source: DGFT (2024).

mostly used in mixing with other pulses by traders, ultimately facilitating their sales at higher prices.

India exported less than 1 mt of pulses each year from 2011 to 2023. The exports have mainly been of chickpea (kabuli) while, to some extent, peas (dry) have also been exported (Table 16.10). Pea exports were highest in 2022–23 at 0.16 mt and the lowest in 2015–16 at 0.01 mt. There was no export of peas in most years. The export of chickpea suddenly increased in 2013–14 after a subdued performance for several years. Then, after 2013–14, it started decreasing continuously until 2016–17. After 2016–17, the export of chickpea increased only in 2018–19 and 2022–23. There was no export of mungbean from 2010–11 to 2014–15. The export of lentil, after increasing from 2014–15 to 2015–16, declined in 2017–18. In 2018–19, suddenly export of lentil increased to the highest level compared to other years at 0.15 mt. The figures show that exports of pulses constitute a negligible share in production, but there is scope for exporting kabuli chickpea, peas and lentil.

Imported dry peas practically compete with all the domestically grown pulses in India as they can be easily admixed with chickpea and arhar and sold through retail markets. The price of dry peas in Canada ranges from US$95/ton to US$180/ton, which is equivalent to Rs 8–15/kg in Indian currency (Fig. 16.16). This price is much less as compared to chickpea and arhar (ranging between Rs 50 to 90/kg) grown in India. The possible reasons for the huge price difference in the Indian and Canadian markets are as follows:

- **Production Costs and Efficiency:** Canadian agriculture is known for its high efficiency and low production costs, largely due to mechanization, advanced agricultural technologies and large-scale farming practices. In contrast, agricultural production in India, especially for crops such as chickpea and arhar, has traditionally been more labour-intensive and less mechanized, leading to higher production costs.

- **Subsidies and Government Policies:** Canadian farmers have benefitted from more substantial government subsidies and support programmes, which can lower production costs and thus market prices. Indian farmers, despite receiving some government support, did not have the same level of subsidies, especially in the context of chickpea and arhar.

- **Climatic and Environmental Factors:** Canadian dry peas are grown in optimal climatic conditions conducive to high yields, whereas chickpea and arhar in India have been subject to more variable and challenging climatic conditions, impacting yields and thus prices.

- **Transportation and Logistics Costs:** The cost of transportation and logistics from Canada to international markets have

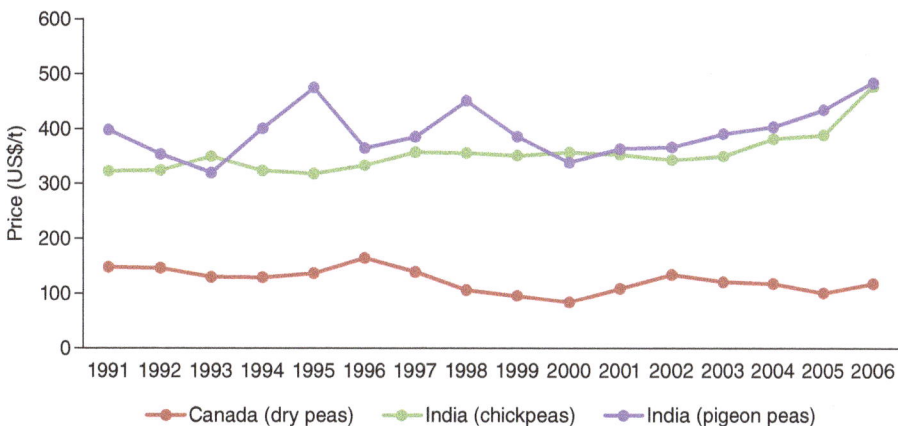

Fig. 16.16. Producer price (US$/t). Source: DGFT (2024) and FAOSTAT(2024). (Author's own figure.)

been lower due to efficient supply chain and logistics systems, compared to potentially higher costs in India. For example, transport costs from Indore to Delhi are much higher than those from Canada to Mumbai port, which makes imports cheaper.

Owing to the escalating prices of pulses, the Indian government has placed pulses under Open General License (OGL) since 2002, allowing for their free import with nominal customs duties, sometimes even zero duty. This strategy aims to ensure the availability of pulses in the domestic market at more affordable prices. However, unlike edible oils, where 60% of India's domestic consumption is met through imports, there is no significant international production of pulses, necessitating an increase in domestic production to meet the growing demand (Fig. 16.17). Although imports surged to 6.5 mt during 2016–17, the Indian government subsequently prioritized boosting domestic production. This effort substantially increased, with production reaching 25–27 mt by 2022–23. Nevertheless, imports still constitute approximately 2–3 mt annually. While the total availability has increased to 28–29 mt, demand continues to surpass production, resulting in rising prices. Addressing this challenge requires a focus on reducing the cost of production for pulses through increased investments in research and development, specifically in developing high yielding varieties (HYVs) that are resistant to pests and diseases and developing matching crop production and protection technologies.

16.10.1 Changing international trade partners

In the first decade of this century, large pulse exporters to India were Canada, Australia, the USA, Myanmar, Tanzania, Thailand and Russia. During recent years, Myanmar, Mozambique, China and Ethiopia have joined the ranks of pulse exporters to India. India primarily imports peas, chickpea, mungbean, urdbean, lentil and arhar (Table 16.11). Each pulse type is predominantly supplied by one country among the major trade contributors: peas (Canada 60.97%), chickpea (Australia 74.4%), mungbean and urdbean (Myanmar 70.37%), lentil (Canada 89.58%), and arhar (Myanmar 46.35%) (Table 16.8). While Myanmar plays a leading role as an exporter of mungbean and urdbean, Mozambique has become a significant exporter of pigeon pea (arhar). This diversity in pulse suppliers leads to varied negotiation dynamics for India. Countries such as Canada exert considerable influence in price negotiations, holding a monopoly on exporting peas to India as the single largest exporter. This is similar to chickpea imports from Australia. In contrast, negotiations with Myanmar to import arhar may give India a stronger position. This diversity underscores the

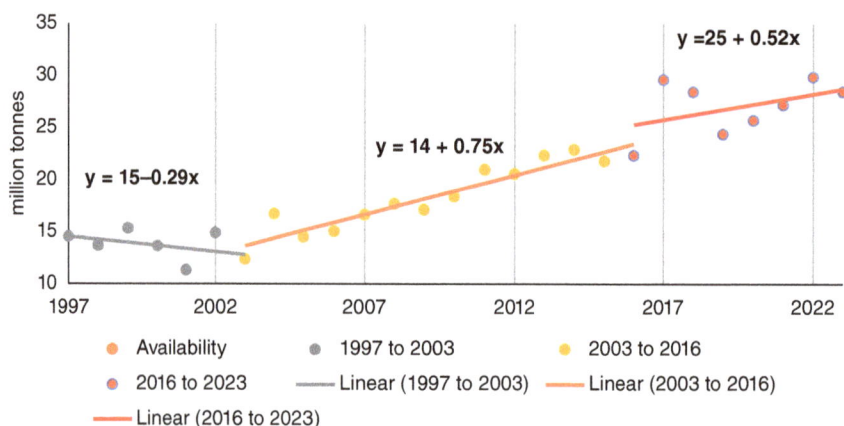

Fig. 16.17. Availability of pulses (million tonnes). Source: Agricultural Statistics at a Glance (2023). (Author's own figure.)

Table 16.11. India's major trade partners during the triennium ending 2021. (Author's own table.)

HS Code	Pulses	Top five import sources
7131000	Peas	Canada (60.97%), Russia (14.26%), USA (6.96%), France (5.36%), Lithuania (4.15%)
7132000	Chickpea	Australia (74.4%), Russia (16.49%), Tanzania (2.79%), Myanmar (0.92%), USA (0.74%)
7133100	Mungbean/ urdbean	Myanmar (70.37%), Kenya (7.43%), Australia (6.32%), Tanzania (3.15%), Uzbekistan (2.6%)
7134000	Lentil	Canada (89.58%), USA (7.47%), Australia (2.88%), Mozambique (0.03%)
7136000	Arhar	Myanmar (46.35%), Tanzania (18.71%), Mozambique (15.36%), Malawi (12.56%), Sudan (3.36%)

Source: DGFT, 2024

complex and dynamic nature of India's pulse trade landscape. It allows for the coexistence of ad-hoc imports in spot markets and the potential for long-term government-to-government contracts, as seen in the current arrangements with Mozambique and Myanmar.

16.11 Government Support for Promotion of Pulses Production and Consumption

The following are some of the important schemes implemented for promotion of pulses (Ministry of Agriculture and Farmers' Welfare, 2024).

NFSM-Pulses: To increase domestic pulses production, the Department of Agriculture and Farmers Welfare (DAFW), Government of India, has implemented the National Food Security Mission Programme (NFSM)-Pulses in identified nationwide districts. The NFSM-Pulses are being implemented in 644 districts of 28 States and two Union Territories of the country (Jammu & Kashmir, and Ladakh). Under NFSM-Pulses, assistance is given to farmers for interventions like cluster demonstrations on an improved package of practices, demonstrations on cropping systems, distribution of seeds of HYVs/hybrids, improved farm machinery/ resource conservation machinery/ tools, efficient water application tools, plant protection measures, nutrient management/soil ameliorants, processing and post-harvest equipment besides imparting cropping system based training to the farmers through State Governments. In addition, the certified seeds of HYVs/ hybrids of pulses are also provided in the form of seed mini kits free of cost to the farmers. Payments made to the farmers under government schemes support the farmers in purchasing seed, as well as purchase of *Rhizobium* and phosphate solubilizing bacteria. Further, 150 'Seed Hubs on Pulses' have been set up under NFSM since 2016–17 to augment the availability of quality seeds. These seed hub centres cumulatively produce >0.1 million quintals of quality seed of pulses every year.

Improved seed availability: In order to increase the productivity potential of pulses crops in the country, ICAR is undertaking basic and strategic research on these crops and conducting applied research in collaboration with State Agricultural Universities (SAU) for developing location-specific high yielding varieties and matching production packages. During 2014–2023, 343 high yielding varieties/hybrids of pulses have been notified for commercial cultivation in the country. During the past five years (2018–19 to 2022–23), cumulatively, about 86,015 q breeder seed of pulse crops was produced against a total demand of about 68,499 q and supplied to various public/private seed agencies for their conversion into certified seed for the farmers. However, there is still a gap in reaching isolated and neglected areas, which needs to be addressed by seed hubs as a priority.

Crop Diversification Programme (CDP): Under *Rashtriya Krishi Vikas Yojana* (RKVY), the governments of the original Green Revolution States, namely Haryana, Punjab and western Uttar Pradesh, have been implementing a Crop Diversification Programme (CDP) since 2013–14 to promote alternative crops such as pulses, oilseeds, coarse cereals, nutri cereals, cotton, etc. CDP aims at demonstrating alternative crops in the farmers' fields. A pilot project for 'Diversification of 4.85 million hectares' in five years (2023–24 to 2027–28) in 75 identified districts of 17 states/ union territories is being implemented.

***Pradhan Mantri Annadata Aay Sanrakshan Yojna* (PM–AASHA)**: Implemented since 2018 by the Government of India, this programme has three sub-schemes: (i) the Price Support Scheme (PSS); (ii) the Price Deficiency Payment Scheme (PDPS); and (iii) the Private Procurement & Stockist Scheme (PPSS). The three components are separate from all the other existing schemes of the Department of Food and Public Distribution (DFPD) for procurement of rice and wheat. The programme targets existing gaps in these procurement schemes, by providing a menu of additional compensation mechanisms. Under the PSS, state governments take a proactive role in the physical procurement of pulses from farmers, which is led by the central nodal agencies and the central government fully funds it. Under the pilot of PPSS, states also have the option to roll out private sector participation in procurement operations of pulses. However, the scheme is not yet popular. PDPS is successful in Madhya Pradesh and has a scope for upscaling to other states.

Maintaining strategic buffer stocks of pulses: Strategic buffer stocks are maintained to stabilize prices by discouraging hoarding and unscrupulous speculation. Pulses from the buffer are released to the market to lower prices and are also supplied on an ongoing basis to the States/Union territories (UTs) for distribution under welfare and nutrition schemes. Pulses from the Price Stabilization Buffer and PSS are made available to UTs for distribution under various welfare schemes, including the Public Distribution System (PDS), as per their requirement.

Distribution of pulses to State/Union territories for welfare schemes: The scheme started in 2019 and is still underway. The scheme is meant for the disposal of pulses procured under the PSS by way of distribution to States/ UTs for utilization under various welfare schemes such as Mid-day Meal (MDM), Integrated Child Development Services (ICDS), PDS, etc. The subsidy

supports the disposal of pulses procured under the PSS by offering a subsidy of Rs 15/kg over the issue price to State/UTs. The responsibility of State/UT Governments is to receive pulses from warehouses at issue price minus central subsidy, get it transported, processed or upgraded to the Food Safety and Standards Authority of India (FSSAI) standards and ensure distribution/ usage in welfare schemes.

Price stabilization: Stocks of chickpea and mungbean from the PSS and Price Stabilization Fund (PSF) buffer are continuously released to the market at moderate prices. Pulses are also supplied to the States at a discount of Rs 15/kg for welfare schemes. Further, the government has also introduced the mechanism to convert raw pulses stock into *dal* for retail disposal under the brand name of '*Bharat Dal*' at a highly subsidized rate of Rs 60/kg for a 1-kg pack and Rs 55/kg for a 30-kg pack in order to make pulses available to consumers at affordable prices, even though the market price is hovering around Rs 100/kg. *Bharat Dal* is distributed through NAFED, NCCF, Kendriya Bhandar and Safal retail outlets. The chickpea dal, under this arrangement, is also made available to state governments for supplies under their welfare schemes, police, jails, and also for distribution through the retail outlets of state government-controlled cooperatives and corporations.

Under National Food Security Mission (NFSM) Programme, Targeting Rice Fallow Area (TRFA)-Pulses: This is being implemented in 11 States of the country, which gives emphasis to land that remains underutilized after harvesting of kharif paddy crops and aims to bring a change in the cropping pattern during rabi season by introducing appropriate varieties of pulses that can be cultivated using available moisture. It is being implemented through a combination of innovative technological interventions and provision of essential agri-inputs, including extension services in 11 TRFA implementing States.

Postharvest: The postharvest pulses losses range between 6.36% and 8.41% (Kumar *et al.*, 2017) and 5.65% and 6.74% (Nathulapati *et al.*, 2022). The Ministry of Food Processing Industries through the implementation of various schemes under the *Pradhan Mantri Kisan Sampada Yojana*

(PMKSY) create post harvest infrastructure and processing facilities, to boost the overall development of the food processing sector including a reduction in post harvest losses, enhancing value addition, etc.

16.12 Export–Import Policy Review

In 2006, import duty was eliminated, and the export of pulses was prohibited. The government directive specified a temporary six-month duration for the export ban. Notably, the prohibition did not apply to *kabuli* chana, exports to Bhutan and the Maldives, as well as the export of organic pulses. Additionally, the export of roasted chickpeas in consumer packs of 1 kg was allowed (Table 16.12).

The export ban on all varieties of pulses was lifted in 2017. However, in 2018, certain pulse exports faced renewed restrictions. Throughout the period from the 2006 export ban to the 2017 resumption of exports, 15 official orders were issued concerning the imports and exports of pulses. This pattern indicated a need for a consistent and long-term policy rather than the adoption of ad-hoc measures in response to short-term challenges. Similarly, in July 2018, the import of peas was permitted, only to be reclassified as restricted in an August 2018 order. In the year 2020, a Minimum Import Price (MIP) for peas was introduced. Between 2006 and 2022, a total of 29 official orders were issued regarding the export–import dynamics of pulses. However, enduring policies are needed, such as long-term government-to-government import–export contracts (Reddy, 2004; Mu *et al.*, 2022). Such policies would instill confidence in the private sector, encouraging investment and ensuring sustained profits.

16.12 Outlook for Pulses: A Holistic Approach

Commodity outlook models employed to portray a country's food demand and supply dynamics are sophisticated, multi-sectoral, multi-regional and multi-dimensional. These models encompass a comprehensive analysis of various sectors, offering insights into the performance and trends of food markets. Through inter-linkages, they assess key variables related to the food balance sheet. As crucial tools, these models provide information on the demand, supply, trade and prices of major agricultural commodities. Furthermore, commodity outlook models play a pivotal role in policy formulation and analysis. They are extensively utilized as simulation tools to assess the potential impacts of alternative policy decisions on the considered commodities. This application allows policy makers to gauge the potential consequences of different policy choices, aiding in informed decision making within the realm of agricultural commodities.

To understand the entire value chain of pulses, it is imperative to delve into four key aspects. The Producer Core System encompasses crucial elements such as the cultivation area, yield equation, production equation and supply equation, providing insights into the cultivation and supply dynamics of pulses. On the Consumer Core System front, the focus shifts to analysing the food and feed demand equation, the other uses demand equation, and the total demand equation, offering a comprehensive view of the consumer-driven demands for pulses in various sectors. The Trade Core System investigates the export and import equations, unravelling the intricacies of international trade and its impact on the pulse market. Lastly, the price linkage equation explores the factors influencing pulse prices, including production, demand, trade and external market conditions. This holistic approach ensures a thorough comprehension of the pulse value chain, aiding stakeholders in making informed decisions and crafting effective strategies for the pulse sector.

16.13 Policy Recommendations

In the realm of government initiatives aimed at promoting pulses, a noteworthy contradiction persists – a policy bias against pulses that necessitates immediate attention and a shift in orientation. It is imperative to address this bias and realign policies to actively foster the cultivation and development of pulses. The subsequent sections delve into a comprehensive set of recommendations strategically designed to rectify the neglect of pulse crops and propel their cultivation forward

Table 16.12. Summary of various pulse policies implemented by the Government of India. (Author's own table.)

1	08.06.2006	Import duty on pulses (HS Code 0713) is reduced to zero to halt rising prices.
2	27.06. 2006	Export of pulses prohibited. This prohibition shall remain for a period of six months from the date of issue of notification.
3	03.07.2006	The period of validity of prohibition on exports of Pulses extended up to 31.3.2007.
4	09.03.2007	The period of validity of prohibition on exports of Pulses (except Kabuli Chana) extended up to 31.3.2008.
5	01.04.2008	The period of validity of prohibition on exports of Pulses (except Kabuli Chana) extended up to 31.3.2009.
6	15.05.2008	The prohibition on export of pulses shall not apply to export of 1000 tons of Pulses to Bhutan.
7	27.03.2009	The period of validity of prohibition on exports of Pulses (except Kabuli Chana) extended up to 31.3.2010.
8	23.03.2011	The period of validity of prohibition on Pulses exports is extended until 31.3.2012. This prohibition will not apply to export of (1) Kabuli Chana (2) 10,000 t of organic pulses during 2011–12.
9	27.3.2012	Prohibition on the export of pulses was extended by one more year, from 31.03.2012 to 31.03.2013. This prohibition will not apply to export of (1) Kabuli Chana (2) 10,000 t/annum of organic pulses.
10	30.05.2012	Export of 73 t of pulses (2012–13) and 80 t (2013–14) to the Republic of Maldives permitted through government channel.
11	27.03.2014	Pulse export to the Republic of Maldives allowed for 2014–15 (87.85 tons), 2015-16 (96.63 tons), and 2016–17 (106.29 tonnes).
12	31.03.2014	The ban on pulse exports is extended, with exceptions for kabuli chana and organic pulses/lentils (up to 10,000 mt annually), subject to conditions per Notification No. 51(RE-2010)/2009-2014 dated 03.06.2011, amending para 3(i) of Notification No. 38(RE-2013)/2009-2014 dated 25.03.2013.
13	13.06.2016	Bhutan is exempted from export bans and quantitative restrictions on Pulses, superseding Notifications No. 87 (05.12.2011) and No. 104 (05.03.2012).
14	20.01.2016	Export of roasted gram (whole/split) in consumer packs up to 1 kg has been permitted.
15	15.02.2016	As per Notification No. 78(RE-2013)/2009-14 dated 31.03.2014, the export prohibition on pulses does not apply to roasted grams (whole/split) in consumer packs up to 1 kg. This exemption is specified under Sl. No. 54 in Chapter 7 of Schedule 2 of ITC(HS) Classification.
16	14.08.2017	Under the 2017–18 bilateral trade agreement, India allowed the export of 122.23 tons of pulses to the Republic of Maldives from April 2017. Such exports are exempted from any present or future restrictions/prohibitions on export.
17	15.09.2017	Export Policy of Pulses - removal of prohibition on export pulses (arhar, mungbean and urdbean) till further orders.
18	22.11.2017	Exports allowed for all varieties of pulses, including organic pulses.
19	10.04.2018	Free export of Kabuli Chana.
20	20.06.2018	Under the 2018–19 bilateral trade agreement, India permitted the export of 31.59 tons of pulses to the Republic of Maldives, exempting it from any existing or future export restrictions/prohibitions.
21	02.07.2018	Extension granted for the import of peas (Exim Code 0713 10 00, including yellow peas, green peas, dun peas and kaspa peas) until 30.09.2018, for three months.
22	30.08.2018	Import of Peas classified under Exim Code 0713 10 00 (including yellow peas, green peas, dun peas, and kaspa peas) is 'Restricted' until 31.09.2018
23	28.09.2018	Import restrictions are until 31.12.2018 for peas (Exim Code 0713 10 00, including yellow peas, green peas, dun peas, and kaspa peas).
24	17.05.2019	In 2019–20, India permitted the export of 161.65 t of Dal to the Republic of Maldives under a bilateral trade agreement, exempting it from current and future export restrictions.

Continued

Table 16.12. Continued.

25	01.01.2020	Pea imports are restricted to an annual quota of 1.5 lakh tons as per procedure notified by DGFT, with a Minimum Import Price (MIP) of Rs 20/- CIF per kg, applicable through Kolkata sea port—exceptions for government commitments under agreements.
26	11.08.2020	During 2020–21, India allowed the export of 185.90 t of pulses to the Republic of Maldives under a bilateral trade agreement, exempting it from existing or future export restrictions.
27	29.03.2022	Urdbean and mungbean import is 'Free' up to 31.03.2023.
28	18.07.2022	For the fiscal years 2022–26, India has established an online procedure for importing pulses under MoUs with Myanmar (250,000 t of urdbean and 100,000 t of arhar), Malawi (50,000 t of ahhar) and Mozambique (200,000 t of arhar).
29	28.12.2022	Mungbean and urdbean import is 'Free' up to 31.03.2024.

Source: DGFT (2024) and Department of Consumer Affairs (2024).

by recognizing and capitalizing on their positive externalities.

One pivotal recommendation involves advocating for compensating positive externalities associated with pulse cultivation through government incentives. This initiative aims to acknowledge and reward the benefits of pulses to the agricultural landscape, encompassing factors such as reduced fertilizer and pesticide use, less water consumption, lowered GHG emissions, and their role in enhancing soil health through N fixation. To provide evidence-based policy recommendations, this chapter quantifies these benefits for pulses and non-pulse crops to determine appropriate levels of incentives.

Another crucial facet of the recommendation centres on neutralizing biased subsidies, particularly those entrenched in fertilizers like urea. The proposal suggests substituting these biased subsidies with direct money transfers, aiming to level the playing field for pulses compared to high-input-intensive crops such as paddy and wheat. This strategic shift seeks to create a fairer agricultural landscape, promoting pulses without the distortion caused by preferential subsidies. Additionally, the recommendations emphasize enhancing market linkages and infrastructure under the 'one-district one-product' initiative. This approach aims to bolster the pulse industry by improving access to domestic and international markets, fostering the growth of pulse-based value chains.

Lastly, the recommendations stress the inclusion of pulses in various government schemes targeted at poverty alleviation. By incorporating

pulses into these initiatives, the aim is to ensure that the economic benefits of these schemes directly reach pulse farmers, contributing to their overall economic wellbeing and enhancing nutritional security for the poorest of the poor in society, especially women and children. Another policy recommendation is a stable long-term import/export policy so that the private sector will participate in both the import and export value chain and reduce costs and enhance quality of pulses to benefit both farmers and consumers. Collectively, these recommendations are crafted to establish a supportive policy environment for pulse cultivation, recognizing and addressing the unique advantages and positive externalities associated with these crops.

16.14 Conclusions

This chapter underscores the critical need for enhanced policy focus supporting pulses, specifically promoting pulse production and consumption in India. It brings attention to the disproportionate allocation of incentives favouring paddy and wheat, which has resulted in a decline in the relative profitability of pulses, primarily owing to skewed subsidies. Despite initial perceptions based on market prices that suggest lower profitability for pulses, a holistic examination incorporating social cost–benefit analysis – factoring in subsidies and externalities – reveals the substantial macro-level benefits associated with pulse cultivation at the macro-level in the Indian context. Pulses, characterized by their non-market attributes such as N fixation, less GHG emissions, lower dependence on fertilizers and pesticides, and water efficiency, contribute significantly to environmental sustainability. Furthermore, pulses play a pivotal role in sustainable agriculture by fostering soil fertility, addressing human protein malnutrition, serving as the main ingredient of animal feed, and offering broader societal and nutritional benefits. In light of these comprehensive implications, the chapter advocates for formulating balanced policies by policy makers. These policies are envisioned to support farmers and prioritize the long-term wellbeing of society and the environment, recognizing the multifaceted positive impacts of pulse cultivation.

References

Agriculture at a Glance (2023) Agriculture At A Glance, Ministry of Agriculture & Farmers Welfare, Government of India, New Delhi. Available at: https://agriwelfare.gov.in/en/Agricultural_Statistics_at_a_Glance (accessed 10 January 2024).

Akber, N., Paltasingh, K.R. and Mishra, A.K. (2022) How can public policy encourage private investments in Indian Agriculture? Input subsidies vs. public investment. *Food Policy* 107, 102210.

Ali, M. and Gupta, S. (2012) Carrying capacity of Indian agriculture: pulse crops. *Current Science* 102, 874–881.

Ali, M., Joshi, P.K., Pande, S., Asokan, M., Virmani, S.M., Kumar, R. and Kandpal, B.K. (2000) *Legumes in the Indo-Gangetic Plain of India*. In: *Legumes in Rice and Wheat Cropping Systems of the Indo-Gangetic Plain - Constraints and Opportunities*. International Crops Research Institute for the Semi-Arid Tropics, Patancheru, Andhra Pradesh, India, pp. 35–70.

Anwar, F., Latif, S., Przybylski, R., Sultana, B. and Ashraf, M. (2007) Chemical composition and antioxidant activity of seeds of different cultivars of mungbean. *Journal of Food Science* 72, 503–510.

Bhutia, D.T. (2014) Protein energy malnutrition in India: the plight of our under five children. *Journal of Family Medicine and Primary Care* 3(1), 63–67. DOI: 10.4103/2249-4863.130279

CACP (2023) Commission for Agricultural Costs and Prices. Available at: https://cacp.da.gov.in (accessed 31 May 2024).

Chand, R., Raju, S.S. and Reddy, A.A. (2015) Assessing performance of pulses and competing crops based on market prices and natural resource valuation. *Journal of Food Legumes* 28(4), 335–340.

Deepthi, R., Anil, N.S., Narayanaswamy, D.M., Sathiabalan, M., Balakrishnan, R. and Lonimath, A. (2023) Recommended dietary allowances, ICMR 2020 guidelines: A practical guide for bedside and community dietary assessment - A review. *Indian Journal of Forensic Community Medicine* 10, 4–10.

Department of Consumer Affairs (2024) Ministry of Consumer Affairs, Food & Public Distribution, Government of India, New Delhi. Available at: https://fcainfoweb.nic.in/reports/report_menu_web.aspx (accessed 10 January 2024).

DGFT (2024) Director General Foreign Trade, Department of Commerce, Ministry of Commerce and Industry, Government of India. Available at: https://tradestat.commerce.gov.in/eidb/default.asp (accessed 10 January 2024).

Directorate of Economics and Statistics (2023a) Index Numbers for Agricultural Production, Department of Agriculture & Farmers Welfare, Ministry of Agriculture and Farmers Welfare, Government of India. Available at: https://desagri.gov.in/statistics-type/index-numbers/ (accessed 10 January 2024).

Directorate of Economic and Statistics (2023b) Advance Estimates of Agricultural Production, Department of Agriculture & Farmers Welfare, Ministry of Agriculture and Farmers Welfare, Government of India. Available at: https://desagri.gov.in/statistics-type/advance-estimates/# (accessed 10 January 2024).

Directorate of Economic and Statistics (2023c) Plot Level Summary Data Under the Cost of Cultivation Scheme, Department of Agriculture & Farmers Welfare, Ministry of Agriculture and Farmers Welfare, Government of India. Available at: https://eands.da.gov.in/Plot-Level-Summary-Data.htm (accessed 31 May 2024).

FAOSTAT (2024) FAO Statistics. Available at https://www.fao.org/statistics/en (accessed 10 January 2024).

Hedley, C.L. (2001) *Carbohydrate in Grain Legume Seeds*. CAB International, Wallingford, UK, pp 1–11.

Hemming, D.J., Chirwa, E.W., Dorward, A., Ruffhead, H.J., Hill, R. *et al.* (2018) Agricultural input subsidies for improving productivity, farm income, consumer welfare and wider growth in low-and lower-middle-income countries: a systematic review. *Campbell Systematic Reviews* 14(1), 1–153.

ICRISAT (2023) District Level Data Base. Available at: http://data.icrisat.org/dld/ (accessed 10 January 2024).

Indian Parliament (2023) Lok Sabha Unstarred Question No.3447 Answered On 23rd March 2022 accessed on 10.01.2024. Available at: https://sansad.in/getFile/loksabhaquestions/annex/12/AU3447.pdf?source=pqals (accessed 10 January 2024).

Joshi, P.K. and Rao, P.P. (2017) Global pulses scenario: status and outlook. *Annals of the New York Academy of Sciences* 1392(1), 6–17.

Kumar, R., Prasad, N., Gautam, V., Tripathi, C.M., Singh, R., Kumar, R. and Praharaj, C.S. (2019) Pulse based bio-village sustainable models through participatory demonstrations for livelihood security. *Journal of Food Legumes* 32(1), 45–48.

Kumar, S., Mohapatra, D., Kotwaliwale, N. and Singh, K.K. (2017) Vacuum hermetic fumigation: A review. *Journal of Stored Products Research* 71, 47–56.

Materne, M. and Reddy, A.A. (2007) Commercial cultivation and profitability. In: *Lentil: An Ancient Crop for Modern Times*. Springer, Dordrecht, the Netherlands, pp. 173–186.

Ministry of Agriculture and Farmers Welfare (2024) Farmers Welfare Schemes. Available at: https://agriwelfare.gov.in/en/FarmWelfare (accessed 10 January 2024).

Mishra, P., Al Khatib, A.M., Lal, P., Anwar, A., Nganvongpanit, K. *et al.* (2023) An overview of pulses production in India: retrospect and prospects of the future food with an application of hybrid models. *National Academy Science Letters* May 24, 1–8.

Mu, L., Hu, B., Reddy, A.A. and Gavirneni, S. (2022) Negotiating government-to-government food importing contracts: A Nash bargaining framework. *Manufacturing & Service Operations Management* 24(3), 1681–1697.

National Family Health Survey (2007) Indian Institute for Population Sciences (IIPS) and MoHFW. Key Indicators for India from NFHS-3. Vol. 18. Available at: http://rchiips.org/nfhs/pdf/India.pdf (accessed 10 January 2024).

National Family Health Survey (2017) Indian Institute for Population Sciences (IIPS) and MoHFW. National Family Health Survey - 4. 2017. Available at: http://rchiips.org/nfhs/pdf/NFHS4/India.pdf (accessed 10 January 2024).

NSSO (2013) National Sample Survey Organisation Consumption Expenditure 68th round (2011–12) and various rounds. Ministry of Statistics and Programme Implementation, Government of India, New Delhi.

NSSO (2019) Situation Assessment of Agricultural Households and Land and Livestock Holdings of Households in Rural India, 2019, NSS Report No.587, National Sample Survey Organisation, Ministry of Planning and Implementation, Government of India.

Nuthalapati, C.S., Dev, S.M. and Sharma, R. (2022) Supply-side Problems in Food Loss and Waste: Issues in Mitigation through Cold Chain. *Economic and Political Weekly* 57(14), 51–61.

Parappurathu, S., Kumar, A., Kumar, S. and Jain, R. (2014) A partial equilibrium model for future outlooks on major cereals in India. *Margin: The Journal of Applied Economic Research* 8 (2), 155–192.

Pratap, A. (2022) Humble yet significant role of mungbean in achieving self-sufficiency in pulses. *Journal of Food Legumes* 35, 229–231.

Rao, P.P., Birthal, P.S., Bhagavatula, S. and Bantilan, M.C.S. (2010) Chickpea and Pigeonpea Economies in Asia: Facts, Trends and Outlook. International Crops Research Institute for the Semi-Arid Tropics, Andhra Pradesh, India.

RBI (2024) Handbook of Statistics on Indian Economy. Available at: https://dbie.rbi.org.in/#/dbie/home (accessed 10 January 2024).

Reddy, A.A. (2004) Consumption pattern, trade and production potential of pulses. *Economic and Political Weekly* 39, 4854–4860.

Reddy, A.A. (2009) Pulses production technology: status and way forward. *Economic and Political Weekly* 44(52), 73–80.

Reddy, A.A. (2013) Strategies for reducing mismatch between demand and supply of grain legumes. *Indian Journal of Agricultural Sciences* 83(3), 243–259.

Reddy, A.A. and Kumara, C. (2016) *Pulses in India - Consumption, Demand and Supply Conditions*. Jaya Publishers, New Delhi.

Saini, S. and Gulati, A. (2017) Price Distortions in Indian Agriculture. International Bank for Reconstruction and Development (IBRD/ The World Bank, Washington, DC.

Sheehy, T., Carey, E., Sharma, S. and Biadgilign, S. (2019) Trends in energy and nutrient supply in Ethiopia: a perspective from FAO food balance sheets. *Nutrition Journal* 18, 1–12.

Singh, J.M., Kaur, A., Chopra, S., Kumar, R., Sidhu, M.S. and Kataria, P. (2022) Dynamics of production profile of pulses in India. *Legume Research - An International Journal* 45(5), 565–572.

Singh, N.P. and Pratap, A. (2016) Food legumes for nutritional quality and health benefits. In: Singh U., Singh, S.S., Praharaj, C.S. and Singh, N.P. (eds) *Biofortification of Food Crops*. Springer India, New Delhi. DOI 10.1007/978-81-322-2716-8_4

Singh, N.P. and Pratap, A. (2018) Pulses for zero hunger and zero malnutrition. *Indian Farming* 68, 33–36.

Index

CABI – who we are and what we do

This book is published by **CABI**, an international not-for-profit organisation that improves people's lives worldwide by providing information and applying scientific expertise to solve problems in agriculture and the environment.

CABI is also a global publisher producing key scientific publications, including world renowned databases, as well as compendia, books, ebooks and full text electronic resources. We publish content in a wide range of subject areas including: agriculture and crop science / animal and veterinary sciences / ecology and conservation / environmental science / horticulture and plant sciences / human health, food science and nutrition / international development / leisure and tourism.

The profits from CABI's publishing activities enable us to work with farming communities around the world, supporting them as they battle with poor soil, invasive species and pests and diseases, to improve their livelihoods and help provide food for an ever growing population.

CABI is an international intergovernmental organisation, and we gratefully acknowledge the core financial support from our member countries (and lead agencies) including:

Discover more

To read more about CABI's work, please visit: **www.cabi.org**

Browse our books at: **www.cabi.org/bookshop**,
or explore our online products at: **www.cabi.org/publishing-products**

Interested in writing for CABI? Find our author guidelines here:
www.cabi.org/publishing-products/information-for-authors/

www.ingramcontent.com/pod-product-compliance
Lightning Source LLC
Chambersburg PA
CBHW040137200326
41458CB00025B/6298